Laboratory Text
for Organic Chemistry

John Bob

Laboratory Text
for Organic Chemistry

a source book of chemical and physical techniques

DANIEL J. PASTO

Professor of Chemistry
University of Notre Dame

CARL R. JOHNSON

Professor of Chemistry
Wayne State University

PRENTICE-HALL, INC., Englewood Cliffs, New Jersey 07632

Library of Congress Cataloging in Publication Data

Pasto, Daniel J.
   Laboratory text for organic chemistry.

   Includes bibliographies and index.
   1. Chemistry, Organic—Laboratory manuals.   I. Johnson, Carl R., joint author.   II. Title
QD261.P36   547′.0028   78-16528
ISBN 0-13-521302-9

© 1979 by PRENTICE-HALL, INC.
Englewood Cliffs, New Jersey 07632

Printed in the United States of America
10   9   8

PRENTICE-HALL INTERNATIONAL, INC., *London*
PRENTICE-HALL OF AUSTRALIA PTY. LIMITED, *Sydney*
PRENTICE-HALL OF CANADA, LTD., *Toronto*
PRENTICE-HALL OF INDIA PRIVATE LIMITED, *New Delhi*
PRENTICE-HALL OF JAPAN, INC., *Tokyo*
PRENTICE-HALL OF SOUTHEAST ASIA PTE. LTD., *Singapore*
WHITEHALL BOOKS LIMITED, *Wellington, New Zealand*

Our thanks to

Bret Watkins

who, at the age of 13,
during the supreme challenge of his life,
assisted with the compilation
of the tables contained in Chapter 9.

Contents

Part II

ABSORPTION SPECTROSCOPY AND MASS SPECTROMETRY

APPENDICES

Preface

This text has evolved from a previous book, *Organic Structure Determination* (OSD), by the same authors. *OSD*, first published in 1969, has continued to be widely used. In its first few years, the most significant utilization of *OSD* was as a text and lab manual for modern variations of a traditional course in organic qualitative analysis. Over the years, many schools have phased out separate courses in this subject and integrated the material, in part, into various other courses, especially beginning organic laboratory courses and lecture courses on spectroscopic methods in organic chemistry. In the last several years, a number of schools have used *OSD* as a core text for the organic laboratory. Although the authors did not specifically design the book for this purpose, it appeared to serve very well. Now with multiple purposes in mind we have revised *OSD*. We believe that we have provided a textbook of exceptional versatility, a textbook that will serve the student in a variety of classes beginning with the introductory organic laboratory and continuing through the first year of graduate school and thereafter as a reference book.

In the present text, Part I provides a detailed discussion on background material and specific instructions on techniques used in running chemical reactions, isolation and separation of products, and characterization of organic compounds. Part II provides a thorough introduction to the major spectroscopic methods used by the organic chemist. Part III integrates spectroscopic and chemical methods used in organic qualitative analysis. Tables of physical properties and derivatives of many common organic compounds are included, as is a chapter on searching the chemical literature.

This book has been written with the following applications in mind:

1. *An undergraduate organic laboratory textbook.* The key ingredients that are missing are the actual experiments. There are none of the traditional melting point, distillation, extraction, simple synthetic transformations, etc., experiments that are available in the scores of laboratory manuals on the market! Why not? Almost without exception, when a traditional laboratory manual is adopted, a significant percentage of the experiments performed are from "handouts." There are two main reasons for this. First, each school's facilities and equipment pose special limitations and/or unique advantages. Second, each teacher or group of teachers have their own favorite experiments. The use of relatively few experiments in a given manual and the duplication of descriptive chemistry found in any modern organic text makes the usual laboratory manual a rather poor investment for the student.

With the present text plus an organic textbook, the students have the necessary information on background and techniques to do any experiment that they are likely to be assigned. For each experiment, the students will require a handout giving specific instructions in as much detail as is appropriate along with references to theories and techniques provided in this text and their organic text.

2. *A textbook for a senior or first-year graduate course in organic applications of spectroscopy.* Part II discusses uv, ir, proton and carbon-13 magnetic resonance spectroscopy, and mass spectrometry at a level appropriate for a survey course. Numerous spectra and tables of spectral data are provided. Part III correlates the various spectral methods plus chemical tests for the identification of organic compounds. This book and a spectral problem workbook would be ideal for a survey course in organic spectroscopy.

3. *A textbook for a modern qualitative organic analysis course.* This text integrates spectroscopic methods with chemical methods for the identification of organic compounds. Nineteen tables of data on physical properties and derivatives are included.

4. *A reference book for the practicing organic chemist.*

To all who have helped with the preparation and production of this book we express our deep gratitude. We thank those users of *OSD* who provided perceptive criticisms and suggestions. We are indebted to Professor Maurice Shamma of Pennsylvania State University, Professor Leon Stock of The University of Chicago, and Professor Christopher S. Foote of the University of California, at Los Angeles, who reviewed this text. Don Schifferl of the University of Notre Dame kindly recorded various FT-NMR spectra. Our thanks go to the Prentice-Hall staff, particularly Linda Mihatov, who ably guided the book through production.

The authors would be pleased to receive corrections, comments, and suggestions from our users for improvement of later editions.

Daniel J. Pasto

Department of Chemistry
University of Notre Dame
Notre Dame, Indiana 46556

Carl R. Johnson

277 Chemistry
Wayne State University
Detroit, Michigan 48202

OSHA TENTATIVE LIST OF CARCINOGENS

The Occupational Safety and Health Administration (OSHA) has issued a list of chemicals for which there is some evidence of carcinogenicity. *Special care should be taken to minimize exposure to these substances.* The list that follows contains compounds selected from the complete OSHA tentative Category I carcinogen list (*Chemistry and Engineering News,* July 31, 1978, p. 20) on the basis of their common occurrence in organic laboratories. Of particular importance is the appearance of the solvents *benzene, carbon tetrachloride, chloroform,* and *dioxane* on the list. In many procedures toluene can be substituted for benzene and dichloromethane can be used in place of chloroform or carbon tetrachloride. Tetrahydrofuran or 1,2-dimethoxymethane can often replace dioxane.

Compounds Selected from the OSHA
Tentative Carcinogen List

Acetamide
Asbestos
Aziridine
Benzene
Benzidine
4-Biphenylamine
Bis(2-chloroethyl) sulfide
Bis(chloromethyl) ether
Carbon tetrachloride
Chloroform
Chromic oxide
Coumarin
Diazomethane
1,2-Dibromo-3-chloropropane
1,2-Dibromoethane
Dimethyl sulfate
p-Dioxane
Ethyl carbamate
Ethyl diazoacetate
Ethyl methanesulfonate

1,2,3,4,5-Hexachlorocyclohexane
Hydrazine and its salts
Lead(2+) acetate
Methyl methanesulfonate
N-Methyl-N-nitrosourea
4-Methyl-2-oxetanone
1-Naphthylamine
2-Naphthylamine
4-Nitrobiphenyl
1,2-Oxathiolane 2,2-dioxide
2-Oxetanone
Phenylhydrazine and its salts
Polychlorinated biphenyl
Thioacetamide
Thiourea
o-Toluidine
Trichloroethylene
Vinyl chloride

General classes of compounds for which the carcinogenic risk is apparently high include:

—Alkylating reagents
—Arsenic and its compounds
—Azo compounds
—Beryllium and its compounds
—Cadmium and its compounds
—Chromium and its compounds
—Estrogenic and androgenic steroids
—Hydrazine derivatives
—Lead(II) compounds
—Nickel and its compounds
—Nitrogen mustards (β-halo amines)
—N-Nitroso compounds
—Polychlorinated substances
—Polycyclic aromatic amines
—Polycyclic aromatic hydrocarbons
—Sulfur mustards (β-halo sulfides)

part I

ORGANIC LABORATORY TECHNIQUES

1

Introduction to the Organic Laboratory

1.1
SUCCESS IN THE ORGANIC LABORATORY

The first step to success in an organic laboratory course is to be prepared *before* you enter the laboratory! Carefully study each experiment before you come to the lab. Outline in a schematic way in your laboratory notebook (Sec. 1.4) the reactions and/or operations you are expected to carry out. Include in your outline a sketch or brief description of any special apparatus that will need to be assembled. When you enter the lab, you should be prepared to begin your experimentation both rapidly and efficiently. The organic chemistry laboratory is perhaps the most expensive space on a square-foot basis that you will occupy during your academic studies. Laboratory time is valuable and limited. The laboratory is neither the time nor the place to be wondering about what to do or to be reading an experiment for the first time!

The second step to success is to be aware of *what* you are doing and *why* at all times. There is nothing more frustrating to the laboratory supervisor nor more humiliating to the student than to have conversations such as the following occur:

SUPERVISOR: What reaction are you running now?

STUDENT: Uh—I don't remember—just a second, let me look in my laboratory instructions.

Such students are not likely to become accomplished chemists, surgeons, or cooks.

Neatness and organization are important. Keep your laboratory area and locker neat and organized. Keep your glassware clean (Sec. 1.6) and ready to use.

Use odd moments—while a reaction mixture is refluxing or a solution filtering, etc.—to clean up your glassware. It is especially important that you allow enough time at the end of each session to clean your equipment so that you may begin experimentation promptly in the next session.

Record your operations and observations in your laboratory book as you go along. Observe good safety practices at all times (Sec. 1.2).

1.2
LABORATORY SAFETY

As in most things we do, there is a certain risk associated with working in an organic laboratory. This risk may be immediate—toxic effects of chemicals, fire, explosions, burns by highly acidic or caustic materials, cuts by broken glassware— or insidious—exposure to chemicals that may cause allergic reactions or may be carcinogenic. Since there is no substitute for hands-on experience in the organic laboratory, risks must be minimized by strict adherence to safety practices.

The first thing a chemist should do when beginning work in a new laboratory is to learn the locations and methods of use of the emergency facilities:

exits	fire blankets
eye wash facilities	gas masks
fire extinguishers	first aid supplies
safety showers	

Of equal importance is knowing how to secure help fast when it is needed.

1.2.1 Eye Protection

1. Wear recommended safety glasses or goggles at all times in the laboratory. If you normally wear prescription lenses, consult with your instructor as to their suitability. (Contact lenses are *not* eye protection devices; in fact, in an accident they may increase the degree of injury to the eye. It is recommended that contact lenses not be worn in the laboratory or that full eye protection be used in conjunction with them at all times.)
2. Never look directly into the mouth of a flask containing a reaction mixture, and never point a test tube or reaction flask at yourself or your neighbor.
3. Avoid measuring acids, caustics, or other hazardous materials at eye level. Place a graduated cylinder on the bench and add liquids a little at a time.

1.2.2 Fire

1. Use flames only when absolutely necessary. Before lighting a flame, make sure there are no highly flammable materials in the vicinity. Promptly extinguish any flame not being used.

2. Learn the location and use of fire extinguishers. For wood, paper, or textile fires, almost any kind of extinguisher is suitable. For grease or oil fires, avoid the use of water extinguishers—they simply spread the burning material. For fires involving electrical equipment, use carbon dioxide or dry chemical extinguishers. For fires involving active metals or metal hydrides, use dry chemical extinguishers or sand.

   To put out a fire, first cool the area immediately surrounding the fire with the extinguishers to prevent the spread of the flames; then extinguish the *base* of the blaze. Remember to aim the extinguisher at the base of the fire and not up into the flames.

3. When clothing is afire, the victim should not run any distance. This merely fans the flames. Smother the fire by wrapping the victim in a fire blanket. Use a coat or roll the victim on the floor if a fire blanket is not readily available, or douse the flames under the emergency shower.

1.2.3 Handling of Chemicals

1. Be cautious at all times when handling chemicals, especially those about which you know little.
2. Work in the hood as much as possible. Keep chemical vapors, particularly of solvents, to a minimum in the laboratory. Handle all chemicals that produce corrosive, toxic, or obnoxious vapors in an efficient hood.
3. Avoid direct contact with organic chemicals. Use plastic or rubber gloves when handling hazardous materials.
4. A lab coat or apron can protect you as well as your clothing.
5. Handle compressed gas cylinders with care. Always move them with a cart and strap them in place.
6. Never pour large quantities of volatile solvents into the sink.
7. Use special precautions with sealed glass vials of hazardous materials. Never subject a sealed glass container to severe thermal shock, e.g., do not place a commercial vial of methylamine, etc., into a Dry Ice bath.
8. Never use your mouth to pipette dangerous liquids. Always use a rubber safety bulb for pipetting.
9. When working with any potentially dangerous reaction, use adequate safety devices—safety shields, gloves, goggles, etc.
10. Do not smoke, eat, or drink in the laboratory.

1.3
FIRST AID

Although severe injuries seldom occur in the chemistry laboratory, it is wise for all chemists to be familiar with such important first-aid techniques as stoppage of severe bleeding, artificial respiration, and shock prevention. Why not consult a first-aid manual today?

1.3.1 Treatment of Chemical Injuries to Eyes

The most important part of the treatment of a chemical injury to the eye is that done by the victim himself in the first few seconds. Get to the eye wash fountain or any source of water immediately, and wash the injured eye thoroughly with water for at least 15 min. Thorough and long washing is particularly important in the case of alkaline materials. A physician should be consulted at once.

1.3.2 Burns from Fire and Chemicals

Chemical burns of all types should be immediately and thoroughly washed with water. Ethanol may prove more effective in removing certain organic substances from the skin.

For simple thermal burns, ice cold water is a most effective first aid measure. If cold water or a simple ice pack is applied until the pain subsides, healing is usually more rapid.

For extensive burns, place the cleanest available cloth material over the burned area to exclude air. Have the victim lie down and call a physician and/or ambulance immediately. Keep the head lower than the rest of the body, if possible, to prevent shock. Do not apply ointments to severe burns.

1.3.3 Cuts and Wounds

The most common minor laboratory accident involves cuts on the hand. Such cuts can usually be treated by applying an antiseptic and a bandage. If the cut is deep and possibly contains imbedded glass, a physician should be consulted.

For severe wounds—don't waste time! Use pressure directly over the wound to stop bleeding. Use a clean cloth over the wound and press with your hand or both hands. If you do not have a pad or bandage, close the wound with your hand or fingers. Raise the bleeding part higher than the rest of the body unless broken bones are involved. Never use a tourniquet except for a severely mangled arm or leg. Keep the victim lying down to prevent shock. Secure professional help immediately.

1.4
NOTEBOOKS AND REPORTS

A bound notebook with numbered pages should be obtained for use as a laboratory notebook. In keeping a laboratory notebook, whether for class or research purposes, there is one cardinal rule: When one makes an observation, it should be written down immediately. Neatness and order, though important, are secondary. Chemists should never get into the habit of recording experimental observations on loose sheets of paper to be transcribed later into a bound notebook. Loose pages tend to get lost, and one's immediate impressions are often tempered with time. One poor recall can be very costly in terms of time, materials, and reputation. The laboratory notebook should be kept at your side in the laboratory at all times. Observations should be recorded when they are made, and plenty of space should be allowed in your notebook for comments, additions of later information, and computations. Adequate references for the procedures used and data cited should be included.

Students should get into the habit of coding all samples, spectra, analyses, etc. with their initials and notebook page numbers. For example, the code RAF-2-147-D appearing on an infrared spectrum weeks, months, or years after the spectrum was originally run indicates that the spectrum is that of compound D described on page 147 of laboratory notebook 2 of chemist RAF. This system will allow the chemist or anyone else to look up the history and source of the compound immediately. Label all samples that you are to turn in or that you intend to store until later. *Never leave unlabeled chemicals in your locker.*

Your laboratory supervisor will probably have specific instructions concerning the format of your notebook; the format used, in part, will depend on the nature and purpose of the experiment. For example, in an experiment involving the synthesis of a particular compound, the format illustrated on the following page is useful.

1.5
CALCULATION OF YIELD

For practical reasons related to costs, reaction rates, positions of equilibria, side reactions, etc., organic reactions are most frequently run with reactants in non-stoichiometric quantities. One reactant is chosen to be the *limiting reagent* which ultimately restricts the theoretical amount of product that can be formed. For example, the oxidation of cyclohexanol to cyclohexanone with sodium dichromate in aqueous sulfuric acid proceeds according to the following *balanced* equation:

$$3 \; \text{(cyclohexanol, mol. wt. 100)} + \text{Na}_2\text{Cr}_2\text{O}_7 \cdot 2\,\text{H}_2\text{O} + 5\,\text{H}_2\text{SO}_4$$

mol. wt. 298   mol. wt. 98

$$\longrightarrow 3 \; \text{(cyclohexanone, mol. wt. 98)} + \text{Cr}_2(\text{SO}_4)_3 + 2\,\text{NaHSO}_4 + 9\,\text{H}_2\text{O}$$

Before Coming to Lab

TITLE   DATE

BALANCED EQUATIONS FOR ALL REACTIONS*

DATA ON REACTANTS, CATALYSTS, AND SOLVENTS

> *Molecular weight, grams, moles, solvent volume, useful physical constants*

PRODUCT DATA

> *Molecular weight, theoretical yield, literature value of melting point or boiling point*

APPARATUS

> *Sketch or brief description*

BRIEF SCHEMATIC OUTLINE OF EXPERIMENT

REFERENCES

During the Lab

RUNNING ACCOUNT OF WHAT YOU DO AND WHAT YOU OBSERVE

SIGN YOUR NOTEBOOK

After the Lab

CALCULATION OF YIELD (SEC. 1.5)

DISCUSSION OF RESULTS

ANSWERS TO QUESTIONS

---

* Review the procedure for balancing equations in your organic textbook.

If a reaction mixture contains 20 g (0.20 mol) of cyclohexanol, 21 g (0.07 mol) of sodium dichromate dihydrate, and 20 mL (37 g, 0.37 mol) of sulfuric acid, it is clear that the cyclohexanol is the limiting reagent (0.07 mol of dichromate is capable of oxidizing 0.21 mol of cyclohexanol). The theoretical yield of cyclohexanone is then 0.20 mol or 19.6 g. If an experimental yield of 15 g is obtained, this corresponds to 15 g/19.6 g × 100 = 76.5% yield.

In a reaction where some reactant is recovered unchanged after termination of the reaction, it is sometimes useful to compute the percent conversion. If during the workup of the above reaction 4.0 g of cyclohexanol were recovered, the consumption of cyclohexanol would have been 20 g − 4 g = 16 g (0.16 mol). That amount of cyclohexanol could lead to 0.16 mol × 98 g/mol = 15.7 g of cyclohexanone; a yield of 15 g would correspond to 15 g/15.7 g × 100 = 96% conversion.

1.6
GLASSWARE

Most laboratories use glassware with standard-taper ground glass joints (Fig. 1.1). Such joints have the advantage of rapid assembly and disassembly and of providing secure seals. (The principal disadvantage of ground glass joints is their expense; each component of a joint is worth two to three dollars.)

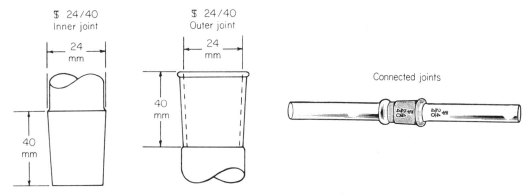

Fig. 1.1. Standard-taper (℥) ground glass joints. The most common sizes are 24/40, 19/38, and 14/20 (diameter/length in mm).

Ground glass joints are obviously made to fit together snugly. Sometimes the joints freeze and disassembly becomes difficult or impossible; this can be avoided by following proper procedures. If joints are kept clean, they normally can be assembled without the use of a lubricant. Disassemble the apparatus immediately after use; if the apparatus is hot, disassemble it before it cools, if possible. Lubricants should always be used when the apparatus is to be evacuated with a vacuum pump or when strongly basic solutions (which tend to etch glass) are used. Lubricants should be used by sparse but even application to the top of the male (inner) joint. Grease lubricants are preferred, as the joints can be easily cleaned by washing with organic solvents after disassembly. Silicone lubricants are almost impossible to remove completely by any means; about the best that can be done is to wipe them with a towel.

Occasionally a joint will become frozen. Gentle tapping will sometimes loosen it; otherwise, after removing all organic material from the apparatus, the outer joint can be gently heated with a Bunsen flame to cause expansion and release.

Glassware that will not become clean simply by rinsing with water can usually be cleaned with the aid of warm water, detergents, and a brush. Organic residues can frequently be removed with the aid of technical acetone. Sometimes it is necessary to soak and scrub stubborn residues with technical acetone and a brush. Wet glassware can be allowed to dry by standing, or more completely, by use of a heated oven. If a glass vessel is needed for use immediately after washing, a final rinse should be made with a little acetone; the vessel may then be dried quickly by placing a hose from an aspirator into the vessel or by using a heat gun. Compressed air lines are not very useful for drying glassware, since dirt and oil are often blown from the lines.

1.7
TEMPERATURE CONTROL

During the course of conducting reactions or their workup it is often necessary to maintain a temperature different from room temperature. Heating is achieved with electrical devices, steam, or gas flames. Cooling is achieved with water, ice, Dry Ice, liquid nitrogen, or mechanical refrigeration. In this section the more common methods of temperature control are surveyed.

1.7.1 Heating Devices

1.7.1a Heating Mantles

Heating mantles have resistance heating wire buried either in glass fabric or ceramic. The temperature is adjusted with fair precision by use of a variable

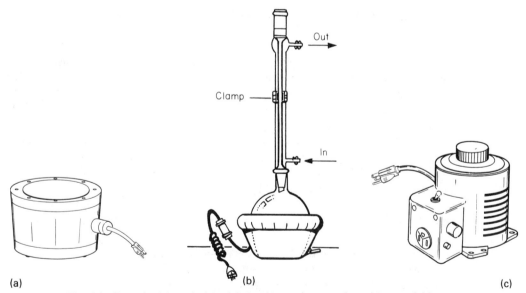

Fig. 1.2. Ceramic (*a*) and glass fabric (*b*) heating mantles with a variable transformer (*c*). (From Addison Ault, *Techniques and Experiments for Organic Chemistry*, 2nd ed. Copyright © 1976 by Allyn and Bacon, Inc. Reprinted by permission.)

transformer (Variac); the temperature limit is about 450°C. Mantles find common use in research laboratories because of their safety feature of having no exposed heating coil and their convenience; however, there are several disadvantages. The sizes and shapes of the mantle and flask must match, so for general use a large range of mantles must be available (at a cost of ten to twenty-five dollars each). The flask must never be allowed to go dry, for overheating of the flasks and any residue occurs rapidly when using mantles. Mantles are useful for refluxing reaction mixtures but are not recommended for distillations of small quantities of materials.

1.7.1b Infrared Lamps

Infrared lamps offer fair safety (provided three-wire plugs are used) at a modest cost; however, they provide a rather low temperature limit, are fragile, and the intense light makes continuous visual observation of the flask contents difficult.

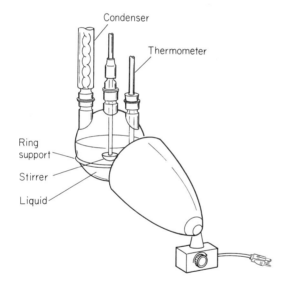

Fig. 1.3. Infrared lamp assembled from an infrared bulb (red), aluminum swivel reflector, dimmer switch, heavy-duty switch box, three-wire lamp cord and plug, and aluminum mounting rod.

1.7.1c Oil Baths

An oil bath consists of a container of oil with an immersed electrical heating coil. An adequate unit can easily be fashioned with a porcelain crucible, a coil made of 10 ft of 26-gauge Nichrome wire, two screw clamps to hold the coil in place, a power cord, and a variable transformer. A more substantial bath can be assembled from a metal pan and a commercially available immersion heater. Magnetic stirring of the oil (Sec. 1.10) can be used to assure uniform heating. Inexpensive mineral oil can be used at temperatures below 150°C, but presents a fire and smoke hazard above that temperature. More expensive silicone oil (e.g., Dow

550) can be used to about 250°C. A polyethylene glycol, Carbowax 600, which solidifies near room temperature and is water-soluble, can be used to a limit of 175°C. Oil baths have the following advantages: rather precise temperature control; bath temperature easily determined with a common thermometer; a range of flasks that can be used with a given bath; and reduced fire hazard. On the other hand, they can be dangerous if overheated, contamination by water can result in splattering, and they can be messy and easily spilled. Oil baths are highly recommended for conducting small- to medium-scale distillations.

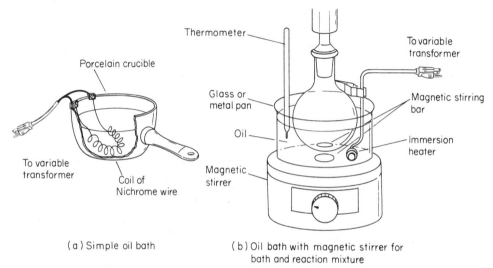

(a) Simple oil bath

(b) Oil bath with magnetic stirrer for bath and reaction mixture

Fig. 1.4. Oil baths. (From Addison Ault, *Techniques and Experiments for Organic Chemistry,* 2nd ed. Copyright © 1976 by Allyn and Bacon, Inc. Reprinted by permission.)

1.7.1d Hot Plates

Electric hot plates are often used to warm solvents, solutions, or mixtures in flat-bottomed containers, particularly Erlenmeyer flasks (whose sloping sides are designed to prevent vapor loss), during dissolution or recrystallization. Highly flammable solvents, particularly ether, carbon disulfide, and low-boiling hydrocarbons, should *never* be warmed on a hot plate. Water, dichloromethane, chloroform, and carbon tetrachloride can safely be used. Aqueous mixtures such as ethanol-water can be used if carefully watched and not allowed to reflux.

Hot plates are sometimes used to heat water or oil baths, which, in turn, are used to control reaction or distillation temperatures.

1.7.1e Steam Baths

Steam baths offer the highest safety factor at the lowest cost. Heat exchange from steam is reasonably efficient. It is not necessary to submerge the flask in the bath. The flow of steam should be adjusted to maintain the desired heating rate, but there is no need to blast steam out into the room; this wastes energy and can make the laboratory uncomfortable. *Caution should be used when first opening a steam valve, as the lines usually contain condensed hot water.* When using a steam bath it may be necessary to protect the sample from moisture.

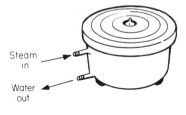

Steam in

Water out

Fig. 1.5. Steam bath. Remove rings to accommodate flasks of different sizes. (From Addison Ault, *Techniques and Experiments for Organic Chemistry*, 2nd ed. Copyright © 1976 by Allyn and Bacon, Inc. Reprinted by permission.)

1.7.1f Bunsen Burners

Open flames should be used only when no other choice is available. Flames should never be employed when flammable solvents are in use by you or *your neighbors* in the laboratory. When a flask is heated with a flame, a wire gauze should be used under the flask to aid in heat distribution. In a modern laboratory, flame use is usually limited to glass-blowing operations and to the flame-drying of apparatus (Sec. 1.11) to protect moisture-sensitive materials.

1.7.1g Heat Guns

These devices, which are much like electric hair dryers, can be used to heat a flowing stream of air up to 600°C. They can be used to heat reactions, and are convenient for use in drying glassware. The heating elements become red-hot, so

Fig. 1.6. Heat gun.

these devices should not be used when vapors of highly inflammable solvents are present.

1.7.2 Cooling Devices

Temperatures below ambient can be maintained by use of circulating tap-water baths (10 to 30°C, seasonal), ice-water baths (~ 0°C), ice-salt baths (33 g of sodium chloride/100 g of crushed ice: ~ −20°C), refrigerators (~ 5°C), and freezers (−20 to −90°C). Very low temperatures can be maintained by the combination of organic solvents and liquid nitrogen or Dry Ice. A wide-mouth vacuum flask (Dewar flask) is convenient for this purpose (if the outside of a glass vacuum flask is exposed, it should be wrapped with electrical tape for safety in the event the flask implodes). Liquid nitrogen (bp −196°C) is poured into a solvent with stirring until a slush is formed; the slush is maintained by periodically adding liquid nitrogen. Otherwise, crushed Dry Ice is added to the solvent until the excess Dry Ice is coated with frozen solvent; small quantities of Dry Ice are added periodically. Some useful combinations are listed in Table 1.1.

Table 1.1.   Low Temperature Cooling Baths

Bath	Temperature (°C)
Liquid nitrogen	−196
Isopentane–liquid nitrogen	−160
Diethyl ether–Dry Ice	−100
Toluene–liquid nitrogen	−95
Acetone–Dry Ice	−78
Trichloroethylene–liquid nitrogen	−73
Chlorobenzene–liquid nitrogen	−45
Carbon tetrachloride–Dry Ice	−23
Sodium chloride (1 part)–ice (3 parts)	−20

1.8
REFLUXING

Refluxing refers to boiling a solution and condensing the vapors in a manner that allows return of the condensate to the reaction flask. The technique of refluxing is used in preparative organic chemistry in order to maintain a reaction mixture at a constant and appropriate temperature. A simple reflux apparatus is pictured in Fig. 1.7. *Boiling chips or magnetic stirring (Sec. 1.10) should be used to prevent bumping!*

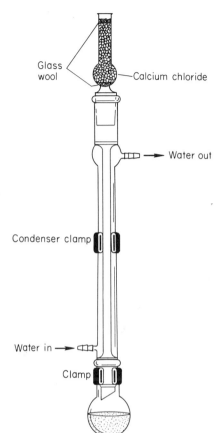

Glass wool

Calcium chloride

Water out

Condenser clamp

Water in

Clamp

Fig. 1.7. Simple reflux apparatus including a drying tube to protect reaction mixture from atmospheric moisture.

1.9
ADDITIONS TO REACTION MIXTURES

1.9.1 Addition of Liquids

A reaction that requires a controlled rate of addition of a liquid is usually run in a three-necked, round-bottomed flask equipped with an addition funnel (dropping funnel) (Fig. 1.8). The simplest addition funnel is similar to a separatory funnel equipped with a male standard-taper joint at the outlet; during dropwise addition of a liquid reactant or solution, it is used with the cap off. Pressure equilibrated addition funnels can be used with the stopper in place; these are convenient for inert atmosphere operations (Sec. 1.11). Teflon stopcocks are preferred on addition funnels so that use of stopcock grease, which can be leached by organic liquids, is avoided.

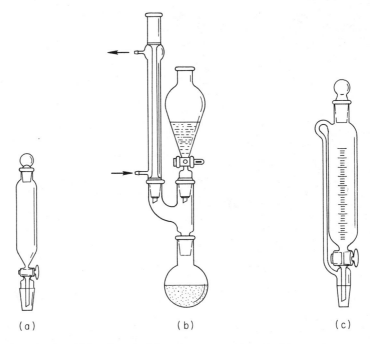

(a)   (b)   (c)

Fig. 1.8. Simple addition funnels (*a*) and separatory funnels (*b*) must have stoppers removed during additions to reactions flasks. Pressure-equilibrating additions funnels (*c*) can be used with stoppers in place.

In many small-scale reactions, and particularly with inert atmosphere operations, it is often convenient to make additions of liquids with a syringe. The needle of the syringe is inserted through a rubber septum fitted securely into one neck of the reaction flask (Fig. 1.9).

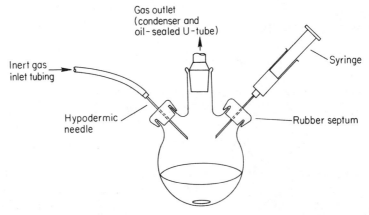

Gas outlet
(condenser and
oil-sealed U-tube)

Inert gas
inlet tubing

Syringe

Hypodermic
needle

Rubber septum

Fig. 1.9. Addition with a syringe. Rubber septums that fit **T** (standard-taper) joints are available from Aldrich Chemical Company, Inc.

1.9.2 Addition of Solids

Solids can often be added directly to a reaction flask by briefly removing a stopper. If the reaction mixture has been heated to near reflux, heating should be ceased and the reaction mixture allowed to cool slightly before the addition of solids is made; otherwise, boilover is likely to occur. For inert atmosphere operations, solids may be added with the aid of Gooch tubing. This is a thin flexible rubber tubing that will fit over a neck of the reaction flask and over the neck of an Erlenmeyer flask. Solids can be periodically poured through the tubing into the reaction flask; when not in use, the Erlenmeyer flask is allowed to hang in order to close the tubing (Fig. 1.10). The rate of addition of solids may be difficult to control, and caution must be exercised in making such additions in highly exothermic reactions. Addition of solutions is usually preferable to addition of solids.

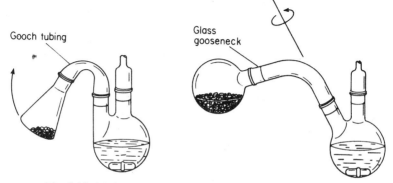

Fig. 1.10. Methods for the addition of a solid to a reaction mixture.

1.9.3 Addition of Gases

A gas may be added to a reaction mixture through a glass tube (often fitted with a glass dispersion frit) or hypodermic needle that dips below the surface of the liquid. If the gas is quite reactive or highly soluble, it may only be necessary or desirable to lead the gas to the surface of a stirred reaction mixture. If the gas reacts to form a precipitate, special provision should be made to avoid clogging of the inlet tube.

1.10
STIRRING AND MIXING

The most commonly used device for stirring and mixing in the organic laboratory is the magnetic stirrer. This simple device consists of a variable speed motor, which spins a magnet (Fig. 1.11). This is placed under the vessel, and a glass or

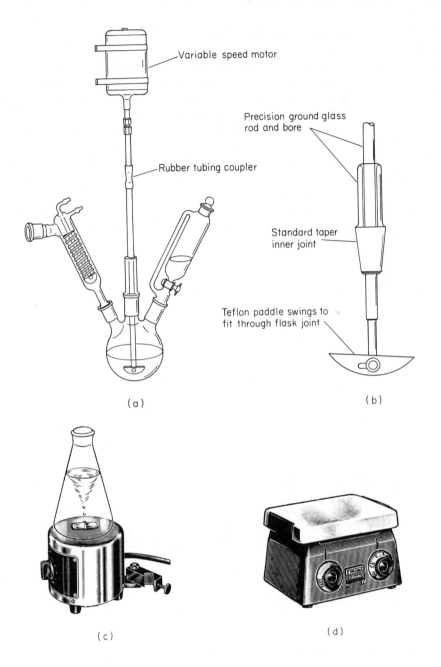

Fig. 1.11. Stirring devices. (*a*) Variable speed motor and paddle stirrer. (*b*) Precision-bore glass stirrer with Teflon paddle. (*c*) Magnetic stirrer with magnetic stirring bar. (*d*) Combination hot plate and magnetic stirrer.

inert plastic-covered magnetic bar is placed inside the vessel. This magnetic system is ideal for stirring reaction mixtures that are not too viscous or that do not contain heavy precipitates.

For reaction mixtures that are more difficult to stir or that need very vigorous mixing, a variety of direct-drive stirrers are available. One commonly used device is the Trubore stirrer, which consists of a precision-ground glass shaft and housing with a male standard-taper joint to fit the flask and a glass or Teflon paddle (which can be tilted to fit into the flask (Fig. 1.11)). This device is rather expensive. In use it should be carefully aligned and secured; the glass bearing should be lubricated with silicone or mineral oil.

There are numerous other types of stirring devices for special application, including mercury-sealed stirrers for inert gas operations and very high-speed stirrers (up to 10,000 rpm) with water-cooled bearings.

1.11
INERT ATMOSPHERES

It is frequently necessary or desirable to protect reaction mixtures from moisture or oxygen, or both. The degree of protection necessary depends on the reactivity of the reagents. During the course of reactions run under reflux, some protection

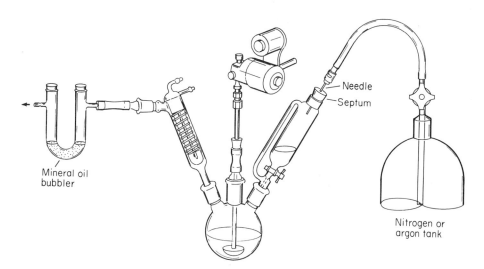

Fig. 1.12. Apparatus for carrying out reactions in an inert atmosphere.

of the reaction mixture is afforded by displacement of air from the system by the solvent vapor. Moderate protection can be achieved in the following manner. A clean flask and auxiliary equipment (addition funnel, stirring bar, etc.) with one neck open are heated with a blue Bunsen flame until just too hot to touch.

(Moisture condensate can usually be observed as it is driven from the glass walls.) A dry condenser capped with a drying tube filled with calcium chloride or Drierite is fitted into the open neck. As the system cools, the air that enters the system passes over the drying agent. When the flask has cooled sufficiently, it is charged with the appropriate solvents and reagents.

More elaborate protection is usually achieved by running reactions under a slight positive pressure of an inert gas; dry nitrogen is usually sufficient, but occasionally argon must be used (e.g., metallic lithium will react slowly with nitrogen). A typical setup is illustrated in Fig. 1.12. The entire system is assembled; however, the condenser should not contain water. The U-tube should contain just enough mineral oil to fill the bottom and to form a liquid seal. The system is flame-dried, while a small steady stream of dry inert gas is flowing. During the course of the reaction a slight positive pressure is maintained, as denoted by the slow release of gas at the U-tube. The inert gas inlet can be glass, or a septum and short hypodermic needle can be used. (Rubber laboratory tubing ($\frac{3}{16}$ in. diam) will fit securely over the large end of a steel hypodermic needle.)

1.12
CONTROL OF EVOLVED NOXIOUS GASES

All reactions that involve volatile, toxic or irritating substances should be run in an efficient hood. Toxic or irritating gases (e.g., HCl, HBr, SO_2, and NH_3) that may be evolved from a reaction can be trapped and kept out of the atmosphere using the apparatus shown in Fig. 1.13. For acidic gases the beaker should contain dilute sodium hydroxide; for ammonia, dilute sulfuric acid is used.

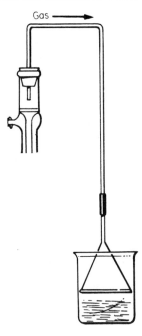

Gas ⟶

Fig. 1.13. Trap for control of water-soluble noxious gases. The funnel should not touch the solution in the beaker. (From Addison Ault, *Techniques and Experiments for Organic Chemistry,* 2nd ed. Copyright © 1976 by Allyn and Bacon, Inc. Reprinted by permission.)

1.13
SOLVENT REMOVAL

It is often necessary to concentrate a solution or completely remove solvents in order to obtain a desired product from a reaction mixture, an extract, a chromatographic fraction, or from mother liquors from a recrystallization. This may be done by simple or fractional distillation techniques at atmospheric pressure or under reduced pressure (Sec. 2.3). Sometimes atmospheric-pressure distillation is used to remove the bulk of the solvents, and last traces are then removed at reduced pressure. In any event, due consideration must be given to the stability and volatility of the desired product. Care must be exercised to prevent product and/or by-product decomposition due to overheating of the residue. This can usually be avoided by use of water, steam, or oil baths (Sec. 1.7.1). The residue should end up in a flask of appropriate size; when small amounts of product are present in dilute solutions, sequential additions of portions of the solution to be concentrated to the distillation flask are often convenient. (Note: Erlenmeyer flasks should not be used for reduced pressure evaporations.)

Rotary evaporators (Fig. 1.14) provide one of the most efficient methods for the removal of volatile solvents. The flask filled to half capacity or less is immersed in a water bath, pressure in the system is reduced by an aspirator, and the flask is rotated. The rotation not only prevents bumping, but continuously wets the warmed walls of the flask from which evaporation is rapid.

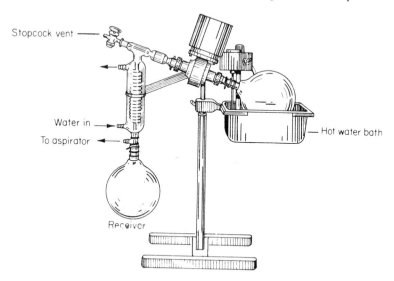

Stopcock vent

Water in

To aspirator

Hot water bath

Receiver

Fig. 1.14. Rotary evaporator.

Figure 1.15 illustrates a simple but reasonably effective apparatus for solvent removal. A boiling stick or stone should be included in the flask, and the flask should be swirled as frequently as is convenient.

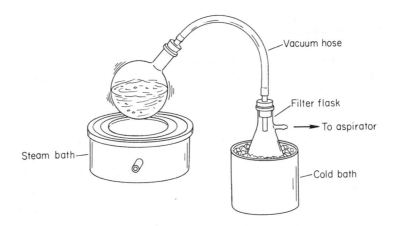

Fig. 1.15. Simple apparatus for removal of small amounts of volatile solvent under reduced pressure.

1.14
DRYING PROCEDURES

1.14.1 Drying Liquids and Solutions

Small amounts of water can be removed from liquids and solutions by allowing them to stand in direct contact with a drying agent for a suitable period of time followed by decantation or filtration. Some common drying agents that function by forming hydrates, e.g.,

$$Na_2SO_4 + 10 H_2O \rightleftharpoons Na_2SO_4 \cdot 10 H_2O$$

are briefly reviewed below.

$CaSO_4$ Sold under the trade name Drierite. Low capacity (forms a hemihydrate), fast acting, efficient general utility. Colorless form is used for drying liquids. The blue indicating form (pink when hydrated) is used in drying tubes, etc.

$CaCl_2$ Generally fast and efficient. Eventually forms a hexahydrate. Can react or form complexes with carboxylic acids, amines, alcohols, and, occasionally, carbonyl compounds.

Na_2SO_4 Neutral drying agent. Slow, but of high capacity (forms a decahydrate). Best for initial drying of rather wet samples.

$MgSO_4$ Slightly acidic drying agent. Not as rapid or intense as $CaSO_4$ or $CaCl_2$. Capacity lower than that of Na_2SO_4.

Dry liquids can be kept water-free by storage over molecular sieves (3A or 4A).

Molecular sieves are synthetic zeolites containing cavities of uniform size which are capable of incorporating small molecules. Water molecules fit into the 3 Å and 4 Å pores of the type 3A or 4A sieves. Small quantities of solvents or reagents can be dried effectively by passing through a short column of activated alumina. Ultra-dry inert solvents (hydrocarbons and ethers) are often prepared by allowing the solvents to stand over CaH_2, $LiAlH_4$ or sodium (usually as a wire or dispersion). After hydrogen evolution is complete, they may be distilled. All of this latter group react vigorously with water, and there is a high explosion and fire risk if not handled properly. *Do not use active metal or metal hydrides as drying agents unless you have received proper instructions from your laboratory supervisor.* Never use active metals or metal hydrides to dry acidic or halogenated liquids. Never use these agents with any solvent before predrying with ordinary drying agents, unless assured of a low water content.

The low-boiling toluene-water azeotrope[1] (bp 85°C) (Sec. 2.2) can be used to advantage to drive water out of wet samples or to remove water from an equilibrium reaction. For example, consider the following equilibrium reaction:

$$\bigcirc\!\!=\!O + H\!-\!N\bigcirc \underset{\longleftarrow}{\overset{H^+}{\rightleftharpoons}} \bigcirc\!\!-\!N\bigcirc + H_2O$$

This reaction can be carried out in toluene at reflux in the apparatus shown in Fig. 1.16. As the toluene-water azeotrope condenses and falls into the water separator (Dean–Stark trap), water separates as a lower phase (the trap is calibrated in mL). Toluene fills the trap and spills back into the reaction flask.

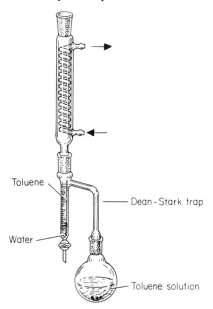

Toluene

Dean-Stark trap

Water

Toluene solution

Fig. 1.16. Apparatus for removing water by azeotropic distillation.

---

[1] Benzene has been most frequently used for this purpose [benzene-water azeotrope (bp 69°C)]; however, due to the high toxicity of benzene, toluene is now preferred for this purpose.

1.14.2 Drying Solids

For some purposes it is possible to dry solids simply by allowing them to stand in an open container (an evaporating dish, Petri dish, or beaker, but not an Erlenmeyer flask). More rigorous drying can be achieved by use of a desiccator (Fig. 1.17). Drying agent is placed in the bottom of the desiccator, and the solid in an open container is placed on a porous plate above the desiccant. Greater efficiency is obtained by evacuation of the desiccator. Drying agents used in desiccators include phosphorous pentoxide, magnesium perchlorate, concentrated sulfuric acid, and calcium sulfate (Drierite). Large solid samples can be dried in vacuum ovens. Very small samples for microanalysis, etc., are usually dried by placing the sample in a small open vial in the inner chamber of a drying pistol [called an Abderhalden (Fig. 1.17)]. The inner chamber containing sample and desiccant is evacuated, and the system is heated to the desired temperature by the refluxing of an appropriate solvent.

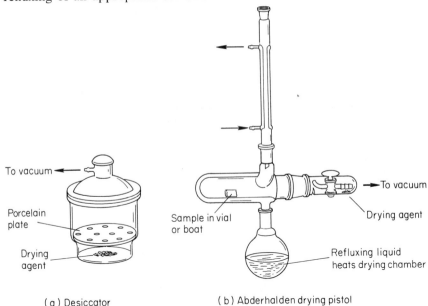

(a) Desiccator   (b) Abderhalden drying pistol

Fig. 1.17. Devices for drying solids.

1.14.3 Drying Gases

Water vapor can be removed from a gas by passing the gas through a drying tower filled with pellets of a desiccant (Drierite, $CaCl_2$, or KOH), or through a gas-wash bottle containing concentrated sulfuric acid (Fig. 1.18). Reactions or containers are often protected from atmospheric moisture by use of drying tubes. Indicating Drierite is often used for this purpose, as is calcium chloride.

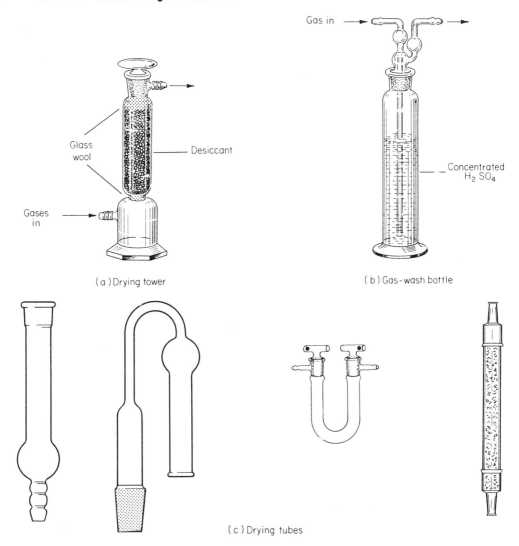

(a) Drying tower

(b) Gas-wash bottle

(c) Drying tubes

Fig. 1.18. Devices for drying gases.

1.15
COMMON ORGANIC SOLVENTS

Organic solvents are the organic substances that are worked with in the largest quantities in the organic laboratory. There are many risks associated with their indiscriminate use. In the compilation that follows, we have listed the properties, advantages, and disadvantages of the solvents that you most frequently encounter in the practice of and reading about organic chemistry.

1.15.1 Acetone

CH_3COCH_3

bp 56°C   Highly flammable.
fp −95°C   Miscible with water.
density 0.78   Health hazard: low.

Technical grade is useful as an inexpensive, water-miscible, highly volatile solvent for cleaning glassware in the organic laboratory. Acetone-Dry Ice baths are used to maintain a temperature of −78°C.

Acetone is a moderately reactive solvent. Mild bases (e.g., alumina) or acids (e.g., $MgSO_4$) can cause aldol condensation. Acetone is a good medium for S_N2 reactions (provided the nucleophilic component does not react with acetone). Acetone is often used as a cosolvent with water in chromic acid oxidations.

Reagent grade acetone is of high purity. Water content can be reduced to less than 0.005% with 4A molecular sieves or Drierite.

1.15.2 Acetonitrile

CH_3CN

bp 82°C   Flammable.
fp −44°C   Miscible with water.
density 0.78   Health hazard:
                                         modestly toxic.

Acetonitrile is a relatively polar, non-hydrogen-bonding material of high solvent power. It is an excellent solvent for many ionic and inorganic substances.

Acetonitrile may be purified by distilling fairly dry samples from phosphorous pentoxide (<5 g/L).

1.15.3 Benzene

C_6H_6

bp 80°C   Flammable.
fp 5.5°C   Immiscible with water.
density 0.87   Health hazard: high.
                                         Cumulative poison.
                                         Is claimed to be carcinogenic.

Because of the leukogenic risk, benzene should be replaced by other solvents (e.g., toluene, hexane) in routine solvent applications. If its use is necessary, it should be used only in an excellent hood.

Benzene, a non-polar solvent, is used as a basic starting material in aromatic chemistry. It has been frequently used for the azeotropic removal of water; the benzene-water azeotrope boils at 69°C and contains 9% water.

High purity grades are readily available. Drying may be accomplished by distillation from phosphorus pentoxide or by refluxing for several days over calcium hydride, followed by slow distillation.

1.15.4 Carbon Disulfide

CS_2

bp 46°C
fp −112°C
density 1.2

Highly flammable.
Immiscible with water.
Health hazard: high.

CAUTION. The high vapor pressure, low minimum ignition temperature (120°C), high flammability (about 1% in air), toxicity, and skin and mucous membrane irritating properties make carbon disulfide a hazardous substance.

Carbon disulfide is occasionally used in infrared spectroscopy and as a solvent for Friedel–Crafts reactions. It reacts readily with nucleophiles such as amines, alkoxides, etc.

1.15.5 Carbon Tetrachloride

CCl_4

bp 77°C
fp −23°C
density 1.58

Non-flammable.
Immiscible with water.
Health hazard: acute and chronic toxicity. Possibly careinogenic.

As a general rule, halogenated hydrocarbons are fairly toxic. They can enter the body by inhalation, ingestion, or absorption through the skin. From a health viewpoint, the concentration of carbon tetrachloride in the air is too high if the odor is detectable.

Carbon tetrachloride is a non-polar solvent; it is useful in infrared and magnetic resonance spectroscopy. It is readily available in high purity and can be dried over small quantities of phosphorus pentoxide.

1.15.6 Chloroform

$CHCl_3$

bp 61°C
fp −6.4°C
density 1.48

Non-flammable.
Immiscible with water.
Health hazard: toxic.
Possibly carcinogenic.

Chronic exposure may cause kidney and liver damage and heart irregularities. When exposed to air and light, chloroform can slowly oxidize to the highly toxic phosgene. Chloroform is irritating to the skin.

Chloroform is an excellent solvent for most organic compounds, including many ammonium, sulfonium, and phosphonium salts. Chloroform is widely used as a solvent in infrared spectroscopy. Deuterochloroform is often used as a solvent in magnetic resonance spectroscopy. Chloroform forms an azeotrope with water (bp 61°C).

CAUTION. Contact of chloroform with basic reagents and solutions should be avoided. Bases can cause the decomposition to dichlorocarbene. The decomposition can become violent or, at the least, cause product contamination by chlorine-containing by-products. Do not use chloroform as a solvent for amines or to extract highly basic solutions.

Chloroform contains small amounts of water and about 0.75% ethanol as an antioxidant. Small quantities *for immediate use* can be purified by passing through an activated alumina column (~ 100 g of alumina/200 mL of chloroform).

1.15.7 Dichloromethane (Methylene Chloride)

CH_2Cl_2 bp 40°C   Poorly flammable.
                           fp -95°C   Immiscible with water.
                                                Health hazard: low.
                                                    Least toxic of the
                                                    chlorinated methanes.

Dichloromethane is an excellent solvent for most organic substances, including many ionic compounds and is recommended for use instead of chloroform. It is highly volatile and its solutions can be concentrated with ease. Dichloromethane is inert to most common reagents but will react with strong bases (e.g., butyl-lithium).

Dichloromethane can be purified by washing with water, then by sodium carbonate solution, followed by drying over calcium chloride and fractionally distilling.

1.15.8 Diethyl Ether (Ether)

$CH_3CH_2OCH_2CH_3$ bp 35°C   Highly flammable.
                          fp -116°C   Immiscible with water.
                          density 0.70   Health hazard: low except
                                                   for explosibility.

Diethyl ether vapor concentrations in the range of 4% to 7% (by volume) produce anesthesia; 10% concentrations can be fatal. Repeated-dose inhalation can cause serious physical- and mental-health problems. Ether is flammable at concentrations as low as 2% and ether in air mixtures can ignite at temperatures as low as 174°C.

All ethers form peroxides with prolonged exposure to oxygen. Residues from the distillation or evaporation of ethers present high explosive hazards. Never distill an ether to near dryness unless peroxides are known to be absent.

1.15.8a Peroxide test:

Place 10 mL of the ether in a test tube and add 1 mL of a freshly prepared 10% aqueous potassium iodide solution. Acidify with several drops of dilute sulfuric acid. Add a drop of starch indicator. A blue color is indicative of the presence of peroxides. (Starch-iodide paper is not reliable for this test.)

Diethyl ether is a highly volatile solvent. The anhydrous grade (sometimes called absolute ether) is used as solvent for many organometallic and related reactions (e.g., Grignard and organolithium reactions, lithium aluminum hydride reductions) and as a cosolvent in column chromatography. Other grades, which may contain ethanol, are often used in extractions.

To purify diethyl ether, first test a small sample in a test tube with lithium aluminum hydride ($LiAlH_4$). *Do not use $LiAlH_4$ without proper supervision!* If a vigorous reaction occurs (hydrogen evolution), the water content should be reduced with conventional drying agents before proceeding. Diethyl ether containing small amounts of water can be purified by refluxing for 12 to 24 hr over $LiAlH_4$, followed by distillation (not to dryness because of the thermal instability of $LiAlH_4$). The $LiAlH_4$ will also destroy any peroxides present. The distilled solvent can be stored over molecular sieves and under nitrogen. The $LiAlH_4$ residues should be destroyed with great care by the slow and cautious addition of ethyl acetate or 10% sodium hydroxide. *Do not add water directly.*

1.15.9 Dimethylformamide (DMF)

$CH_3CON(CH_3)_2$

bp 153°C
fp −60°C
density 0.94

Flammable.
Miscible with water.
Health hazard:
    moderate.

The vapors of dimethylformamide may cause nausea, headaches, etc. The liquid and vapor are absorbed through the skin.

Dimethylformamide has high solvent power for a wide variety of organic and inorganic compounds. It is a good dipolar, aprotic medium for nucleophilic reactions. DMF will react with Grignard reagents, alkyllithiums, lithium aluminum hydride, etc.

Molecular sieves can be used to reduce the water content from 1% down to a few ppm.

1.15.10 Dimethyl Sulfoxide (DMSO)

$$\begin{matrix} & O \\ & \parallel \\ CH_3 & \!\!\! S \!\!\! & CH_3 \end{matrix}$$

bp 189°C
fp 64°C (4 mm)

Flammable.
Miscible with water.
Health hazard: low toxicity.

CAUTION. Although dimethyl sulfoxide itself has low toxicity, it readily penetrates the skin and can carry toxic solutes with it into body fluids. When working with DMSO, many people experience a strange and persistent taste caused by the absorption of DMSO liquid and/or vapors through the skin.

Dimethyl sulfoxide has extraordinary solvent power for organic and inorganic substances. DMSO is an excellent solvent for SN2 reactions and reactions of stabilized carbanions (e.g., Wittig reactions). *DMSO should not be used with oxidizing or reducing reagents.* (Numerous explosions have been reported with perchlorates, etc.) DMSO reacts with sodium hydride or other strong bases to produce methylsulfinyl carbanion, which is of use as a base and a reagent. Because it is difficult to remove, DMSO is usually a poor choice as a recrystallization solvent.

Reagent-grade dimethyl sulfoxide can be used as is for most purposes. Drying can be achieved by stirring overnight with calcium hydride and *distilling at reduced pressure.* Distillation should be conducted below 90°C to prevent disproportionation to the sulfide and sulfone.

1.15.11 *Ethanol (Absolute Ethanol)*

CH_3CH_2OH

bp 78.5°C   Flammable.
fp −114°C   Miscible with water.
density 0.78   Health hazard: very
                   low.

Ethanol is often used as a solvent, a cosolvent, or reactant. It is often used in combination with sodium ethoxide (prepared *in situ* by addition of sodium) or potassium hydroxide in base-catalyzed or base-promoted reactions. Both polar and non-polar compounds have reasonable solubility in ethanol. It is an excellent solvent for many recrystallizations. The less expensive 95% *ethanol* (ethanol-water azeotrope, bp 78.2°C) should be used for all applicants where the water is not a problem.

Traces of water can be removed from commercial absolute ethanol (water content 0.01% or less) by use of magnesium turnings. To dry 1 liter, about 5 g of Mg turnings is placed in a 2L round-bottomed flask, and 60 mL of ethanol and several drops of carbon tetrachloride are added. The mixture is refluxed until the Mg is vigorously reacting. The remaining ethanol (940 mL) is added to the flask, and the reflux is continued for 1 hr. The ethanol is distilled into a receiver while being protected from atmospheric moisture.

1.15.12 *Ethyl acetate*

$CH_3COOCH_2CH_3$

bp 77°C   Flammable.
fp −84°C   Immiscible with water.
density 0.90   Health hazard: low.
                  Vapors can cause dizziness.

Ethyl acetate is a useful solvent in thin-layer and column chromatography, particularly in combination with hexane or other hydrocarbons. It is sometimes used in recrystallizations, and it is a good solvent for polymers and resins. Commercial material usually has a purity of 99.5% or better.

1.15.13 Hexamethylphosphoramide (HMPA)

$[(CH_3)_2N]_3P{=}O$

bp 233°C — Flammable.
fp 7.2°C — Miscible with water.
density 1.0 — Health hazard: may be a
       potent carcinogen.

CAUTION. *Hexamethylphosphoramide has been shown to cause cancer of the nasal passage in rats exposed to its vapors.* Its use should be avoided whenever possible. It should be used only with extreme safety precautions.

Hexamethylphosphoramide is a dipolar aprotic solvent that, even as a cosolvent in small quantities, has been found to have profound effects on certain nucleophilic and strong base reactions.

1.15.14 Hexane

$CH_3(CH_2)_4CH_3$

bp 69°C — Flammable.
fp −95°C — Immiscible with water.
density 0.65 — Health hazard: low.

The volatile, saturated alkanes are all highly flammable, have anesthetic properties by inhalation, and are skin irritants.

Hexane and related hydrocarbon solvents are useful in extraction, recrystallization, and chromatography of non-polar organic compounds. Because they are transparent to ultraviolet and visible light, saturated hydrocarbons are often used as solvents in ultraviolet spectroscopy.

The principal impurities in saturated hydrocarbons are alkenes and aromatics. Saturated hydrocarbons can be purified by stirring or shaking several times with concentrated sulfuric acid, and then with 0.1 M potassium permanganate in 10% sulfuric acid. The hydrocarbon layer is then washed with water, dried over calcium chloride, refluxed over calcium hydride or sodium wire, and distilled.

1.15.15 Methanol

CH_3OH

bp 65°C — Flammable.
fp −98°C — Miscible with water.
density 0.79 — Health hazard: low by
       inhalation, high by
       ingestion.

Ingestion of methanol can lead to blindness and death. Death from ingestion of less than 30 mL has been reported.

Methanol is a useful solvent for the recrystallization of a wide range of compounds. It is often used in combination with sodium methoxide or sodium hydroxide in base-promoted reactions.

Methanol can be purified by refluxing with magnesium turnings (10 g/L) for 4 hr followed by distillation. The distilled material can be stored over 3A molecular sieves.

1.15.16 Pentane

$CH_3(CH_2)_3CH_3$

bp 36°C
fp −128°C
density 0.62

Highly flammable.
Immiscible with water.
Health hazard: low.

See comments under hexane.

1.15.17 Petroleum Ether

Highly flammable.
Immiscible with water.
Health hazard: low.

See comments under hexane. The petroleum ethers are mixtures of hydrocarbons and can be used interchangeably with the pure substances for many solvent purposes. Petroleum ethers are sold in various boiling ranges: 20 to 40°C, 30 to 60°C, 30 to 75°C, 60 to 70°C, 60 to 110°C, etc.

1.15.18 2-Propanol (Isopropyl Alcohol)

$CH_3CHOHCH_3$

bp 82°C
fp −88°C
density 0.78

Flammable.
Miscible with water.
Health hazard: low.

Solvent properties of 2-propanol are much the same as those of ethanol. Dry-Ice and 2-propanol baths are often used to maintain temperatures near −80°C.

1.15.19 Tetrahydrofuran (THF)

bp 66°C
fp −109°C
density 0.9

Flammable.
Miscible with water.
Health hazard:
moderately toxic.

Tetrahydrofuran is an excellent solvent for organometallic and metal hydride reactions. Grignard reagents can be prepared from vinyl halides and aryl chlorides in this solvent, whereas they do not form in diethyl ether.

Tetrahydrofuran forms peroxides very readily. (See peroxide test under diethyl ether.) Small amounts of peroxide can be removed by shaking TMF with cuprous chloride. THF containing large amounts of peroxide should be discarded by flushing down the drain with tap water. For instructions and precautions concerning the purification of THF, see *Org. Syn.*, Coll. Vol. **5,** 976 (1973). Many of the explosions reported with THF are probably due to the exothermic cleavage of the THF by reaction with excess metal hydride at high temperature zones produced on dry portions of the wall of the flask. Purified THF should be stored over molecular sieves and under nitrogen.

1.15.20 Toluene

$C_6H_5CH_3$

bp 111°C	Flammable.
fp −95°C	Immiscible with water.
density 0.86	Health hazard: moderate. More acutely toxic than benzene.

Although toluene is more acutely toxic than benzene, exposure to low concentrations of its vapor probably involves less chronic risks (benzene has been claimed to be carcinogenic).

Toluene is an excellent, non-polar solvent for low-temperature operations. Toluene and water form an azeotrope (bp 85°C) that contains 20% water.

1.16 REFERENCES

GENERAL

1. GORDON, A. J., and FORD, R. A., *The Chemist's Companion. A Handbook of Practical Data, Techniques, and References.* New York: Wiley-Interscience, 1972.

INERT ATMOSPHERE OPERATIONS

2. BROWN, H. C., *Organic Syntheses via Boranes.* New York: Wiley-Interscience, 1975.

3. LANE, C. F., and KRAMER, G. W., "Handling Air-Sensitive Reagents," *Aldrichimica Acta*, **10,** 11 (1977).

4. SHRIVER, D. F., *The Manipulation of Air-Sensitive Compounds.* New York: McGraw-Hill, 1969.

SOLVENTS

5. RIDDICK, J. A., and BUNGER, W. B., *Organic Solvents. Physical Properties and Methods of Purification*, 3rd ed. New York: Wiley-Interscience, 1970.

2

Separation
and Purification
Techniques

2.1
FILTRATION

Filtration is the separation of insoluble solids from a liquid by use of a porous barrier known as a filter; the liquid passes through the filter while the solid is retained by it. The liquid can pass through the filter by gravity (gravity filtration) or by use of suction (suction filtration). The process of filtration is used to clarify a solution and/or to collect a solid. The filters used most often in organic chemistry are filter paper and sintered glass frits.

2.1.1 Gravity Filtration

The apparatus for gravity filtration consists simply of a sheet of filter paper and a conical funnel. The circle of filter paper can be used folded in quarters; however, the flow rate is maximized by using fluted paper (Fig. 2.1). Fluted filter paper can be purchased or folded from plain filter paper. A size of filter paper should be chosen so that the paper cone projects just above the top of the funnel; if a smaller size is used, be sure the solution to be filtered is not allowed to fill the funnel above the paper level. Depending on the relative size of the funnel and collection flask, it may be advisable to support the funnel in an iron ring. If the funnel is not supported independently, some other provision may be necessary to prevent the filled funnel from sealing the top of the flask, as flow may be

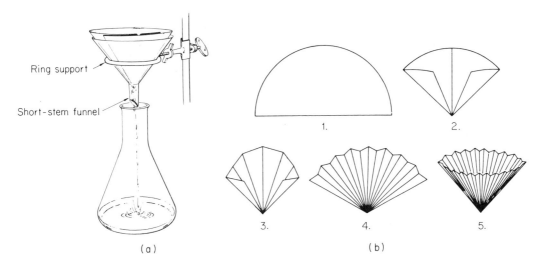

Ring support

Short-stem funnel

(a)

1.

2.

3.

4.

5.

(b)

Fig. 2.1. (*a*) Simple filtration apparatus. (*b*) Fluted filter paper. To flute plain filter circles, first fold into quarters; then by alternating the direction of each fold and limiting each fold to two thicknesses of paper, continue folding the paper into 32nds.

retarded by a pressure differential as the collection flask fills. Filtration is accomplished by pouring, in portions if necessary, the sample to be filtered into the filter paper cone. For filtering hot solutions from which crystallization is likely to occur upon cooling, a wide-bore short-stem funnel is used, and the solution in the collection flask is refluxed to keep the funnel warm.

2.1.2 Suction Filtration

In suction filtration, pressure is reduced below the filter by means of an aspirator; atmospheric pressure then forces the liquid through the filter. Following a recrystallization (Sec. 2.2), the crystals are usually collected by suction filtration.

For suction filtration three types of funnels are in general use—Büchner, Hirsch, and sintered glass funnels (Fig. 2.2). Büchner and Hirsch funnels require circles of filter paper of correct diameter to just cover the perforations at the bottoms. The filtrate collection devices are filter flasks, which are shaped like Erlenmeyers with thick walls and a side arm, and filter bells (Fig. 2.3). Filter flasks smaller than 1 liter should be clamped because the assembled apparatus is top-heavy and is easily tipped over. Filtration is accomplished by placing a circle of filter paper in the funnel, starting a gentle vacuum, wetting the paper with a little of the solvent used in the solution to be filtered (this seals the paper in place), and pouring the solution to be filtered into the funnel.

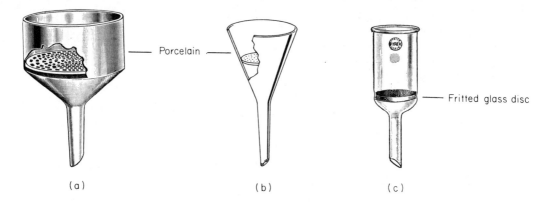

— Porcelain —

— Fritted glass disc

(a)   (b)   (c)

Fig. 2.2. (*a*) Büchner funnel. (*b*) Hirsch funnel. (*c*) Sintered glass funnel.

2.1.2a Trap for Water Aspirator

If water pressure decreases while a system is under aspirator vacuum, water may be sucked into the system. Although many aspirators have one-way check valves designed to prevent this, these valves may fail. To prevent water from entering the vacuum device, a trap should be placed in the aspirator line (Fig. 2.4). The vacuum should always be released before the water flow is turned down by disconnecting the vacuum tubing or by allowing air to enter through the stopcock on the trap.

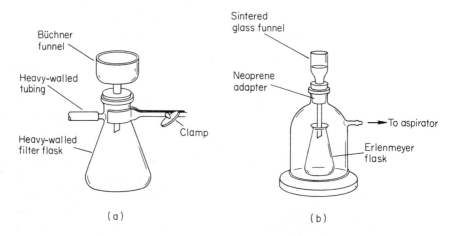

(a)   (b)

Fig. 2.3. (*a*) Filter flask with Büchner funnel. (*b*) Filter bell with sintered glass funnel. Rubber adapters are used to seal funnels.

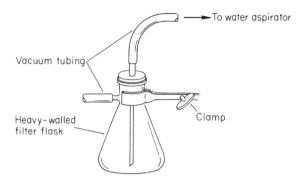

Fig. 2.4. Trap for a water aspirator. If the tube connected directly to the aspirator leads almost to the bottom of the flask, as shown, the trap will be self-cleaning.

2.1.3 Filter Aids

When the object is to clarify a solution by removal of finely suspended particles, the use of a more efficient filter than filter paper may be necessary. An effective filter can be created in a Büchner funnel by depositing a pad of some porous, inert filter aid on top of the paper prior to filtration of the suspension. Such a filter pad is not easily clogged by fine particles. The filter aid most often used is a diatomaceous earth product sold under the trade name Celite.

To prepare a filter pad, place the filter paper in the Büchner, wet with solvent, add filter aid as a heavy slurry in the solvent (use enough filter aid to create a pad about 0.5 cm thick), turn on the aspirator to remove the excess solvent, disconnect the aspirator and rinse the collection flask, reconnect the aspirator, and gently pour the solution to be filtered into the Büchner funnel in order to cause minimum disturbance to the filter pad.

2.2
RECRYSTALLIZATION

Crystallization is the deposition of crystals from a solution or melt of a given material. During the process of crystal formation, a molecule will tend to become attached to a growing crystal composed of the same type of molecules because of a better fit in a crystal lattice for molecules of the same structure than for other molecules. If the crystallization process is allowed to occur under near-equilibrium conditions, the preference of molecules to deposit on surfaces composed of like molecules will lead to an increase in the purity of the crystalline material. Thus the process of recrystallization is one of the most important methods available to the chemist for the purification of solids. Additional procedures can be incorporated into the recrystallization process to remove impurities. These include filtration to remove undissolved solids and adsorption to remove highly polar impurities.

Recrystallization depends on the differential solubility of a substance in a hot and a cold solvent. It is desirable that the solubility of the substance be high in the hot solvent and low in the cold solvent to facilitate the recovery of the starting material. The solution remaining after crystals have deposited is known as the *mother liquor.* The proper choice of solvent is critical and may require trial tests with small quantities of the material in a variety of solvents or solvent pairs (combinations of two solvents).

Solvents used in recrystallizations should be relatively low-boiling so that solvent adhering to the crystals can be readily removed by evaporation. If structural characteristics of the compound to be recrystallized are known, the adage "like dissolves like" should be kept in mind; polar compounds are more soluble in polar solvents, and non-polar compounds are more soluble in non-polar solvents. Some common solvents used in recrystallization are listed in Table 2.1.

Table 2.1. Common Solvents* for Recrystallization Listed in Order of Decreasing Polarity

Solvent	bp °C	Useful for	Common Cosolvents
Water	100	Salts, amides, some carboxylic acids	Acetone, methanol, ethanol
Methanol	65	General use	Water, diethyl ether, benzene
Ethanol	78	General use	Water, diethyl ether, benzene
Acetone	56	General use	Water, hexane
Ethyl acetate	77	General use	Hexane
Dichloromethane	40	General use, low-melting compounds	Ethanol, hexane
Diethyl ether	35	General use, low-melting compounds	Methanol, ethanol, hexane
Chloroform†	62	General use	Ethanol, hexane
Toluene	111	Aromatic compounds	
Benzene†	80	Aromatic compounds	Ethanol, diethyl ether, ethyl acetate
Hexane (or petroleum ether)	69	Hydrocarbons	All solvents on this list except water and methanol

* For additional information on organic solvents see Sec. 1.15.
† Not recommended for general use because of health hazard (Secs. 1.15.3 and 1.15.6).

2.2.1a Recrystallization procedure

The solvent, or solvent pair, to be used in the recrystallization of a substance is chosen in the following manner. A few milligrams of the substance is placed in a small test tube and a few drops of solvent are added. In general, one should use the least polar solvent first, for example, hexane or petroleum ether, progressing to the more polar solvents, for example, an alcohol. Should the sample completely dissolve, chill the solution to see whether crystals will form. If no crystals appear, the material

is too soluble in that solvent, and that solvent should not be used for the recrystallization. Other solvents should be tested until one is found in which the sample does
not completely dissolve at room temperature, but undergoes solution on heating. If
crystals reappear on cooling, the solvent is suitable for use. If no single solvent
provides suitable results, a mixture of two solvents can be employed, one of the
solvents being a good solvent for the sample, and the other being a poor solvent for
the sample. The correct proportion of the two solvents must be determined by trial
and error.

Once the proper solvent has been chosen, the remainder of the sample is
recrystallized.

The material to be recrystallized is placed in a suitable container, either an
Erlenmeyer flask or a centrifuge tube (depending on the amount of material to be
recrystallized), and solvent is added slowly, maintaining a gentle reflux of the solvent
in the container, until no more sample dissolves in the solution. Occasionally, highly
insoluble materials may be present in the sample, which, regardless of the amount of
solvent added, will not undergo solution. If either the solution or starting crystals
shows evidence of colored impurities, it may be desirable to add activated carbon to
absorb highly polar contaminants. The addition of activated carbon must be carried
out with great care, for if the saturated solution has become superheated, the
addition of carbon will induce violent boiling with the possible loss of material; the
solution should be at a temperature below its boiling point before addition of
activated carbon. After the addition of the absorbent, the mixture is boiled gently for
about 30 sec, with rapid stirring to avoid bumping, and then the solids are allowed to
settle to the bottom of the container. Any change in the color of the supernatant
liquid is noted, and an additional amount of the adsorbent is added, if needed. The
process is repeated until no further change is noted in the color of the solution.
(Caution must be exercised when dealing with highly polar or colored compounds in
order to avoid a great excess of the adsorbent. The use of a great excess of the
adsorbent may cause considerable loss of material.) The hot, saturated solution is
then filtered through a hot, solvent-saturated filter, which is kept warm by refluxing
solvent to prevent the premature formation of crystals (Fig. 2.5). The filter paper is
washed with a small portion of hot solvent. The volume of the solvent in the

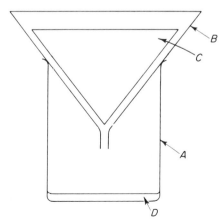

Fig. 2.5. Filtration setup for filtering a
hot saturated solution during a recrystallization process: (*A*) beaker, (*B*) glass
funnel, (*C*) filter paper, (*D*) refluxing
solvent. The entire setup is heated on a
steam bath or a hot plate (never over an
open flame!).

collection flask is then reduced by boiling until a saturated solution is again attained. Occasionally it is necessary to use a filter aid, for example, Celite, to completely remove the small granules of carbon. Suction filtration techniques, using a Büchner funnel and suction flask, are generally employed when such a filter aid is used (Sec. 2.1.3).

The rate of crystal growth from the saturated solution is critical. Too rapid a precipitation rate does not allow equilibrium conditions to exist, and the very slow precipitation with the formation of large crystals often leads to extensive solvent inclusion in the crystals. The hot, saturated solution should be allowed to cool at a rate between the two extremes. This can generally be accomplished by allowing the solution to cool on the bench top, followed by chilling in an ice bath. Occasionally, with low boiling solvents such as ether or pentane, it may be desirable to chill the sample in a Dry Ice-acetone bath to induce crystal growth. In such cases the collection funnel should be chilled prior to filtration by pouring a portion of chilled solvent through the filter funnel.

Some materials tend to "oil out" as a liquid instead of forming crystals. This usually occurs when the melting point of the material, or of the liquid phase formed, is below the temperature at which a saturated solution is attained. Additional solvent should be added, maintaining a clear solution, until crystals finally begin to form. If crystals are reluctant to form, scratching the side of the container with a glass stirring rod or adding a seed crystal will generally induce crystallization. The crystals are then collected by suction filtration and washed with a small portion of cold solvent to remove any adhering mother liquor. When using filter paper to collect the sample, care should be taken in removing the sample from the paper to avoid dislodging any fibers of the paper, which may contaminate the sample. These fibers may interfere with subsequent elemental or spectral analysis.

The recrystallization of milligram quantities of materials can best be carried out in small test tubes or centrifuge tubes. Centrifuge tubes are the most convenient in that the crystalline mass can be forced to the bottom of the tube and the mother liquors removed by a capillary pipette (Fig. 2.6). The crystalline mass can then be washed with several small quantities of cold solvent and then removed from the tube to be dried.

The recrystallization of highly insoluble compounds can be accomplished by use of a Soxhlet extractor (see Sec. 2.5 on extraction), in which the material is placed in

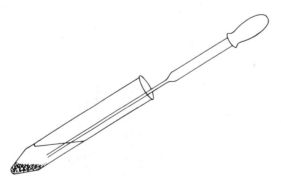

Fig. 2.6. Removing the solvent from a recrystallization in a centrifuge tube with a capillary pipette.

the extraction shell, and the purified material is recovered in the solvent distillation flask.

The crystalline material obtained in the procedure outlined above is thoroughly dried to remove adhering solvent. Many compounds can be dried by allowing the samples to set in the open air (care must be exercised to protect the sample from contamination). Hygroscopic compounds or compounds recrystallized from high-boiling solvents must be dried in a vacuum drying apparatus (e.g., a vacuum dessicator) (Sec. 1.14.2). The entire recrystallization procedure should be repeated until a constant melting point of relatively narrow range is obtained (see Sec. 3.2 on melting points for the details of melting-point determinations).

2.3
DISTILLATION

Distillation can be defined as the partial vaporization of a liquid with transport of these vapors and their subsequent condensation in a different portion of the apparatus. Distillation is one of the most useful methods for the separation and purification of liquids. The successful application of distillation techniques depends on several factors. These include the difference in vapor pressure (related to the difference in the boiling points) of the components present, the size of the sample and the distillation apparatus (as well as the type of apparatus employed), the occurrence of codistillation or azeotrope formation, and the care exercised by the experimentalist. Distillation relies on the fact that the vapor above a liquid mixture is richer in the more volatile component, the composition being controlled by Raoult's Law.

Raoult's Law states that the partial pressure (P_A) of component A in an ideal solution at a given temperature is equal to the vapor pressure (at the same temperature) of pure A (P_A^o) multiplied by the mole fraction of A (N_A) in solution. Consider an ideal solution of A and B:

$$N_A = \frac{\text{moles of } A}{\text{moles of } A + \text{moles of } B} \qquad N_B = \frac{\text{moles of } B}{\text{moles of } A + \text{moles of } B}$$

$$N_A + N_B = 1$$

$$P_A = P_A^o N_A \qquad P_B = P_B^o N_B \tag{2.1}$$

$$P_T \text{ (total vapor pressure)} = P_A + P_B$$

The boiling point of the solution is reached when P_T is equal to the pressure applied to the surface of the solution [Eq. (2.1)].

Phase diagrams are helpful in illustrating simple and fractional distillation. The diagram shown in Fig. 2.7 plots the equilibrium composition of vapor and liquid phases against temperature. It can be seen that at 105°C, liquid with composition

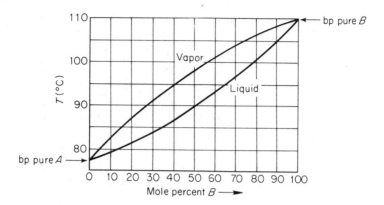

Fig. 2.7. Boiling point phase diagram for an ideal solution of two liquids: *A* (bp 76°C) and *B* (bp 110°C).

90% *B* and 10% *A* is in equilibrium with vapor of composition 73% *B* and 27% *A*. If this vapor were condensed, the condensate, with composition 73% *B* and 27% *A*, would be enriched in the lower-boiling component. This is exactly what happens in a simple distillation. However, if the (73% *B*)/(27% *A*) condensate were allowed to reach equilibrium with its vapor at 97°C, and if this vapor were condensed, this new condensate would be further enriched in *A*, whereas the residual solution would become enriched in *B*. In a fractional distillation of the *A–B* mixture, this vaporization-condensation process is repeated many times in a single apparatus. Early portions coming out of the distillation column will be essentially pure *A*, intermediate portions will be a mixture of *A* and *B* (the size and composition of intermediate portions depending on the difference in the boiling points of *A* and *B* and the efficiency of the apparatus), and the latter portions will be pure *B*.

In considering the distillation of ideal solutions, the following points are pertinent:

1. The boiling points of mixtures are always intermediate between those of the highest- and lowest-boiling constituents.
2. The closer the boiling points of the pure constituents of a mixture, the more difficult their separation by distillation.
3. A mixture of two or more substances with the same boiling point will have the same boiling point as the constituents and cannot be separated by distillation.

A simple distillation will never lead to the complete separation of two volatile substances. The use of efficient fractionating columns may produce more effective separations when sufficient quantities of the sample are available. When the sample size is too small to carry out an effective distillation, the use of chromatographic techniques is recommended, particularly preparative gas or liquid chromatography (Sec. 2.6).

There are a large number of liquid mixtures that do not obey Raoult's Law but form an *azeotrope* (also called an azeotropic mixture). An azeotrope is a constant-boiling mixture of definite composition. The boiling point of an azeotrope is *always outside the range* of the boiling points of its components. For example, the minimum-boiling azeotropic mixture of 68% benzene and 32% ethanol boils at 68°C; its pure components boil at 80°C and 78.5°C respectively. Other examples of azeotropes are listed in Table 2.2.

Table 2.2. Some Common Azeotropes

Components	Percent by Weight	bp Pure Component (°C)	bp Azeotrope (°C)
Acetone	88	56	56
Water	12	100	
Benzene	91	80	69
Water	9	100	
Benzene	68	80	68
Ethanol	32	78.5	
Chloroform	93	61	59
Ethanol	7	78.5	
Chloroform	97	61	56
Water	3	100	
Dichloromethane	98	40	39
Water	2	100	
Ethanol	68	78.5	77
Toluene	32	111	
Ethanol	96	78.5	78
Water	4	100	
Toluene	80	111	85
Water	20	100	
Benzene	74	80	
Ethanol	19	78.5	65
Water	7	100	
Toluene	51	111	
Ethanol	37	78.5	74
Water	12	100	

2.3.1 Simple (Non-fractional) Distillation

If a sample is known to contain essentially only one volatile component, a simple, non-fractional distillation can be used to effect a suitable purification of the component. A suitable apparatus for such a distillation is illustrated in Fig. 2.8 and employs standard-taper 14/20 precision-ground glassware (the correct handling of precision-ground glassware is discussed in Sec. 1.6). The apparatus shown in Fig. 2.8 can be used for atmospheric or reduced pressure (vacuum) distillations and is suitable for sample sizes down to 0.5 g.

A small portion of the sample should be subjected to a micro boiling-point determination (Sec. 3.4) to determine the approximate boiling point of the sample and to determine whether the sample is stable under the conditions of the distillation. In general, if the boiling point of the sample is above 180°C, a reduced pressure distillation should be employed. This can be accomplished by the use of a water aspirator capable of producing pressures down to 12 to

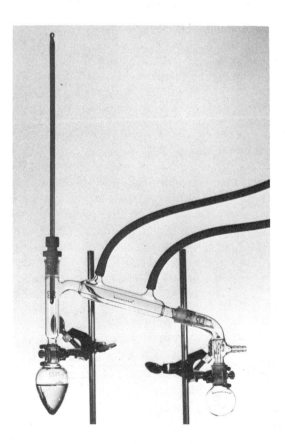

Fig. 2.8. Simple distillation apparatus employing 14/20 precision-ground glassware. (Photograph by courtesy of the Kontes Glass Company, Vineland, N.J.; photograph by Bruce Harlan, Notre Dame, Ind.)

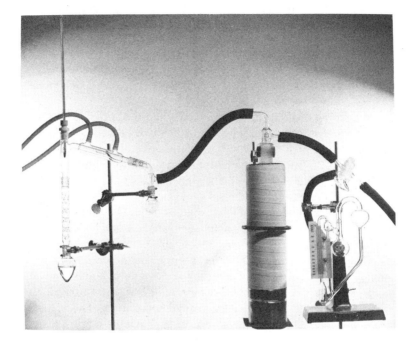

Fig. 2.9. Typical setup for a vacuum distillation including a fractionating column, cold trap, and vacuum gauge. (Photograph by courtesy of the Kontes Glass Company, Vineland, N.J.; photograph by Bruce Harlan, Notre Dame, Ind.)

15 mm-Hg, or a vacuum pump capable of producing pressures down to 0.01 mm-Hg. Figure 2.9 shows the recommended setup for a vacuum distillation, including a vacuum gauge for determining the pressure at which the distillation is being carried out, a cold trap (to be used only when a vacuum pump is employed) to prevent vapors being drawn into the vacuum pump, and an air bleed to introduce air into the system to bring the internal pressure back to atmospheric pressure before disassembling the apparatus. The vacuum should never be released by the removal of a flask or the thermometer. A safety shield should be routinely used to shield the experimentalist carrying out a vacuum distillation. The comparison of boiling points obtained at different pressures is readily accomplished by use of a boiling-point nomograph (Appendix I).

The sample is placed in the distilling flask; a boiling chip or a boiling stick is then added to promote a smooth and continuous boiling action. In many cases, boiling chips or sticks do not provide sufficient agitation of the sample to promote a smooth and continuous boiling action. In such cases it is necessary to provide the necessary agitation by means of a slow passage of nitrogen or argon through a capillary into the liquid. A magnetic stirrer can also be used. The distillation flask should be heated by means of an oil or wax bath: never with a heating mantle or the flame of a Bunsen burner. Adequate control of the temperature of the distillation flask generally cannot be maintained with the latter two methods of heating.

2.3.2 Fractional Distillation

Fractional distillation is employed when the separation of two or more volatile components is required. The principle of fractional distillation is based on the establishment of a large number of theoretical vaporization-condensation cycles (Sec. 2.3). A fractionating column that allows equilibrium of the descending condensed liquid with the ascending vapors is used, thus producing the effect of a multiple vaporization-condensation cycle.

The length and type of fractionating column required depends on the boiling points of the components to be separated. Suitable separations of components differing in boiling points by 15 to 20°C can be accomplished by means of a Vigreux column (Fig. 2.10). For separations of components with closer boiling points, packed columns or spinning-band-type columns can be used (Fig. 2.10).

Equilibrium conditions must be maintained in the fractionating column at all times in order to achieve a successful separation. The ratio of distillate to the amount of condensed material returning to the distillation flask (referred to as the reflux ratio) should always be much larger than 1, generally in the region of 5 to 10 for relatively easily separated components. Maintaining the reflux ratio in this region requires a very careful control of the amount of heat applied to the distillation flask. Flooding of the column should be avoided at all times. Fractional distillations carried out under reduced pressures should involve the use of fraction cutting devices, as shown in Fig. 2.10. These devices do not require the breaking of the vacuum, which would in turn destroy the equilibrium conditions established on the fractionating column.

2.3.3 Microdistillation

Frequently the chemist is faced with the problem of distilling milligram quantities of a liquid. Small quantities of liquids recovered from preparative chromatographic separations require a final distillation to remove solvents or high-boiling residues (such as the liquid stationary phases used in gas chromatography which are continually eluted from the column) prior to elemental analysis or recording of the physical properties of the sample. Such microdistillations can be accomplished in several ways.

A common method in research laboratories involves a Kugelrohr (German: bubble tube) apparatus (Fig. 2.11). The glass bubble tube can be blown from a single tube or assembled in sections using ground glass joints. The material to be distilled is placed in the end bulb (not more than one-third full). With the bubble tube held horizontal, the filled bulb is placed in a heated air bath. The bulbs are

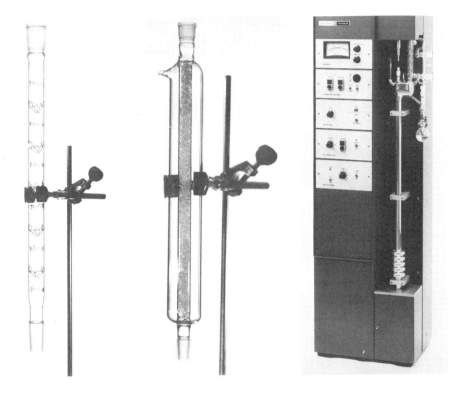

Fig. 2.10. Typical fractionating columns. (*a*) Vigreux column. (*b*) Glass helices-packed, vacuum-jacketed fractionating column (Photographs by Bruce Harlan, Notre Dame, Ind.) (*c*) Spinning band column. A tightly-fitted Teflon spinning spiral contacts the glass walls of the column, wiping the descending condensate into a thin film and exposing almost each molecule to the heated ascending vapors. The still shown has 200 theoretical plates and is capable of separating materials with boiling point differences of less than 0.5°C. (Photograph by courtesy of Perkin-Elmer Corporation, Norwalk, Conn.)

gently rocked by an oscillating motor, which speeds distillation and avoids bumping. The distillation is carried out under reduced pressure (as low as 10^{-5} mm), and the distillate is collected in the adjacent air- or water-cooled bulb.

Other devices can be used to achieve such simple "trap-to-trap" distillations. A simple double-U tube still is shown in Fig. 2.12. The sample is introduced into one of the U sections, and that end of the tube is sealed. The portion of the U tube containing the sample is cooled by means of an ice-water or Dry Ice–acetone bath, depending on the boiling point of the sample, and a vacuum is applied at the other end. In this manner, volatile solvents can be removed effectively. The cold trap is removed from the portion of the double-U tube containing the sample, and the other (empty) portion of the U tube (the receiver) is placed in the cold bath.

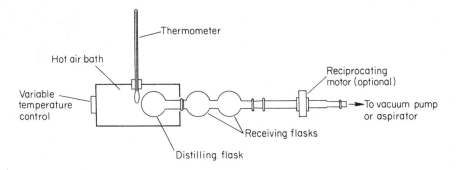

Fig. 2.11. Kugelrohr distillation apparatus. Samples from 50 mg to 5 g can be quickly distilled. A light plug of glass wool added to the distilling flask will help to prevent bumping. (A unit is available from Aldrich Chemical Company Inc., Milwaukee, Wis.)

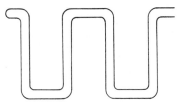

Fig. 2.12. Microdistillation tube.

Heat is then applied to the portion of the U tube containing the sample, and the sample condenses in the U tube immersed in the cold bath, leaving behind the non-volatile impurities.

A more sophisticated apparatus is illustrated in Fig. 2.13. The sample is placed in the inner sample space; it is vaporized, condensed on the cold finger, and collected in the small flask at the bottom of the apparatus. The collection flask can also be immersed in a cold bath. The value of this apparatus lies in the greater control of the heating source and pressure. Sample sizes from 10 to 200 mg can be conveniently distilled in this apparatus. Distillation in this apparatus is essentially a molecular distillation, and very little fractionation can be accomplished, as is true with all microdistillations.

2.3.4 Steam Distillation

A *heterogeneous* mixture of two liquids (A and B) does not obey Raoult's Law, but each component exerts a partial vapor pressure (P_A° or P_B°) that is the same as that for the pure substance at any given temperature. In other words, the partial pressure of each component of a heterogeneous mixture is dependent only on

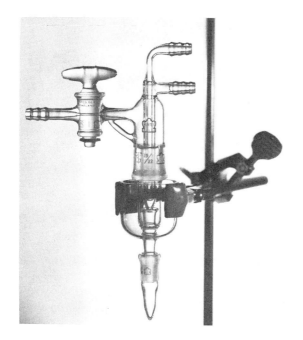

Fig. 2.13. Micromolecular distillation apparatus. (Marketed by and photograph by courtesy of the Kontes Glass Company, Vineland, N.J.; photograph by Bruce Harlan, Notre Dame, Ind.)

temperature. When $P_A^o + P_B^o$ equals the applied pressure, the mixture boils. Since P_A^o and P_B^o are additive, the boiling point of the mixture is always *below* the boiling point of the lowest-boiling component (cf., ideal homogeneous solutions, Sec. 2.3). The boiling point of the mixture and the composition of the distillate will remain constant as long as there is some of each constituent present. Since the molar concentration of each component in the vapor is proportional to its vapor pressure, the composition of the distillate is given by the following expression, where P_A^o and P_B^o are the partial pressures of the components at the temperature at which distillation occurs [Eq. (2.2)].

$$\frac{\text{moles of } A}{\text{moles of } B} = \frac{P_A^o}{P_B^o} \quad \text{or} \quad \frac{\text{wt. } A}{\text{wt. } B} = \frac{P_A^o \cdot \text{mol. wt. } A}{P_B^o \cdot \text{mol. wt. } B} \tag{2.2}$$

If water is employed as one phase in the distillation of a two-phase immiscible system, the method is called steam distillation. A practical application of steam distillation is provided by the example of bromobenzene. Bromobenzene (bp 156°C) and water (bp 100°C) are, for all practical purposes, insoluble in each other. At 95°C the combined vapor pressures of bromobenzene (120 mm) and water (640 mm) equal 760 mm, and the mixture boils. The composition of the distillate, as long as some bromobenzene remains in distillation flask, is given by Eq. (2.3).

$$\frac{\text{wt. C}_6\text{H}_5\text{Br}}{\text{wt. H}_2\text{O}} = \frac{P_{\text{C}_6\text{H}_5\text{Br}} \cdot \text{mol. wt. C}_6\text{H}_5\text{Br}}{P_{\text{H}_2\text{O}} \cdot \text{mol. wt. H}_2\text{O}}$$

$$= \frac{120 \cdot 157}{640 \cdot 18} = \frac{1.64 \text{ g of bromobenzene}}{\text{g of water}}$$

(2.3)

Steam distillations are very useful for separating volatile components from relatively non-volatile components, particularly when the more volatile component possesses a very high boiling point and may be subject to decomposition if a direct distillation is attempted. This technique is also useful if the presence of other lesser volatile impurities causes extensive destruction of the desired fraction under normal distillation conditions. Steam distillation can often be used to separate isomeric compounds, particularly when one isomer is capable of extensive intramolecular hydrogen bonding; hence, that isomer is made more volatile, whereas the other isomers can participate only in intermolecular hydrogen bonding. A mixture of o- and p-nitrophenol can be separated by steam distillation. The ortho isomer, which exhibits intramolecular hydrogen bonding, readily steam distills, whereas the para isomer remains in the distillation flask.

A typical steam distillation setup is illustrated in Fig. 2.14. Small quantities of materials can be steam-distilled by the direct distillation of a mixture of the sample and water, deleting the external steam source and a water entrainment trap. The organic material is then recovered from the distillate by extraction with an organic solvent.

2.4
SUBLIMATION

Many solid substances pass directly into the vapor phase when heated and, on cooling, pass directly back to the solid phase. Such a process is termed *sublimation*

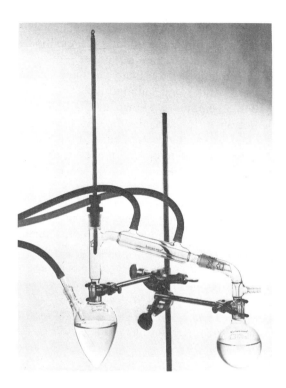

Fig. 2.14. Steam distillation setup with an external steam source. (Photograph by courtesy of the Kontes Glass Company, Vineland, N.J.; photograph by Bruce Harlan, Notre Dame, Ind.)

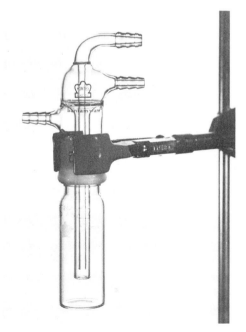

Fig. 2.15. Vacuum sublimation apparatus. (Photograph by courtesy of the Kontes Glass Company, Vineland, N.J.; photograph by Bruce Harlan, Notre Dame, Ind.)

and, as such, provides a useful method for the purification of materials. Unfortunately, however, relatively few organic substances undergo sublimation at atmospheric pressure. Generally, sublimations are carried out under reduced pressures. Types of compounds that readily sublime include many α-amino acids, ketones, carboxylic acids, and most acid anhydrides and quinones.

A typical vacuum sublimation apparatus is shown in Fig. 2.15. The finely divided sample is placed in the bottom of the cup, and the cold-finger condenser is then inserted into the container cup. A vacuum is applied, and the temperature of the sample is slowly increased until sublimation occurs; however, the temperature of the sample *must* be kept well below the melting point of the sample. The flow of coolant to the condenser is shut off, the condenser is removed, and the sample is scraped from the cold finger. Repeated sublimations may be required to obtain suitably pure material. If the apparatus such as that shown in Fig. 2.15 is not available, the sample can be placed in a side-arm test tube, with a small test tube inserted to act as the condenser. The inner tube can be filled with ice or constructed as a condenser.

2.5
EXTRACTION

The distribution of a substance between two immiscible phases is the basis of extraction. The distribution equilibrium constant K is defined in Eq. (2.4),

$$K = \frac{[C_A]}{[C_B]} \tag{2.4}$$

where $[C_A]$ is the equilibrium concentration of a substance in phase A and $[C_B]$ is the equilibrium concentration in phase B. The distribution coefficient can be calculated using the limiting equilibrium solubilities of the material in the two phases. In actuality, the equilibrium constant expression may be more complex than is illustrated in Eq. (2.4) in that the substance may exist in dimeric or higher polymeric forms in one or both of the phases, necessitating the introduction of coefficients and exponents in Eq. (2.4) to adequately describe the concentration dependences in the two phases.

In extraction procedures normally employed in organic chemistry, one phase is aqueous and the other phase a suitable organic solvent. A change in solvent will change the solubility in that phase and thus alter the distribution coefficient. In general, the rule of "like dissolves like" is a suitable guideline in choosing the appropriate solvent. The organic solvent should be relatively volatile for ease of removal after the extraction has been accomplished.

2.5.1a Procedure for simple extraction

A separatory funnel supported by a ring is charged with the cool aqueous solution to be extracted. Add the organic solvent. (Do not add volatile solvent to a hot solution or fill the separatory funnel to more than three-quarters full.) Place the stopper in the funnel. Remove the funnel from the supporting ring. Grasp the funnel with both hands so that the stopper and stopcock are held firmly in place. (The stopper should be against the palm of one hand, and the fingers of the other hand should be around the handle of the stopcock.) Shake the funnel once or twice so that the phases mix. With the stopcock end pointed upward and away from all persons, vent the stopcock to release excess pressure. Repeat several times, making sure that the phases have been vigorously intermixed. Replace the funnel in the support ring, remove the stopper, and allow the phases to separate. Drain the lower phase slowly through the stopcock, and just as the interface of the two phases reaches the stopcock, close it. Be sure that you can distinguish between the two phases so you do not discard your product. In aqueous extractions, organic layers heavier than water will be the lower phase, and those lighter than water will be the upper phase (Table 2.3). A simple test that will usually allow one to distinguish between an organic phase and an aqueous phase is to place a drop of each phase on the edge of a circle of filter paper; the spot that is wet with water will tear easily while the other spot will resist tearing.

Table 2.3 Common Solvents for Extraction of Aqueous Solutions

Lighter than Water	Heavier than Water
Ether	Dichloromethane
Pentane, hexane, petroleum ether	Chloroform
Benzene, toluene	Carbon tetrachloride

In any extraction, each phase becomes saturated with respect to the other phase. When extracting an aqueous phase with an organic phase, it is necessary to remove the dissolved water before recovering the extracted material by evaporation of the solvent. Generally most of the dissolved water in the organic phase can be removed by washing the organic phase with a saturated aqueous solution of sodium chloride. Final drying of the organic phase is accomplished by allowing the organic phase to stand for a period of time over an anhydrous inorganic salt (see Drying Agents, Sec. 1.14.1).

Various materials can be added to either phase, to either increase or decrease the solubility of a substance in that phase. The solubility of a material in one phase can be increased by the addition of a reagent that is capable of forming a stable, soluble complex in that phase. For example, many reactive alkenes and polyenes can be extracted into an aqueous phase in the presence of silver ion, or chelating agents can be extracted either into or from an aqueous phase in the presence of a complexing metal ion. These complexes can be subsequently degraded, and the desired material recovered.

The solubility of a substance can be decreased, particularly in aqueous phases, by the addition of neutral salts that reduce the solubility of the substance in that phase. Various salts can be used, sodium chloride probably being the most widely used. Other salts that can be similarly used are sodium sulfate, potassium carbonate, and calcium chloride.

The distribution coefficient for acidic or basic materials can be greatly altered by changing the pH of one phase. For acidic materials, one can show that the distribution coefficient between two phases, one of which is the aqueous phase with a hydrogen ion concentration of $[H^+]$, can be represented by an effective distribution coefficient K_{eff} [Eq. (2.5)], where K_D is the distribution coefficient between an organic phase and pure water, and K_A is the ionization constant of the acid.

$$K_{eff} = \frac{K_D K_A}{[H^+]} \tag{2.5}$$

This approach is the basis for the facile separation of strong acids from weak acids by extraction with aqueous bicarbonate. After extraction of the acid or base from an organic phase, the compound can be recovered from the aqueous phase by acidification or basification.

In many situations, however, it is not possible to obtain a favorable distribution coefficient by any of the foregoing procedures. When an unfavorable distribution coefficient is encountered, it can be shown by Eq. (2.6) that several extractions with small volumes of the extracting phase will be more efficient than one extraction utilizing all of the extracting phase.

$$[C_A] = [C_A^\circ]\left(\frac{V_A}{KV_B + V_A}\right)^n \tag{2.6}$$

$[C_A]$ and $[C_A^\circ]$ are the final and original concentrations of the material in phase A, K is the distribution coefficient, V_A is the volume of phase A, V_B is the volume of the extracting phase used in each extraction step, and n is the number of extraction steps.

Continuous liquid-liquid extractions may be required when particularly unfavorable distribution coefficients are encountered. Continuous liquid-liquid extractors suitable for use with lighter- and heavier-than-water organic phases are shown in Figs. 2.16 and 2.17.

Another type of liquid-liquid extraction is termed *countercurrent distribution*. Countercurrent distribution involves two immiscible phases that are in contact with each other as they flow from tube to tube in opposite directions. The apparatus is elaborate and its use is normally restricted to specialized separations. The extracting phase is introduced at one end of the apparatus and encounters the phase to be extracted progressing from the other end.

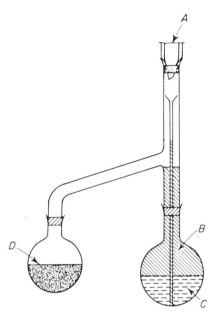

Fig. 2.16. Diagram of a simple continuous extractor for use with lighter-than-water extracting phase: (A) condenser, (B) lighter-than-water extracting phase, (C) water phase being continuously extracted, and (D) distillation reservoir. The solvent is vaporized from D, condensed in the condenser A, and conducted to the bottom of the extracting flask by the small inner tube. The organic phase percolates up through the water phase and returns to the distillation flask, where the extracted material concentrates.

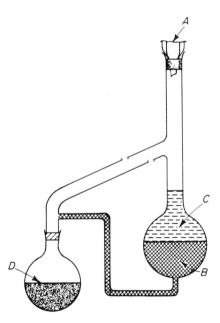

Fig. 2.17. Diagram of a continuous extractor for use with heavier-than-water extracting phases: (A) condenser, (B) heavier-than-water extracting phase, (C) water phase being continuously extracted, and (D) distillation reservoir. The extracting solvent is vaporized from the distillation flask, condensed in the condenser, the drips down through the phase being extracted. The extracting phase returns to the distillation flask by the small return tube, and the material extracted concentrates in the distillation flask.

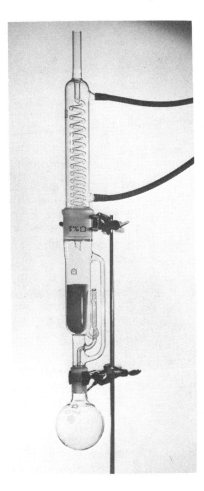

Fig. 2.18. Soxhlet extractor for continuous liquid-solid extractions. (Photograph by courtesy of the Kontes Glass Company, Vineland, N.J.; photograph by Bruce Harlan, Notre Dame, Ind.)

Continuous liquid-solid extractions can be carried out by the Soxhlet extractor (Fig. 2.18). The finely ground solid is placed in the extraction shell, and fresh solvent is allowed to percolate through the sample, the extracted material accumulating in the solvent distillation flask.

2.6
CHROMATOGRAPHY

2.6.1 Introduction

Chromatography can be defined as the separation of a mixture into various fractions by distribution between two phases, one phase being stationary and essentially two-dimensional (a surface), and the remaining phase being mobile. The underlining principle of chromatography is that different substances have different partition coefficients between the stationary and mobile phases. A compound that interacts weakly with the stationary phase will spend most of its time in the mobile phase and move rapidly through the chromatographic system. Compounds that interact strongly with the stationary phase will move very slowly. In the ideal case, each component of a mixture will have a different partition coefficient between the mobile and stationary phases, and consequently each will move through the system at a different rate, resulting in complete separations. Various types of chromatography are possible, depending on the physical states of the phases. Employing a gas as the mobile phase is termed *gas chromatography (gc)* or *vapor phase chromatography (vpc)*. Separations using gas chromatography involve vapor phase *vs.* adsorption and/or solution equilibria. *Liquid chromatography (lc)* refers to any chromatographic process that employs a mobile liquid phase.

Liquid chromatography is inherently more versatile in that it is not limited by sample volatility, and separations are the result of specific interactions of sample molecules with *both* the stationary and mobile phases. (Such specific interactions are absent in the mobile phase of gc.) There are four basic types of liquid chromatography. *Liquid-liquid*, or *partition chromatography*, involves a stationary phase of an immiscible liquid coated on particles of an inert support, or, in some special cases, a stationary phase chemically bonded to the support surface. *Liquid-solid*, or *absorption chromatography*, involves partition between solution in the mobile liquid phase and adsorption on high-surface-area particles acting as the stationary phase. *Gel*, or *exclusion chromatography*, employs stationary phases that possess cavities or pores of rather uniform size into which molecules of appropriate size can enter and be retained. Molecules of size inappropriate to the cavities move rapidly with the mobile phase. The two basic types of materials used in exclusion chromatography are molecular sieves (inorganic silicates with pore sizes of 3 Å to 10 Å) and organic polymeric gels (e.g., Bio-Gels-P and Sephadex Gels, which have an operating range of molecular weights of 200 to 400,000). Gel chromatography is an important technique in biochemistry. The stationary phase in *ion-exchange chromatography* is an organic resin substituted with sulfonic or carboxylic acid, amine, or quaternary ammonium groups that act as ion-exchange donor groups. The position of ion-pairing equilibria involving ionic samples and the ionic groups on the resin is controlled by changing the pH or some ion concentration in the mobile phase. Ion exchange resins find common use in water purification, in water-softening operations, in amino acid analyses, and in inorganic separations.

All types of chromatography are useful for analytical purposes. Under appropriate conditions, all types of chromatography can be used for preparative scale separations, In every type of chromatography there are three elements to be considered:

> Load (sample size)
> Resolution (relative separation of components)
> Speed

Of course, it would be ideal if all three elements could be maximized so that complete separation of samples of any desired size could be quickly achieved. In practice, generally any two of these elements can be maximized at the expense of the third. For routine analytical work, resolution and speed are maximized at the expense of load. In preparative scale separations load and speed can be maximized, but then separations are usually incomplete. Complete separations of large samples can be achieved, but the overall operation is likely to be slow and tedious. Keep these elements in mind as we examine the chromatographic techniques of greatest interest in organic chemistry in the sections that follow.

2.6.2 Adsorption Chromatography

The separation of the components of a mixture by adsorption chromatography depends on adsorption-desorption equilibria between compounds adsorbed on the surface of the solid stationary phase and dissolved in the moving liquid phase. The extent of adsorption of a component depends on the polarity of the molecule, the activity of the adsorbent, and the polarity of the mobile liquid phase. The actual separation of the components in a mixture is dependent on the relative values of the adsorption-desorption equilibrium constants for each of the components in the mixture.

In general, the more polar the functional groups of a compound, the more strongly it will be adsorbed on the surface of the solid phase. Table 2.4 lists typical classes of organic compounds in the order of elution (increasing polarity). Minor alterations in this series may occur if a functional group is highly sterically protected by other less polar portions of the molecule.

Table 2.4. Compound Elution Sequence

Saturated hydrocarbons
Alkenes
Aromatic hydrocarbons,
    halogen compounds
Sulfides
Ethers
Nitro compounds
Aldehydes, ketones, esters
Alcohols, amines
Sulfoxides
Amides
Carboxylic acids

The activity of the adsorbent depends on its chemical composition, the particle size, and the porosity of the particles. Table 2.5 lists the more common adsorbents in the usual order of increasing activity.

Table 2.5. Adsorbents for Adsorption Chromatography

Cellulose Starch Sugars Silicic acid (silica gel) Florisil (magnesium silicate) Aluminum oxide (alumina) Activated charcoal	General order of increasing activity

The choice of the proper adsorbent will depend on the types of compounds to be chromatographed. Cellulose and starch are used primarily with very labile, polyfunctional plant and animal products. Silica gel is the adsorbent with the widest variety of organic applications. It is available over a large range of particle sizes (10 to 150 μm) and porosities. An advantage of a silica gel column is that it is translucent in chloroform and in chloroform containing up to 5% ethanol. Colorless compounds can often be observed as opaque bands as they move down the column. Above 5% ethanol in chloroform the column becomes opaque and this advantage is lost. Many chromatograms can be started with pure chloroform and eluted with chloroform containing increasing amounts of ethanol. Florisil has properties similar to those of silica gel, and it can be used for the separation of most functional classes, including carboxylic acid derivatives and amines.

Alumina, a widely used adsorbent, can be obtained in three forms: acidic, basic, and neutral. Acidic alumina is an acid-washed alumina giving a suspension in water with pH of approximately 4. This alumina is useful for the separation of acidic materials such as acidic amino acids and carboxylic acids. Basic alumina (pH of approximately 10) is useful for the separation of basic materials such as amines. Neutral alumina (pH of approximately 7) is useful for the separation of nonacidic and nonbasic materials.

In addition to varying the acid-base properties of the adsorbent, the activity of alumina can be varied by controlling the moisture content of the sample. For example, alumina can be prepared as activity grades I through V (Brockman scale), in which the activity decreases as the activity grade number increases. In general, the more active grades of alumina should be used when the separation of the components of a mixture is difficult. Detailed procedures are available for the preparation and standardization of the various types and grades of alumina in ref. 9, p. 86, in Sec. 2.8. Acidic, basic, and neutral activity I aluminas are commercially available[1] and can be converted to the other activity grades by the addition of 3% (activity II), 6% (III), 10% (IV), and 15% (V) by weight of water.

The use of the more active adsorbents, particularly alumina, may lead to the destruction of certain types of compounds during chromatography. Owing to the

---

[1] Woelm alumina, distributed by Waters Associates, Inc., 61 Fountain Street, Framingham, Massachusetts 01701.

presence of water in the lower-activity grades of alumina, many esters, lactones, and acid halides may undergo hydrolysis, the resulting carboxylic acid being very strongly adsorbed. The low-molecular-weight aldehydes and ketones undergo extensive aldol and ketol condensations on the surface of the more active adsorbents. The correct choice of adsorbent will be contingent on some knowledge of the types of compounds to be separated. Even then, a small trial chromatographic separation on a small portion of the sample is recommended to determine the proper conditions for separation of the remainder of the sample.

In addition to controlling the pH of the adsorbent, the properties of the adsorbent can be altered by coating the adsorbent with various chemicals. For example, silver nitrate-coated alumina, prepared by dissolving 20% by weight silver nitrate in aqueous methanol and slurrying with the alumina, followed by removal of the solvent, is particularly useful for the separation of alkenes. A stationary liquid phase, for example, water, ethylene glycol, or a low-molecular-weight carboxylic acid, can be applied to some adsorbents. Separations employing such preparations are termed liquid-liquid chromatographic separations.

The recovery of the material from a chromatogram can be accomplished in either of two ways. The individual components of the mixture can be separated and developed into distinct bands on the column by allowing a solvent, or solvent mixture, of sufficient polarity to proceed down the column affecting the separation. If the individual bands can be discerned by their color, fluorescence under the influence of ultraviolet light, or reaction with a colored indicator, the developed chromatogram can be extruded from the column and the separated components recovered by leaching cut portions of the column with a solvent of high polarity. The general procedure is, however, to continually flush the column and, if necessary, with solvents of increasing polarity, to remove each component individually. This type of procedure is generally referred to as elution chromatography.

The solvents generally used as eluents are listed in Table 2.6 in the order of increasing polarity. Often a pure solvent or single solvent system (e.g., 5% diethyl ether in hexane) can be used to elute all components. In other cases gradient solvent systems are used. Since the entire separation is dependent on the establishment and maintenance of equilibrium conditions, in gradient elution the polarity of the solvent system is gradually increased by the slow increase of the concentration of a more polar solvent. The rate of change of solvent polarity will depend upon the similarity or dissimilarity of the components in the mixture to be separated. If closely related compounds are to be separated, for example, isomeric alkenes or alcohols, the change in solvent composition should be very gradual. The separation of compounds having distinctly different functional groups can often be accomplished by a much more rapid change in solvent composition.

The choice of eluent solvents will depend on the type of adsorbent used and the nature of the compounds to be chromatographed. The solvents must be of high purity. The presence of traces of water, alcohols, or acids in the lesser polar solvents will alter the adsorption activity of the adsorbent.

Table 2.6.   Eluotropic Series

Petroleum ether	
Cyclohexane	
Carbon tetrachloride	
Benzene	
Dichloromethane	
Chloroform (alcohol free)	
Diethyl ether	Increasing
Ethyl acetate	polarity
Pyridine	
Acetone	
1-Propanol	
Ethanol	
Methanol	
Water	
Acetic acid	↓

In many respects, the application of chromatographic separation procedures is more an art than a science. Experience gained in the use of these techniques leads to a more efficient use of these techniques in future work. When faced with a separation problem involving a totally unknown mixture, it is usually best to carry out a crude and rapid trial chromatographic separation with a small portion of the mixture, using the information and experience gained in the trial run to carry out a more efficient separation on the remainder of the material. Such a procedure usually results in a saving of time and material. Thin-layer chromatography (Sec. 2.6.3) can often serve as an excellent guide to the conditions for column chromatography. For a discussion of the relationship between thin-layer and column chromatography see Sec. 2.6.3b.

2.6.2a   Techniques of Column Chromatography

The success of an attempted chromatographic separation depends on the care exercised in the preparation of the column and in the elution procedure. Figure 2.19 shows a typical adsorption chromatography column. The container is usually a glass tube provided with some means of regulating the flow of eluent through the column (stopcock preferably made of Teflon). If a ground-glass stopcock is used to control the rate of flow of solvent, all lubricating grease should be removed. Stopcock grease is quite soluble in many of the organic solvents used for elution purposes and may be leached from the stopcock, contaminating the material being eluted from the column.

The actual packing of the column is very important. The adsorbent must be uniformly packed, with no entrainment of air pockets. A plug of glass wool or cotton is placed in the bottom of the column, and a layer of sand is added on top of the plug to provide a square base for the adsorbent column. Failure to provide

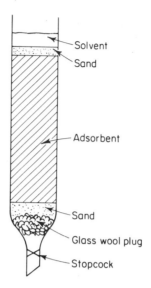

Fig. 2.19. Construction of a chromatography column.

a square base for the column may well lead to elution of more than one fraction at a time if the separation is not very great, as is illustrated in Fig. 2.20.

The packing of the column can be accomplished in several ways. The adsorbent can be dry-packed, the solvent being added after the adsorbent. However, this method is generally not suitable because air pockets become lodged within the column and are difficult to remove; in addition, some adsorbents undergo an expansion in volume when wetted with a solvent, thus bursting the column (for example, silica gel and many ion-exchange resins). The recommended method involves filling the container with petroleum ether, or the least polar solvent to be used, and then adding the adsorbent in a continuous stream until the desired amount of adsorbent has been added. The tube should be maintained in a vertical position to promote uniform packing. Adsorbents that are not of uniform

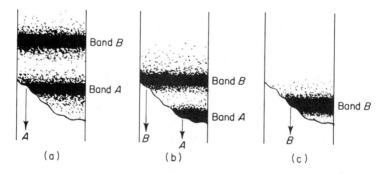

Fig. 2.20. Elution of successive bands from a chromatography column that is not square at the bottom. (*a*) Elution of *A*. (*b*) Elution of *A* and *B*. (*c*) Elution of *B*.

size and possess a low density tend to produce columns in which layering of large and small particles occurs when filled in this manner. Florisil and silica gel are notorious in this respect. In such cases, the adsorbent is slurried with the solvent, and the slurry is rapidly added to the empty column. This usually prevents segregation of particle sizes in the column. A layer of sand is added to the top of the column to protect the top of the column from being disturbed during the addition of solvent during the elution process.

The quantity of adsorbent and the column size required will depend on the type of separation to be carried out. Generally 20 to 30 g of adsorbent is required per gram of material to be chromatographed. Ratios as high as 50 or 100 to 1 may be required in cases when the components of a mixture are all quite similar. The height and diameter of the column are also important. Too short a column may not provide a sufficient length of column to effect the separation. In general, a height-to-diameter ratio of 8:1 to 10:1 is recommended.

After the column has been prepared, the solvent level is lowered to the top of the column by draining from the bottom of the column. The material to be separated is dissolved in the least polar solvent where solubility can be attained (10 to 15 mL/g of material) and carefully added to the top of the column. Elution is begun with a solvent system corresponding to the composition of the solvent used to dissolve the sample. Occasionally some highly polar components present in mixtures may require a solvent system of such highly polar character for complete solubility that the lesser polar components are immediately eluted from the column with no separation being achieved. In such cases, the sample may have to be added to the column as a two-phase system; however, this may lead to problems of column congestion.

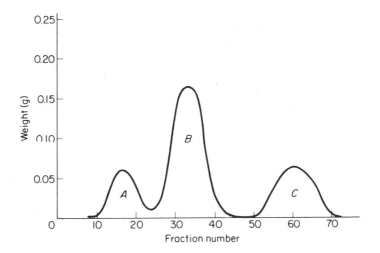

Fig. 2.21. Typical weight *vs.* fraction-number chromatogram.

The rate of change of the composition of the eluting solvent will depend on the type of mixture being separated. Elution with one solvent composition should continue until a decrease in the amount of material being eluted is noted. The fractions should be evaporated as soon as possible, and the weight recovered in each fraction recorded and plotted versus fraction number. A typical plot of a chromatographic separation appears in Fig. 2.21. A decrease in the weight per fraction indicates the approaching end of a component band. The fractions appearing under one peak can be combined for subsequent purification and identification steps. In general, one should recover 90 to 95% of the material placed on the chromatographic column.

2.6.3 Thin-Layer Chromatography

Thin-layer chromatography (TLC) is a special application of adsorption chromatography in which a thin layer of adsorbent supported on a flat surface is utilized instead of a column of adsorbent. Elution or, more properly, development of the chromatogram is accomplished by capillary movement of the solvent up the thin layer of adsorbent. Unfortunately, one is restricted to the use of a single solvent system; however, after use of one solvent system, the chromatogram can be dried and developed further by use of a second solvent system, either in the same direction or at right angles to the direction of the first development.

The most common adsorbents used in TLC in order of importance are silica gel, alumina, keiselguhr, and cellulose. The first two are far more important for general use than the last two. The adsorbents for use in TLC are more finely divided than those for column chromatography, and they usually contain plaster of Paris as a binding agent. For these reasons, only the commercially available adsorbents especially prepared for TLC should be employed.

The choice of the best eluent for TLC will depend on the type of adsorbent and the components in the mixture to be separated. The same general rules for column adsorption chromatography also apply to TLC. However, since only one solvent system is generally used in TLC, trial chromatographic separations with a variety of solvents should be carried out to determine which solvent leads to the best separation. A series of small individual plates (microscope slides are excellent for this purpose) are prepared, spotted, and developed in the different solvents.

Visualization of the chromatogram to locate the position of each of the components in the separated mixture will depend on the type of molecules present in the original mixture. If all the components are colored, then visual inspection is sufficient to locate the spots. In TLC the most common methods for visualizing compounds are visualization by iodine vapors, the use of ultraviolet light on phosphor-containing layers, and charring with sulfuric acid.

The most commonly used indicator is iodine vapor. The developed and dried plate is placed in a closed container containing a few crystals of iodine. The iodine vapor is adsorbed into the areas of the plate containing organic compounds;

brown spots due to iodine charge-transfer complexes appear on a white background. Almost all organic compounds can be detected with this technique. The method is usually nondestructive. The spots can be marked and the plate gently warmed to allow the iodine to sublime out of the layer, leaving the compounds unchanged.

Inert, inorganic phosphors can be added to the slurry prior to the preparation of TLC plates. Such phosphors are frequently supplied with the commercial plaster-of-Paris-containing adsorbents especially prepared for TLC. These phosphors emit visual light when excited with ultraviolet radiation. Many organic compounds quench this emitted light and a dark spot will appear on a bright layer. The method has the advantages of simplicity and high sensitivity as well as being nondestructive.

Many organic compounds char when heated with sulfuric acid. A small amount of sulfuric acid can be incorporated into the slurry prior to the formation of the plates, or the developed chromatogram can be sprayed with sulfuric acid. Charring is then accomplished by heating the developed plate over a hot plate or in a drying oven at 100 to 110°C.

Many reagents used for qualitative functional tests can also be used for the visualization of TLC plates. For example, acid-base indicators for acidic or basic compounds, ferric chloride solution for phenols and enols, 2,4-dinitrophenyl-hydrazine for ketones and aldehydes, dichromate reagents, etc. can be used as reagents. The use of more than one developing reagent may be required to detect all the components present in any given mixture. This can be carried out on a single chromatogram or, if one indicator interferes with subsequent tests, on different plates.

A value of TLC lies in the relatively small amount of time required per analysis. With thin-layer chromatography on microscope slides, the entire operation from the preparation of the layer to the visualization of the spots can be accomplished in about 10 min. TLC is very useful for monitoring the progress of reactions, in detecting intermediates in reactions, in analyzing crude products or unknown mixtures to determine the number of components, and for checking the efficiency of purification processes.

TLC can be used for relatively small preparative separations employing thicker layers of absorbent (~ 2 mm) and applying the sample as a band instead of spots. Only a narrow section of the chromatogram is developed by use of indicators to locate the bands of material. The bands are then scraped from the plate, and the material is leached from the adsorbent.

TLC can also be used as a tentative means of identification. If plate conditions are kept constant, a compound in several initial spots will progress up the plate at the same rate relative to an added standard or to the solvent front.
displacement is referred to as

$$R_x = \frac{\text{displacement of compound}}{\text{displacement of standard}} \tag{2.7}$$

or

$$R_f = \frac{\text{displacement of compound}}{\text{displacement of solvent front}} \qquad (2.8)$$

For identification purposes, a comparison of a known and unknown should be carried out with a number of different solvent systems on several adsorbents. R_x and R_f values from the literature are not reliable enough for use in final identification; they should be used only as general guides. Whenever possible, direct comparison of a known and unknown should be made. It is wise to spot the known, the unknown, and a mixture of the two on a single chromatogram. Even with direct comparison, a final unambiguous identification is not possible in that different compounds may have identical or very similar adsorption properties.

2.6.3a Techniques of Thin-Layer Chromatography

Chromatography on Microscope Slides

The chemist involved in problems of organic structure determination will find TLC on microscope slides (Fig. 2.22) constantly useful in his work. The remarkable utility and convenience of this method for monitoring reactions and checking composition and purity cannot be overemphasized.

Preparation of the plates.   The slides for TLC are most conveniently prepared by dipping the slides into a slurry of commercial TLC adsorbents in organic solvents. For the slurry use 35 g of silica gel G or 60 g of alumina in 100 mL of chloroform-methanol (2 : 1 v/v). Shake or stir the slurry for about 2 min before use. The slurry can be stored for some time in tightly sealed containers. The plates are prepared as follows:

1. Dip two *clean, dry* microscope slides held back-to-back into the slurry. Remove the slides slowly and allow them to drain on the edge of the container. If the layers are thin and grainy, the slurry should be thickened by the addition of more adsorbent.
2. Separate the two slides and allow them to dry on the desk for 5 min. Remove the excess adsorbent from the edges. Silica gel and alumina plates prepared in this manner are ready for use. If they are not to be used immediately, store them in a dry atmosphere.

Application of the sample.   The most convenient applicators are made by pulling the center of an open melting-point capillary to a very fine capillary and breaking at the center. Apply small spots of a dilute solution of the sample in a volatile solvent about 10 mm from the end of the slide. If more than one application is necessary, allow the spot to dry before reapplying more sample. With care, up to three samples can be placed on one slide. Allow the spotting solvent to evaporate before placing the slide in the developing chamber.

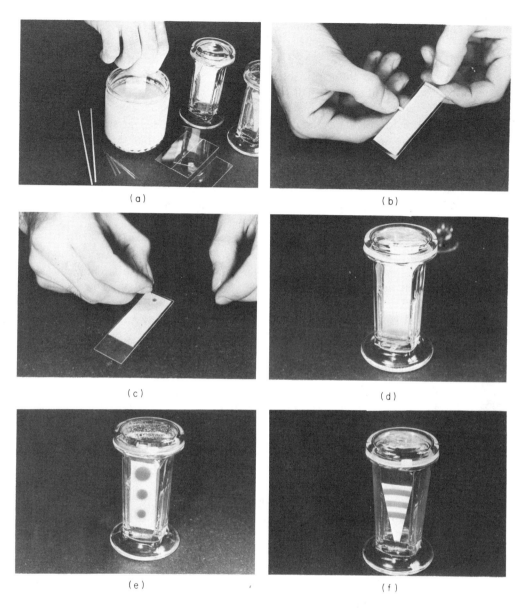

| (a) | (b) |

| (c) | (d) |

| (e) | (f) |

Fig. 2.22. TLC on microscope slides. (*a*) Two microscope slides, back-to-back, are dipped into the slurry, separated, and allowed to dry. (*b*) The edges are smoothed by removing a small amount of the adsorbent. (*c*) The sample is applied with a small capillary. Up to three samples can be applied on a single slide. (*d*) The chromatogram is developed in a small closed chamber for about 5 min. Jars designed for slide-staining make convenient development chambers. (*e*) The chromatogram is visualized by placing it in a closed chamber containing a few crystals of iodine. (*f*) Chromatography on shaped layers results in the formation of narrow bands rather than round spots. Only one sample can be placed on a slide, but the method often results in improved separation of the components and facilitates the detection of trace components of low R_f value. (Photographs by L. Pepoy and D. Vegh.)

Development of the plate. Solvent systems are usually chosen by trial and error by using the eluotropic series shown in Table 2.6 as a general guide. The use of mixed solvent systems will often be most valuable.

The spotted plate is placed in a small jar or beaker fitted with a cover (aluminum foil or glass plate). The solvent level in the jar should be below the level of the spots on the plate. Development, such that the solvent progresses about three-fourths of the way up the plate, usually requires about 5 min. The plates are then removed from the development chamber and allowed to air dry. The spots are visualized by the methods discussed earlier. If the spots have an R_f of less than 0.4, the chromatogram can be redeveloped in the same or a different solvent system for increased resolution.

Recording of data. R_x or R_f values can be calculated, or a picture of the chromatogram can be drawn in the laboratory notebook. One can also use transparent tape to place the layer directly in the notebook. The tape is pressed onto the layer, and the layer will adhere to the tape when it is removed. An additional piece of tape is placed on the back of the layer, completely sealing it between the tapes.

Chromatography on Larger Plates

The commonly used larger plates are 5 by 20 cm or 20 by 20 cm. (Prepared layers are commercially available on either glass plates or plastic sheets. The latter have the advantage in that they can be easily cut to any desired size but are flexible and must be supported during the development.) The preparation of these larger plates is best accomplished by means of a commercially available apparatus. One can, however, prepare them by spreading a slurry by means of a glass rod. An aqueous slurry is prepared from commercial TLC adsorbent according to the directions on the container. The slurry is spread evenly across the surface of the plate(s) by pulling a glass rod that has been wrapped at both ends with masking tape to the desired thickness of the layer across the surface of the plate(s), forcing the slurry ahead of the glass rod. Alternately an edge of tape of the desired thickness can be placed down each side of the glass plate and the slurry spread between the tapes by means of a glass rod. The plates are air dried and placed in an oven at 105°C, for 1 hr or more to activate them for use in adsorption chromatography. The spotting and development follow from the directions given for the microscope slide plates.

2.6.3b The Relationship between Column and Thin-Layer Chromatography

Thin-layer chromatography can be used as a model for column chromatography. TLC on microscope slides will suggest adsorbent and solvent systems for use in column chromatography and provide details about the complexity of the mixture to be separated. For an effective transition from TLC to column chromatography, it is important that the TLC and column adsorbent come from the same manufacturer. The following procedure can be recommended:

1. Find an appropriate solvent system for clean separation on TLC. Such a system should consist of two liquids—one polar and the other less polar.
2. Reduce the polarity of the solvent system by the addition of the less polar solvent until the R_f values of the components are below 0.3 on TLC. If there is only one desired component of the mixture and this component has a high R_f value, reduction of the solvent polarity may not be necessary.
3. Use the less polar solvent system from step 2 to make a slurry and pack the column.
4. Place the sample on the column and develop with the modified solvent system. Analyze the fractions by TLC using microscope slides.

2.6.4 High-Pressure Liquid Chromatography

Column chromatography in which a liquid mobile phase is forced over the stationary phase by pressure is known as *high-pressure liquid chromatography* (hplc). By placing the mobile phase under pressure, column efficiency per unit time is markedly increased.

Compared to traditional column chromatography, high-pressure liquid chromatography may offer major advantages in convenience, speed, versatility, and ability to achieve difficult separations and lower repetitive costs (solvents and adsorbents). Hplc involves the same principles as traditional column chromatography, but the pressure flow allows the use of much finer stationary phases that are totally impractical with gravity-flow columns. For example, preparative columns filled with the same mesh of silica as used in TLC perform excellently under pressure, whereas flow is negligible with gravity. Pressure chromatographic equipment varies from analytical instruments that operate at 5000 psi and take loads in the sub-milligram range to preparative systems that operate with pressures as low as 25 psi but can handle loads of 10 or more grams. A schematic diagram of a high-pressure liquid chromatograph is shown in Fig. 2.23.

The two most common types of detectors used are uv/visible photometers and differential refractometers. With the former detector the sample must adsorb in the uv or visible range of the instrument, and the solvent must be transparent in the region of operation. A differential refractometer detector continuously monitors the difference in refractive index between a reference mobile phase and the eluent from the column. Such detectors respond to all solutes under proper operating conditions, but they are not exceptionally sensitive and are difficult to use with gradient elution. In preparative work a detector is often omitted from the system. Fractions are collected and analyzed by evaporation and/or by thin-layer chromatographic examination.

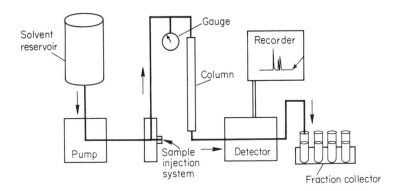

Fig. 2.23. Schematic of a high-pressure liquid chromatograph.

Preparative pressure columns are more expensive and more difficult to pack than gravity columns; however, once packed, under proper use preparative pressure columns can be used many times. After each use residual materials can be removed from the column by back-flushing with a polar solvent. After back-flush is complete, the columns can be reactivated by flushing with solvents of lower polarity. The useful lifetime of a preparative column can be greatly increased if tars and highly polar by-products are removed by passing the sample through a "pre-column" or scrubber column before the sample is placed on the main column. The scrubber may be a short column built into the pressure system, a short external gravity column, or simply a pad of adsorbent in a Büchner or sintered glass funnel.

What about the safety of high-pressure liquid chromatography? Liquids are difficult to compress, and, therefore, very little energy is stored up at the pressures used. The pressures used in hplc pose very few hazards. The main hazards come from the flammability and/or toxicity of the solvents and samples. The usual safety measures—good ventilation, safety glasses, etc.—should be used.

2.6.5 Paper Chromatography

Paper chromatography is somewhat similar to thin-layer chromatography except that high-grade filter paper is used as the adsorbent or solid stationary phase. In actual fact, paper chromatography is not strictly adsorption chromatography but a combination of adsorption and partition chromatography. A partitioning of the solute occurs between the water of hydration of the cellulose and the mobile organic phases. Paper chromatography is used primarily when extremely polar or polyfunctional compounds, for example, sugars and amino acids, are to be separated. Such materials cannot be chromatographed on more active adsorbents. The method is used as an analytical procedure much like analytical thin-layer chromatography; sample sizes for paper chromatography are usually in the range of 5 to 300 μg.

The selection of solvent systems for paper chromatography is very important. For a discussion of solvent systems, refer to the books listed in Sec. 2.8. In most cases the solvent system should contain some water. The solvent development can be accomplished by suspending the paper strip so that the end of the strip is immersed in a container of solvent (ascending paper chromatography) or by immersing the top of the strip in a small trough of solvent and allowing the solvent movement to occur by a combination of capillary action and gravity (descending paper chromatography). It is very important in all paper chromatography procedures that the development of the chromatogram occurs in an atmosphere saturated with solvent vapors. Otherwise, the solvent will evaporate from the paper faster than it is replaced by capillary action and separation of the components will not be achieved. After development is complete, the position of the solvent front is noted and the paper strip is allowed to dry. The paper is then sprayed with a visualizing reagent. R_f and R_x values are calculated.

2.6.5a Test-Tube Technique
for Paper Chromatography

As in thin-layer chromatography, both elaborate and simple techniques and equipment can be employed in paper chromatography. By far the simplest technique involves the use of an ordinary test tube as the developing chamber; the method can be recommended for most identification work or, at least, for exploratory work in determining conditions for using larger strips or sheets.

A 6- or 8-in. test tube can be used. The paper strip cut from Whatman No. 1

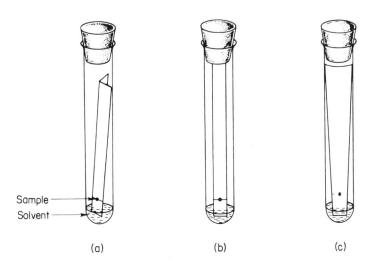

Sample
Solvent

(a)   (b)   (c)

Fig. 2.24. Paper chromatography in test tubes. (*a*) Paper suspended in tube by folding. (*b*) Paper suspended from slit in rubber stopper. (*c*) Paper beveled to suspend from sides of tube at the broad end.

filter paper can be suspended in the test tube by suspending it from a slit cut in the stopper, by putting a fold in the paper, or by cutting the paper with a broad end at the top, which will serve to suspend the paper in the tube (Fig. 2.24). Handle the paper only at the extreme top edge since fingerprints will contaminate the chromatogram.

Place a light pencil mark about 1 cm from the bottom of the strip. The sample is applied as a small spot with a micropipette (made by pulling a capillary) at the center of the line and allowed to dry. The sample size should be about 10 μg. A small amount of solvent is placed in the bottom of the tube by means of a pipette in such a manner that the sides of the tube above the surface of the solvent remain dry. The spotted and dried paper strip is carefully inserted into the tube so that the end dips into the solvent, but the surface of the solvent is below the pencil line. The tube is stoppered and allowed to stand until the solvent ascends near the top of the strip. This process may take up to 3 hr. The paper is removed with forceps and the solvent front marked. The strip is allowed to dry and the compounds are visualized with an appropriate spray.

2.6.6 Gas Chromatography

Gas chromatography (gc), sometimes called vapor phase chromatography (vpc), utilizes a moving gas phase with a solid or liquid stationary phase. With a solid stationary phase, the separation of the components of a mixture is dependent on the establishment of adsorption equilibria as in column adsorption chromatography. When a liquid stationary phase is supported on the surface of an inert solid, the separation is dependent on solubility equilibria between the components in the gas phase and liquid phase. Herein lies the great utility of gc. Proper choices of the stationary liquid phases allow the experimentalists to greatly vary the sequence of elution of compounds in a mixture from the column.

Gas chromatography has developed into one of the most powerful analytical tools available to the chemist. The technique allows separation of extremely small quantities of material, of the order of 10^{-4} to 10^{-6} g. The quantitative analysis of mixtures can be readily accomplished (as discussed in greater detail later in this section). The possibility of using long columns, producing a great number of theoretical plates, increases the efficiency of separation. The technique is applicable over a wide range of temperatures (-70 to $+500°C$), making it possible to chromatograph materials covering a wide range of volatilities. Finally, gas chromatographic analysis requires very little time compared with most other analytical techniques.

The instrumentation involved may be quite simple or may contain a high degree of sophistication and complexity. The basic components required are illustrated in Fig. 2.25. A relatively high-pressure gas source, 30 to 100 psi, is required to provide the moving gas phase. The gas is introduced into a heated injector block. The sample is introduced into the gas stream in the injector block by means of a microsyringe forced through a syringe septum. Liquids can be injected directly, whereas solids are generally dissolved in a volatile solvent. The injector block is maintained at temperatures ranging from room temperature to approximately 350°C. The operating temperature is chosen to ensure complete vaporization of all components present in the mixture. The temperature of the injector, or the column, need not be higher than the boiling point of the least volatile substance. The partial vapor pressure of most substances, in the presence of the high-pressure carrier gas, is sufficient to ensure complete vaporization. It must be kept in mind that high injector block temperatures may lead to decomposition of heat-sensitive compounds.

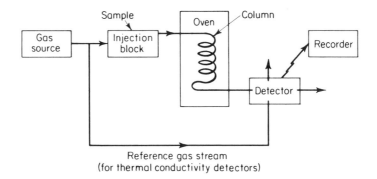

Fig. 2.25. Block schematic of a simple gas chromatograph.

The gas stream, after leaving the injector block, is conducted through the chromatographic column, which is housed in an oven. If all the retention times (see later discussion for definition) of the components are quite similar, the column can be maintained at a constant temperature (isothermal operation). If the retention times are greatly different, it may be desirable to slowly increase the temperature of the column during the chromatographic process. Instruments are available for constant-temperature and/or variable (programmed)-temperature operations. The length of the column required will depend on the difficulty of separation of the components in a given mixture; this is usually determined by trial runs. Packed columns are usually 0.25 in. in diameter and 2 to 50 ft in length. Normal column lengths are usually 6 to 10 ft; the longer columns are required for the more difficult separations. Larger diameter columns, up to 2 in., are used for preparative purposes (collection of the individual fractions). Capillary columns of up to 300 ft in length are also available.

The effluent gas stream containing the separated components is conducted through a detector, which measures some physical phenomenon and sends an electronic signal to a strip chart recorder. Many types of detectors are available. These include thermal-conductivity devices and ionization devices. The actual interpretation of the data recorded by the strip chart recorder depends on the

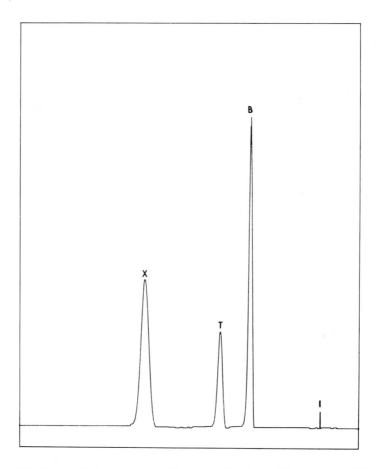

Fig. 2.26. Gas-liquid chromatogram of benzene (*B*), toluene (*T*), and xylene (*X*) on a 5½ ft, 20% Carbowax 20M column at 140°C. The small peak labeled *I* indicates the point of injection of the sample.

type of detector used. The types of detectors and the method of handling the data will be discussed in detail in the following paragraphs.

The strip chart recorder provides a chromatogram (not a spectrum!) such as that illustrated in Fig. 2.26. The first question to be asked is: What do the individual peaks in the chromatogram represent? First of all, the peak represents the detection of some material being eluted from the column. The peak can be characterized by a retention volume or retention time. The *retention volume* is the total volume of gas passing through the column to effect elution of the compound responsible for a given peak. The retention volume is a function of the volume of stationary liquid phase, the void volume of the column (volume of gas in the column at a given time), the temperature, and the flow rate. The experimentalist generally does not determine retention volumes but, instead, characterizes the peaks with respect to their *retention times*, the elapsed time between injection and elution. In the comparison of retention times, constant column conditions are implied, those conditions being the functions of retention volume cited before.

The identification of an unknown giving rise to a peak in a chromatogram can be tentatively inferred by the comparison of retention times of the unknown with a possible known on a number of columns differing widely in polarity. Additional evidence can be obtained by recording the chromatogram of an admixture of the unknown and known samples. A single peak whose appearance (symmetry) is the same as the unknown and known samples indicates the possible identity of the unknown and known samples. A rigorous identification of an unknown cannot be made by comparison of retention times since it is very conceivable that many other materials may possess similar elution properties.

Rigorous identification of the material giving rise to a peak can be accomplished by collecting the fraction as it emerges from the exit port and then characterizing the material by its physical properties. However, it must be remembered that isomerizations and fragmentations can and do occur in the injector block and column and that the material going through the detector may not be the same as that which was injected. Gas chromatographs have been developed specifically to take advantage of the fragmentation of compounds. Injection of highly nonvolatile materials into a pyrolysis injector results in extensive cracking of the material, and the chromatogram obtained is referred to as a *cracking pattern*.

Gas chromatographs can also be used in conjunction with other types of instruments. The emerging gas stream can be directed through an infrared cell of a fast-scan infrared spectrometer, or directly into a mass spectrometer. Such procedures eliminate the necessity of collecting the individual fractions.

The area of the peak of a gas chromatogram represents a quantitative measurement of some physical property of the material going through the detector. For a given material, the response will be linear with concentration; however, the physical constant for a series of different molecules may not have the same absolute value, and the relative response of one compound with respect to another will not be directly proportional to the partial pressures of the materials in the eluent gas stream. The relative response of different compounds depends on the type of detector used, and appropriate corrections in calculations must be made if quantitative data are to be derived.

2.6.6a Detectors

Detectors generally fall into one of two classes: thermal conductivity devices and ionization detectors.

The thermal conductivity device measures changes in the thermal conductivity of the eluent gas stream. The detector usually consists of four heated filaments acting as the four resistances of a Wheatstone bridge. Two of the filaments are immersed in a reference gas stream, a small stream of gas bypassing the injector and column in Fig. 2.25 and entering the detector, to provide a constant reference of resistance. The two remaining filaments are immersed in the eluent gas stream. With a constant flow of pure carrier gas past the filaments, a constant rate of heat transfer from the filaments to the gas stream will occur. As a compound is eluted from the column, the thermal conductivity of the gas stream will change and more or less heat will be transferred from the filament to the gas stream. The change in temperature of the filament results in a change in the resistance of the filament, which is noted by a change in current flow. This signal is relayed to the recorder and plotted on a time scale. Therefore, with the thermal conductivity detector, the chromatogram represents the change in the thermal conductivity of the eluent gas stream.

Table 2.7 lists the thermal conductivities of several potential carrier gases and organic compounds in the gas phase at two different temperatures. The thermal conductivity of a gaseous mixture is equal to the sum of the products of the mole fraction and thermal conductivity. Of the four potential carrier gases listed in Table 2.7, helium and hydrogen provide the greatest sensitivity of measurement, sensitivity meaning the greatest change in the thermal conductivity of the gas stream with respect to change in concentration. Helium is the most widely used carrier gas in thermal conductivity devices, owing to its greater chemical inertness. Argon and nitrogen are not suitable carrier gases for use with thermal conductivity devices, since "negative" peaks may occur with compounds having a thermal conductivity larger than that of argon or nitrogen.

Table 2.7. Thermal Conductivities

of Various Substances $\left(\dfrac{g\ cal}{sec\ cm^2}\Big/\dfrac{°C}{cm}\right)$

Substance	0°C	100°C	Ratio $\dfrac{100°C}{0°C}$
Helium	33.60	39.85	1.18
Hydrogen	39.60	49.94	1.26
Argon	3.88	5.09	1.31
Nitrogen	5.68	7.18	1.26
Carbon dioxide	3.35	4.96	1.48
Methane	7.16	12.3	1.72
Ethane	4.26	7.77	1.83
Pentane	2.94	—	—
Benzene	2.10	4.17	1.99
Ethanol	—	5.03	—

The conversion of the peak areas of a chromatogram into relative weights, or moles, requires having available the thermal conductivities of the components present in the mixture at the temperature of the detector. Such information is extremely limited, and the measurement of such values would be a prohibitive task reducing the utility of gas chromatography as a quantitative analytical tool. To circumvent this problem, one can determine relative response ratios of the components present in the mixture. This is accomplished by recording the chromatogram of a carefully weighed mixture of the pure components and calculating the (weight ratio)/(area ratio) of each component relative to a single component or to an added internal standard. The area ratios, relative to the same component or internal standard used in the known composition mixture, determined from the unknown mixture chromatogram, are converted to relative weight ratios by multiplication of the area ratios by the (weight ratio)/(area ratio) values (illustrated in the sample calculations that follow).

Inspection of the data presented in the last column of Table 2.7 indicates that the thermal conductivities of compounds in the gas phase are not constant with temperature. Secondly, the temperature dependence, which is nonlinear, varies from compound to compound. Therefore, the relative response ratios calculated at a given detector temperature can be used *only* at that temperature. Should an analysis be carried out at a different detector temperature, the relative response ratio must be redetermined at that temperature. It is also apparent from Table 2.7 that the difference in thermal conductivities increases with temperature. To obtain maximum sensitivity with the use of the thermal conductivity device, the temperature of the detector is maintained at a relatively high value of 250 to 350°C.

Quantitative yield determinations of volatile components can also be determined by the use of gas chromatography. This can be accomplished by either of two procedures. An additional weighed amount of one of the components present in the mixture is added to a weighed aliquot of the mixture. The gas chromatograms of the mixture and admixture are recorded. The absolute weights or moles present in the mixture can then be calculated (see the sample problem below). An alternative procedure involves the addition of a weighed amount of an internal standard to an aliquot of the mixture and then the recording of the gas chromatogram. For this method, one needs the relative response ratio of one of the components with respect to the internal standard. Yields should not be calculated without the use of an internal standard unless *all* products appear as peaks in the chromatogram and there has been *no* mechanical loss of material.

2.6.6b Sample calculations

Reaction of A (2.50 g, 134 mol. wt.) with reagent X gave 2.40 g of a product mixture containing A and two products B (148 mol. wt.) and C (162 mol. wt.). The gas chromatogram of the product mixture showed peaks for A, B, and C, with areas of 77, 229, and 276, respectively. Addition of 36.4 mg of B to a 120.2-mg aliquot of the reaction mixture produced a gas chromatogram with areas of 111, 567, and 396 for A, B, and C. A mixture of 52.7 mg of A, 47.3 mg of B, and 63.2 mg of C produced a gas chromatogram with areas of 287, 275, and 423, respectively.

Calculate the yields of B and C based on reacted starting material according to the following equation:

$$A + X \longrightarrow B + C$$

Solution:

$$\text{Response ratio } \frac{A}{B} = \frac{\text{wt. A/wt. B}}{\text{area A/area B}} = \frac{52.7/47.3}{287/275} = \frac{1.113}{1.044} = 1.067 \left(\frac{A}{B}\right)$$

$$\text{Response ratio } \frac{C}{B} = \frac{\text{wt. C/wt. B}}{\text{area C/area B}} = \frac{63.2/47.3}{423/275} = \frac{1.337}{1.540} = 0.867 \left(\frac{C}{B}\right)$$

Area of B in admixture
chromatogram due to added $B = \text{area B}_{admix} - \text{area A}_{admix} \cdot \dfrac{\text{area B}_{orig}}{\text{area A}_{orig}}$

$$= 567 - 111 \cdot \frac{229}{77} = 567 - 330 = 237$$

From the proportion

$$\frac{\text{wt. B}_{added}}{\text{area B}_{added}} = \frac{\text{wt. B}_{orig}}{\text{area B}_{orig}}$$

we have

$$\frac{36.4 \text{ mg B}}{237} = \frac{\text{wt. B}_{orig}}{330}$$

$$\text{wt. B}_{orig} \text{ in aliquot} = \frac{36.4 \cdot 330}{237} = 50.7 \text{ mg}$$

$$\text{wt. of B in total sample} = \frac{50.7 \text{ mg}}{120.2 \text{ mg}} \cdot 2.40 \text{ g} = \underline{1.014 \text{ g of B}}$$

$$\text{wt. of A in sample} = \frac{\text{area A}}{\text{area B}} \cdot \text{response ratio} \left(\frac{A}{B}\right) \cdot \text{wt. B}$$

$$= \frac{77}{229} \cdot 1.067 \cdot 1.014 \text{ g A} = \underline{0.364 \text{ g of A}}$$

$$\text{wt. of C in sample} = \frac{\text{area C}}{\text{area B}} \cdot \text{response ratio} \left(\frac{C}{B}\right) \cdot \text{wt. B}$$

$$= \frac{276}{229} \cdot 0.867 \cdot 1.014 \text{ g C} = \underline{1.060 \text{ g of C}}$$

The percent yields are calculated in the usual manner.
In this particular case, the entire sample proves to be volatile. This may not be true in all cases.
A similar approach would be used if an internal standard had been added instead of employing additional B.

The second category of detectors can be generally classified as ionization detectors. The eluent stream is subjected to partial ionization; the ions being formed then migrate to a polarized electronic grid, producing a current. The current is amplified and recorded, producing the chromatogram. The peaks of chromatograms derived from ionization detectors represent the number of ions formed. The extent of ionization varies with the type of ionization detector employed and the chemical nature of the eluent.

The first of this type of detector to be discussed is the flame ionization detector. The eluent stream is mixed with a stream of hydrogen and oxygen and is then subjected to combustion. During this process, the compounds eluted from the chromatographic column undergo partial combustion producing ionic fragments, or undergo ionization induced by the energy produced from the combustion of the hydrogen. The extent of ionization of individual compounds depends on the number and types of functional groups contained in the molecule. To derive quantitative information from the chromatogram, the approach described earlier for the thermal conductivity detector must be used.

Instead of employing a hydrogen flame to induce ionization, a source of low energy β or γ rays can be used. The mechanism of the ionization process depends on the carrier gas employed. Direct ionization of eluted materials occurs when helium is used as the carrier gas. For each primary electron (β particle) of approximately 1 mev energy, 100 to 200 ionizations/cm of path occur, resulting in a total ionization level of $10^{-6}\%$. The relatively high energies involved do not appear to lead to discrimination, and the response is generally linear with respect to the number of molecules of a substance present. Such a detector is referred to as a cross-section ionization detector and employs Ra, ^{90}Sr, ^{90}Y, or ^{3}H as the ionizing particle source.

Substitution of the helium carrier gas by argon results in a change in the ionization mechanism. The primary particle excites an argon atom to a metastable state with a potential of 11.6 ev. The excited argon atom transfers this energy upon collision with the other molecules in the eluent gas stream producing ionization. The response of this type of detector depends on the ionization potential of the materials being chromatographed. Molecules with ionization potentials greater than 11.6 ev will be "transparent." Most organic compounds possess ionization potentials in the 8 to 11 ev region and thus most organic molecules will be detectable. The level of ionization is approximately 1%, providing an increase in the sensitivity of detection by a factor of 10^6 over that when helium is used as the carrier gas.

The use of nitrogen as a carrier gas in an ionization detector results in the ejection of an electron from a molecule of nitrogen. The secondary electron thus produced attaches itself to a molecule of the material being eluted. The negatively charged sample ions are then counted. The response is dependent on the electron affinities of the compounds being chromatographed. Since electron affinities can vary even more greatly than gas phase thermal conductivities, it is necessary to employ relative response ratios in quantitative determinations.

Several other types of detectors have been described in the literature but do not appear to be as generally applicable as those just described. In general, the use of ionization detectors provides a much more sensitive method of detection than is obtainable with the thermal conductivity device. They are not, however, applicable in preparative separations without use of stream splitters.

2.6.6c Column Stationary Solid and Liquid Phases for Gas-Liquid Chromatography

The choice of solid phases for gc is quite limited. Adsorbents that have been used include finely divided alumina, silica gel, charcoal, and molecular sieves. The utility of such phases rests in the temperature range available for use, approximately -70 to $+500°C$. The type of samples that can be subjected to gas chromatography using these solid phases is limited. Highly polar compounds generally cannot be successfully chromatographed.

The great utility of gc lies in the great number of combinations of solid supports and stationary liquid phases available. Table 2.8 lists a number of solid supports available with a short description of their properties. The solid supports also can be subjected to specific treatments to improve their utility in certain cases. These treatments include acid washing to remove traces of metal salts and basic materials that may otherwise interfere in the separations. The solid support is generally treated with concentrated hydrochloric acid for a period of time and is then washed with water until neutral. Treatment with base may be required when acid-sensitive compounds must be chromatographed. Such treatment greatly

Table 2.8.   Solid Supports

Name	Description
Chromosorb P	Highly adsorptive, pink calcined diatomaceous silica, pH 6 to 7.
Chromosorb W	Medium adsorptive, white diatomaceous silica, which has been flux-calcined with sodium carbonate, pH 8 to 10.
Chromosorb G	Dense, low adsorptivity, flux-calcined material, pH 8.5 (maximum liquid phase loading, 5%).
Fluoropak 80 Haloport F Chromosorb T	Fluorocarbon polymer, maximum usable temperature 260°C, reduces tailing of highly polar compounds.
Glass beads	Used only with very low amounts of liquid phase (0.5% and less).
Poropak	Polymeric porous beads used without a liquid stationary phase.

reduces tailing of highly polar or basic compounds. The support is washed with dilute methanolic potassium hydroxide and is then washed with water until the desired pH is obtained. Occasionally the support is coated with a layer of potassium hydroxide by evaporation of the methanol used as solvent under reduced pressure. To obtain a nonreactive support, the solid support, which has been acid- and base-washed, is stirred for a short period of time with a 5% solution of dichlorodimethylsilane in toluene. The support is washed with toluene and methanol. Such a process is referred to as silanizing.

Tables 2.9 and 2.10 list a number of liquid phases and the classes of compounds that can be separated. The list is by no means exhaustive. It should also be pointed out that the success of a given column substrate depends on the care in preparation, length of use, and on the gross structure of the components being separated. Although Table 2.10 includes recommended liquid phases for a

Table 2.9. Examples of Liquid Stationary Phases

Number	Abbreviation	Trade Name or Description	Maximum Temperature, °C
1	Apiezon L	Hydrocarbon grease	300
2	Apiezon M	Hydrocarbon grease	275
3	Carbowaxes (400–6000)	Polyethylene glycols	100–200
4	Carbowax 20M	Polyethylene glycol	250
5	DC-200	Dow Corning methyl silicone fluid	225
6	DC-550	Dow Corning phenyl methyl silicone fluid	225
7	DC-710	Dow Corning phenyl methyl silicone fluid	250
8	DEGS or LAC 728	Diethyleneglycol succinate	225
9	LAC 446	Diethyleneglycol adipate	225
10	Flexol Plasticizer 10-10	Didecyl phthalate	175
11	Nujol	Paraffin oil	200
12	QF-1-6500	Dow Corning fluorinated silicone rubber	260
13	SE-30	G.E. methyl silicone rubber	300
14	SE-52	G.E. phenyl silicone rubber	300
15	SF-96	G.E. fluoro silicone fluid	300
16	—	Silicone gum rubber	375
17	—	Silver nitrate-propylene glycol	75
18	TCP	Tricresyl phosphate	125
19	TCEP	1,2,3-Tris(2-cyanoethoxy)propane	180
20	THEED	Tetrakis(2-hydroxyethyl)ethylene-diamine	130
21	Ucon polar	Polyalkylene glycols and derivatives	225
22	Ucon non-polar	Polyalkylene glycols and derivatives	225
23	—	4,4'-Dimethoxyazobenzene	120–135*

*The stationary phase exists as liquid crystals, referred to as nematic or smectic phases depending on crystal orientations. The temperature range is limited to the regions where these phases exist. Other alkoxyderivatives of azobenzene have also been used [M. J. S. Dewar and J. P. Schroeder, *J. Am. Chem. Soc.*, **86**, 5235 (1964)].

Table 2.10.   Suggested Column Uses

Class of Compound	Column Number (from Table 2.9)
Acetates	8, 10, 12, 19, 20, 21
Acids	6 (on acid-washed support), 16
Alcohols	3, 4, 5, 6, 16, 19, 20, 21, 22
Aldehydes	3, 4, 10, 11
Amines	3, 4, 6, 20 (on fluoropak or haloport or potassium hydroxide-coated support)
Aromatics	1, 2, 10, 16, 18, 19, 20, 21
Esters	3, 4, 8, 9, 10, 12, 16, 19, 20
Ethers	6, 9, 16, 18, 19
Halides	5, 10, 11, 18, 19
Hydrocarbons	1, 2, 6, 10, 16, 19, 20
Ketones	3, 4, 6, 10, 16, 18, 20
Nitriles	9, 18, 20
Alkenes	10, 17, 19, 20
Phenols	4, 12
Isomeric aromatics	23

given class of compounds, the experimentalist may find that none of these suggestions is suitable, and he may have to search for a different column. The selection of the proper liquid phase and column conditions, temperature and flow rate, is many times a trial-and-error procedure, but the end results are usually worth the time and effort expended.

The column packings are prepared by dissolving the liquid phase in a volatile solvent, usually methylene chloride or methanol, and slurrying with the solid support. The solvent is evaporated under reduced pressure, with continuous agitation of the slurry. The dried packing is poured into the column and firmly packed by vibration. Before use, the column must be purged by a stream of gas, while maintained at the maximum column-operating temperature until equilibrium conditions are reached, as indicated by the production of a steady base line. After each use, the column should be purged at the maximum temperature to remove any relatively nonvolatile materials that may have been left on the column. Capillary columns are coated with the stationary liquid phase by slowly evaporating a solution of the stationary phase inside the column.

2.7
SEPARATION OF MIXTURES

The separation of a complex mixture poses a challenging problem to the chemist, regardless of his area of chemistry. In the previous sections of this chapter, we have discussed several techniques for the purification of individual compounds and some techniques applicable to the separation of complex mixtures.

The proper choice of the method of separation depends on the physical and chemical properties of the mixture (the number of phases present, the solubility and volatility of the various components, the type of functional groups present, the quantity of material present, etc.). The three most important techniques for the separation of mixtures, used singly or in combination, are fractional distillation, extraction, and chromatography. Specific separation procedures applicable to all cases cannot be prescribed because each mixture presents an individual challenge. The chemist must use his experience and knowledge to successfully achieve the desired separation.

Prior to the start of a separation of a complex mixture, the infrared spectrum and gas-liquid chromatogram or thin-layer chromatogram of a representative sample of the mixture should be recorded.

The physical state of the sample will dictate the individual procedure to be employed in the attempted separation. If the sample is a homogeneous liquid, a fractional distillation can be attempted. A fractional distillation generally requires 2 to 3 mL of sample in a small fractionating column to achieve a suitable separation. For smaller samples, one must employ chromatographic techniques. A small portion of the sample is tested for thermal stability up to a maximum temperature to be employed (approximately 180°C is recommended). If the sample appears to be stable, for example, if there is no dramatic color change, evolution of a gas, or alteration of viscosity, the entire sample can be subjected to fractional distillation (Sec. 2.3.2). Care should be exercised not to heat the distillation pot above the temperature tested for the thermal stability of the sample (use an oil bath with a thermometer for heating purposes). If the sample does not distill up to 180°C, a reduced pressure distillation should be attempted. Care must be exercised to avoid reducing the boiling point of any fraction present to below the temperature of the condenser, in which case that fraction will disappear into the cold trap protecting the vacuum pump (or down the drain with the water when a water aspirator is used!). It is always a good practice to clean the cold traps before and after each distillation so that recovery of such volatile fractions is possible.

The volume collected *vs.* temperature is recorded so that a distinction between fractions can be noted. The infrared spectrum and gas-liquid chromatogram of each fraction, and the pot residue, if any remains, should be recorded. If the degree of separation in the fractional distillation was not sufficient, each fraction can be individually redistilled or one must revert to the use of extraction or chromatographic techniques.

Heterogeneous samples, containing two liquid phases or one liquid and a solid phase, can also be subjected to distillation (it is generally not advisable to filter the solid phase since a considerable quantity of the solid phase will undoubtedly be soluble in the liquid phase, and vice versa, only a minor degree of separation being accomplished).

Extraction techniques should be attempted if fractional distillation techniques do not prove satisfactory. Prior to attempting an extractive separation, the solubility of the sample in the two extracting phases should be determined. Water is almost always one of the solvents employed; however, it presents the greatest difficulty of recovery with highly polar compounds that may be contained in the mixture. The solubility of a small weighted portion of the unknown in a small volume of water should be determined before proceeding with the extraction. If the mixture is soluble, the aqueous phase should be extracted several times with small portions of ether or dichloromethane to determine the amount of material that is recoverable. If only a small portion of the material is recovered, extraction techniques may not prove satisfactory. (The unknown sample can be recovered by evaporating the water solution.) Compounds that show great water solubility and low organic solvent solubility include polyfunctional compounds, e.g., glycols, aminoalcohols, and amino acids. The solubility of such compounds in water can be substantially reduced by saturating the water phase with sodium chloride; however, this then forbids recovery of the compounds remaining in the aqueous phase by simple evaporation of the water. The separation of such highly polar or polyfunctional compounds is most readily accomplished by ion-exchange techniques

Samples that display low water solubilities may be subjected to an extractive separation. A trial run on a small portion of the sample is carried out first; this avoids the possibility of losing the entire sample if the separation is not successful. A small weighted portion of the sample (300 to 500 mg) is dissolved in 25 mL of ether. The ether layer is first extracted with 10-mL portions of 5% sodium bicarbonate (CAUTION: The formation of carbon dioxide may generate considerable pressure in the separatory funnel) until the aqueous phase remains slightly basic (this usually requires only one extraction on this scale). The aqueous phase will contain the salts of the strongly acidic compounds (carboxylic acids, sulfonic acids, etc.). The free acids are regenerated by careful acidification (carbon dioxide evolution!) with concentrated hydrochloric acid. The compounds can be recovered by filtration, if solid, or by extraction with ether. Occasionally the acids can be recovered by precipitation as the calcium or barium salts (particularly with sulfonic acids). The original ether layer is then extracted with a 10-mL portion of 5% sodium hydroxide to remove the weakly acidic compounds (phenols, etc.). The free acids are regenerated by acidification with concentrated hydrochloric acid and recovered by extraction. The aqueous phase remaining after extraction of the weakly acidic fraction is carefully neutralized by the slow addition of 20% sodium hydroxide, carefully observing the solution near the neutralization point. A cloudiness, or a change in color, which develops and disappears on further basification indicates the possible presence of an amphoteric compound, for example, an aminophenol. The solution should be carefully neutralized to the point of maximum cloudiness or to the point of color change, and repeatedly extracted with ether or an ether-chloroform mixture.

The original ether phase is finally extracted with a 10-mL portion of 5% hydrochloric acid until the aqueous phase remains acidic after extraction. The aqueous phase will contain all the basic components originally in the mixture as their conjugate acids. The acid phase is made strongly basic by the addition of 20% sodium hydroxide, and the free bases are recovered by extraction with ether.

The resulting final ether phase contains the neutral component(s) of the mixture that are recovered by cvaporation of the ether.

The weight of each fraction is determined, and the total weight of recovered material is compared with the weight of the initial sample. Total recoveries should range in the 85 to 95% region. The amounts of each of the components in the mixture do not have to be the same; in fact, reaction mixtures are generally composed of one or two major components and one or more minor components.

Each fraction obtained in the separation sequence is analyzed for purity by gas-liquid or thin-layer chromatography. If any fraction still contains more than one component, further separation by chromatographic techniques must be carried out. Chromatographic techniques are outlined in the foregoing sections on chromatography and will not be discussed further here.

It is important for the chemist to realize that the scheme discussed above may not always provide the desired results and that modifications may be necessary. Furthermore, during the extraction sequence, the unknown mixture comes in contact with both acid and base, and the components present in the mixture can undergo chemical reactions under these conditions, for example, the acid-catalyzed hydrolysis of acetals, ketals, esters, or acid chlorides, the base-catalyzed hydrolysis of acid derivatives, and the solvolysis of reactive organic derivatives. If all the infrared peaks appearing in the separated fractions are present in the original infrared spectrum of the mixture and if no additional peaks appear, it can be assumed that it is then safe to invest the remainder of the sample through the separation scheme, being careful to reproduce the exact procedures carried out in the trial run with the smaller portion of the sample.

2.8
REFERENCES

GENERAL

1. BERG, E. W., *Physical and Chemical Methods of Separation.* New York: McGraw-Hill Book Company, 1963.

2. SIXMA, F. L. J., and WYNBERG, H., *A Manual of Physical Methods in Organic Chemistry.* New York: John Wiley & Sons, Inc., 1964.

3. WIBERG, K. B., *Laboratory Technique in Organic Chemistry.* New York: McGraw-Hill Book Company, 1960.

4. WEISSBERGER, A., *Technique of Organic Chemistry*, Vol. III. New York: Interscience Publishers, Inc., 1956.

LIQUID CHROMATOGRAPHY

5. BOBBITT, J. M., SCHWARTING, A. E., and GRITTER, R. J., *Introduction to Chromatography.* New York: Reinhold Book Corp., 1968.

6. BROWN, P. R., *High Pressure Liquid Chromatography*: *Biochemical and Biomedical Applications*. New York: Academic Press, 1973.

7. HEFTMANN, E., ed., *Chromatography: A Laboratory Handbook of Chromatographic and Electrophoretic Methods*. New York: Van Nostrand Reinhold, 1975.

8. KIRKLAND, J. J., ed., *Modern Practice of Liquid Chromatography*. New York: Wiley-Interscience, 1971.

9. LEDERER, E., and LEDERER, M., *Chromatography*, 2nd ed. Amsterdam: Elsevier Publishing Company, 1957.

10. PERRY, S. G., AMOS, R., and BREWER, P. I., *Practical Liquid Chromatography*. New York: Plenum Press, 1972.

11. SNYDER, L. R., and KIRKLAND, J. J., *Introduction to Modern Liquid Chromatography*. New York: Wiley-Interscience, 1974.

12. STOCK, R., and RICE, C. B. F., *Chromatographic Methods*, 3rd ed. London: Chapman and Hall, 1974.

THIN-LAYER CHROMATOGRAPHY

13. BOBBITT, J. M., *Thin-Layer Chromatography*. New York: Reinhold Publishing Corporation, 1963

14. RANDERATH, K., *Thin-Layer Chromatography*. New York: Academic Press, Inc., 1963

15. STAHL, E., ed., *Thin-Layer Chromatography*. *A Laboratory Handbook*, 2nd ed. New York: Springer, 1969.

16. TRUTER, E. V., *Thin-Film Chromatography*. New York: Interscience Publishers, 1966.

PAPER CHROMATOGRAPHY

17. BLOCK, R. J., DURRUM, E. L., and ZWEIG, G., *Paper Chromatography and Paper Electrophoresis*. New York: Academic Press, Inc., 1958.

18. HAIS, I. M., and MACEK, K., eds., *Paper Chromatography*. New York: Academic Press, Inc., 1963.

GAS CHROMATOGRAPHY

19. BURCHFIELD, H. P., and STORRS, E. E., *Biochemical Applications of Gas Chromatography*. New York: Academic Press, Inc., 1962.

20. CRIPPEN, R. C., *Identification of Organic Compounds with the Aid of Gas Chromatography*. New York: McGraw-Hill, 1973.

21. GRANT, D. W., *Gas-Liquid Chromatography*. New York: Van Nostrand Reinhold, 1971.

22. HARDY, C. J., and POLLARD, E. H., "Review of Gas-Liquid Chromatography" in *Chromatographic Reviews*, Vol. I, M. Lederer, ed. Amsterdam: Elsevier Publishing Company, 1960.

23. KAISER, R., *Gas Phase Chromatography*. London: Butterworths Scientific Publications, 1963.

24. KEULEMANS, A. I. M., *Gas-Chromatography*, 2nd ed., C. G. Verver, ed. New York: Reinhold Publishing Corporation, 1959.

25. LITTLEWOOD, A. B., *Gas Chromatography, Techniques and Applications.* New York: Academic Press, Inc., 1962.

26. NOGARE, S. D., and JUVET, R. S., Jr., *Gas-Liquid Chromatography.* New York: Interscience Publishers, 1962.

27. PURNELL, H., *New Developments in Gas Chromatography.* New York: Wiley-Interscience, 1973.

3

Physical Characterization

3.1
INTRODUCTION

The physical properties of a substance—melting and boiling points, density, refractive index, solubilities, optical rotation, and spectral properties (to be considered in Part II of this text)—along with the empirical formula (calculated from its elemental analysis) and molecular weight provide information useful in making a structural assignment. The physical properties of materials are dependent on the purity of the sample; therefore, to derive the maximum utility out of the measured data, the samples must be of a high degree of purity. Chapter 2 was devoted to discussions of separation and purification techniques that can be applied to obtain pure samples. This chapter is devoted to discussions concerning the measurement and interpretation of the physical properties of a material.

3.2
MELTING POINTS

The melting point of a solid, or conversely, the freezing point of a liquid, can be defined as the temperature at which the liquid and solid phases are in equilibrium at a given pressure. Implied in this statement is the more quantitative definition of melting point or freezing point as that temperature at which the vapor pressures of the solid and liquid phases are equal.

Although the intersection of the vapor-pressure curves for the liquid and solid phases gives a single point temperature for the melting point of a solid, in an actual experimental determination this is rarely observed owing to the experimental methods employed. In general, very small samples of solids are placed in a capillary tube in an oil bath, or between glass plates on a hot stage, whose temperature is continuously raised by 1 to 2°C/min. Since a finite time is required for complete melting to occur, a finite temperature range will be traversed during this process. Therefore, it is necessary to report a temperature range for the melting point of any substance, for example, benzoic acid—121.0 to 121.5°C. Handbooks and tables of melting points list a single temperature, usually an average of several reported, indicating where a sample of the pure substance should melt. However, in reporting the melting point of a specific sample, a melting-point range should be reported, that is, the temperature at which the first melting is observed and the temperature at which the last crystal disappears. Both the temperature and range of an observed melting point are important attributes. Both reflect the purity of the sample.

The addition of a nonvolatile soluble substance to a liquid produces a decrease in the vapor pressure of the liquid phase (an application of Raoult's Law). From the definition of melting or freezing point, a decrease in the vapor pressure of the liquid phase will result in a lowering of the melting or freezing point (Fig. 3.1).

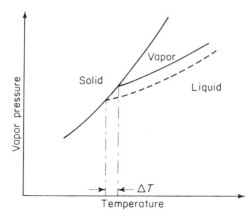

Fig. 3.1. Vapor pressure *vs.* temperature diagram. Solid lines represent the pure substance *A*, and the dotted lines represent a solution of a nonvolatile material in *A*.

The vapor pressure of the solid in equilibrium with the liquid will, for all purposes, be unaffected, since the solvent in the solid phase will tend to be present as pure small crystalline units with crystalline solute dispersed throughout. The solute will not, in general, be incorporated into the crystal lattice of the crystalline solvent. Therefore, if the solute remains soluble in the liquid phase, the solute concentration will decrease as the amount of liquid solvent increases during the melting process. The net result is that the melting point of the mixture will increase as melting proceeds, giving a melting range much greater (several degrees) than that observed for a pure substance.

During the purification of a solid material, the melting range should decrease, and the melting point should increase as the impurities are removed during successive stages of purification. When no further change in the melting range and point is observed on further purification, the material is probably as pure as can be obtained using that purification technique.

Several methods can be used to determine melting points. The capillary-tube method is one of the most common methods employed. A portion of a finely ground sample is introduced into a fine glass capillary, 1 mm by 100 mm, sealed at one end. (Glass capillaries are commercially available, or the student can prepare his own by drawing a *clean* piece of 8 to 12 mm soft glass tubing.) Enough sample is placed in the capillary tube and firmly packed until a column of sample 1 to 2 mm is obtained. Occasionally it is difficult to introduce the sample into the capillary and get the sample firmly packed in the closed end of the capillary. Small quantities of material can be forced into the open end of the capillary and then induced to move down the capillary to the sealed end by vibrations. This can be accomplished by firmly grasping the capillary tube and tapping against the bench or table top, by allowing the capillary to fall down a glass tube and bounce on the bench top, or by drawing a file across the side of the capillary.

The capillary is attached to a thermometer by means of a small rubber band cut from $\frac{1}{4}$-in. rubber tubing, and the thermometer and sample are placed in a heating medium. The sample must be maintained at the same level as the mercury bulb of the thermometer. Mineral oil is usually employed as the heating medium and can be safely used up to 225 to 250°C. For higher temperatures, a metal block device can be used.

A variety of liquid heating bath devices can be employed. Figure 3.2a illustrates the proper use of the Thiele tube. The thermometer and sample are placed in the Thiele tube so that the sample and sensing bulb of the thermometer are slightly below the bottom of the top side-arm neck of the tube. A Bunsen burner is used to heat the side arm of the Thiele tube gently, the circulation of the heating medium occurring by convection. More elaborate setups can be employed, which contain internal resistance wire heaters and mechanical stirrers. A magnifying lens may be used to observe the sample during the melting process. Figure 3.3 illustrates a popular commercial apparatus for melting-point determinations.

Figure 3.2b illustrates the use of a small round-bottom flask as the heating bath vessel; however, maintaining a uniform temperature throughout the heating medium is quite difficult unless some means of stirring is provided, for example, a stream of air bubbles introduced by means of an air tube.

The Fisher–Johns melting-point apparatus (Fig. 3.4) is a typical hot-stage melting-point apparatus. A small portion of the sample is placed between two 18-mm microscope slide cover glasses, which are then placed in the depression of the electrically heated aluminum block. The rate of temperature rise is regulated by means of a rheostat and the sample is observed with the aid of a magnifying lens. The Fisher–Johns apparatus is useful for determining melting points between

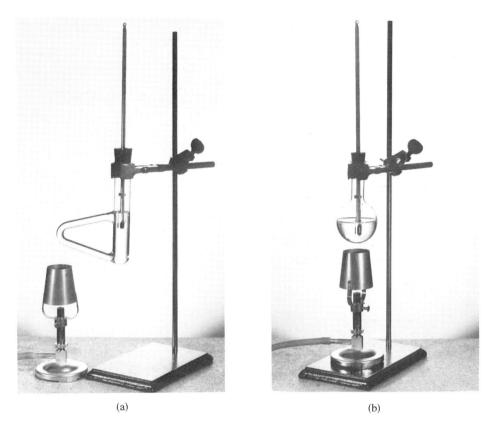

(a)   (b)

Fig. 3.2. Setups for the determination of melting points. (*a*) Employing a Thiele tube. (*b*) Employing a small round-bottomed flask. (Photographs by Bruce Harlan, Notre Dame, Ind.)

30 and 300°C. This apparatus must be calibrated periodically using a set of melting-point standards.

The rate of heating of a sample in any method of melting-point determination should be approximately 1°C/min while traversing the melting-point range. Greater heating rates can be used to raise the temperature of the heating bath to about 10 to 15°C below the expected melting-point temperature. Many compounds display melting ranges that depend on the rate of heating of the sample. Examples of compounds that display such behavior are those that undergo decomposition, chemical isomerization (thermally induced), or changes in crystal form.

The sample should be carefully observed during the melting process for changes in crystal form, decomposition with the formation of gaseous products, or decomposition with general charring. Such observations may be very useful in deriving structural information. Changes in crystal form may be due to a simple change of one allotropic crystal modification to another of the same substance, or they may be due to a thermally induced rearrangement to an entirely new

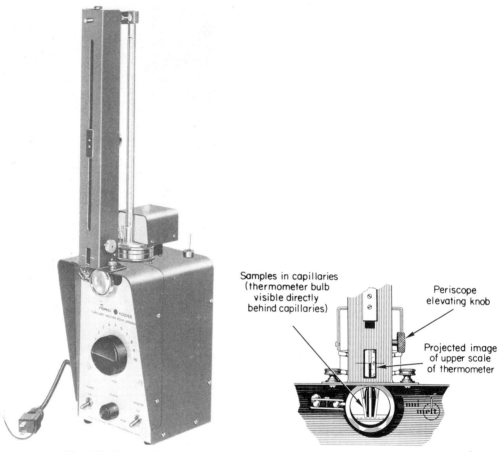

Samples in capillaries
(thermometer bulb
visible directly
behind capillaries)

Periscope
elevating knob

Projected image
of upper scale
of thermometer

Fig. 3.3. Thomas-Hoover melting-point apparatus. Periscopic thermometer reader allows sample observation and temperature readings to be made at the same eye level. (Reproduced by courtesy of Arthur H. Thomas Company, Philadelphia, Pa.)

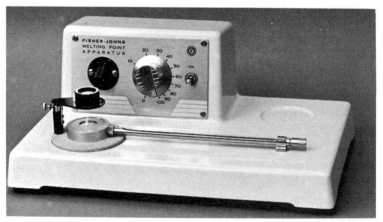

Fig. 3.4. Fisher-Johns hot-stage melting-point apparatus. (Reproduced by courtesy of Fisher Scientific Company.)

compound, where the final melting point then corresponds to the rearranged material and not to the original sample. The evolution of a gas is generally the result of a thermal decarboxylation of β-carboxy carbonyl compounds, although the loss of small molecules such as water, hydrogen halides, and hydrogen cyanide may occur with some compounds. Should such phenomena occur during the melting point determination, the thermally induced reaction should then be carried out on a larger portion of the sample and the product(s) identified. General decomposition with charring is typical of polyfunctional compounds, such as many sugars and very high-melting compounds.

The melting points of substances that undergo decomposition will vary with the rate of temperature rise of the sample; hence, reliable comparisons of melting or decomposition points are not possible. Determination of the instantaneous melting or decomposition point, accomplished by dropping small quantities of the material on a previously heated hot stage and by increasing the temperature of the hot stage until the sample melts or decomposes upon striking the surface, gives a more reliable melting or decomposition temperature.

3.2.1 Correction of Melting Points

In all the foregoing procedures for the determination of melting points, only a very small portion of the thermometer is in contact with the heating medium. Owing to a difference in the coefficient of expansion between mercury and glass, a stem correction must be applied to give a true or "corrected" melting point. This correction is calculated by means of Eq. (3.1),

$$\text{Correction in }°C = N(t_1 - t_2)0.000154 \tag{3.1}$$

where N equals the degrees of mercury column above the level of the heating medium, t_1 is the observed melting point, and t_2 is the average temperature of the mercury column. Some research journals recommend the reporting of corrected melting points. If the melting-point apparatus is calibrated with standards, the stem correction is not to be applied.

3.2.2 Mixture Melting Points

The phenomenon of melting-point depression has many very useful applications in structure determinations. Qualitatively, melting-point depressions, or really, the lack of such depression, can be used for the direct comparison of an unknown with possible knowns. For example, chemical and physical properties may limit the number of possibilities for the structure of an unknown to two or three. The melting point of the mixture of the unknown with a known that has the same structure as the unknown does not lead to a depression in the melting point, whereas mixtures of the unknown with knowns that differ in structure from the unknown show depressions in their mixture melting points. The mixtures are prepared by taking approximately equivalent amounts of unknown and known and then melting the mixture until homogeneous. The melt is allowed to crystallize, and the melting point of a small, finely ground portion of the mixture is determined. This procedure has limited utility in that it requires the availability of suitable known compounds. Data derived from such mixture melting point experiments must be used with caution, however, since some mixtures may not display a depression, or may display only a very small depression, due to compound or eutectic formation. Some mixtures may display an elevation of melting point. This can occur when compound formation occurs or when one has a racemic mixture (see following paragraph).

Mixture melting point data can also be used to distinguish between racemic mixtures [eutectics of equal amounts of crystals of (−)- and (+)-enantiomers], racemic compounds [crystals containing equal quantities of (−)- and (+)-molecules in a specified arrangement], and racemic solid solutions [crystals containing equal amounts of (−)- and (+)-molecules arranged in a random manner in the crystal].[1] The addition of a pure enantiomer (1) to a racemic mixture results in an elevation of the melting point, (2) to a racemic compound results in a depression of the melting point, and (3) to a racemic solid solution results in no effective change in the melting point.

The quantitative aspects of melting-point or freezing-point depression can be applied in the determination of molecular weights (see Sec. 3.8 on determination of molecular weights).

3.3
FREEZING POINTS

The freezing points of liquids can be used for purposes of characterization; however, freezing points are much more difficult to determine accurately and freezing point data for liquids are much more limited than are melting and boiling point data. In addition, the quantity of sample required to determine a freezing point is much greater than that required for a melting point determination. One to two milliliters of the sample is placed in a test tube and cooled by immersion in a cold bath (the temperature of the cold bath should not be too far below the freezing point of the sample in order to avoid extensive supercooling). The sample is continuously stirred, and the temperature of the sample is recorded periodically. The cooling curve will usually reach a minimum (supercooling) and rise to a constant temperature as crystal growth occurs. This temperature is the freezing point of the sample.

3.4
BOILING POINTS

The boiling point of a sample is generally determined during purification of the sample by distillation. Difficulties arise in comparison of boiling points, however,

---

[1] E. L. Eliel, *Stereochemistry of Carbon Compounds*, McGraw-Hill Book Company, New York, 1962, pp. 43–46.

when a boiling point is determined at an applied pressure that is different from the literature reference value. Conversion of a boiling point at one pressure to another pressure can be accomplished by the use of boiling point nomographs for different classes of compounds (Appendix I).

The determination of a boiling point by distillation requires the availability of sufficient sample to attain temperature equilibrium in the distillation apparatus. Frequently, insufficient quantities of sample are available to determine boiling points by distillation techniques, and one must use semimicrotechniques. One such technique involves the heating of 0.3 to 0.4 mL of a liquid, with a small boiling chip, in a test tube into which a thermometer is suspended just above the liquid level. This sample is gently heated (usually with a small Bunsen flame) until a continuous reflux of sample from the thermometer and the wall of the test tube is attained. If possible, the reflux line should reach the calibration line of the thermometer. The temperature of the vapors is the boiling point of the sample. Care must be taken not to overheat the sample, since superheating of the liquid and vapors will yield erroneous results.

A second microprocedure requires only two to three drops of sample. The sample is placed in a micro test tube, and a very fine capillary, with one end sealed, is placed open-end into the sample. The test tube is suspended in an oil bath and gently heated until a steady stream of bubbles issues from the end of the capillary. The temperature of the oil bath is allowed to decrease slowly until the bubbling stops; this temperature is the boiling point of the sample. The process should be repeated several times until a reproducible boiling point is obtained. This technique is relatively difficult to master, but it should give boiling points within a few degrees of the boiling point that would be obtained during distillation.

3.5
DENSITY MEASUREMENTS

Determination of the density of a substance provides valuable information for the identification of compounds that are not readily converted into derivatives.

Densities are usually determined by a direct comparison of the weights of equal volumes of sample and water at a given temperature, t, and correcting to the density of water at 4°C as given in Eq. (3.2).

$$D_t = \frac{\text{weight of sample}}{\text{weight of water at } t} \cdot D_{\text{H}_2\text{O}} \quad \text{at} \quad t \tag{3.2}$$

Densities are determined using vessels referred to as pycnometers. Pycnometers with capacities of 1 to 25 mL are commercially available; however, since the structural chemist usually has only limited amounts of sample available, he must occasionally make a pycnometer of suitable size for his use. Suitable micropycnometers can be constructed from 2 to 4 mm soft glass tubing or capillaries as shown in Fig. 3.5. Pycnometers prepared from 3 mm glass tubing require 0.2 to 0.6 mL of sample and from capillaries, 0.02 to 0.05 mL. Density measurements should be carried out using as large a sample size as possible to reduce weighing errors. Densities should be calculated to four significant places, requiring the weighings to be precise to the same degree.

Fig. 3.5. Micropycnometer.

The pycnometer illustrated in Fig. 3.5 is cleaned, completely filled with water, and thermostated at a given temperature. After temperature equilibrium has been attained, the liquid level of the sample in the pycnometer is adjusted to the level of the ring on the right side of the pycnometer by removing liquid from the capillary end with a piece of adsorbent paper. The pycnometer is removed from the constant temperature bath, carefully dried, and weighed. The pycnometer is emptied, dried, and filled with the sample, and the process is carried out with the sample. The sample can be recovered uncontaminated for further use.

3.6
REFRACTIVE INDEX

The refractive index, as normally determined, is the ratio of the velocity of light in air to the velocity in the substance being determined. The refractive index can be precisely determined, usually to five significant figures, and is one of the most important physical constants used to describe a liquid.

Refractive indexes are determined using refractometers, of which there are several types available. Figure 3.6 shows the Abbé refractometer, which is the most widely used. The Abbé refractometer consists of a light source, a pair of movable, hinged, water-cooled prisms between which the sample is placed, two movable Amici prisms, which compensate for differences in the degree of refraction of light of different wavelengths, a telescope for observing the refraction field, and a scale for reading the refractive index. If the liquid is quite mobile, a dropping pipette is used to introduce the sample into the small sample depression between the prisms. For viscous samples, the prisms are opened, and one drop of the sample is cautiously applied to the bottom prism; then the prisms are gently closed. Care must be exercised to avoid touching the surface of the prism with the dropper. The light is adjusted to obtain maximum illumination, and the cross hairs in the telescope are brought into focus by adjustment of the eyepiece of the telescope. The Amici prisms are adjusted to remove colors from the field, and the

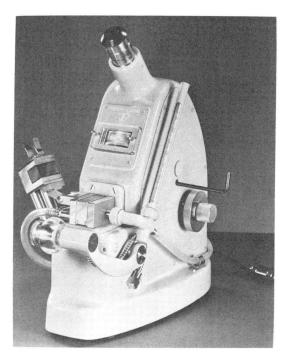

Fig. 3.6. Abbé refractometer. (Courtesy of Bausch and Lomb, Incorporated, Rochester, N.Y.)

position of the borderline between the black and white fields is adjusted to intersect at the intersection of the cross hairs by adjustment of the coarse- and fine-adjustment knobs. The refractive index is read from the drum to three decimal places, and the fourth place is estimated. The temperature of the prism assembly is recorded since the refractive index is a function of the temperature. The variation of refractive index with temperature is dependent on the types of functional groups present, but it can be approximated by using a correction of 0.00045/°C, the refractive index decreasing as the temperature rises. Thus refractive indexes determined at one temperature can be reliably compared with values determined at other temperatures. The refractive index n_D^T, for example, for water at 20°C, is reported as

$$n_D^{20}\ 1.3333$$

where 20 is the temperature at which the refractive index was measured, and D indicates the wavelength of the light employed as that of the sodium D line.

3.7
OPTICAL ROTATION

Conducting a beam of plane polarized light through an optically active medium results in an angular displacement of the plane of the polarized light. The principal components of a polarimeter (Fig. 3.7) used to determined optical rotations are a light source, a polarizing Nicol prism, a sample tube compartment, an analyzing Nicol prism, an observation lens, and a scale indicating the optical rotation of the sample.

Fig. 3.7. Polarimeter. A top-filling polarimeter tube is shown beside the instrument. (Reproduced by courtesy of Rudolph Research, Fairfield, N.J.)

The purified sample, approximately 200 to 500 mg, is accurately weighed and is dissolved in an appropriate volume of solvent, the volume depending on the volume of the sample tube to be used. Suitable solvents include water, methanol, ethanol, chloroform, or mixtures of water and an alcohol. If the rotation of the sample is low, higher concentrations can be used. Liquid samples can also be run as neat samples. The sample solution must be free of small particles (dust, lint, etc.) and should also be colorless if possible. The sample is placed in the sample tube, generally a 10 cm tube with glass end plates held firmly in place with rubber washers and a screw cap, by removing the screw cap and end plate from one end of the tube, holding the tube nearly vertical, and adding the sample by means of a capillary dropper. The sample is added until it rises slightly above the end of the sample tube. The glass end plate is carefully placed over the end of the tube so that no air bubbles are trapped in the sample. If bubbles have been trapped, the end plate must be removed and more sample added until the sample level extends above the end of the tube; then the glass end plate is again put in place. When no air bubbles have been trapped, the rubber washer is placed on top of the glass end plate, and the screw cap is firmly secured. One must be careful not to apply too much pressure in screwing the end cap in place since anomalous rotations may be produced. Top-filling polarimeter tubes, which provide for a somewhat easier filling operation, are also available (Fig. 3.7).

The *blank* or *zero* reading of the polarimeter is determined before the sample tube is placed in the light beam. The light field is observed through the observation lens. If the orientations of the two Nicol prisms are not the same, the light field will appear as two halves, one dark and one light half, depending on the relative angular displacement from 0°. The observation lens is adjusted until maximum sharpness between the two fields is obtained. The movable prism is then adjusted until no distinction can be made between the two halves of the light field. Finally, the rotation is read from the scale on the movable prism mount. This scale is usually divided into 0.25° divisions with a vernier capable of reading to the nearest 0.01°. Several adjustments of the movable prism are made, and readings are taken and averaged to obtain the rotation. The *blank* or *zero* reading may or may not coincide with the 0° mark on the scale, and an appropriate correction must be applied to the observed rotation of the sample. The sample tube is then placed in the light path and the rotation is determined as outlined earlier in determining the instrument *zero* point.

The specific rotation $[\alpha]$ is given by Eqs. (3.3) and (3.4),

$$\textit{For solutions:} \qquad [\alpha]_D^T = \frac{\alpha \cdot 100}{l \cdot c} \qquad\qquad (3.3)$$

$$\textit{For pure liquids:} \qquad [\alpha]_D^T = \frac{\alpha}{l \cdot d} \qquad\qquad (3.4)$$

where α is the observed rotation of the sample in degrees, l is the length of the sample tube in decimeters, c is the concentration of solute in g/100 mL of solution, d is the density of the pure liquid, T is the temperature of the sample at which the rotation was measured, and D specifies the wavelength of the light used as that at the sodium D line (otherwise the wavelength is given in nanometers). The solvent, as well as the concentration, should be specified. The rotation of the sample is given, for example, as $[\alpha]_D^{20} +65.2°$ $(c\,1.0, H_2O)$. Occasionally materials may possess very large specific rotations, and a single measurement will not allow one to distinguish between, for example, a moderate negative specific rotation and a large positive specific rotation. *Specific rotations can be compared only when using the same solvent, since specific rotations may vary with solvent.*

In quantitative comparisons of compounds of different molecular weights *molecular rotations*, $[\phi]$, as defined by Eq. (3.5), are often used,

$$[\phi] = [\alpha]\frac{M}{100} \qquad\qquad (3.5)$$

where M is the molecular weight.

The measurement of the optical rotation as a function of the wavelength is termed *optical rotatory dispersion* (ORD), and the measurement of the unequal absorption of right and left circularly polarized light is termed *circular dichroism* (CD). The two phenomena are intimately related; the same qualitative information is available from either type of measurement. Although only ORD curves are discussed in this text, the application of CD data parallels the application of ORD data.

Figure 3.8 illustrates two of many possible types of optical rotatory dispersion curves. Optical rotatory dispersion curves can exhibit maxima, minima (extrema), or inflections. Curve *A* (−) of Fig. 3.8 is referred to as a normal negative dispersion curve. The curve is assigned a negative sign because the molecular rotation increases in the negative direction in going to shorter wavelengths. Curve *B* (---) in Fig. 3.8 is referred to as a plain positive dispersion curve, even though it starts out with negative rotations at longer wavelengths. (Actually, both curves are plain dispersion curves; however, the term *normal* implies no extrema, inflections, or crossings of the zero rotation line.)

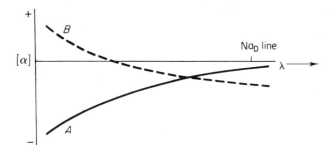

Fig. 3.8. Examples of typical optical rotatory dispersion curves: curve *A*, normal negative curve; curve *B*, plain positive curve.

Figure 3.9 illustrates two complex anomalous dispersion curves. Curve *A* (—) is a positive, single, Cotton effect curve. The various points on the curve are designated as peak (*P*) (instead of maximum, to avoid confusion with the ultraviolet maximum, which occurs at a different wavelength) and trough (*T*). The vertical distance (*a*) between the peak and trough is the amplitude, and the horizontal distance (*b*) is the breadth. A positive Cotton effect curve is one in

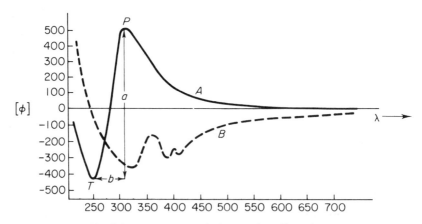

Fig. 3.9. Example of typical anomalous dispersion curves.

which the peak occurs at a longer wavelength than does a trough. If an optical rotatory dispersion curve displays several peaks, troughs, or inflections, it is referred to as a multiple Cotton effect curve [see curve *B* (---) in Fig. 3.9].

Empirical and theoretical correlations of the sign of the dispersion curve with absolute configuration and structure have been developed for a wide variety of classes of compounds including cyclic ketones, α,β- and β,γ-unsaturated cyclic ketones, lactones, alkenes and dienes, α-amino- and α-hydroxyacids, benzoate derivatives of alcohols and glycols, and carboxylic acid derivatives. Such correlations make the assignment of absolute configurations of chiral centers considerably easier than by purely chemical methods. The reader is referred to the references listed at the end of this chapter for further information.

3.8
DETERMINATION OF MOLECULAR WEIGHTS

The molecular weights of substances can be determined in many ways. Mass spectrometric methods (Sec. 7.4.1) are generally the easiest and most accurate, although the necessary instrumentation is very expensive. Molecular weights of volatile compounds can be determined by use of the ideal gas law, in which a known weight of material is vaporized at a given temperature, and the volume and pressure are measured. The high molecular weights of biological polymers are often estimated by gel exclusion chromatography (Sec. 2.6.1) and ultracentrifuge methods. Many methods for molecular weight determination are based on Raoult's Law, i.e., changes in vapor pressure upon addition of a nonvolatile solute to a pure solvent. These methods include boiling-point elevation, vapor-pressure equalization (isopiestic and osmometric methods), freezing-point depression, and melting-point depression (Rast method). The latter method is unique in that it requires only the simplest of equipment—a melting-point apparatus.

3.8.1 Rast Method

The addition of a nonvolatile solute to a solid material results in a reduction of the vapor pressure of the liquid phase in equilibrium with the solid phase (see Sec. 3.2 on melting points), and ultimately in a lowering of the melting point of the solid. The magnitude of the melting-point depression for a given concentration of the sample is a unique physical constant of the pure solid phase and is referred to as the *molal melting-point depression constant*. The molecular weight is calculated from Eq. (3.6)

$$\text{mol. wt.} = \frac{K \cdot w \cdot 1000}{\Delta t \cdot W} \tag{3.6}$$

where K is the molal melting-point depression constant, w is the weight in grams of the solute, Δt is the depression in the melting point, and W is the weight of the solid solvent (or liquid solvent in the case of freezing-point depression determinations).

Table 3.1 lists various solid solvents and their molal freezing-point depression constants, which can be used in Rast molecular-weight determinations. The choice of the solid solvent depends on several factors. The solute must be soluble in the liquid phase of the solvent. There must be no chemical or physical (complex formation) reaction of the solute with the solvent that would change the nature of the solvent or the solute. The solute must also be chemically stable at the temperature of the melting point of the mixture. Finally, the solid solvent should be chosen to provide the maximum melting-point depression for similar concentrations of various solutions to increase the accuracy of the determination.

Table 3.1   Solid Solvents for Rast Molecular-Weight Determinations

	Melting Point, °C	Molal Depression Constant, °C
Benzophenone	48	9.8
Borneol	202	35.8
Bornylamine	164	40.6
Camphene	49	31.08
Camphoquinone	190	45.7
Camphor	178	40.0
Cyclopentadecanone	65.6	21.3
Biphenyl	70	8.0
Naphthalene	80.2	6.9
Perylene	276	25.7

3.8.1a Procedure

An accurately weighed sample of the unknown substance is added to an accurately weighed portion of the solid solvent (the concentration of the solute must be kept below 10 and preferably 5 mol%). The mixture is melted and mixed until homogeneous. The melting points of the mixture and pure solid solvent are determined (the melting point being the temperature at which the last crystals disappear into the liquid phase), preferably employing capillary melting-point tubes. The samples are allowed to cool and solidify, and the melting points are redetermined. This process is repeated until consistent results are obtained.

Precautions must be exercised with compounds that undergo decomposition giving two or more lower molecular-weight fragments, for example, chlorohydrins and solvates. In such cases, anomalously low values of molecular weights are derived.

3.9
REFERENCES

GENERAL

1. CHERONIS, N. D., "Micro and Semimicro Methods" in *Technique of Organic Chemistry*, Vol. VI, Part I, A. Weissberger, ed. New York: Interscience Publishers, 1954.

2. KING, J. F., "Applications of Dissociation Constants, Optical Activity, and Other *Physical Measurements" in Technique of Organic Chemistry*, Vol. XI, Part I, New York: Interscience Publishers, 1963.

3. OVERTON, K. H., "Isolation, Purification, and Preliminary Observations" in *Technique of Organic Chemistry*, Vol. XI, Part I, A. Weissberger, ed. New York: Interscience Publishers, 1963.

OPTICAL ROTATION

4. CRABBÉ, P., *ORD and CD in Chemistry and Biochemistry.* New York: Academic Press, Inc., 1972.

5. DJERASSI, C., *Optical Rotatory Dispersion.* New York: McGraw-Hill Book Company, 1960.

6. ELIEL, E. L., *Stereochemistry of Carbon Compounds.* New York: McGraw-Hill Book Company, 1962.

7. VELLUZ, L., LEGRAND, M., and GROSJEAN, M., *Optical Circular Dichroism.* New York: Academic Press, Inc., 1965.

8. WILEN, S. H., *Tables of Resolving Agents and Optical Resolutions.* Notre Dame: University of Notre Dame Press, 1972.

part II

ABSORPTION SPECTROSCOPY AND MASS SPECTROMETRY

4

Ultraviolet Spectroscopy

4.1
GENERAL INTRODUCTION TO SPECTROSCOPY

The determination of molecular structure has been greatly facilitated by recent advances in spectroscopy, both in the interpretation and correlation of spectral data and in the design of the various spectrometers. These techniques are based on the fact that molecules are capable of absorbing radiant energy by undergoing a variety of different excitation processes. The modes of excitation available to most molecules include electronic excitation, vibrational excitation, rotational excitation, and nuclear spin inversion. The various modes of excitation require vastly different quantities of energy; hence, absorption of energy by molecules occurs over the entire electromagnetic spectrum. Table 4.1 illustrates the regions, wavelengths, energies, and the types of transitions occurring in each region of the spectrum. Below the ultraviolet region, the energy available exceeds the ionization potential of molecules and results in ion formation.

The energies involved in the various transitions are quantized; that is, a given transition in a specific molecule can be effected only by radiation of energy corresponding to the energy difference between the two states of the molecule involved in the transition. This may lead to a sharp absorption line at a wavelength corresponding to the energy involved; however, in many instances, the absorption lines are broadened into bands due to interaction with other types of transitions occurring in the same molecule. These instances will be discussed in the individual sections.

107

Table 4.1. Regions of the Electromagnetic Spectrum

Region	Wavelength*	Energy of Excitation	Type of Excitation
Vacuum ultraviolet	100–200 nm	286–143 kcal	Electronic
Quartz ultraviolet	200–350 nm	143–82 kcal	Electronic
Visible	350–800 nm	82–36 kcal	Electronic
Near infrared	0.8–2.0 μm	36–14.3 kcal	Overtones of bond deformations
Infrared	2–16 μm	14.3–1.8 kcal	Bond deformations
Far infrared	16–300 μm	1.8–1 kcal	Bond deformations
Microwave	~ 1 cm	~ 10^{-4} kcal	Rotational
Radio frequency	~ meters	~ 10^{-6} kcal	Electron and nuclear spin transitions

* In the past, the dimensions of wavelength in the ultraviolet and visible and infrared regions were different. Nanometers (nm) was previously millimicrons (mμ) and micrometers (μm) was microns (μ). The reader must be aware of these changes when reading the literature prior to about 1974.

The energy gap between similar transitions in various molecules is a function of the environment of the chromophore undergoing excitation. Correlations of the effects of the molecular environment on the energies of transitions in the various spectral regions have been made and are extremely useful in recognizing certain functional units within a molecule. The following sections will discuss the origin of the transitions, the effects of the molecular environment on the transition energies, and correlations found to be useful in structure determination.

Certain definitions of terms and equations must be presented before discussing the individual spectral regions. These are given in Table 4.2. Variances in the use of the terms will be noted. A thorough understanding of the meaning and the usage of each of the terms is necessary.

Table 4.2. Definition of Terms and Equations

Term	Symbol	Equation	Dimensions
Wavelength	λ	—	Å (Ångström)(10^{-8} cm)
		—	μm/micrometers
		—	nm/nanometers
Frequency	ν	$\dfrac{c}{\lambda}$	Hz (cycles per second)
Wave number	n	$\dfrac{1}{\lambda}$	cm^{-1}
Energy	E	$h\nu, \dfrac{hc}{\lambda}$	Depends on the units of h

c = velocity of light $(2.9979 \times 10^{10}$ cm-sec$^{-1})$; h = Planck's constant $(6.6237 \times 10^{-27}$ erg-sec$)$

Not all these physical methods will lend themselves to application in every structural problem. Microwave spectroscopy, in general, does not lend itself to applications in structural problems, except with very simple molecules, due to the complexity of the spectral data. Information from one spectral region alone will hardly ever provide sufficient information to derive the complete structure; instead, an integration of data from several regions, along with chemical data, is usually necessary. To repeat, *it cannot be emphasized too strongly that these physical methods by themselves may not lead to the complete and correct structure assignment; an integration of both physical and chemical data is necessary.* Experience has shown that the chemist who tries to rely only on spectral or chemical information wastes a great deal of time and often increases the chances of error in the final assignment of structure. The sections on chemical methods of structure determination (Chapter 9) will candidly evaluate the utility of physical *vs.* chemical methods for the detection of the individual functional groups.

4.2
INTRODUCTION TO ULTRAVIOLET
AND VISIBLE SPECTROSCOPY

The absorption of energy in the ultraviolet and visible regions of the spectrum results in the excitation of an electron from an occupied molecular orbital or a nonbonding atomic orbital to a higher energy, unoccupied molecular orbital. As molecules may contain σ and π molecular orbitals and nonbonded (n) atomic orbitals, the type of transition occurring on absorption of energy is conveniently indicated by designating the orbital from which the electron is excited and to which it goes. For example, the excitation of an electron from a σ molecular orbital to a σ^* molecular orbital is designated as a $\sigma \rightarrow \sigma^*$ transition. Other transitions commonly encountered are the $\pi \rightarrow \pi^*$, $n \rightarrow \pi^*$, and $n \rightarrow \sigma^*$ transitions. As the energy gap between σ and σ^* molecular orbitals is almost always greater than that between π and π^*, and n and π^*, $\sigma \rightarrow \sigma^*$ transitions occur at shorter wavelengths than do $\pi \rightarrow \pi^*$ transitions, which in turn appear at shorter wavelengths than do $n \rightarrow \pi^*$ transitions.

The ultraviolet portion of the spectrum extends from 100 to 350 nm, while the visible region extends from 350 to 800 nm. The ultraviolet region is divided into two regions: the "far" or "vacuum" ultraviolet region, extending from 100 to 200 nm, and the "quartz" region, extending from 200 to 350 nm. Below 200 nm, oxygen and carbon dioxide in air and the sample containers absorb strongly, rendering impossible the detection of absorption at wavelengths shorter than 200 nm without the use of vacuum techniques. Such techniques are difficult to employ, and absorption data from this region are not routinely used. The $\sigma \rightarrow \sigma^*$ and $\pi \rightarrow \pi^*$ transitions of alkanes, alkenes, alkynes, and simple aldehydes and ketones, and some $n \rightarrow \pi^*$ transitions occur at wavelengths less than 200 nm. The $\pi \rightarrow \pi^*$ transitions of conjugated polyenes, carbonyl compounds, and aromatics absorb at wavelengths longer than 200 nm, making ultraviolet spectroscopy particularly useful for detecting and characterizing such systems.

Ultraviolet and visible spectral data are obtained as plots of absorbance or extinction coefficient *vs.* wavelength (Figs. 4.2 to 4.4). The absorbance and extinction coefficients are calculated according to Eq. (4.1), in which I_0 is the intensity of the incident light, I is the intensity of the light transmitted through the sample, ε is the extinction coefficient, l is the cell length in centimeters, and c is the concentration in moles per liter. Extinction coefficients range from 1 to 100,000 depending on the nature of the absorbing chromophore. Unless

$$\text{Absorbance} = \log \frac{I_0}{I} = \varepsilon c l \tag{4.1}$$

direct comparison of features is desired, only the wavelength at maximum intensity of absorption is noted as $\lambda_{max}^{solvent}$, followed by the extinction coefficient or log of the extinction coefficient in parentheses, i.e., $\lambda_{max}^{C_2H_5OH}$ 227 nm (log ε 3.72). Occasionally the wavelengths at which other features appear are also listed, such as inflections ($\lambda_{infl}^{solvent}$) and minima ($\lambda_{min}^{solvent}$). The solvent used must always be specified, as in many cases the nature of the solvent has a profound effect on the position of maximum absorption.

4.3
SPECTRAL DATA
AND STRUCTURE CORRELATIONS

4.3.1 Hydrocarbon Derivatives

Table 4.3 lists the wavelengths of typical $n \rightarrow \sigma^*$ absorption bands of saturated hydrocarbon derivatives.

Table 4.3. Absorption Maxima of Saturated Systems Containing Heteroatoms

Type of Compound	λ_{max}(nm)	Approximate Log ε
Amines	190–200 Also occasionally a longer wavelength shoulder	3.4
Alcohols	180–185	2.5
Ethers	180–185	3.5
Epoxides	~170	3.6
Mercaptans	190–200 and 225–230 (shoulder)	3.2 and 2.2
Sulfides	210–215 and 235–240 (shoulder)	3.1 and 2.0
Disulfides	~250	2.6
Chlorides	170–175	2.5
Bromides	200–210	2.6
Iodides	255–260	2.7

4.3.2 Alkenes, Alkynes, and Polyenes

Nonconjugated alkenes and alkynes. The $\pi \rightarrow \pi^*$ band of ethylene occurs at 165 nm. Each sequential attachment of an alkyl group to the C=C produces a bathochromic shift (a shift to longer wavelength) of 3 to 5 nm. As even the λ_{max} of tetraalkyl-substituted alkenes occurs below 200 nm, the λ_{max} of such chromophores cannot generally be observed. However, the absorption band widths are sufficiently broad that a portion of the absorption band extends beyond 200 nm (termed end absorption), and the presence of the C=C function can be detected in the absence of other strongly absorbing groups. Similarly, the intense $\pi \rightarrow \pi^*$ band of alkynes appears below 200 nm; however, in many cases a very weak band is also observed near 223 nm.

Conjugated polyenes. Conjugated dienes and enynes have $\pi \rightarrow \pi^*$ absorptions that occur at wavelengths longer than 200 nm. Spectra-structure correlations have been developed which are useful in predicting the type and degree of substitution on the conjugated system. The empirical correlations initially proposed by R. B. Woodward and later extended by L. and M. Fieser are known as the Woodward Rules or Woodward–Fieser Rules.

A parent system is defined, and increments characteristic of the functions directly attached to the chromophore are added to the base absorption of the parent. The parent system is the acyclic diene, or heteroannular diene, system **1,** in which the two double bonds are not contained within a single ring. The base absorption of **1** is at 214 nm, while the homoannular diene, system **2**, has a base absorption at 253 nm. In addition to differences in the base absorption wavelengths, the heteroannular dienes absorb more intensely than do the homoannular diene systems. It should be noted that in the homoannular diene system the

1   **2**

λ_{max} 214 nm (log ε 3.9 − 4.3) λ_{max} 253 nm (log ε 3.7 − 3.9)

two carbon atoms of the six-membered ring attached to the ends of the diene chromophore **boldface** in **2**) are not included in the base absorption value. Homoannular dienes contained in other size rings possess different base absorption values, for example, cyclopentadiene (228 nm) and cycloheptadiene (241 nm). Cross-conjugated polyenes, represented by the partial structure **3,** in general do not correlate well.

Table 4.4 lists the substituent increments. In general, the magnitude of the increment is not a function of the position of the substituent on the diene chromophore.

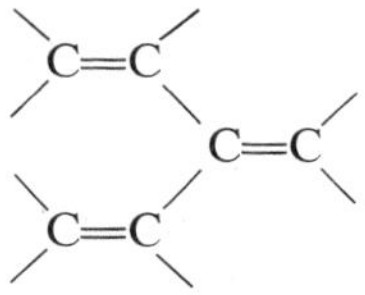

3

A few examples will illustrate the use of these rules. In structure **4**, the parent diene is of the homoannular type, with one of the double bonds exocyclic to ring A and one exocyclic to ring C, exocyclic being defined as having only one of the carbon atoms of the double bond belonging to a ring. In structure **5**, the parent diene is of the heteroannular type in which the double bond in ring B is exocyclic to ring A. Structure **6** is an example of a "through" conjugated triene system,

Table 4.4. Substituent Constants for Calculation of λ_{max}'s of Substituted Dienes

Function	Increment (nm)
Alkyl (per group)	+5
—OCOR*	0
—OR	+6
—SR	+30
—Cl, —Br	+5
—NR$_2$	+60
Extended conjugation (per double bond)	+30
Exocyclic double bond	+5

* R represents any alkyl group.

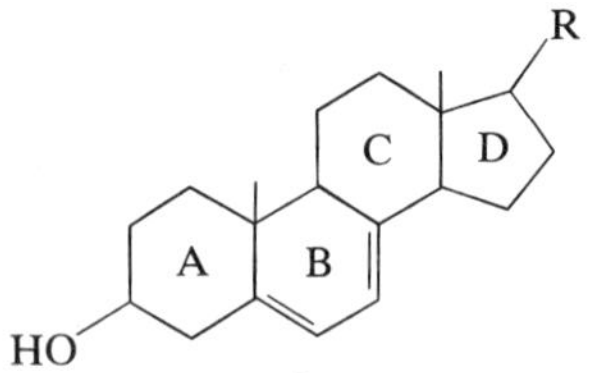

4

Parent diene	homoannular	253	nm
Substituents	4(×5)	20	
Exocyclic double bonds	2(×5)	10	
Calculated λ_{max}		283	nm
Observed		282	nm

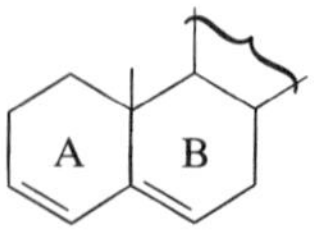

5

Parent diene	heteroannular	214	nm
Substituents	3(×5)	15	
Exocyclic double bonds	1(×5)	5	
Calculated λ_{max}		234	nm
Observed		234	nm

and it presents us with a choice of selecting either the homo- or heteroannular diene base. In general, one selects the longest wavelength base, i.e., that of the homoannular diene. In counting the number of substituents on the chromophore in **6**, all substituents on the entire triene system are included; it should be pointed out that C_{10} (the circled carbon atom in **6**) of the steroid nucleus is a substituent on two different carbon atoms of the triene system and is counted as two alkyl groups.

Parent diene	homoannular	253	nm
Extended conjugation	$1(\times 30)$	30	
Substituents	$5(\times 5)$	25	
Exocyclic double bonds	$3(\times 5)$	15	
Calculated λ_{max}		323	nm
Observed		324	nm

6

Parent diene	homoannular		heteroannular	
	253	nm	214	nm
Substituents	25	nm	20	
Exocyclic double bonds	15	nm	10	
Calculated λ_{max}	293	nm	244	nm
Observed		285 nm		

7

Structure **7** contains a cross-conjugated triene. Calculations employing both the homo- and heteroannular diene bases are shown, neither of which is in acceptable agreement.

It must be pointed out that in systems where the π-electron system is twisted or otherwise distorted owing to the geometry of the molecule, the preceding rules may not satisfactorily predict the position of maximum absorption. For example, 1,2-dimethylenecyclohexane (**8**) absorbs at 220 nm instead of at the calculated

8

value of 234 nm. The use of two-dimensional representations of molecules may not always give a true indication of the distortion within a molecule. Molecular models are more effective in revealing steric distortions of a π-system.

Conjugated enynes and diynes. Conjugated enynes and diynes also possess $\pi \rightarrow \pi^*$ absorption bands that appear at greater than 200 nm; however, no correlation of absorption band positions with structure has been found possible. Most compounds of these types display several closely spaced absorption bands owing to vibrational interactions. The most intense peak of selected examples is given below the following examples.

$$CH_3CH_2 \quad\quad C\!\equiv\!CH \quad\quad CH_3CH\!=\!CH\!-\!C\!\equiv\!CCH_3 \quad\quad CH_3\!-\!C\!\equiv\!C\!-\!C\!\equiv\!C\!-\!CH_3$$

$$\underset{H}{\overset{}{\diagdown}}C\!=\!C\underset{H}{\overset{}{\diagup}}$$

λ_{max} 221 nm λ_{max} 223 nm λ_{max} 227 nm

4.3.3 Carbonyl Compounds

Nonconjugated carbonyl derivatives. The $\pi \rightarrow \pi^*$ absorption band of all nonconjugated carbonyl derivatives occurs at wavelengths below 200 nm. The $n \rightarrow \pi^*$ excitation band, however, occurs at wavelengths longer than 200 nm, the position of which is somewhat characteristic of the type of carbonyl function (Table 4.5).

Table 4.5.   Absorption Maxima of Nonconjugated Carbonyl Compounds

Function	$\lambda_{max}(n \rightarrow \pi^*, \text{nm})$	Log ε
Aldehydes	290	0.9–1.4
Ketones	280	1.0–2.0
Carboxylic acids	206	1.6–2.0
Esters (in water)	206	1.6–2.0
(in non-protic solvents)	212	1.6–2.0
Amides	210	1.0–2.0
Acid chlorides	235	1.0–2.0
Anhydrides	225	≈ 1.7

Conjugated carbonyl compounds. Similar to conjugated dienes, conjugated carbonyl compounds possess $\pi \rightarrow \pi^*$ absorption bands at wavelengths longer than 200 nm, and spectra-structure correlations have been developed. In contrast to the conjugated dienes, the substituent increment depends on the position of the substituent on the unsaturated carbonyl chromophore. The parent base absorption for the various types of conjugated carbonyl compounds, along with the substituent parameters for the general chromophore **9**, is given in Table 4.6.

$$\gamma \text{ or higher} \left\{ \underset{}{\overset{\gamma}{\diagdown}}C\!=\!C \diagup \quad \overset{}{\diagdown}C\!=\!O \right.$$

$$\underset{\beta \quad\quad \alpha}{\diagup C\!=\!C \diagdown}$$

9

It should be noted that in contrast to the conjugated dienes, the position of maximum absorption of the conjugated carbonyl compounds is dependent on the

Table 4.6.   Base Absorptions and Substituent Parameters for Calculation of λ_{max}'s of Conjugated Carbonyl Compounds in Ethanol Solution

α,β-Unsaturated Parent	Base Absorption (nm)
Acyclic or six-membered or higher-ring ketone	215
Five-membered ring ketone	205
Aldehydes	210
Carboxylic acids and esters	195
Extended conjugation	+30
Homodienic component	+39
Exocyclic double bond	+5

Substituent	Substituent Parameters		
	α	β	γ
Alkyl	+10	+12	+18
Hydroxyl	+35	+30	+50
Alkoxyl	+35	+30	+17 (+31 for δ)
Acetoxyl	+6	+6	+6
Dialkylamino		+95	
Chloro	+15	+12	
Bromo	+25	+30	
Alkylthio		+85	

Solvent	Solvent Correction (nm)
Water	−8
Methanol	0
Chloroform	+1
Dioxane	+5
Ether	+7
Hexane	+11
Cyclohexane	+11

nature of the solvent. Also, in contrast to the change in position of the $\pi \rightarrow \pi^*$ band with increasing solvent polarity, the $n \rightarrow \pi^*$ band shifts to lower wavelength with increasing polarity of the solvent and in the presence of hydrogen-bond donors.

The following examples illustrate the use of these rules.

Parent base	215 nm
Extended conjugation	30
Substituents	
beta	12
delta	18
Exocyclic double bond	5
Calculated λ_{max}^{EtOH}	280 nm
Observed	284 nm

10

Parent base	215 nm
Extended conjugation	
(2 × 30)	60
Homodienic component	
(ring B)	39
Substituents	
beta	12
3γ or higher	54
Exocyclic double bond	5
Calculated λ_{max}^{EtOH}	385 nm
Observed	388 nm

A homodienic component is included for compound **11** to account for the presence of two double bonds in the B ring. Compound **12** illustrates an application to a cyclopentenone derivative. Both the carbonyl group and the double bond must be in the five-membered ring:

Parent base	215 nm
Substituents	
alpha	10
2 beta	24
Five-membered ring	−10
Calculated λ_{max}^{EtOH}	239 nm
Observed	237.5 nm

Parent base		215 nm
Substituents	2 beta	24
Aldehyde		−5
Calculated λ_{max}^{EtOH}		234 nm
Observed		235 nm

Parent base		215 nm
Substituents	2 beta	24
Acid		−20
Calculated λ_{max}^{EtOH}		219 nm
Observed		216 nm

Compounds **13** and **14** illustrate the application of these rules to aldehydes and acids. The correspondence between calculated and observed positions is much poorer in the application to unsaturated acids (±5 nm), and certain caution must be exercised. The accuracy of the application of the solvent corrections to carboxylic acids has not been investigated.

The partial structure **15** illustrates a cross-conjugated system in which the chromophore giving rise to the longest wavelength maximum is to be used.

Eneone system		$\Delta^{1,2}$		$\Delta^{4,5}$
Parent base		215 nm		215 nm
Substituents	beta	12	2 beta	24
Exocyclic double bond		═══		5
Calculated λ_{max}^{ETOH}		227 nm		244 nm
Observed				244 nm

The ultraviolet spectra of derivatives of carbonyl compounds are also quite useful. Oximes absorb at approximately the same wavelength as the original carbonyl compounds, whereas semicarbazones display maxima that are shifted to longer wavelengths by 30 to 40 nm, with an average increase in the extinction coefficient of 10,000 compared with the original carbonyl compound. 2,4-Dinitrophenylhydrazones of saturated ketones absorb in the 360 to 365 nm region, whereas similar derivatives of α,β-unsaturated ketones absorb in the 375 to 385 nm region.

Quinones. Quinones comprise a family of compounds that contain an α,β-unsaturated ketone system; however, because of the nature of the chromophore, they give rise to distinctly different types of absorption spectra.

There are two classifications of quinones: *para*-quinones and *ortho*-quinones. Distinction between the two categories of quinones is readily accomplished on the basis of their ultraviolet absorption spectra. *Para*-quinones give rise to three absorption bands in the ultraviolet-visible region. These bands occur in the 240 to 260 nm (arbitrarily assigned a band I designation), 280 to 340 nm (band II), and near 430 nm (band III). For example, *p*-benzoquinone absorbs at 244 nm (log ε 4.3), 279 nm (log ε 2.8), and 430 nm (log ε 1.3) in ethanol solution. The introduction of alkyl groups produces (1) a bathochromic shift in band I of 4 nm per alkyl group, (2) substantial bathochromic shifts in band II (however, no consistent trend is noted), and (3) effectively no change in the position of band III. Very minor solvent effects are noted. The ultraviolet spectra of *para*-quinones recorded in hydrocarbon or chloroform solution display bands in the same region as in ethanol; however, bands I and III generally appear as two and three distinct maxima and/or shoulders in nonpolar solvents.

Hydroxy-*para*-quinones usually display only two absorption bands in ethanol solution. These bands appear in the 250 to 310 nm (band I) and 350 to 435 nm (band II) regions, for example, hydroxy-*para*-quinone absorbs at 260 nm (log ε 4.0) and 369 nm (log ε 3.1). The introduction of additional hydroxy groups results in bathochromic shifts of approximately 15 nm and 20 nm in bands I and II, respectively. The introduction of alkyl groups in hydroxy-*para*-quinones results in a rather consistent bathochromic shift in band I of 4 nm per alkyl group, with no apparent effect on the position of band II. Alkoxy-*para*-quinones give rise to two intense bands in the 250 nm (log ε 4) and 360 nm (log ε 3.2) regions; however, no consistent trend with substitution is apparent.

Ortho-quinones display three bands in the ultraviolet and visible region near 260 nm (log ε 3.3), 400 nm (log ε 3.0), and 560 nm (log ε 1.6). The two longer wavelength bands are particularly useful in the characterization of *ortho*-quinones. There appear to be no obvious trends of the wavelength of absorption with substitution.

4.3.4 Aromatic Compounds

Aromatic compounds may display several absorption bands in the ultraviolet region. All benzenoid compounds display what are commonly termed E and B bands representing $\pi \rightarrow \pi^*$ excitations. For example, the E band of aromatics occurs in the 180 to 210 nm region with an ε of 2–6 × 10^3, while the B band appears in the 250 to 295 nm region with an ε of 10^2 to 10^3. In simple substituted aromatics the B band quite often appears as several closely spaced peaks (see the uv spectrum of benzene, Fig. 4.2, p. 124). Aromatic compounds that contain an unsaturated function in conjugation with the aromatic ring display a band in the 220 to 250 nm region with an ε of 1–3 × 10^4, which is termed the K band. Finally, if an attached function possesses a nonbonded pair of electrons, such as an alkoxy group, $n \rightarrow \pi^*$ transitions are possible, giving rise to absorption bands in the 275 to 330 nm region with ε's in the range of 10^2 to 10^3, which are termed R bands. Quite often the B and R absorption bands are obscured by the more intense K band. Examples of compounds displaying these bands are given in Table 4.7.

The introduction of substituents onto the aromatic ring produces several changes in the appearance of the absorption spectrum. With substituted aromatics, the E band shifts to longer wavelength, usually into the accessible region, regardless of the electronic properties of the substituent. The absorption maximum of the B band moves to longer wavelength in cases where the substituent is electron donating or capable of conjugation. On the other hand, with electron-withdrawing substituents, practically no change in the maximum position is observed (Table 4.8). The intensity of the B band generally increases as the electron-donating character of the substituent increases.

Table 4.7. Typical Absorption Maxima of Aromatic Compounds

Compound	Band λ_{max} nm(ε)			
	E	K	B	R
Benzene	198(8000)	—	255(230)	—
t-Butylbenzene	208(7800)	—	257(170)	—
Styrene		244(12,000)	282(450)	—
Acetophenone		240(13,000)	278(1100)	319(50)

Table 4.8.   UV Absorption Maxima of Substituted Benzenes

Function	E Band (Log ε)	B or K Band (log ε)	R Band (Log ε)
	Electron-donating substituents		
—CH_3	207 (3.85)	261(2.35)	
—$C(CH_3)_3$	207.5(3.78)	257(2.23)	
—OCH_3	217 (3.81)	269(3.17)	
—OH	210.5(3.79)	270(3.16)	
—O^-	235 (3.97)	287(3.42)	
—NH_2	230 (3.93)	280(3.16)	
—SH	236 (4.00)	269(2.85)	
	Electron-withdrawing substituents		
—F	204 (3.80)	254(3.00)	
—Cl	210 (3.88)	257(2.23)	
—Br	210 (3.88)	257(2.23)	
—I	207 (3.85)	257(2.85)	
—NH_3^+	203 (3.88)	254(2.20)	
	π-Conjugating Substituents		
—CHO	244 (4.18)	280(3.18)	328(1.30)
—$COCH_3$	240 (4.11)	278(3.04)	319(1.70)
—CO_2H	230 (4.00)	270(2.93)	
—CO_2^-	224 (3.94)	268(2.75)	
—CN	224 (4.11)	271(3.00)	
—NO_2	252 (4.00)	270(2.90)	330(2.10)
—CH=CH_2	244 (4.08)	282(2.65)	
—C_6H_5	246 (4.30)		

Inspection of the data given in Table 4.8 should point out the fact that no accurate correlation with structure is possible. This is also true of the disubstituted benzenes. The various disubstituted isomers do show slight differences in the positions of the absorption maxima and their intensities; however, the changes are so slight, and the absorption maxima lie in the same general region as monosubstituted benzenes so that no useful correlation can be made. The careful use of model compounds (see later discussion on model compounds) is recommended in these cases, or, better yet, refer to infrared and nuclear magnetic resonance data.

It is important to note the wavelength shifts involved in going from phenol to phenoxide and from aniline to anilinium ion. The bathochromic shift observed on going from phenol to phenoxide anion is typical of enol-enolate anion systems and is useful as a diagnostic tool to detect enol systems, not only in aromatics but also with nonaromatics. Similarly, the hypsochromic shift observed in the aniline-to-anilinium ion transformation is typical of the spectral changes due to the protonation of basic sites.

The fusion of additional rings to the benzene nucleus, for example, forming naphthalene, anthracene, etc., results in substantial bathochromic shifts. The *B* band of benzene occurs at 255 nm, whereas the similar bands in naphthalene and anthracene occur at 314 nm and 380 nm, respectively.

The introduction of heteroatoms into the aromatic ring, for example, pyridine, quinoline, etc., produces ultraviolet spectra very similar to those obtained from the related benzenoid compounds. The spectra of the heteroaromatic compounds display less fine structure than the benzenoid aromatics; for example, compare the spectrum of pyridine (Fig. 4.3, p. 125) with that of benzene (Fig. 4.2, p. 124). Absorption data for these and other heterocyclic aromatic systems appear in Table 4.9.

Table 4.9.   Absorption Maxima of Heterocyclic and Benzenoid Aromatics

		λ_{max} (nm)	ε
Benzene	(*B* band)	255	230
Pyridine	(*B* band)	252	2090
Pyridazine	(*B* band)	246	1150
Pyrimidine	(*B* band)	243	2950
Pyrazine	(*B* band)	261	6000
Naphthalene	(*B* band)	314	250
Quinoline	(*B* band)	313	2500
Anthracene	(*B* band)	380	9000
Furan		205	6300
Pyrrole		211	10000
Thiophene		235	4500
Imidazole		207	5000
Pyrazole		210	3100
Isoxazole		211	4000
Thiazole		204	4000

4.3.5 Miscellaneous Chromophores

Table 4.10 contains a list of miscellaneous functional groups that absorb in the ultraviolet region.

4.4
THE USE OF MODEL COMPOUNDS

Often the partial structure of a complex molecule can be deduced by comparison of the spectral properties of the compound with those of simpler molecules thought to contain the same chromophore. This is particularly true in cases in which the geometry of the molecule causes a distortion of the chromophore, resulting in shifts of the maxima to different wavelengths. Electronic interaction

**Table 4.10. Absorption Data
for Miscellaneous Functional Groups***

Functional Group	λ_{max} nm(ε)
Allenes	175–185(~10,000)
Cumulenes (butatriene)†	241(20,300)
Nitriles	~170(weak) and ~340(120)
Nitro	~210(~16,000)$(\pi \to \pi^*)$ and 270–280(~200)$(n \to \pi^*)$
Nitrate	$\pi \to \pi^*$ end absorption and ~260–270(shoulder)(150)
Nitrite	~350(~150)‡
Azo	$\pi \to \pi^*$ end absorption and ~350(weak)$(n \to \pi^*)$
Diazo	~400(~3)‡
Sulfoxide	210–215(~1600)
Sulfone	no absorption above 208
Vinyl sulfone	~210(~300)

* Functional groups contained in aliphatic systems only.
† Cumulenes containing more double bonds absorb at correspondingly longer wavelengths.
‡ Composed of a number of bands for which the approximate center is given.

with a neighboring functional group may also lead to anomalous shifts in the band positions. The simpler molecules used for such comparisons are called model compounds. Model compounds are used extensively in natural products chemistry and for highly strained ring compounds.

Although one generally uses the positions of the maxima for correlation purposes, similar changes in the positions of maxima throughout a series of compounds may be very helpful. The chromophores may not be completely identical, but may be similar enough to allow a correlation to be made. The

308 nm $\Delta\lambda = 12$ nm   320 nm $\Delta\lambda = 73$ nm   393 nm

328 nm $\Delta\lambda = 12$ nm   340 nm $\Delta\lambda = 74$ nm   414 nm

similarity in the shifts observed in the ultraviolet spectra of the following steroidal compounds and seven-membered ring compounds indicates a similarity in the chromophores.

4.5
ADDITIVITY OF CHROMOPHORES

Many molecules can contain more than one absorbing chromophore. In such cases, the observed ultraviolet absorption spectrum will be the sum of the absorption bands of the individual chromophores, provided there is no electronic interaction between the chromophores. If the absorption maxima of the two chromophores lie close to one another, resulting in a great deal of overlapping of the bands, we can subtract an expected absorption of one of the known chromophores to derive the absorption spectrum of the other chromophore. Experimentally this can be readily accomplished by placing a reference compound containing the desired chromophore into the reference cell of the spectrometer, with the sample in the sample cell. The resulting spectrum will be the difference spectrum between the sample and reference compound. The concentration of the reference compound may have to be adjusted so that the net absorptivity of the reference chromophore equals that of the same chromophore in the sample. A great deal of caution must be exercised in the selection of model compound because the chromophores must be identical.

4.6
PREPARATION OF SAMPLES

The primary consideration in the preparation of samples for recording their ultraviolet spectra is that of concentration. The concentration must be adjusted such that the absorption peak remains on the recording scale of the instrument and, furthermore, in the more accurate region on the recording scale. The region of greatest accuracy is between 0.2 and 0.7 absorbance units. If we specify that the maximum absorbance should be near 0.7 and we have a general idea of what type of chromophore is present and thus a general region for the value of the extinction coefficient, we can use Eq. (4.1) to calculate the concentration required for use in a 1 cm cell. For conjugated diene chromophores with extinction coefficients in the general region of 8000 to 20,000, the concentration should be near 4×10^{-5} mol/L, whereas for the $n \rightarrow \pi^*$ transition of a carbonyl group (ε of 10 to 100), the concentration should be near 10^{-2} mol/L. Preparation of very dilute samples cannot, in general, be accomplished by a direct weighing technique. The quantity of material would be quite small, in the submilligram range, and large weighing errors would be introduced. In general, such sample concentrations are obtained by successive dilutions of more concentrated solutions.

The choice of solvent rests on several factors. The most important criterion is that the solvent be transparent in the region over which you wish to record the absorption spectrum. Figure 4.1 lists several solvents and the cutoff point below

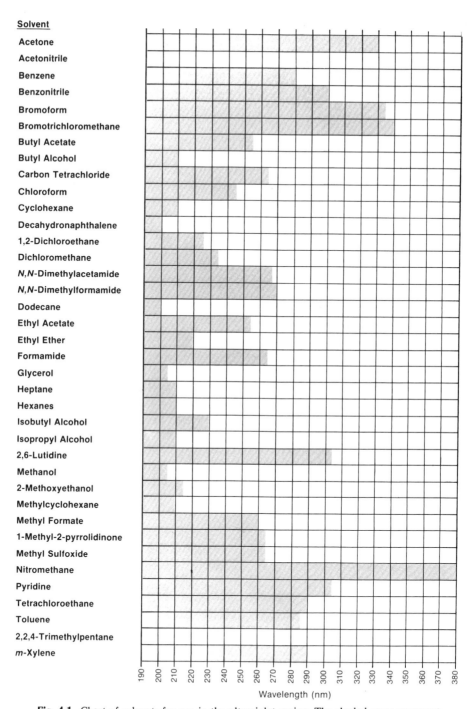

Fig. 4.1. Chart of solvents for use in the ultraviolet region. The shaded areas represent absorbances greater than unity in a 1-cm pathlength cell, prohibiting use as a solvent in those regions. (Reproduced by courtesy of Eastman Kodak Company, Rochester, N.Y.)

which they absorb too strongly to be useful as a solvent. Secondly, the solvent should be a good solvent for dissolving the material. Even though the concentrations are quite small, many compounds may be only partially soluble or may exist as aggregates in solution. Finally, a solvent should be chosen so that it does not react chemically with the sample.

Care should be taken in the purification and drying of the samples. The presence of trace amounts of materials having absorption maxima with large extinction coefficients may give rise to maxima that might be interpreted as belonging to the predominant material. The solvent used should be of high purity, generally referred to as "spectro grade" by distributors. Care should be taken to keep lint and dust from contaminating the final solutions.

The sample solution is placed in the sample cell and compartment of the instrument, and pure solvent, or a reference compound if a difference spectrum is to be determined, is placed in the reference cell and compartment. The description of the actual recording of the spectrum will not be explained here; the reader is referred to the individual operation manual available with each instrument.

4.7
EXAMPLES OF ULTRAVIOLET SPECTRA

Figures 4.2, 4.3, and 4.4 show the ultraviolet spectra of benzene, pyridine, and 4-methyl-3-penten-2-one (mesityl oxide) in ethanol solution.

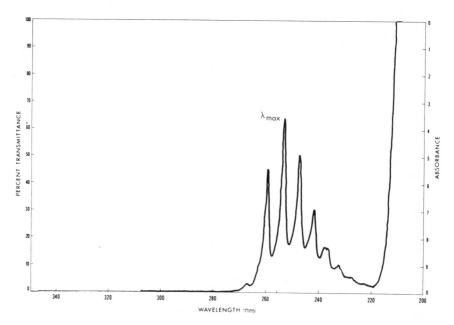

Fig. 4.2. Ultraviolet spectrum of benzene in 95% ethanol: concentration 2.92×10^{-3} m/L.

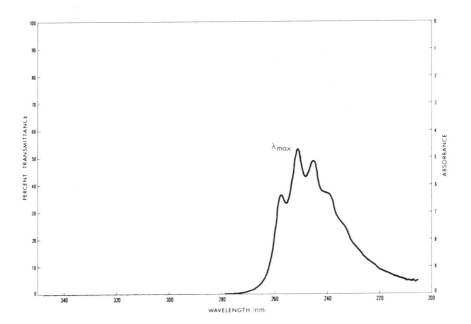

Fig. 4.3. Ultraviolet spectrum of pyridine in 95% ethanol: concentration 1.98×10^{-4} m/L.

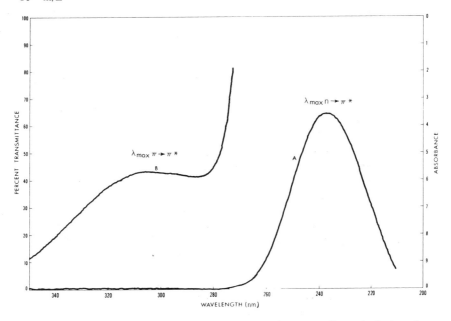

Fig. 4.4. Ultraviolet spectrum of mesityl oxide in 95% ethanol. Trace A displays the $\pi \to \pi^*$ band, concentration 6.29×10^{-5} m/L, and trace B displays the $n \to \pi^*$ band, concentration 6.29×10^{-3} m/L.

1. Calculate the λ_{max}'s for each of the following compounds.

a.

$$CH_3-\overset{\overset{\displaystyle O}{\|}}{C}-CH=CHN(CH_3)_2$$

b.

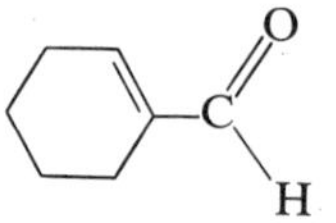

c. $CH_2=C(CH_3)-CH=CH-CO_2CH_3$

d.

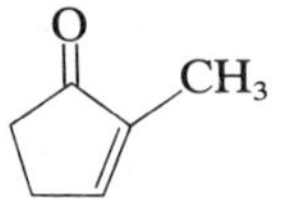

e.

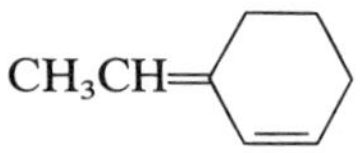

$CH_3CH=$

f. $CH_2=CHCH_2CH_2CCH_3$ (with carbonyl O)

g.

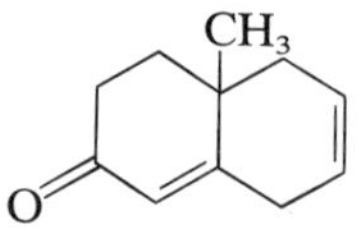

h. $CH_3O_2CCH=C(CH_3)_2$

i.

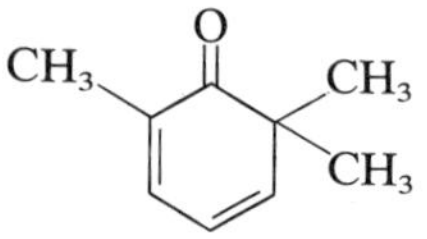

j.

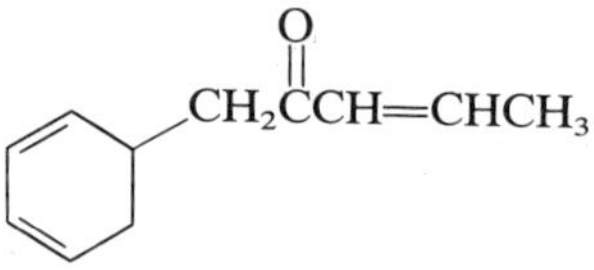

2. Suggest structures for each of the following that are compatible with the given spectral data (all in ethanol solution).

a. A $C_7H_{10}O$ compound (λ_{max} 259 nm), which undergoes acid-catalyzed hydrolysis to a C_6H_8O compound with λ_{max} 227 nm.

b. A C_7H_9N compound with λ_{max} 207 and 257 nm.

c. A C_4H_6O compound with λ_{max} 222 nm, which undergoes reduction with sodium borohydride to give a product that displays only end absorption.

d. A C_5H_8O compound with λ_{max} 290 (log ε 1.2) and end absorption, which does not readily add hydrogen in the presence of a catalyst.

e. A $C_8H_{10}O$ compound with λ_{max} 216 and 270 nm, which possesses the same spectral properties when dissolved in 0.1 M ethanolic sodium hydroxide.

f. A $C_9H_{12}O$ compound with λ_{max} 207 and 260 nm, which reacts with sodium dichromate in the presence of sulfuric acid to give a compound with λ_{max} 207, 261, and 280 (weak) nm.

g. A C_9H_{10} compound (λ_{max} 244 and 285 nm), which undergoes addition of hydrogen in the presence of a catalyst to produce a compound with λ_{max} 207 and 260 nm.

h. A C_3H_8OS compound with λ_{max} 212 nm.

i. A C_6H_{10} compound with λ_{max} 224 nm, which on reaction with one equivalent of ozone in methylene chloride-pyridine solution gives, as one of the products, a compound with λ_{max} 234 nm.

j. A C_7H_9NO compound with λ_{max} 228 and 281 nm, $\lambda_{max}^{0.1\ M\ base}$ 228 and 281 nm, and $\lambda_{max}^{0.1\ M\ acid}$ 202 and 255 nm.

4.9
LOCATING ULTRAVIOLET AND VISIBLE
ABSORPTION DATA IN THE LITERATURE

Several compilations of ultraviolet and visible absorption data have been published. H. M. Hershenson has compiled literature references in two volumes entitled *Ultraviolet and Visible Absorption Spectra* (ref. 11) covering the years 1930 to 1954 and 1955 to 1959. Nine volumes entitled *Organic Electronic Spectral Data* (ref. 13) have been published, which give spectral data and original references covering the years 1946 to 1973.

Reproductions of ultraviolet and visible spectra have been compiled by the American Petroleum Institute Research Project 44 (ref. 16), Sadtler Research Laboratories (ref. 15), as well as in the *UV Atlas of Organic Compounds* (ref. 17), and the compilation of UV spectra of aromatic compounds (ref. 10).

The *Handbook of Ultraviolet Methods* (ref. 18) is a bibliography of references of ultraviolet analytical procedures dealing with specific compounds or substances.

Many other sources, too numerous to mention here, give spectral data for individual, specific classes of compounds.

General references pertaining to the general theory and interpretation of ultraviolet and visible spectra are given in refs. 1 to 9.

4.10
REFERENCES

GENERAL

1. FIESER, L. F., and FIESER, M., *Steroids.* New York: Reinhold Publishing, 1959, pp. 15–21.

2. GILLAM, A. E., and STERN, E. S., *An Introduction to Electronic Absorption Spectroscopy in Organic Chemistry.* London: Edward Arnold Publishers, 1957.

3. JAFFÉ, H. H., and ORCHIN, M., *Theory and Applications of Ultraviolet Spectroscopy.* New York: John Wiley & Sons, Inc., 1962.

4. MASON, S. F., "Molecular Electronic Absorption Spectra," *Quart. Revs.,* **15,** 287 (1961).

5. MATSEN, F. A., "Applications of the Theory of Electron Spectra" in *Technique of Organic Chemistry,* A. Weissberger, ed. New York: Interscience Publishers, Inc., 1956.

6. RAO, C. N. R., *Ultra-Violet and Visible Spectroscopy.* London: Butterworths and Co., Ltd., 1961.

7. "Report on the Notation for the Spectra of Polyatomic Molecules," *J. Chem. Phys.,* **23,** 1997 (1955).

8. SANDORFY, C., *Electronic Spectra and Quantum Chemistry.* Englewood Cliffs, New Jersey: Prentice-Hall, Inc., 1964.

9. WOODWARD, R. B., *J. Am. Chem. Soc.,* **64,** 72 (1952).

SPECTRAL DATA AND REFERENCES

10. FRIEDEL, R. A. and ORCHIN, M., *Ultraviolet Spectra of Aromatic Compounds.* New York: John Wiley & Sons, 1951.

11. HERSHENSON, H. M., *Ultraviolet and Visible Absorption Spectra.* New York: Academic Press, index for 1930–1954 published in 1956, index for 1955–1959 published in 1961.

12. HIRAYA, K., *Handbook of Ultraviolet and Visible Absorption Spectra of Organic Compounds.* New York: Plenum Press Data Division, 1967.

13. *Organic Electronic Spectral Data.* New York: Interscience Publishers, Inc., nine volumes covering the years 1946–1973.

14. SCOTT, A. I., *Interpretation of the Ultraviolet Spectra of Natural Products.* New York: Pergamon Press, 1964.

15. *Standard Ultraviolet Spectra.* Compiled by the Sadtler Research Laboratories, 1517 Vine St., Philadelphia, PA.

16. *Ultraviolet Spectral Data.* Compiled by the American Petroleum Institute Research Project 44, Chemical Thermodynamics Properties Center, Department of Chemistry, Texas A & M University, College Station, Texas.

17. *UV Atlas of Organic Compounds.* London: Five volumes published by Butterworths, 1966–1971.

18. WHITE, R. G., *Handbook of Ultraviolet Methods.* New York: Plenum Press, 1965.

5

Infrared Spectroscopy

The positions of atoms within molecules do not remain constant, but undergo continual periodic movement (vibrations) relative to each other, the position of the center of mass of the molecule remaining constant during the vibrational processes. The energies associated with the vibrational processes within a molecule can assume only specific values, i.e., the vibrational energy levels are quantized. Transitions between vibrational energy levels can be induced by the absorption of electromagnetic radiation in the infrared region.

A molecule containing n number of atoms will possess $3n - 6$ independent vibrational motions called *normal modes* (linear molecules possess $3n - 5$ normal modes). Although at first glance it might appear that the infrared spectrum of a molecule containing a large number of atoms would contain a large number of absorption bands, and thus would be hopelessly complex, many of the molecular vibrations are highly localized in specific functions of the molecule. As a result, vibrations that are highly concentrated in a given type of function give rise to characteristic absorption bands in fairly well-defined and limited portions of the infrared region, allowing for the development of spectra-structure correlations.

The vibrational motions within molecules fall into two categories: (1) those involving the vibration of a bond along the bond axis (a stretching deformation designated by ν); and (2) those involving a vibration of a bond perpendicular to the bond axis (a bending deformation of the bond in a plane or perpendicular to an internal plane of symmetry in the function and designated by σ). If we view the

$$A \longleftrightarrow B \qquad A \xrightarrow{\ \ } B$$

stretching deformation   bending deformation

vibrational motions of a bond as a classical harmonic oscillator (such as a spring), the vibrational frequency, ν_{osc}, in a system A—B, in which the mass of A is much larger than that of B, is given by Eq. (5.1), in which k is the force constant (i.e., a measure of the energy required to deform the oscillator), and m is the mass of B.

$$\nu_{osc} = \frac{1}{2\pi}\sqrt{\frac{k}{m}} \tag{5.1}$$

In bonded systems k increases with increasing bond order, causing ν_{osc} to increase. For example, the vibrational frequencies increase in the order $\nu_{C-C} < \nu_{C=C} < \nu_{C\equiv C}$. When the masses of A and B are comparable Eq. (5.2) is applicable, in which μ is the reduced mass of the system ($m_A \times m_B/m_A + m_B$). If one applies Eq. (5.2) to a C—H bond in a complex molecule of mol. wt. 100, increasing the mol. wt. to 114 (the next higher homolog) results in very little change in ν_{osc}, which is consistent with the observation that bands arising from a given function tend to appear in narrow portions of the infrared region.

$$\nu_{osc} = \frac{1}{2\pi}\sqrt{\frac{k}{\mu}} \tag{5.2}$$

Not all of the $3n - 6$ normal modes of a molecule necessarily absorb in the infrared region. In order for a normal mode to absorb in the infrared, a bond must possess a dipole moment that changes in magnitude on vibrational excitation (vibrationally excited states generally have slightly longer bond lengths and, thus, larger dipole moments than the ground vibrational state).[1] As an example, let us consider the linear molecule CO_2, which possesses $3 \times 3 - 5 = 4$ normal modes of which two are stretching (ν_1 and ν_2) and two are bending (σ_1 and σ_2) deformations (Fig. 5.1). The stretching deformation ν_1 does not result in a change

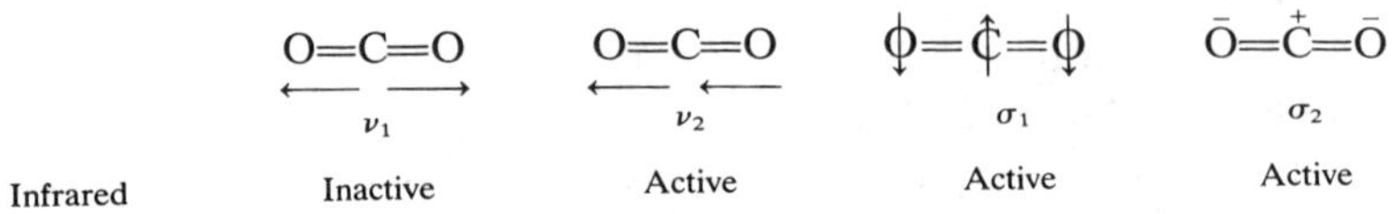

Fig. 5.1. Normal modes of carbon dioxide. Arrows indicate the direction of movement of the atoms in the plane of the page; (+) and (−) indicate movement forward and backward, respectively, out of the plane of the paper.

---

[1] This is an extremely simplified statement of the requirements for infrared activity. The reader is referred to F. A. Cotton, *Chemical Applications of Group Theory*, Interscience Publishers, New York, 1963, Chapter 9, and H. Jaffe and M. Orchin, *Symmetry in Chemistry*, John Wiley & Sons, Inc., New York, 1965, for more detailed and rigorous discussions of the requirements for infrared and Raman activity.

in the dipole moment on vibrational excitation, and thus does not give rise to an infrared absorption band. The deformations ν_2, σ_1, and σ_2 do result in a change in dipole moment on vibrational excitation; hence, they absorb infrared radiation.

In addition to the normal modes (fundamentals), other types of absorption bands may appear in the infrared region. These include overtone, combination, coupled, and Fermi resonance bonds. Overtone bands are high frequency harmonics of the fundamental normal modes and appear near integral multiples of the fundamental absorption frequencies. These are referred to as the first, second, etc. overtones, and they decrease markedly in intensity as the order of the overtone increases. First overtones of many intense bands appear in the infrared region and are generally the most noticeable.

Combination bands are relatively weak bands appearing at frequencies equal to the sum or difference of two or more fundamental frequencies. Coupling bands occur when two absorption bands of the same portion of the molecule interact, causing the absorption bands to be shifted out of the expected region for each chromophore independently. Fermi resonance occurs when an overtone or combination band lies close to a fundamental absorption band, resulting either in an enhancement of the intensity of the overtone or combination band or in a splitting of the bands.

Certain functions give rise to characeristic group deformation bands which are extremely useful in differentiating between related functions. In an A—B system only a single stretching deformation is possible. In an A$\genfrac{}{}{0pt}{}{\nearrow B}{\searrow B}$ system, however, two stretching deformations are possible, one being a symmetric stretching deformation and the other being an antisymmetric deformation. Thus two bands are found in the infrared when such groups are present. For example, this difference provides a ready means for distinguishing between primary and

Symmetric stretching deformation   Antisymmetric stretching deformation

secondary amines, the former giving rise to two N—H bands and the latter to one N—H stretching band. Other functions readily distinguished are: primary and secondary amides, nitroso (—N=O) and nitro $\left(-\overset{+}{N}\genfrac{}{}{0pt}{}{\nearrow O}{\searrow O^-}\right)$, and sulfinyl $\left(-\overset{O}{\underset{}{S}}-\right)$ and sulfonyl $\left(-\overset{O}{\underset{O}{\overset{\|}{S}}}-\right)$ functions. AB$_3$ systems, such as the methyl group, similarly give rise to two stretching deformations: one symmetric and one antisymmetric. Similar combinations of bending deformations are also possible (Fig. 5.2, p. 136).

The infrared region of the spectrum extends from the upper end of the visible region, approximately 0.75 μm, to the microwave region near 400 μm. The portion of this region generally used by the organic chemist for structural work is from 2.5 to 16 μm. This is primarily due to instrument design and cost and to the fact that most of the useful information can be derived in this region. The region from 0.75 to 2.5 μm is referred to as the near-infrared region and contains absorption bands due to overtone and combination bands. The region extending from 16 to 400 μm is referred to as the far-infrared region. The normal and far-infrared regions contain absorptions due to fundamental, overtone, and combination bands.

The position of absorption in the infrared region can be expressed in the wavelength [micrometer (μm)] of the absorbed radiation, or in terms of the wave number (cm^{-1}, the reciprocal of the wavelength in centimeters). Prior to the early 1970's, wavelengths were commonly given in microns (μ); 1 μ = 1 μm. The wave number, or incorrectly termed the frequency, convention is the most commonly used. In the following portions of this text, absorption positions will be given in wave numbers followed by wavelength in μm in parentheses.

Infrared spectrometers that record linear in wave number or linear in wavelength are available. The use of linear-in-wave numbers instruments results in a considerable expansion of the more important high-wave number end of the infrared region, resulting in an increased ability to resolve bands and define their position. The use of linear-in-wavelength instruments results in a considerable compression of the high-wave number region, resulting in a loss of resolution.

The intensities of the infrared absorption bands vary from strong to very weak (eventually disappearing if there is little or no change in dipole moment on excitation). The intensity of an infrared absorption band cannot be expressed as a unique constant as was possible in ultraviolet and visible spectroscopy. This is due to the fact that the slit widths (fine slits controlling the amount of infrared radiation passing through the instrument) used in most instruments are of the same order as the typical infrared bandwidths, which causes the measured optical density to be a function of the slit width. Despite this problem, infrared spectroscopy can be used as a quantitative analytical tool with certain precautions. Instead of determining the true molar extinction coefficient, one can determine an apparent molar extinction coefficient, ε_a, which will be a function of the slit width and even sample concentration. The apparent extinction coefficient is calculated as shown in Eq. (5.3).

$$\varepsilon_a = \frac{\text{absorbance}}{\text{concentration (mol/L} \cdot \text{cell length (cm)}} \qquad (5.3)$$

A series of apparent extinction coefficients *vs.* concentration should be determined, and the data plotted for use in final analysis. The apparent extinction coefficients can be used only with the instrument they have been measured on. Analytical limits are approximately ±5% when these methods are used.

5.2
CHARACTERISTIC ABSORPTION BANDS

Not all the absorption bands appearing in an infrared spectrum will be useful in deriving structural information. Certain portions of the infrared region, for example, the region of 1300 to 1000 cm^{-1} (7.5 to 10 μm), are extremely difficult to interpret due to the variety and number of fundamental absorptions occurring in this region. Certain narrow regions of the infrared spectrum provide most of the important information. In the derivation of information from an infrared spectrum, the most prominent bands in these regions are noted and assigned first.

The characteristic bands of various individual functions will be discussed in the following sections. These discussions will include the effect of molecular structure and electronic effects on the more prominent absorption bands.

5.2.1 Carbon-Hydrogen Absorption Bands

5.2.1a Alkanes

Carbon-hydrogen bond absorption occurs in two regions: the C—H stretching region of 3300 to 2500 cm^{-1} (3 to 4 μm) and the bending region of 1550 to 650 cm^{-1} (6.5 to 15.4 μm). The methyl group gives rise to two stretching bands that occur at 2960 and 2870 cm^{-1} (3.38 and 3.48 μm). The 2960 cm^{-1} (3.38 μm) band is ascribed to the asymmetric stretching deformation, and the 2770 cm^{-1} (3.48 μm) band is ascribed to the symmetric stretching deformation. The methyl group also gives rise to two bands in the bending deformation region at 1460 cm^{-1} (6.85 μm) and approximately 1380 cm^{-1} (7.25 μm). The higher-wave number band is due to the asymmetric bending motion, and the lower-wave number band is due to the symmetric bending. The 1380 cm^{-1} (7.25 μm) region is extremely valuable for the detection of the presence of methyl groups. This region is almost devoid of other types of absorption bands, whereas the other three absorption bands of the methyl group lie extremely close to similar absorption bands arising from methylene and methine C—H. The symmetrical bending absorption band of the methyl group is quite sharp and of medium intensity (Figs. 5.13, 5.14, and 5.16). The position of this band is quite sensitive to the type of substituent attached to the methyl group, and it can be used to detect the possible presence of acetyl and methoxyl groups (Table 5.1).

The presence of two or more methyl groups bonded to a single carbon atom results in a splitting of the symmetric C—H bending absorption bands. For example, the isopropyl group gives rise to two bands at 1397 and 1370 cm^{-1} (7.16 and 7.30 μm), and the tertiary butyl group gives rise to two bands at 1385 and 1370 cm^{-1} (7.22 and 7.30 μm). Confirmation for the presence of these groups can be obtained by observing the C—C skeletal vibrational bands (Sec.

5.2.5).

Table 5.1. Selected Carbon-Hydrogen Absorption Bands*

Functional Group	Wave Number cm^{-1}	Wavelength μm	Assignment[†]	Remarks[‡]
Alkyl				
—CH$_3$	2960	3.38	ν_{as}	s, lu
	2870	3.48	ν_s	m, lu
	1460	6.85	σ_{as}	s, lu
	1380	7.25	σ_s	m, gu
—CH$_2$—	2925	3.42	ν_{as}	s, lu
	2850	3.51	ν_s	m, lu
	1470	6.83	Scissoring	s, lu
	1250	~8.00	Twisting and wagging	s, nu
—C—H	2890	3.46	ν	w, nu
	1340	7.45	σ	w, nu
—OCOCH$_3$	1380–1365	7.25–7.33	σ_s	s, gu
—COCH$_3$	~1360	~7.35	σ_s	s, gu
—COOCH$_3$	~1440	~6.95	σ_{as}	s, gu
	~1360	~7.35	σ_s	s, gu
Vinyl				
=CH$_2$	3080	3.24	ν_{as}	m, lu
	2975	3.36	ν_s	m, lu
	~1420	~7.0–7.1	σ(in-plane)	m, lu
	~900	~11	σ(out-of-plane)	s, gu
C=C H	3020	3.31	ν	m, lu
Monosubstituted	990	10.1	σ(out-of-plane)	s, gu
	900	11.0	σ(out-of-plane)	s, gu
Cis-disubstituted	730–675	13.7–14.7	σ(out-of-plane)	s, gu
Trans-disubstituted	965	10.4	σ(out-of-plane)	s, gu
Trisubstituted	840–800	11.9–12.4	σ(out-of-plane)	m–s, gu
Aromatic				
C—H	3070	3.30	ν	w, lu
5-adjacent H	770–730	13.0–13.7	σ(out-of-plane)	s, gu
	710–690	14.1–14.5	σ(out-of-plane)	s, gu
4-adjacent H	770–735	13.0–13.6	σ(out-of-plane)	s, gu
3-adjacent H	810–750	12.3–13.3	σ(out-of-plane)	s, gu
2-adjacent H	860–800	11.6–12.5	σ(out-of-plane)	s, gu
1-adjacent H	900–860	11.1–11.6	σ(out-of-plane)	m, gu

Table 5.1. (Continued)

Functional Group	Wave Number cm^{-1}	Wavelength μm	Assignment†	Remarks‡
1, 2-, 1,4- and	1275–1175	7.85–8.5	σ(in-plane)	w, lu
1,2,4-substituted	1175–1125	8.5–8.9	(only with 1,2,4-)	
	1070–1000	9.35–10.0	(two bands)	w, lu
1-, 1,3-, 1,2,3-, and	1175–1125	8.5–8.9	σ(in-plane)	
1,3,5-substituted	1100–1070	9.1–9.35	(absent with 1,3,5-)	
	1070–1000	9.35–10.0		
1,2-, 1,2,3- and	1000–960	10.0–10.4	σ(in-plane)	w, lu
1,2,4-substituted				
Alkynyl				
$\equiv$C—H	3300	3.0	ν	m–s, gu
Aldehydic				
	2820	3.55	ν	m, lu
	2720	~ 3.7 ~	Overtone or combination band	m, gu

* Values selected from tables presented in Bellamy (ref. 2) and Nakanishi (ref. 11).

† Assignments are designated as follows: ν, stretching deformation; as, asymmetric; s, symmetric; σ, bending deformation.

‡ Intensities are designated as: s, strong; m, medium; w, weak. Band utilities for structural assignments are indicated as: gu, great utility; lu, limited utility (depends on the complexity of the structure); nu, no practical utility.

Methylene groups give rise to a number of absorption bands corresponding to the types of deformations illustrated in Fig. 5.2. The symmetric and asymmetric-stretching modes, Fig. 5.2a and b, give rise to bands at 2850 and 2930 cm^{-1} (3.51 and 3.42 μm), respectively. Absorption due to the *scissoring* action of the methylene group, Fig. 5.2c, occurs at 1470 cm^{-1} (6.80 μm), very close to the asymmetric bending band of the methyl group. The scissoring of the methylene group varies substantially with changes in the molecular environment. The *rocking* bending deformation, Fig. 5.2d, gives rise to absorption in the 720 cm^{-1} (13.9 μm) region. The *wagging* and *twisting* bending deformations, Fig. 5.2e and f, give rise to absorption in the 1350 to 1180 cm^{-1} (7.4 to 8.5 μm) region. The latter three bending absorptions are not particularly useful in structural identification in molecules of any complexity.

The methine group, —C̶—H, gives rise to single stretching and bending absorptions at 2890 and 1340 cm^{-1} (3.46 and 7.45 μm), which are very weak and usually are of no practical utility in structural identification.

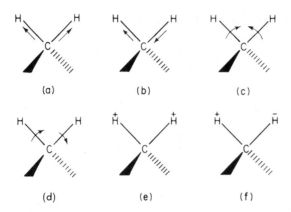

Fig. 5.2. Stretching and bending deformations of the methylene group. Arrows indicate movement in the plane of the page, and the + and − signs indicate movement in the forward and rearward directions.

The wavelengths of absorption of the stretching and bending deformation modes for methyl, methylene, and methine all lie quite close together, with the exception of the symmetrical bending mode of the methyl group. These change position only slightly with change in the molecular environment. The differences in position caused by changes in substitution are not sufficiently great to allow resolution by most of the common infrared spectrophotometers available for general use; hence, the amount of structural information that one can derive is limited.

5.2.1b Alkenes and Aromatic Compounds

The absorption peaks due to stretching modes of excitation of aromatic carbon-hydrogen bonds occurs in the 3180 to 2980 cm^{-1} (3.15 to 3.36 μm) region. The terminal methylene group, $C{=}C\overset{H}{\underset{H}{\diagup}}$, displays two stretching bands, one occurring near 3090 cm^{-1} (3.24 μm) and the other near 2980 cm^{-1} (3.36 μm), corresponding to the asymmetric and symmetric stretching modes. The stretching bands of other types of vinyl carbon-hydrogen bonds occur between 3080 and 3020 cm^{-1} (3.24 and 3.31 μm), while aromatic carbon-hydrogen absorptions occur near 3030 cm^{-1} (3.30 μm). These absorption bands are usually quite weak, and the positions vary only slightly with change in substitution. Owing to the possible congestion in this very narrow region, caution must be exercised in interpreting information derived from this region for distinguishing between aromatic and olefinic compounds. One must rely on the carbon-carbon double bond stretching region between 1700 and 1450 cm^{-1} (5.9 and 6.9 μm) and the fingerprint region from 1000 to 650 cm^{-1} (10 to 15.5 μm).

The absorption bands due to the C—H bending deformations of alkenes and aromatic compounds are particularly useful in that one can determine the extent and position of substitution and the stereochemistry of disubstituted alkenes.

There are two possible bending deformations: the in-plane bending deformation, in which the hydrogen remains in the plane of the $C=C$ or aromatic system, and the out-of-plane deformation, in which the hydrogen bends out of the plane of the molecule. The out-of-plane bending deformations are the most valuable for deriving structural information. These peaks occur in the so-called "fingerprint" region 1000 to 650 cm^{-1} (10 to 15 μm), in which very few other absorptions occur, whereas the in-plane deformations occur in the congested 1400 to 1000 cm^{-1} (7 to 10 μm) region. The in-plane bending region is useful only in relatively simple molecules due to the great number of bands that appear in this region.

Vinyl groups ($—CH=CH_2$) give rise to two out-of-plane bending deformation bands, one band appearing near 900 cm^{-1} (11 μm), and the other near 990 cm^{-1} (10.1 μm). The terminal methylene of 1,1-disubstituted alkenes also appears near 900 cm^{-1} (11 μm). This band is quite intense and usually appears as one of the more prominent bands in a spectrum. *Cis*- and *trans*-disubstituted alkenes absorb near 685 and 965 cm^{-1} (14.2 and 10.4 μm), respectively. The out-of-plane bending deformation of a trisubstituted alkene appears near 820 cm^{-1} (12.1 μm). The position of this band is quite different compared with the corresponding band of the terminal vinyl group, which appears at 990 cm^{-1} (10.1 μm). Typical fingerprint regions of spectra of alkenes are illustrated in Fig. 5.3.

A monosubstituted benzene displays two out-of-plane bending deformation bands near 750 and 710 cm^{-1} (13.4 and 14.3 μm). *Ortho*-disubstituted benzenes display a single band near 750 cm^{-1} (13.4 μm), characteristic of a four-adjacent-hydrogen system; *para*-disubstituted benzenes display a single band near 830 cm^{-1} (12.1 μm), characteristic of a two-adjacent-hydrogen system; *meta*-disubstituted benzenes display bands near 780 cm^{-1} (12.8 μm), characteristic of a three-adjacent-hydrogen system, and 880 cm^{-1} (11.4 μm), characteristic of a one-adjacent-hydrogen system. Several of these types of systems are illustrated at the end of this chapter. These characteristic absorption patterns for the various adjacent hydrogen systems are also observed with substituted pyridines and polycyclic benzenoid aromatics, for example, substituted naphthalenes, anthracenes, and phenanthrenes.

The in-plane and out-of-plane bending deformation band positions are summarized in Table 5.1. These bands are quite intense and are subject to considerable change in wave number upon change of substitution. The presence of electron-donating substituents on an alkene or aromatic ring results in shifts to lower wave number (longer wavelengths), and vice versa for electron-withdrawing substituents. Occasionally a band may shift into the region normally assigned to a different type or degree of substitution. Chemical or other physical data are required in such cases to determine the type of substituent that is present before an accurate assignment of these bands can be made.

The 2000 to 1670 cm^{-1} (5 to 6 μm) region of spectra of aromatic compounds displays several weak bands that are overtone and combination bands. These bands are usually quite weak, and a fairly concentrated solution of the sample

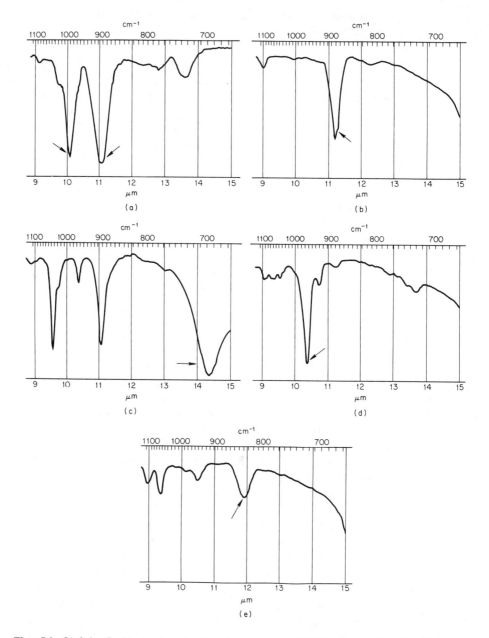

Fig. 5.3. Olefinic C—H bending bands of a monosubstituted alkene (*a*), 1,1-disubstituted alkene (*b*), *cis*-di- (*c*), and *trans*-disubstituted (*d*) alkenes, and a trisubstituted alkene (*e*). Note that some of the spectra also contain other deformation bands in the low-frequency region that are not due to the olefinic C—H.

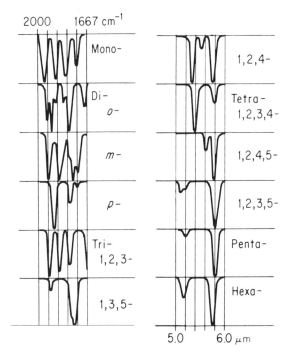

Fig. 5.4. Typical absorption patterns of substituted aromatics in the 2000 to 1670 cm^{-1} (5.0 to 6.0 μm) region.

must be used to record these bands (Fig. 5.9). These bands are also useful in deriving structural information; however, the positions of these bands are sensitive to the types of substituents present in the system, as was noted for the out-of-plane deformations discussed in preceding paragraphs. Figure 5.4 displays typical absorption patterns of substituted aromatics in the 2000 to 1670 cm^{-1} (5 to 6 μm) region.

5.2.1c Acetylenic Compounds

Acetylenic carbon-hydrogen absorbs very close to 3330 cm^{-1} (3.0 μm). The band is quite sharp and intense.

5.2.1d Aldehydic Carbon-Hydrogen

The aldehyde C—H gives rise to two bands at approximately 2820 and 2700 cm^{-1} (3.55 and 3.7 μm). The lower-wave-number band is quite useful in detecting the presence of an aldehyde group.

5.2.2 Oxygen-Hydrogen Absorption Bands

The oxygen-hydrogen stretching deformation occurs in the 3700 to 2500 cm^{-1} (2.7 to 4.0 μm) region. The position and the shape of the absorption bands vary greatly with structure and can be very useful in deriving information concerning the structure of the molecule. The position and shape of the absorption band are also quite sensitive to the type and extent of hydrogen bonding either to another hydroxyl group or another acceptor group.

The nonbonded hydroxyl gives rise to a sharp absorption whose position varies slightly from primary to secondary to tertiary hydroxyl. The difference in band positions varies only about 10 cm^{-1} ($0.01 \ \mu\text{m}$) between primary and secondary, and secondary and tertiary. This difference in position borders on the limits of resolution of spectrometers commonly available in laboratories, and it should not be relied on to decide the environment of the hydroxyl group. Nonbonded hydroxyl absorption can be observed only in dilute solutions in nonpolar solvents. At normal concentrations, approximately 5 to 10% by weight, extensive inter-molecular hydrogen bonding occurs, causing an additional broad absorption at lower wave number (Fig. 5.9). The position and shape of the band depend on the extent of hydrogen bonding, becoming more broad and moving to lower wave number as the strength of the hydrogen bond increases (Table 5.2). Occasionally another functional group may be present in the same molecule as the hydroxyl group to act as an acceptor functional group for intramolecular hydrogen bonding. The shape and intensity of the absorption band are not functions of the concentration of the sample and can be differentiated from intermolecular hydrogen bonding by dilution studies.

The oxygen-hydrogen stretching absorption of phenols occurs at slightly lower frequency than those of alcohols, but usually not sufficiently different to allow distinction between phenols and alcohols.

The oxygen-hydrogen stretching absorption of carboxylic acids appears as a very broad band with maximum absorption at approximately 2940 cm^{-1} ($3.4 \ \mu\text{m}$); it extends to nearly 2500 cm^{-1} ($4.0 \ \mu\text{m}$), owing to the very strong intermolecular hydrogen bonding between carboxyl groups. Since acids exist as dimers in non-polar solvents, the position and shape of this band are little affected by changes in concentration.

The oxygen-hydrogen bending absorption occurs in the 1500 to 1300 cm^{-1} (6.7 to 7.7 μm) region and is of no practical value for analysis.

5.2.3 Nitrogen-Hydrogen Absorption Bands

Nitrogen-hydrogen stretching absorptions occur at slightly lower frequencies than the O—H stretching absorptions do. Primary amines display two bands corresponding to asymmetric and symmetric stretching deformations, whereas secondary amines display a single peak (Table 5.2). Imines, $>$C=NH, display a single peak in the same region. In general, it is easy to distinguish primary from secondary amines, but distinction between various types of $>$N—H is difficult.

Amines display N—H bending deformation absorption bands similar to those of —CH$_2$— and —$\overset{|}{\underset{|}{\text{C}}}$—H, except that they occur at slightly higher frequency.

Table 5.2. Selected X-H Absorption Bands

Functional Group	Wave Number cm^{-1}	Wavelength μm	Remarks
Alcohols			
Primary (nonbonded)	3640	2.72	m, gu, usually determined in dilute solution in nonpolar solvents.
Secondary	3630	2.73	
Tertiary	3620	2.74	
Phenols	3610	2.75	
Intermolecularly	3600–3500	2.78–2.86	m, dimeric, rather sharp
H-bonded	3400–3200	2.94–3.1	s, polymeric, usually quite broad
Intramolecularly H-bonded	3600–3500	2.78–2.86	m-s, much sharper than intermolecular hydrogen bonded OH; is not concentration dependent.
Amines			
RNH$_2$	~3500	~2.86	m, gu, ν_{as}
	~3400	~2.94	m, gu, ν_s
	1640–1560	6.1–6.4	m-s, gu, corresponds to scissoring deformation.
R$_2$NH	3500–3450	2.86–2.90	w-m, gu, ν
ArNHR	3450	2.90	w-m, gu
Pyrroles, indoles	3490	2.86	w-m, gu
Ammonium salts			
NH$_4^+$	3300–3030	3.0–3.3	s, gu
	1430–1390	7.0–7.2	s, gu
—$\overset{+}{\text{N}}$H$_3$	3000	~3.0	s, gu, usually quite broad
	1600–1575	6.25–6.35	s, gu, σ_{as}
	1490	~6.7	s, gu, σ_s
$>\overset{+}{\text{N}}$H$_2$	2700–2250	3.7–4.4	s, gu, ν_{as} and ν_s, usually broad or a group of bands
$>\overset{+}{\text{N}}$H	1600–1575	6.25–6.35	m, gu, σ
	2700–2250	3.7–4.4	s, gu, ν, σ_{NH^+} band is weak and of no practical utility.
Mercaptans			
—SH	2600–2550	3.85–3.92	s, gu band is often very weak and can be missed if care is not exercised.

Protonation of amines, to give the corresponding amine salts, results in the formation of bands similar to those of $-CH_3$, $-CH_2-$, and $-\overset{|}{\underset{|}{C}}-H$, again appearing at higher wave numbers.

Primary amides display N—H stretching bands at slightly higher wave numbers than do amines. The N—H bending deformations of amides occur in the 1610 to 1490 cm^{-1} (6.2 to 6.7 μm) region and are generally referred to as "amide II" bands. These bands are quite intense.

Hydrogen bonding involving N—H gives rise to similar effects noted with hydroxyl absorption bands, although N—H bonds show a lesser tendency to hydrogen-bond than do alcohols.

5.2.4 Sulfur-Hydrogen Absorption Bands

The sulfur-hydrogen stretching absorption occurs near 2630 to 2560 cm^{-1} (3.8 to 3.9 μm), is usually very weak, and is not subject to large shifts due to hydrogen bonding.

5.2.5 Carbon-Carbon Absorption Bands

Carbon-carbon single-bond stretching bands appear from 1350 to 1150 cm^{-1} (7.4 to 8.7 μm), are extremely variable in position, and are usually quite weak in intensity. These absorption bands are of little practical use in structure determination.

Specific alkyl group stretching deformation bands have been characterized. In addition to the characteristic C—H stretching frequencies produced (Sec. 5.2.1a), the tertiary butyl group gives rise to two bands at 1250 cm^{-1} (8.00 μm) and near 1208 cm^{-1} (8.28 μm), and the isopropyl group gives rise to a band at 1170 cm^{-1} (8.55 μm), with a weaker band at 1145 cm^{-1} (8.73 μm). A geminal dimethyl group produces two bands at 1125 and 1195 cm^{-1} (8.23 and 8.37 μm). Groups of two, three, and four methylene groups absorb at 790 to 770 cm^{-1} (12.7 to 13.0 μm), 743 to 734 cm^{-1} (13.5 to 13.7 μm), and 725 to 720 cm^{-1} (13.8 to 13.9 μm), respectively. Very few other characteristic group absorptions have been defined.

Carbon-carbon double-bond stretching absorption occurs in the 2000 to 1430 cm^{-1} (5 to 7 μm) region. The position and intensity of these bands are very sensitive to the degree and type of substitution and, in cyclic alkenes, to ring size. Alkyl substitution results in a shift of the absorption band to higher wave numbers. The shift to higher wave numbers with increasing substitution is least with acyclic alkenes and increases dramatically with decreasing ring size in cyclic alkenes (Table 5.3) Decreasing the ring size of cyclic alkenes from a six- to a four-membered ring results in shifts to lower wave numbers; however, a further decrease in ring size to cyclopropene results in a dramatic shift to higher wave

Table 5.3.   Selected C—C and C—N Absorption Bands

Functional Group	Wave Number cm^{-1}	Wavelength μm		Remarks
Acyclic C=C($\nu_{C=C}$)				
Monosubstituted	1645	6.08	m,	These bands are of little
1,1-Disubstituted	1655	6.04	m,	utility in assigning sub-
Cis-1,2-disubstituted	1660	6.02	m,	stitution and stereo-
Trans-1,2-disubstituted	1675	5.97	w,	chemistry; the C—H
Trisubstituted	1670	5.99	w,	out-of-plane bending
Tetrasubstituted	1670	5.99	w,	bands in the fingerprint
				region are recommended
				for this purpose.
Conjugated C=C($\nu_{C=C}$)				
Diene	1650 and 1600	6.06 and 6.25	s, gu	
With aromatic	1625	6.16	s, gu,	
With C=O	1600	6.25	s, gu,	
Cyclic C=C($\nu_{C=C}$)				
6-membered ring and larger	1646	6.08	m,	Of limited utility due to closeness to the acyclic region.
Monosubstituted	1680–1665	5.95–6.00	m,	
5-m. unsubstituted	1611	6.21	m,	Can be of great utility
Monosubstituted	1660	6.02	m,	although the ring size
Disubstituted	1690	5.92	w,	must be assigned first
4-m. unsubstituted	1566	6.39	m,	for those compounds
3-m. unsubstituted	1640	6.10	m,	whose absorption falls
Monosubstituted	1770	5.65	m,	into regions consistent
Disubstituted	1880	5.32	m,	with other types of
				absorption bands (nu-
				clear magnetic reso-
				nance spectra can be very
				useful in this respect).
Aromatic C=C	1600	6.24	w-s,	In-plane skeletal vibra-
	1580	6.34	s,	tions; the intensities of
	1500	6.67	w-s,	the 1600 and 1500 cm^{-1}
	1450	6.9	s,	may be rather weak.
C—N				
Aliphatic amine	1220–1020	8.2–9.8	m-w, lu, ν_{C-N}	
Aromatic amine	1370–1250	7.3–8.0	s, gu-lu, ν_{C-N}	
C=N				
	1700–1615	5.9–6.2	s, gu, $\nu_{C=N}$	

numbers. These shifts are probably due to changes in hybridization involving the carbon atoms that comprise the carbon-carbon double bond. The effect of conjugation with an aromatic nucleus results in a slight shift to lower wave number, and with another C=C or C=O, a shift to lower wave number of approximately 40 to 60 cm^{-1} (0.15 to 0.20 μm) with a substantial increase in intensity.

The intensity of the C=C stretching absorption band is a function of the symmetry of substitution on the double bond; mono- and trisubstituted alkenes produce more intense bands than do unsymmetrical di- and tetrasubstituted alkenes, with symmetrically substituted di- and tetrasubstituted alkenes generally showing no $\nu_{C=C}$ absorption. Enamines (C=C—NR$_2$) and the related enol ethers (C=C—OR) give rise to very intense absorption bands.

The skeletal C=C vibrations of aromatics give rise to a series of four bands in the 1660 to 1430 cm^{-1} (6 to 7 μm) region. These bands occur very close to 1600 cm^{-1} (6.24 μm), 1575 cm^{-1} (6.34 μm), 1500 cm^{-1} (6.67 μm), and 1450 cm^{-1} (6.9 μm). The first and third bands vary greatly in intensity, generally becoming more intense in the presence of electron-donating groups. Table 5.3 lists the positions of absorption of various C=C systems.

Bending deformation absorptions of C—C and C=C bonds occur below 670 cm^{-1} (15 μm) and are of little utility in structure identification.

5.2.6 Carbon-Nitrogen Absorption Bands

Aliphatic amines display C—N stretching absorptions in the 1220 to 1020 cm^{-1} (8.2 to 9.8 μm) region. These bands are of medium to weak intensity. Owing to the intensity and position of these bands, they are of little practical value for structural work, except in relatively simple molecules. Aromatic amines absorb in the 1370 to 1250 cm^{-1} (7.3 to 8.0 μm) region and give quite intense bands. Carbon-nitrogen double bond absorption occurs in the 1700 to 1615 cm^{-1} (5.9 to 6.2 μm) region and is subject to similar environmental effects as the C=O absorption bands (see following section).

5.2.7 Carbon-Oxygen Absorption Bands

Carbon-oxygen single bond absorption occurs in the 1250 to 1000 cm^{-1} (8 to 10 μm) region. Although both alcohols and ethers absorb in this region, distinction between these classes can be made from information gained from the hydroxyl region.

The position of the carbon-oxygen stretching absorption bands depends on the extent and type of substitution on the carbinol carbon atom. Nakaniski (ref. 11) has developed the following correlation to predict the wave number of absorption of the carbon-oxygen stretching deformation, based on standard wave numbers of absorption for primary, secondary, and tertiary alcohols (structure **1**) of 1050, 1100, and 1150 cm^{-1}, respectively.

C$^\beta$—C$^\alpha$	α-branching	-15 cm^{-1} (per branch in α' and α'')
C—C$^{\alpha'}$—C—OH	α-unsaturation	-30 cm^{-1}
	α, α'-ring formation	-50 cm^{-1}
C—C$^{\alpha''}$	α-unsaturation + α'-branching	-90 cm^{-1}
1	α- and α'-unsaturation	-90 cm^{-1}
	α-, α'-, and α''-unsaturation	-140 cm^{-1}

The effect of ring size on the position of absorption in ethers is indicated in Table 5.4. These absorption bands in the cyclic ethers are due to the antisymmetric C—O—C vibrational modes.

The carbon-oxygen double bond absorbs in the 2000 to 1540 cm^{-1} (5 to 6.5 μm) region, except for ketenes, which absorb near 2200 cm^{-1} (4.65 μm). This is probably the most useful portion of the infrared region because the position of the carbonyl group is quite sensitive to substituent effects and the geometry of the molecule. The band positions are also solvent-sensitive owing to the high polarity of the C=O bond.

An empirical correlation that can be used to adequately predict the positions of most carbonyl bands has been developed. Table 5.4 lists the band positions for a number of different types of compounds that contain the C=O group. For ketones, aldehydes, acids, anhydrides, acid halides, and esters, the wave numbers (wavelengths) cited are for the normal, unstrained (acyclic or contained in a six-membered ring) *parent* compound in carbon tetrachloride solution. The introduction of a double bond or aryl group in conjugation with the C=O results in a reasonably consistent 30 cm^{-1} (0.1 μm) shift to lower wave number. Introduction of a second double bond results in an additional shift of approximately 15 cm^{-1} (0.05 μm) in the same direction.

The position of the C=O band is very sensitive to changes in the C(CO)C bond angle. A decrease in ring size, resulting in a decrease in the C(CO)C bond angle, of ketones and esters results in a reasonably consistent shift of 30 cm^{-1} (0.1 μm) to higher wave number per each decrease in ring size from the six-membered ring. In cyclic structures containing more than six atoms, slight shifts to lower wave numbers are observed. For example, a shift of -10 cm^{-1} ($+0.03$ μm) is observed with seven-membered ring compounds, and an approximate -5 cm^{-1} (0.01 to 0.02 μm) shift with eight- and nine-membered ring systems. Further increases in ring size result in a moderate increase in wave number back to the parent compound position. Highly strained bridge carbonyl compounds, for example, 7-oxobicyclo[2.2.1]heptanes and 6-oxobicyclo-[2.1.1]hexanes, absorb at relatively high wave numbers, in the general region of 1800 cm^{-1} (5.5 μm).

The effects of conjugation and ring size are cumulative and can be used to adequately predict the position of absorption in more complex molecules. For example, the calculated positions of absorption in compounds **2** and **3** in carbon

Table 5.4. Selected Carbon-Oxygen Absorption Bands

Functional Group	Wave Number cm^{-1}	Wavelength μm	Remarks
C—O Single Bonds			
Primary C—O—H	1050	9.52	s, gu, ν_{C-O} (See discussion for substituent effects.)
Secondary C—O—H	1100	9.08	s, gu, ν_{C-O}
Tertiary C—O—H	1150	8.68	s, gu, ν_{C-O}
Aromatic C—O—H	1200	8.33	s, gu, ν_{C-O}
Ethers-acyclic	1150–1070	8.7–9.35	s, gu, antisymmetric ν_{C-O-C}
C=C—O—C	1270–1200 and	7.9–8.3 and	s, gu, antisymmetric ν_{C-O-C}
	1070–1020	9.3–9.8	s, gu, symmetric ν_{C-O-C}
Cyclic ethers			
6-member and larger	1140–1070	8.77–9.35	s, gu,
5-m.	1100–1075	9.1–9.3	s, gu,
4-m.	980–970	10.2–10.3	s, gu,
Epoxides	1250	8.0	s, gu,
Cis-disubstituted	890	11.25	s, gu,
Trans-disubstituted	830	12.05	s, gu,
C—O Double Bonds			
Ketones	1715	5.83	s, gu, $\nu_{C=C}$, unstrained C=O group in acyclic and 6 m. ring compounds in carbon tetrachloride solution. (See discussion for effects of conjugation and ring size.)
α, β-unsaturated	1685	5.93	s, gu, $\nu_{C=O}$. For the s-*cis* 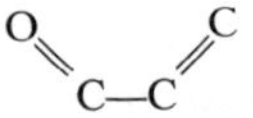 conformation the $\nu_{C=C}$ may appear above 1600 cm^{-1} (below 6.26 μm with an intensity approximately that of the $\nu_{C=O}$. The *trans*-conformation 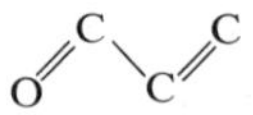 does not show this enhanced intensity of absorption of the C=C.
α- and β-diketones	1720	5.81	s, gu, $\nu_{C=O}$, two bands at higher frequency when in the s-*cis* conformation 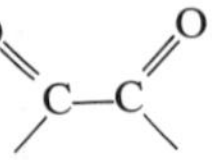
	1650	6.06	s, gu, $\nu_{C=C}$ if enolic, 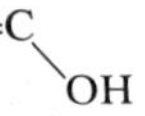 C=C—OH

Table 5.4. (Continued)

Functional Groups	Wave Number cm^{-1}	Wavelength μm	Remarks
Quinones	1675	5.97	s, gu, $\nu_{C=O}$
Tropolones	1650	6.06	s, gu, $\nu_{C=O}$
	1600	6.26	s, gu, $\nu_{C=O}$, if intramolecularly hydrogen bonded as in α-tropolones.
Aldehydes $\left(-C\!\!\!\!\begin{smallmatrix}O \\ H\end{smallmatrix}\right)$	1725	5.80	s, gu, $\nu_{C=O}$ (See discussion for effects of conjugation.)

Carboxylic acids and derivatives

$\left(-C\!\!\!\!\begin{smallmatrix}O \\ OH\end{smallmatrix}\right)$	1710	5.84	s, gu, $\nu_{C=O}$, usually as the dimer in nonpolar solvents, monomer absorbs at 1730 cm^{-1} (5.78 μm) and may appear as a shoulder in the spectrum of a carboxylic acid; for conjugation effects see the discussion.
Esters	1735	5.76	s, gu, $\nu_{C=O}$, acyclic and 6-m. lactones; for conjugation and ring size effects see the discussion.
	1300–1050	7.7–9.5	s, lu, symmetric and antisymmetric ν_{C-O-C} giving 2 bands, indicative of type of ester, for example, formates: 1178 cm^{-1} (8.5 μm) acetates: 1242 cm^{-1} (8.05 μm) methyl esters: 1164 cm^{-1} (8.6 μm) others: 1192 cm^{-1} (8.4 μm) but the distinction generally is not great enough to be of diagnostic value.
Vinylesters $\left(-C\!\!\!\!\begin{smallmatrix}O \\ O-C=C\end{smallmatrix}\right)$	1755	5.70	
Anhydrides	1820 and 1760	5.48 and 5.68	s, gu, $\nu_{C=O}$, the intensity and separation of the bands may be quite variable. (See discussion for general effects of conjugation and ring size.)

Table 5.4.   (Continued)

Functional Groups	Wave Number cm^{-1}	Wavelength μm	Remarks
Acid halide	1800	5.56	s, gu, $\nu_{C=O}$, acid chlorides and fluorides absorb at slightly higher wave number while the bromides and iodides absorb at slightly lower wave number. (See discussion for effects of conjugation.)
Amides	1650	6.06	s, gu, $\nu_{C=O}$, "Amide I" band, this frequency is for the associated amide (see COOH), free amide at 1686 cm^{-1} (5.93 μm) in dilute solution, cyclic amides shift to higher wave number as ring size decreases.
	1300	7.7	s, gu, ν_{C-N}, "Amide III" band, free amide at slightly higher wave number.
α, β-unsaturated	1665	6.01	s, gu, $\nu_{C=O}$
Carboxylate	1610–1550 and 1400	6.2–6.45 and 7.15	s, gu, antisymmetric and symmetric stretching of

tetrachloride solution are illustrated below and are compared with the observed absorptions.

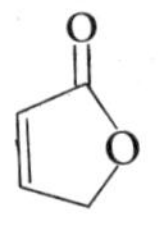

		Wave Number	Wavelength
2	Normal ester (from Table 5.4)	1735 cm^{-1}	5.76 μm
	Ring size correction	+30	−0.10
	Unsaturation	−30	+0.10
	Calculated $\nu_{C=O}$	1735 cm^{-1}	5.76 μm
	Observed $\nu_{C=O}$	1740 cm^{-1}	5.74 μm

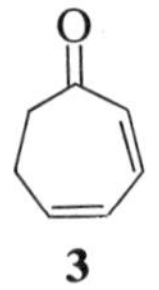

		Wave Number	Wavelength
3	Normal ketone	1715 cm^{-1}	5.83 μm
	Ring size correction	−10	+0.03
	Unsaturation (first C=C)	−30	+0.10
	(second C=C)	−15	+0.05
	Calculated $\nu_{C=C}$	1660 cm^{-1}	6.01 μm
	Observed $\nu_{C=O}$	1666 cm^{-1}	6.00 μm

The C=O stretching absorption band of amides displays a different behavior when conjugated with an unsaturated system than do other types of C=O containing functional groups. Instead of moving to lower wave numbers when α,β-unsaturated, the band actually moves to higher wave numbers owing to the negative inductive effect of the C=C. The amide group is highly resonance-stabilized and does not strongly interact with the additional C=C.

Although anhydrides were considered earlier, this functional group is quite different from other acid derivatives in that two bands appear in the 1850 to 1725 cm^{-1} (5.4 to 5.8 μm) region. The relative intensities of these bands vary greatly, but both are usually easily discernible.

The introduction of electron-withdrawing groups on the α-carbon atom leads to a displacement of the absorption band to higher wave numbers (shorter wavelengths), the magnitude of the shift being a function of the electronegativity, or electron-withdrawing ability, of the group and the dihedral angle in the X—C—C=O group. An example of the latter effect is demonstrated by the α-substituted cyclohexanones in which the axial isomers (**4**) absorb at a lower wave number (longer wavelength) than do the equatorial isomers (**5**).

Many of the listed carbonyl-containing compounds give rise to absorption bands in other portions of the infrared region that are equally important in assigning partial structures, for example, the O—H and N—H stretching absorptions of carboxylic acids and amides, which have been discussed in earlier sections. The identification of these bands, integrated with the results of functional classification tests, should allow the facile assignment of the type of carbonyl group that is present in any given molecule.

5.2.8 Carbon-Carbon Triple Bond Absorption

Carbon-carbon triple bond stretching absorption in terminal acetylenes occurs in the 2140 to 2080 cm^{-1} (4.6 to 4.8 μm) region. The absorption band is relatively weak but quite sharp. A band in this region, along with the C—H stretching absorption band near 3300 cm^{-1} (3.0 μm) is indicative of the presence of a terminal acetylene.

Unsymmetrically disubstituted acetylenes absorb in the 2260 to 2190 cm^{-1} (4.42 to 4.57 μm) region. The intensity of this band is a function of the symmetry of the molecule; the more symmetrical the substitution, the less intense the absorption band.

Conjugation with a C=C causes an increase in the band intensity and a shift to slightly lower wave numbers. The observed shifts are usually less than those observed with alkenes. Conjugation with carbonyl groups does not appreciably alter the position of absorption.

5.2.9 Carbon-Nitrogen Triple Bond Absorption

The carbon-nitrogen triple bond of nitriles absorbs in the 2260 to 2210 cm^{-1} (4.42 to 4.52 μm) region. Saturated nitriles absorb in the higher-wave number portion of the region, while unsaturated nitriles absorb in the lower-wave number end. The C≡N absorption band is of much greater intensity than the C≡C absorption bands.

5.2.10 Absorption of X=Y=Z Functional Groups

Functional groups having the general structure X=Y=Z, for example, allenes, ketenes, ketenimines, isocyanates, carbodiimides, and azides, absorb in the 2500 to 2000 cm^{-1} (4 to 5 μm) region (Table 5.5). The bands are very intense and are of great diagnostic value. Distinction between the possible functional groups falling in this region may require functional classification tests or conversion to suitable derivatives.

5.2.11 Nitrogen-Oxygen Absorption Bands

Nitro groups display two intense absorption bands near 1520 and 1350 cm^{-1} (6.6 and 7.4 μm). Conjugation with an aromatic ring causes a shift to lower wave numbers. Nitrates, RONO$_2$, absorb in several regions, including near 1640 and 1280 cm^{-1} (6.1 and 7.8 μm) (Table 5.6).

Nitroso groups give rise to a single absorption peak in the 1600 to 1500 cm^{-1} (6.2 to 6.7 μm) region. Nitrites, RONO, give rise to two bands, ascribed to the *cis* and *trans* forms, in the 1680 to 1610 cm^{-1} (5.9 to 6.2 μm) region.

Nitrogen-oxygen single bond absorption occurs in the 1320 to 910 cm^{-1} (7.6 to 11 μm) region and may require additional spectral and chemical data for identification purposes (Table 5.6).

5.2.12 Sulfur-Oxygen Absorption Bands

Functional groups containing a sulfur-oxygen single bond display absorption bands in the 910 to 710 cm^{-1} (11 to 14 μm) region. Sulfur-oxygen bonds of the

Table 5.5. Absorption Bands of Y≡Z and X=Y=Z Type Groups

Functional Group	Wave Number cm^{-1}	Wavelength μm	Remarks
Acetylenes			
terminal	2140–2100	4.67–4.76	w, gu, $\nu_{C\equiv C}$, to be used in conjunction with the ν_{C-H} band.
disubstituted	2260–2190	4.42–4.57	v.w, lu, $\nu_{C\equiv C}$, may be completely absent in symmetrical acetylenes.
Nitriles			
alkyl	2260–2240	4.42–4.47	m-s, gu, $\nu_{C\equiv N}$
aryl	2240–2220	4.46–4.51	s, gu, $\nu_{C\equiv N}$
α,β-unsaturated	2235–2215	4.47–4.52	s, gu, $\nu_{C\equiv N}$
Allenes	1950	5.11	m, gu, $\nu_{C=C=C}$, terminal allenes display two bands and ν_{C-H} at 850 cm^{-1} (11.76 μm).
Ketenes	2150	4.65	s, gu
Ketenimines	2000	5.0	s, gu
Cyanates	2275–2250	4.39–4.45	v.s, gu, extremely intense
Carbodiimides	2145–2130	4.66–4.69	v.s, gu, extremely intense
Azides	2160–2120	4.63–4.72	v.s, gu

Table 5.6. Selected N—O Absorption Bands

Functional Group	Wave Number cm^{-1}	Wavelength μm	Remarks
Nitro (R—NO$_2$)	1570–1500	6.37–6.67	s, gu, antisymmetric ν_{NO_2}
	1370–1300	7.30–7.69	s, gu, symmetric ν_{NO_2}, conjugated nitro absorbs in the lower-wave number portions of the cited regions.
Nitrate (RO—NO$_2$)	1650–1600	6.06–6.25	s, gu, antisymmetric ν_{NO_2}
	1300–1250	7.69–8.00	s, gu, symmetric ν_{NO_2}, additional bands at 870–855 cm^{-1} (11.5–11.7 μm) (ν_{O-N}), 763 cm^{-1} (13.1 μm) (out-of-plane bending) and 703 cm^{-1} (14.2 μm) NO$_2$ bending).
Nitramine (N—NO$_2$)	1630–1550	6.13–6.45	s, gu, antisymmetric ν_{NO_2}
	1300–1250	7.69–8.00	s, gu, symmetric ν_{NO_2}

Table 5.6   (Continued)

Functional Group	Wave Number cm^{-1}	Wavelength μm	Remarks
Nitroso (R—N=O)	1600–1500	6.25–6.66	s, gu, $\nu_{N=O}$, conjugation shifts to lower wave numbers in the cited region
Nitrite (R—ONO)	1680–1610	5.95–6.21	s, gu, $\nu_{N=O}$, two bands usually present due to *cis* and *trans* forms.
Nitroamide $\left(-N\begin{smallmatrix}COR\\NO_2\end{smallmatrix}\right)$	1580	6.33	s, gu, $\nu_{C=O}$, appears at 1724 cm^{-1} (5.80 μm).
Nitroasoamide $\left(-N\begin{smallmatrix}COR\\N=O\end{smallmatrix}\right)$	1522	6.57	s, gu, $\nu_{C=O}$, appears at 1740 cm^{-1} (5.75 μm).
Nitrone $\left(\begin{smallmatrix}\\C=\overset{+}{N}\end{smallmatrix}\begin{smallmatrix}R\\\bar{O}\end{smallmatrix}\right)$	1170–1280	7.8–8.6	s, gu, $\nu_{C=N}$, appears at 1580–1610 cm^{-1} (6.2–6.3 μm).
Amine Oxide $\left(\geq\overset{+}{N}-\bar{O}\right)$ aliphatic	970–950	10.3–10.5	v.s, gu, ν_{N-O}
aromatic	1300–1200	7.7–8.3	v.s, gu, ν_{N-O}
Azoxy $\left(N=\overset{+}{N}\begin{smallmatrix}\\O^-\end{smallmatrix}\right)$	1310–1250	7.63–8.00	v.s, gu, ν_{N-O}

type appearing in sulfoxides and sulfones absorb at considerably higher wave numbers. Sulfoxides display a single absorption band near 1050 cm^{-1} (9.5 μm), whereas sulfones give rise to two bands near 1330 and 1140 cm^{-1} (7.5 and 8.7 μm). These absorption bands are little affected by conjugation or ring strain (Table 5.7).

General spectra-structure correlation tables of group wave numbers in the infrared region are presented in Figs. 5.5, 5.6, 5.7, and 5.8. These tables summarize in more general terms the group wave numbers discussed in greater detail in the earlier sections of this chapter.

5.3
ISOTOPE EFFECTS

Substitution of an atom by a heavier isotope, for example, deuterium for hydrogen, results in a shift of the absorption band to lower wave numbers. The

Table 5.7.   Selected S—O Absorption Bands

Functional Group	Wave Number cm^{-1}	Wavelength μm	Remarks
—S—O	900–700	11.1–14.2	s, gu, $\nu_{\text{S}-\text{O}}$
Sulfoxide $\left(\begin{array}{c}\diagdown\\\diagup\end{array}\text{S}{=}\text{O}\right)$	1090–1020	9.43–9.62	s, gu, $\nu_{\text{S}=\text{O}}$
Sulfone $\left(\text{S}\begin{array}{c}\diagup\!\!^{\text{O}}\\\diagdown\!\!_{\text{O}}\end{array}\right)$	1350–1310	7.42–7.63	s, gu, antisymmetric ν_{SO_2}
	1160–1120	8.62–8.93	s, gu, symmetric ν_{SO_2}
Sulfonic acid (—SO$_2$OH)	1260–1150	7.93–8.70	s, gu, antisymmetric ν_{SO_2}
	1080–1010	9.26–9.90	s, gu, antisymmetric ν_{SO_2}
	700–600	14.2–16.6	s, gu, $\nu_{\text{S}-\text{O}}$, may appear outside the general infrared region; in addition, O—H absorption appears.
Sulfonyl chlorides (—SO$_2$Cl)	1385–1340	7.22–7.46	s, asymmetric ν_{SO_2}
	1185–1160	8.44–8.62	s, symmetric ν_{SO_2}
Sulfonate (—SO$_2$OR)	1420–1330	7.04–7.52	s, gu, antisymmetric ν_{SO_2}
	1200–1145	8.33–8.73	s, gu, symmetric ν_{SO_2}
Sulfonamide (—SO$_2$NR$_2$)	1370–1330	7.30–7.52	s, gu, antisymmetric ν_{SO_2}
	1180–1160	8.47–8.62	s, gu, symmetric ν_{SO_2}

position of absorption of the new bonded system can be approximated fairly closely by Eq. (5.4)

$$\nu_{\text{A}-\text{C}} \approx \nu_{\text{A}-\text{B}} \sqrt{\frac{m_{\text{B}}}{m_{\text{C}}}} \tag{5.4}$$

where C is the isotope of B in the bond A—C, and m_{B} and m_{C} are the masses of the two isotopes. Eq. (5.4) is derived from Eq. (5.1), assuming that the force constant k is the same for both systems.

This same technique can be applied reasonably well to bonding systems in which one atom is exchanged for another atom of the same family, for example, O—H to S—H and C=O to C=S. This approach is useful when insufficient information is available in the literature on the position of absorption of a particular functional group.

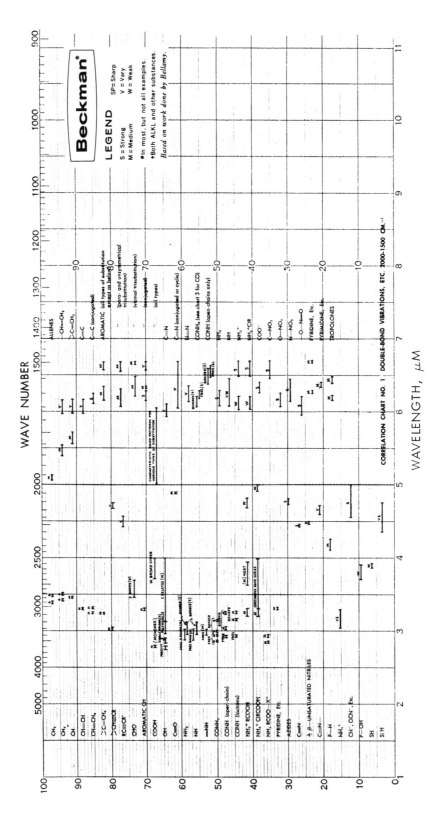

Fig. 5.5. Correlation chart 1. Hydrogen stretching, and double- and triple-bond stretching deformation band positions (3750 to 1500 cm⁻¹). (Reproduced by permission of Beckman Instruments, Inc., Fullerton, Calif.)

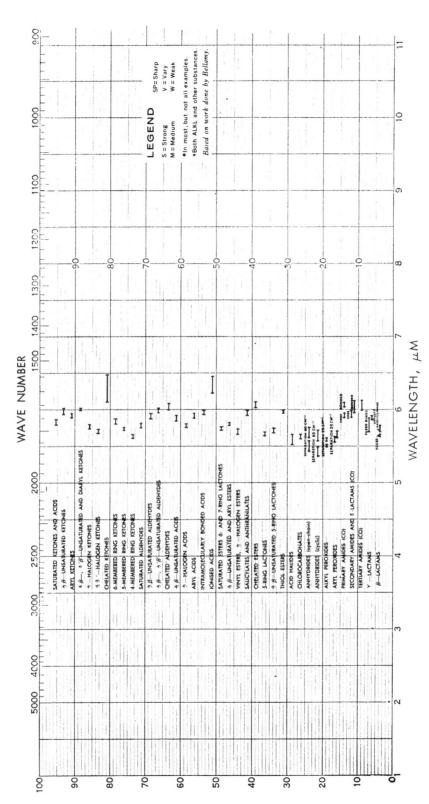

Fig. 5.6. Correlation chart 2. Carbonyl stretching deformation band positions (1900 to 1500 cm⁻¹). (Reproduced by permission of Beckman Instruments, Inc., Fullerton, Calif.)

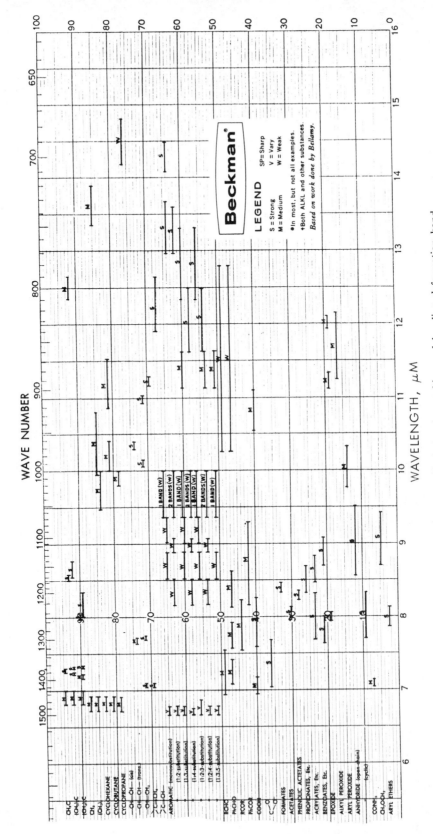

Fig. 5.7. Correlation chart 3. Single-bond stretching and bending deformation band positions (1500 to 650 cm⁻¹). (Reproduced by permission of Beckman Instruments, Inc., Fullerton, Calif.)

156

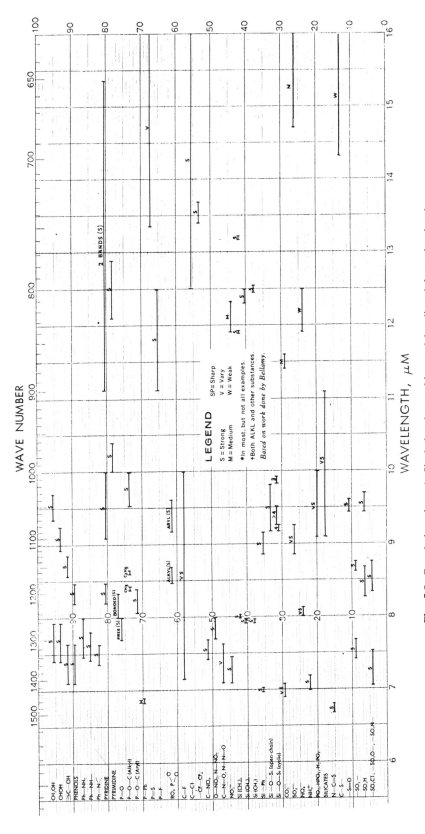

Fig. 5.8. Correlation chart 4. Single-bond stretching and bending deformation band positions (1500 to 650 cm⁻¹). (Reproduced by permission of Beckman Instruments, Inc., Fullerton, Calif.)

5.4
INTERPRETATION OF REPRESENTATIVE
INFRARED SPECTRA

In the analysis of the infrared spectrum of an unknown material certain guidelines are useful. As indicated in the previous sections, there are particular regions that provide unambiguous information concerning the nature of functions present in a molecule. We suggest the following steps in interpreting an infrared spectrum of an unknown.

1. Check the 3700 to 3300 cm^{-1} (2.7 to 3.0 μm) region for the possible presence of O—H or N—H containing functions.
2. Check the 1750 to 1670 cm^{-1} (5.7 to 6.0 μm) region for the possible presence of C=O containing functions.
3. Check the 1660 to 1450 cm^{-1} (6.0 to 6.9 μm) region for the possible presence of aromatic and/or C=C functions. If sharp bands are noted in this region, analyze the fingerprint region 1000 to 650 cm^{-1} (10.0 to 15.4 μm) for C—H bending deformation bands for possible assignment of degree and position of substitution on aromatic rings and double bonds.
4. Finally, look for the most intense band, or bands, in other regions, and attempt assignments based on the molecular formula and chemical properties, and by reference to Figs. 5.5 to 5.8.

In the following, an interpretation of representative infrared spectra is outlined.

The infrared spectrum of 1-phenylethanol is shown in Fig. 5.9. The complete trace was taken of a neat liquid film between sodium chloride plates (see

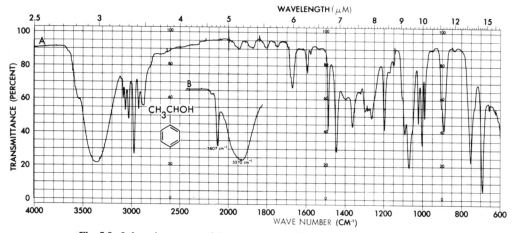

Fig. 5.9. Infrared spectrum of 1-phenylethanol. Trace *A*, neat liquid film; trace *B*, 5% in carbon tetrachloride. Trace *B* is of the 3900 to 3150 cm^{-1} region and has been displaced on the chart paper.

discussion on the preparation of samples, Sec. 5.5). The intense, broad band near 3360 cm^{-1} (3 μm) represents the hydrogen bonded O—H stretch. The non-bonded absorption peak is barely perceptible as a shoulder near 3550 cm^{-1} (2.81 μm). Trace B is of the hydroxyl region of a carbon tetrachloride solution of 1-phenylethanol, which has been displaced on the chart paper for greater clarity, showing the sharp nonbonded band and the broad bonded band. The C—H stretching region shows the aromatic C—H stretch as three bands just above 3000 cm^{-1} (3.33 μm). The band at 2970 cm^{-1} (3.36 μm) represents the asymmetric stretch of the methyl.

The bending deformation bands of the methyl group appear at 1450 and 1365 cm^{-1} (6.9 and 7.3 μm), the former overlapping one of the aromatic skeletal stretching bands. The out-of-plane C—H bending deformation bands occur at 755 and 685 cm^{-1} (13.1 and 14.8 μm), characteristic of a monosubstituted aromatic. The in-plane C—H bending deformation bands appear in the 1100 to 1000 cm^{-1} (9 to 10 μm) region, along with the C—O stretch and other possible skeletal deformation bands, thus making definite assignments rather tenuous. The overtone and combination bands of the out-of-plane and in-plane C—H deformations appear in the 2000 to 1670 cm^{-1} (5 to 6 μm) region. These bands are perceptible in trace A. Note the similarity of these band shapes and positions with those predicted in Fig. 5.4. Only two of the aromatic skeletal stretching bands are readily visible in the 1650 to 1430 cm^{-1} (6 to 7 μm) region, those appearing at 1590 and 1485 cm^{-1} (6.3 and 6.7 μm). The immediate information that a reader should derive from this spectrum, if given to the person as an unknown, is the presence of O—H, a monosubstituted benzene ring, a methyl group, and probably very little other aliphatic C—H.

Figure 5.10 displays the spectrum of phenylacetylene. The acetylenic C—H band appears at 3290 cm^{-1} (3.03 μm) as a very sharp, intense band. The C≡C

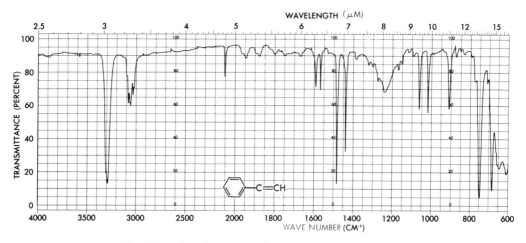

Fig. 5.10. Infrared spectrum of phenylacetylene as a neat liquid film.

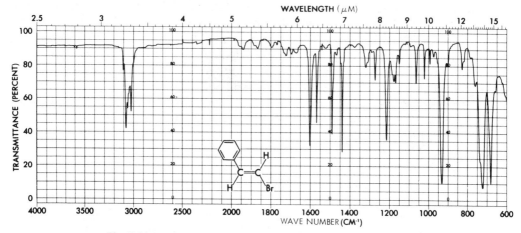

Fig. 5.11. Infrared spectrum of *trans*-β-bromostyrene as a neat liquid film.

stretch, on the other hand, appears as the relatively weak band at 2100 cm^{-1} (4.76 μm). The typical out-of-plane C—H bending bands appear at 745 and 680 cm^{-1} (13.4 and 14.6 μm), with the in-plane C—H bending bands appearing in the 1100 to 900 cm^{-1} (9 to 11 μm) region. The typical pattern of a monosubstituted benzene ring is clearly seen in the 2000 to 1650 cm^{-1} (5 to 6 μm) region. The 1590 and 1570 cm^{-1} (6.28 and 6.37 μm) bands due to the aromatic system are clearly more distinct than in Fig. 5.9, along with the 1480 and 1440 cm^{-1} (6.74 and 6.94 μm) bands.

Figure 5.11 shows the infrared spectrum of *trans*-β-bromostyrene. The

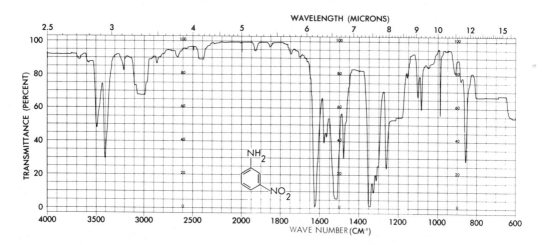

Fig. 5.12. Infrared spectrum of *m*-nitroaniline in chloroform solution.

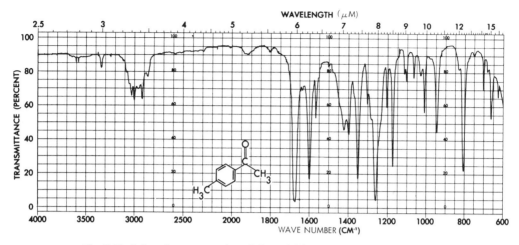

Fig. 5.13. Infrared spectrum of methyl *p*-tolyl ketone as a neat liquid film.

olefinic C—H out-of-plane deformation appears at 930 cm^{-1} (10.7 μm), corresponding to a *trans*-disubstituted alkene. The C=C double-bond stretch overlaps the 1600 cm^{-1} (6.2 μm) aromatic band. The interpretation of the remainder of the spectrum is left to the reader.

Figure 5.12 shows the infrared spectrum of *m*-nitroaniline in chloroform solution. The N—H absorption of the primary amine appears as two bands at 3500 and 3400 cm^{-1} (2.86 and 2.93 μm). The aromatic C—H stretch is barely resolved near 3030 cm^{-1} (3.3 μm). The aromatic C—H out-of-plane deformations occur at 860 cm^{-1} (11.7 μm), characteristic of a single isolated C—H, and at 820 cm^{-1} (12.2 μm). This latter band appears only as a slight shoulder where the solvent (chloroform) absorbs very strongly. The characteristically very intense nitro group bands appear at 1515 and 1340 cm^{-1} (6.6 and 7.4 μm). Note the intense band at 1625 cm^{-1} (6.16 μm) due to in-plane bending of the —NH$_2$.

The infrared spectrum of methyl *p*-tolyl ketone is shown in Fig. 5.13. The C=O stretch appears at 1675 cm^{-1} (5.95 μm), characteristic of a conjugated ketone. The methyl C—H bending deformation peak appears at 1350 cm^{-1} (7.41 μm). The aromatic C—H out-of-plane bending deformation band appears at 805 cm^{-1} (12.2 μm), characteristic of a two adjacent hydrogen system.

The infrared spectrum of 5-hexen-2-one, Fig. 5.14, shows the olefinic C—H stretch at 3080 cm^{-1} (3.24 μm), with C—H out-of-plane deformation bands characteristic of the —CH=CH$_2$ group appearing at 985 and 905 cm^{-1} (10.1 and 11.0 μm). The methyl C—H bending deformation appears at 1355 cm^{-1} (7.39 μm). The C=O stretching band appears at 1720 cm^{-1} (5.83 μm), and the bending band appears at 1155 cm^{-1} (8.65 μm). The terminal C=C stretching deformation band appears at 1635 cm^{-1} (6.11 μm).

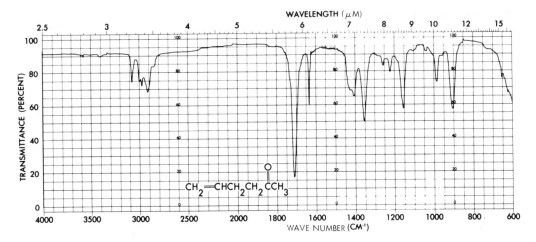

Fig. 5.14. Infrared spectrum of 5-hexen-2-one as a neat liquid film.

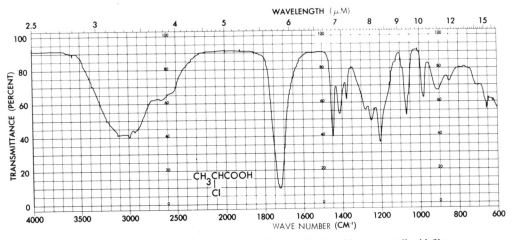

Fig. 5.15. Infrared spectrum of α-chloropropionic acid as a neat liquid film.

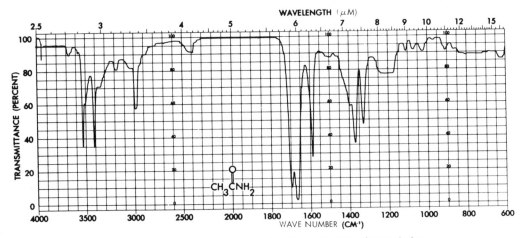

Fig. 5.16. Infrared spectrum of acetamide in chloroform solution.

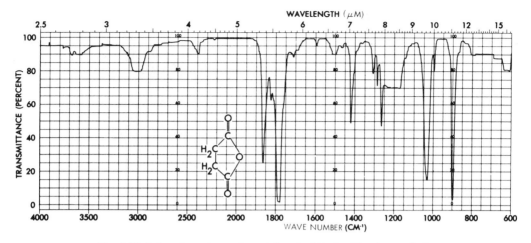

Fig. 5.17. Infrared spectrum of succinic anhydride in chloroform solution.

Figure 5.15 shows the infrared spectrum of α-chloropropionic acid. Note the extreme broadening of the O—H and C=O bands due to strong intermolecular hydrogen bonding of the O—H to the C=O. The C=O absorbs at a slightly higher wave number due to the presence of the electronegative chlorine on the α-carbon.

The infrared spectrum of acetamide in dilute chloroform solution is shown in Fig. 5.16. The N—H stretching absorptions appear as two bands at 3530 and 3410 cm^{-1} (2.83 and 2.93 μm). The "Amide I" band, the C=O stretch, appears at 1670 cm^{-1} (5.97 μm), while the "Amide II" bands, the N—H bending deformations, appear at 1585 cm^{-1} (6.29 μm) and near 1400 cm^{-1} (7.2 μm); the "Amide III" band appears at 1330 cm^{-1} (7.52 μm). The band at 1370 cm^{-1} (7.29 μm) represents the C—H bending deformation of the methyl group. The C=O bending band appears close to 1240 cm^{-1} (8.1 μm) but is broadened and partly obscured by an absorption band of the solvent chloroform.

The infrared spectrum of succinic anhydride appears in Fig. 5.17. Two bands occur in the carbonyl region, which are displaced to higher wave numbers due to the five-membered ring.

5.5
PREPARATION OF THE SAMPLE

The method of handling the sample for recording the infrared spectrum is generally dictated by the physical state of the sample and the region one wishes to record.

Gas samples are placed in gas cells whose internal path length can be greatly multiplied by internal mirrors. The effective length of gas cells may extend from a

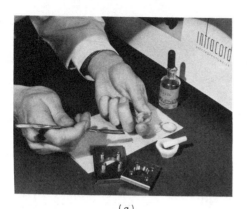

(a)   (b)

Fig. 5.18. (*a*) Preparation of a neat sample for recording the infrared spectrum of a liquid. A drop of the liquid, or mull, is placed between the sodium chloride plates, and the plates are secured between the metal retainers of the cell (bottom of the photograph). (*b*) The filling of a solution cell. (Courtesy of the Perkin-Elmer Corporation, Norwalk, Conn.)

few centimeters up to several meters (generally in multiples of meters). The handling of gases usually requires a vacuum line for storage and transfer to the cell.

Liquid samples can be run either as a neat (pure liquid) sample or in solution. Neat samples can be prepared by placing one or two small drops of the sample between two highly polished pieces of cell material (Fig. 5.18). The thickness of the capillary film is very difficult to control and reproduce, giving spectra with varying absorption intensities. Thin metal foil spacers with thicknesses accurately measured down to 0.001 mm can be used to control the thickness of the sample. A few small drops of the sample are placed in the open area of the spacer on one plate, and the second plate is firmly pressed to the spacer in a cell holder (Fig. 5.18). Solution spectra are obtained by dissolving the sample in an appropriate solvent (see subsequent paragraphs) and placing it in a cell. The concentration range normally employed is 2 to 10% by weight. A variety of cells are available for holding the sample solution. In addition to the sample cell, a reference cell of the same thickness as the sample cell is filled with pure solvent and placed in the reference beam of the instrument. The weak absorption bands occurring in the reference beam offset similar weak absorption bands of the solvent occurring in the sample beam; thus interfering extraneous absorption bands are removed from the spectrum of the compound. Major absorption bands of the solvent absorb so strongly that no effective infrared energy passes through the cell, and no differential absorption due to the sample can be detected.

Solid samples can be run as solutions, mulls, or as a solid dispersion in potassium bromide. The solutions are prepared as described before. Mulls are prepared by suspending finely ground sample particles in Nujol (paraffin or mineral oil) and then recording the spectrum of the mull as a neat sample. Fluorolube can also be used in the preparation of mulls. Not all solids can be mulled successfully. In addition, the Nujol displays intense C—H absorption and renders these regions useless for identification purposes. In general, mulls should be used only if no other method is available.

Solid dispersions of samples, approximately 1% by weight in potassium bromide, are prepared by carefully grinding a mixture of the sample and potassium bromide until composed of very finely ground particles. The finely ground mixture is then pressed into a transparent disc under several tons of pressure. The disc is mounted on a holder, and the spectrum is then recorded. Since potassium bromide is transparent in the infrared region, only bands corresponding to the sample appear in the spectrum. The spectrum obtained thus, as well as with Nujol mull dispersions, will be of the material in the solid phase and may differ from solution spectra owing to restrictions of molecular configurations (in microcrystals) or increased functional group interactions. Broad hydroxyl absorption near 3300 cm^{-1} (3 μm) is usually present, owing to moisture absorbed by the potassium bromide, unless one is careful when preparing the disc.

The sample container is placed in the sample beam of the instrument (Fig. 5.19) along with the appropriate reference beam blank (pure solvent), and the spectrum is recorded. Instructions for operating the instrument are provided in the manufacturer's instruction manual and should be thoroughly understood before individual operation.

5.5.1 Solvents

The choice of solvent depends on the solubility of the sample and the absorption characteristics of the sample and solvent. Since almost all solvents employed in infrared spectroscopy are organic molecules themselves, they will

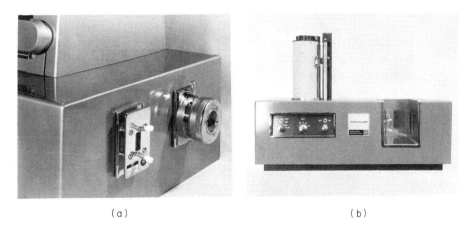

(a)   (b)

Fig. 5.19. (a) The cell compartment of a typical infrared spectrophotometer. The left cell is the sample cell, and the right cell is a variable-thickness cell filled with pure solvent in the reference beam of the instrument. The thickness of the variable-thickness cell is adjustable to match exactly the thickness of the sample cell to balance the solvent-absorption peaks. (b) A typical infrared spectrophotometer. (Courtesy of the Perkin-Elmer Corporation, Norwalk, Conn.)

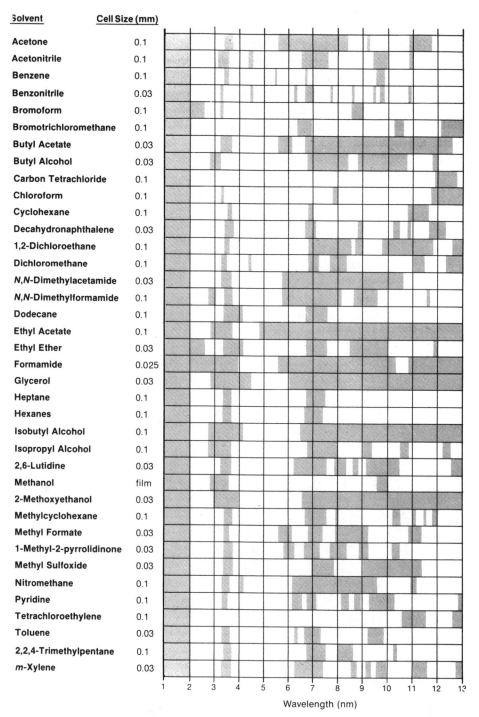

Solvent	Cell Size (mm)
Acetone	0.1
Acetonitrile	0.1
Benzene	0.1
Benzonitrile	0.03
Bromoform	0.1
Bromotrichloromethane	0.1
Butyl Acetate	0.03
Butyl Alcohol	0.03
Carbon Tetrachloride	0.1
Chloroform	0.1
Cyclohexane	0.1
Decahydronaphthalene	0.03
1,2-Dichloroethane	0.1
Dichloromethane	0.1
N,N-Dimethylacetamide	0.03
N,N-Dimethylformamide	0.1
Dodecane	0.1
Ethyl Acetate	0.1
Ethyl Ether	0.03
Formamide	0.025
Glycerol	0.03
Heptane	0.1
Hexanes	0.1
Isobutyl Alcohol	0.1
Isopropyl Alcohol	0.1
2,6-Lutidine	0.03
Methanol	film
2-Methoxyethanol	0.03
Methylcyclohexane	0.1
Methyl Formate	0.03
1-Methyl-2-pyrrolidinone	0.03
Methyl Sulfoxide	0.03
Nitromethane	0.1
Pyridine	0.1
Tetrachloroethylene	0.1
Toluene	0.03
2,2,4-Trimethylpentane	0.1
m-Xylene	0.03

Wavelength (nm)

Fig. 5.20. Chart of solvents for use in the infrared region. The unshaded areas represent useful regions in which transmittance of the pure solvent is $\geq 60\%$. (Reproduced by courtesy of Eastman Kodak Company, Rochester, N.Y.)

give absorption bands characteristic of the types of bonds present. It is always desirable to use a solvent having the least amount of absorption in the infrared region. For example, carbon tetrachloride absorbs strongly only in the 830 to 670 cm^{-1} (12 to 15 μm) region, whereas chloroform absorbs strongly near 3030, 1220, and 830 to 670 cm^{-1} (3.3, 8.2, and 12 to 15 μm); thus carbon tetrachloride is more useful than chloroform, provided suitable concentrations can be achieved in both solvents. Frequently it may be necessary to record the spectrum of a sample in two different solvents to derive all of the available spectral information. For example, the use of carbon tetrachloride as the solvent for alkenes and aromatics allows one to observe all absorption bands out to 900 cm^{-1} (11 μm), but the fingerprint region will be obscured. Recording the spectra of these compounds in carbon disulfide, or acetonitrile, allows one to record the absorption bands appearing in the fingerprint region. Figure 5.20 lists several solvents suitable in the infrared region. The darkened areas indicate regions in which the solvent absorbs most, or all, of the available energy. Slight solvent shifts are noted with some functional groups, particularly those capable of entering into hydrogen bonding either as acceptors or donors.

5.5.2 Cell Materials

Cell window materials must be transparent in the desired portion of the infrared region, must not be soluble in the solvent used, and must not react chemically with the solvent or sample. Materials suitable for use as cell windows are tabulated in Table 5.8. Sodium chloride is the most commonly used cell material owing to its relatively low cost and ease of handling; however, Irtran-2, despite being expensive, is finding extensive use because of its chemical and physical durability.

Certain care must be exercised in the use of these cell materials. The reader must be familiar with the general properties of the cell window materials being used, and he must take proper precautions to protect the windows. Particular care must be exercised in preparing the samples and solvents to be used.

Cells and plates should always be carefully cleaned after each use by thorough washing with purified chloroform or the solvent used in preparing the sample solution, and drying with a stream of dry nitrogen.

5.6
NEAR-INFRARED SPECTROSCOPY

The near-infrared region, 0.75 to 3 μm, contains the overtones of the fundamental bond deformations appearing in the infrared region, as well as various combination bands. The band positions of the overtone bands appear approximately at $2\nu_0$, $3\nu_0$, $4\nu_0$, etc., where ν_0 is the wave number of the fundamental. The

Table 5.8. Cell Materials

Material	Useful Range (μm)	General Properties
NaCl	0.25–16	Soluble in water; slightly soluble in low-molecular-weight alcohols, glycols, amines, carboxylic acids; low cost; easy to polish.
KBr	0.25–30	Soluble in water, low-molecular-weight alcohols, glycols, acids; slightly soluble in amines, ethers; low cost; easy to polish; slightly hygroscopic.
CsBr	1–40	Soluble in water, low-molecular-weight alcohols, glycols, acids; quite expensive; soft; very hygroscopic; used primarily for long wavelength transmission.
CaF$_2$	0.2–10	Insoluble in water; soluble in ammonia and salt solutions; relatively expensive; hard material; useful for high-pressure cells.
BaF$_2$	0.2–12	Insoluble in water; soluble in acids; hard material; sensitive to mechanical and thermal shock.
AgCl	1–22	Insoluble in water; soluble in ammonia and strong bases; should not be extensively exposed to light of wavelength less than 0.6 μm; possesses high reflectivity giving rise to interference fringes (may appear as anomalous peaks).
KRS-5	1–40	Double salt-containing TlBr (48%) and TlI (52%); excellent transmission properties, chemical inertness; sparingly soluble in water, soluble in bases, insoluble in acids; high reflectivity; expensive.
KRS-6	1–22	Double salt-containing TlBr (40%) and TlCl (60%) (TlCl leaches out in water and bases); physical and chemical properties similar to KRS-5; high reflectivity; expensive.
Irtran-2	2–13	Zinc sulfide; extremely durable; insoluble in virtually all organic and inorganic solvents (except with strong oxidizing agents); useful temperature range -200 to $+800°C$; high reflectivity; expensive.
Glass	0.35–2	Useful in the near-infrared region.
Quartz	0.2–4	Useful in the near-infrared region.

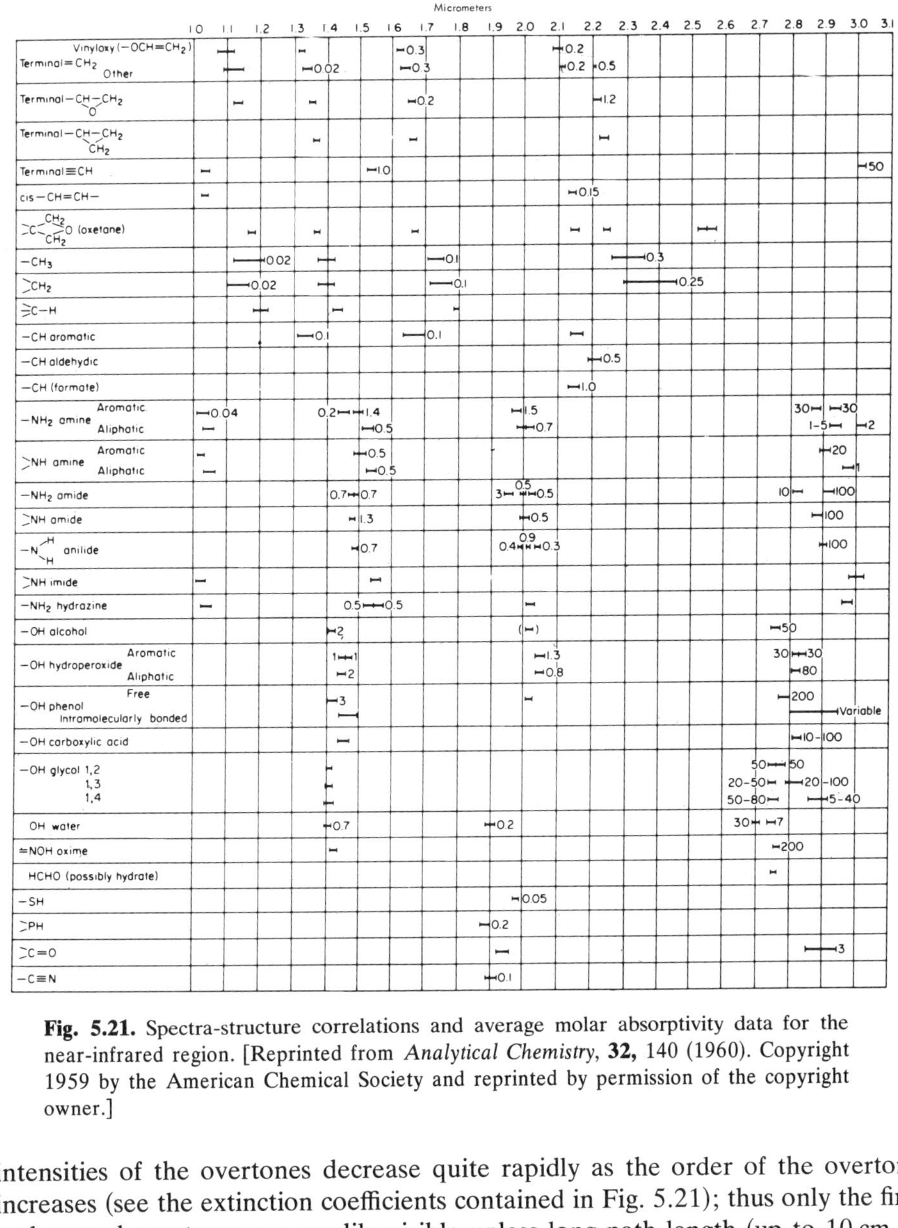

Micrometers

Group	1.0	1.1	1.2	1.3	1.4	1.5	1.6	1.7	1.8	1.9	2.0	2.1	2.2	2.3	2.4	2.5	2.6	2.7	2.8	2.9	3.0	3.1
Vinyloxy (−OCH=CH₂) Terminal=CH₂			⊢		⊢		0.3					0.2										
Other			⊢		0.02		0.3					0.2	0.5									
Terminal −CH–CH₂ (O)				⊢		⊢		0.2					1.2									
Terminal −CH–CH₂–CH₂					⊢			⊢					⊢									
Terminal ≡CH	⊢					1.0																50
cis −CH=CH−	⊢											0.15										
C(CH₂/CH₂)O (oxetane)				⊢		⊢		⊢				⊢	⊢		⊢							
−CH₃			0.02	⊢			0.1						0.3									
CH₂			0.02	⊢			0.1						0.25									
C−H				⊢	⊢			⊢														
−CH aromatic					0.1		0.1			⊢												
−CH aldehydic												0.5										
−CH (formate)										1.0												
−NH₂ amine Aromatic	0.04				0.2 1.4					1.5									30	30		
Aliphatic	⊢				0.5						0.7								1-5		2	
NH amine Aromatic	⊢				0.5														20			
Aliphatic	⊢				0.5																1	
−NH₂ amide					0.7 0.7				3 0.5 0.5								10			100		
NH amide					1.3					0.5										100		
−N(H/H) anilide					0.7				0.4 0.9 0.3											100		
NH imide	⊢				⊢															⊢		
−NH₂ hydrazine	⊢				0.5 0.5				⊢											⊢		
−OH alcohol					2				(⊢)								50					
−OH hydroperoxide Aromatic					1 1					1.3							30 30				
Aliphatic					2					0.8							80					
−OH phenol Free					3				⊢								200					
Intramolecularly bonded					⊢												Variable					
−OH carboxylic acid					⊢												10-100					
−OH glycol 1,2					⊢											50 50						
1,3					⊢										20-50 20-100							
1,4					⊢										50-80 5-40							
OH water					0.7					0.2							30 7					
=NOH oxime					⊢												200					
HCHO (possibly hydrate)																	⊢					
−SH										0.05												
PH										0.2												
C=O										⊢								3				
−C≡N										0.1												

Fig. 5.21. Spectra-structure correlations and average molar absorptivity data for the near-infrared region. [Reprinted from *Analytical Chemistry*, **32**, 140 (1960). Copyright 1959 by the American Chemical Society and reprinted by permission of the copyright owner.]

intensities of the overtones decrease quite rapidly as the order of the overtone increases (see the extinction coefficients contained in Fig. 5.21); thus only the first and second overtones are readily visible unless long path-length (up to 10 cm of pure liquid) cells are used. This severely limits the types of functional groups that will give rise to overtone peaks in the near-infrared region to X—H type groups, for example, C—H, N—H, O—H, S—H, etc., and carbonyl overtones. (The carbonyl fundamental is extremely intense and thus gives rise to more intense overtones.)

The combination bands are often the most intense bands in the near-infrared region. The combination bands represent the sum or the difference of two fundamentals, or a fundamental and a first overtone, generally involving the vibrational and bending deformation modes of the same functional group.

The value of near-infrared spectroscopy lies in the greater resolution obtainable with quartz prism or grating near-infrared spectrophotometers than can be obtained with the salt optics or grating infrared instruments (except in the more refined and expensive instruments). Near-infrared spectrophotometers provide a resolution of approximately 5 nm, allowing one to determine band positions to ± 0.001 μm. This is in contrast to the resolution of infrared instruments, which is about ± 0.02 μm for small laboratory instruments and slightly less than $\pm 0.01 \mu$m for the better grating instruments. With reasonably complex molecules, the congestion in the 3 to 4 μm region limits the amount of useful information that can be obtained from this region, whereas the greater resolution provided in the near-infrared region permits greater distinction of the bands for diagnostic purposes.

Figure 5.21 presents a spectra-structure correlation chart, compiled by Goddu and Delker in 1960, of characteristic group frequencies in the near-infrared region. Near-infrared spectroscopy has been shown to be very useful for the detection of the presence of cyclopropanes.[2] A more complete discussion of individual absorbing chromophores will not be presented here owing to the limited current use of near-infrared spectroscopy; however, the potential utility of near-infrared spectroscopy should not be ignored, and the practicing chemist should be aware of its potential uses.

5.7
FAR-INFRARED SPECTROSCOPY

The far-infrared region contains absorption bands arising from heavy atom bond stretching and bending deformations, functional group skeletal deformations, ring torsional deformations, and lattice mode vibrations (in the solid state only). Considerable research effort is currently being devoted to the study of far-infrared spectra of molecules. Figure 5.22 presents a spectra-structure correlation for the far-infrared region. Reference 19, Sec. 5.10, can be referred to for a more detailed description of far-infrared spectroscopy.

5.8
RAMAN SPECTROSCOPY

An important adjunct to infrared spectroscopy is Raman spectroscopy. Highly symmetrical chromophores, such as the carbon-carbon double bond of ethylene, do not absorb infrared radiation and thus cannot be detected by means of infra-

---

[2] P. G. Gassman and F. V. Zalar, *J. Org. Chem.*, **31**, 166 (1966).

red spectroscopy. To be infrared-active, the chromophore must have a dipole moment; with Raman spectroscopy, however, only a change in the polarizability of the electrons of the system on vibrational excitation is required. In general, bands appearing in the infrared spectrum will also appear in the Raman spectrum, but the intensities will be reversed, i.e., strong bands in the infrared spectrum will appear as weak bands in the Raman spectrum, and vice versa. The principal utility of Raman spectroscopy lies in dealing with highly symmetrical molecules, or chromophores within a molecule, for which sufficient information cannot be gained from the infrared spectrum.

Raman spectroscopy involves the inelastic collision of a photon with a molecule, approximately 1% of the incident light, with transferral of a portion of the energy of the proton to the molecule resulting in vibrational excitation. The scattered photon has had its energy reduced by an amount corresponding to the energy required for the vibrational excitation process, and it will appear at a lower wave number than the incident beam. In general, a number of such lines will appear, resulting from a number of different vibrational excitations occurring within the molecule. These lines are referred to as *Stokes lines*. The wave number separation between the incident beam and the Stokes lines, the Raman wave number $\Delta\nu$, is the wave number of the fundamental of the vibration involved. A second set of much weaker lines appears at higher wave numbers than the incident beam, and these lines are referred to as *anti-Stokes lines*. These lines represent the transfer of energy from the vibrationally excited molecule to a photon of the incident radiation, the molecule returning to the ground vibrational state. Because the anti-Stokes lines are usually very weak, and the information derived duplicates the information gained from the Stokes lines, the anti-Stokes lines are usually ignored.

Laser Raman spectrometers covering the range from 20 to 3500 cm^{-1} are commercially available; for resolution and accuracy comparable to infrared the instrument cost is two to three times higher. Samples may be in the form of liquids, solutions, powders, or even single crystals. The spectrometers are similar to infrared instruments. The source of the radiation is the sample that has been excited by intense monochromatic radiation from a laser. The Raman scattered light is passed into a monochromator, where it is dispersed into its spectrum.

Raman spectra are usually simpler than infrared because of the absence of overtone or combination bands. A wide choice of solvents, including water, can be used. Raman spectroscopy complements infrared spectroscopy. The two methods used together provide a powerful tool for organic structure determination. Vibrations that result in strong infrared peaks are often very weak in the Raman spectrum and conversely. Important Raman group frequencies are given by C=C, C≡C, N=N and S—S stretching modes.

Since the bands appearing in the Raman spectrum correspond to the bands normally occurring in the infrared region, the spectra-structure correlations outlined in the previous sections on infrared spectroscopy also apply to the interpretation of Raman spectra.

BECKMAN FAR-INFRARED VIBRATIONAL FREQUENCY CORRELATION CHART

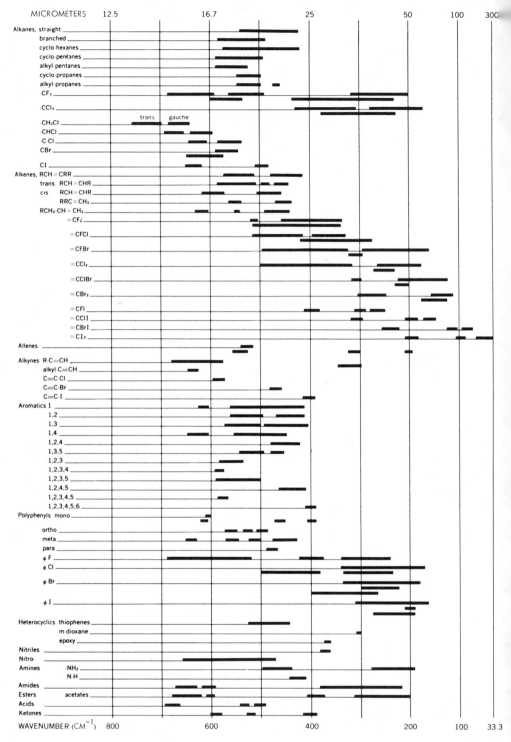

Fig. 5.22. Spectra-structure correlation for the far-infrared region. (Reprinted by permission of Beckman Instruments, Inc., Fullerton, Calif.)

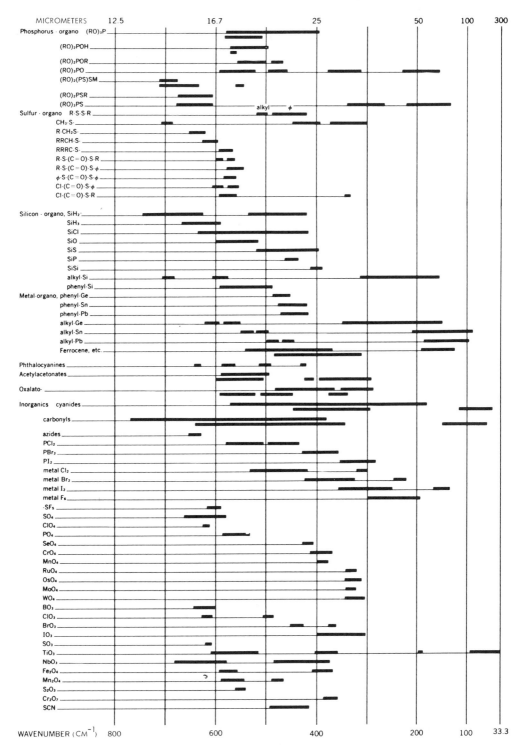

Fig. 5.22 (Continued)

5.9
INFRARED SPECTRAL PROBLEMS

Identify each of the following compounds from their infrared spectra.

1. $C_5H_8O_2$; infrared spectrum of unknown recorded as a neat liquid film.

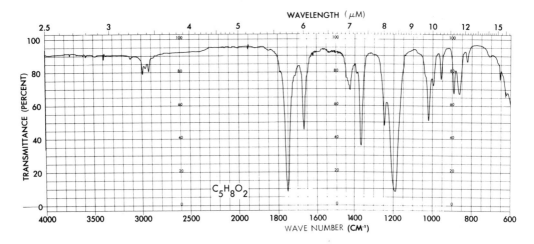

Fig. 5.23.

2. C_5H_8O; infrared spectrum of unknown recorded as a neat liquid film.

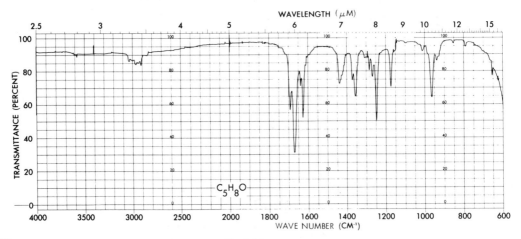

Fig. 5.24.

3. $C_8H_{10}O$; infrared spectrum of unknown recorded in carbon tetrachloride (4000 to 1000 cm^{-1}) and carbon disulfide (1000 to 600 cm^{-1}) solution.

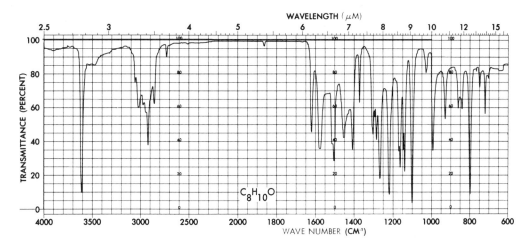

Fig. 5.25.

4. $C_9H_{10}O$; infrared spectrum of unknown recorded in carbon tetrachloride (4000 to 1000 cm^{-1}) and carbon disulfide (1000 to 600 cm^{-1}) solution.

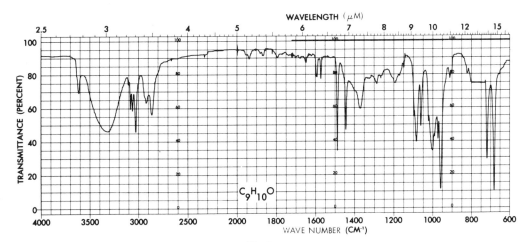

Fig. 5.26.

5. $C_4H_6O_2$; infrared spectrum of unknown recorded as a neat liquid film.

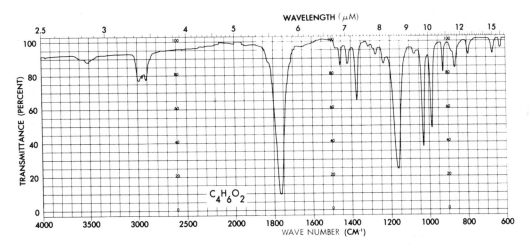

Fig. 5.27.

6. $C_7H_7NO_3$; infrared spectrum of unknown recorded as a neat liquid film.

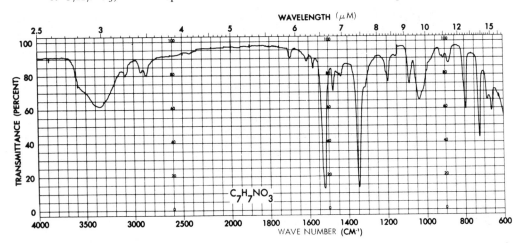

Fig. 5.28.

7. C_4H_7Br; infrared spectrum of unknown recorded as a neat liquid film.

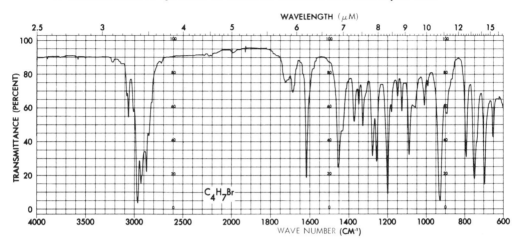

Fig. 5.29.

8. C_6H_8O; infrared spectrum of unknown recorded as a neat liquid film.

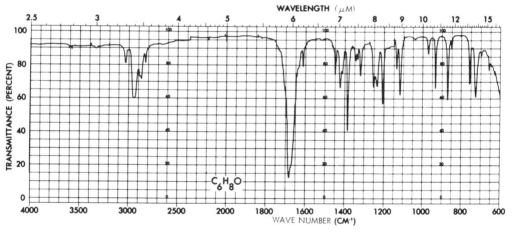

Fig. 5.30.

5.10
LITERATURE SOURCES
OF INFRARED SPECTRAL DATA

The literature covering infrared spectroscopy can be conveniently grouped into three categories: general reference books, specific literature reviews covering limited areas, and specific compound references and spectral compilations.

The reference books by Conley (ref. 4), Bellamy (ref. 2), Lawson (ref. 8), Nakanishi (ref. 11), and Szymanski (ref. 15) are general reference texts on the use and interpretation of infrared spectroscopy. *Infrared—A Bibliography*, by Brown et al. (ref. 3), is a compilation of original literature references on instrumentation, techniques, and general compound classifications covering the years from 1935 to 1951. Specific literature reviews covering limited areas are too numerous to cite in this text. The reader is referred to abstracting and current literature services to locate specific reviews, for example, ref. 7.

Literature references for specific compounds can be found in refs. 6, 7, 10, and 16 in Sec. 5.11. Published spectra of individual compounds can be found in the Aldrich Library of Infrared Spectra (ref. 12), the Sadtler Standard Spectra files (ref. 14), the American Petroleum Institute infrared spectra files (ref. 1), the Manufacturing Chemists' Association Research Project infrared spectra files (ref. 9), and the atlases of infrared absorption spectra of steroids (refs. 5 and 13).

Instructions on the use and applications of the various compilations of literature references and spectra accompany each compilation.

5.11
REFERENCES

INFRARED SPECTROSCOPY

1. *American Petroleum Institute Research Project 44 Infrared Files*, Petroleum Research Laboratory, Carnegie Institute of Technology, Pittsburgh, Pennsylvania.
2. BELLAMY, L. J., *The Infra-red Spectra of Complex Molecules*. New York: John Wiley & Sons, Inc., 1958; and *The Infra-red Spectra of Complex Molecules*, Vol. 1, 3rd ed. New York: Wiley/Halsted, 1975.
3. BROWN, C. R., AYTON, M. W., GOODWIN, T. C., and DERBY, T. J., *Infrared—A Bibliography*. Washington, D.C.: Library of Congress, Technical Information Division, 1954.
4. CONLEY, R. T., *Infrared Spectroscopy*, 2nd. ed., Boston: Allyn and Bacon, Inc., 1972.
5. DOBRINER, K., KATZENELLENBOGEN, E. R., and JONES, R. N., *Infrared Absorption Spectra of Steroids—An Atlas*, Vol. 1. New York: Interscience Publishers, Inc., 1953.
6. HERSHENSON, H. M., *Infrared Absorption Spectra Index*. New York: Academic Press, 1959.
7. *IR, Raman, Microwave Current Literature Service*, Butterworth and Co., Ltd., London, England; and Verlag Chemie, Gmbh., Weinheim, Germany.
8. LAWSON, K. E., *Infrared Absorption of Inorganic Substances*. New York: Reinhold Publishing Corporation, 1961.
9. *Manufacturing Chemists' Association Research Project Infrared Files*, Chemical Thermodynamic Properties Center, Department of Chemistry, Texas A & M University, College Station, Texas.
10. Ministry of Aviation Technical Information and Library Services, ed., *An Index of Published Infra-Red Spectra*, Vols. 1 and 2. London: Her Majesty's Stationery Office, 1960.

11. NAKANISHI, K., and SOLOMON, P. H., *Infrared Absorption Spectroscopy*, 2nd ed. San Francisco: Holden-Day, Inc., 1977.

12. POUCHERT, C. J., *Aldrich Library of Infrared Spectra*, 2nd ed. Milwaukee: Aldrich Chemical Company, Inc., 1975.

13. ROBERTS, G., GALLAGHER, B. S., and JONES, R. N., *Infrared Absorption Spectra of Steroids—An Atlas*, Vol. 2. New York: Interscience Publishers, Inc., 1958.

14. *Sadtler Standard Spectra*, Sadtler Research Laboratories, Philadelphia, Pennsylvania.

15. SZYMANSKI, H. A., *Interpreted Infrared Spectra*, Vols. 1 to 3, New York: Plenum Publishing Company, 1964, 1966, and 1967.

RAMAN SPECTROSCOPY

16. COLTHUP, N. B., DALY, L. H., and WIBERLEY, S. E., *Introduction to Infrared and Raman Spectroscopy*. New York: Academic Press, 1964.

17. TOBIN, M. C., *Laser Raman Spectroscopy*. New York: Wiley-Interscience, 1971.

FAR-INFRARED SPECTROSCOPY

18. MÖLLER, K. D., and ROTHSCHILD, W. G., *Far Infrared Spectroscopy*. New York: Wiley-Interscience, 1971.

19. STEWART, J. E., "Far Infrared Spectroscopy" in *Interpretive Spectroscopy*, Freeman, S. K., ed. New York: Reinhold Publishing Corporation, 1965, p. 131.

REFERENCES CONTAINING CHAPTERS ON INFRARED SPECTROSCOPY

20. GUILLORY, W. A., *Introduction to Molecular Structure and Spectroscopy*. Boston: Allyn & Bacon, 1977.

21. LAMBERT, J. B., SHURVELL, H. F., VERBIT, L., COOKE, R. G., and STOUT, G. H., *Organic Structural Analysis*. New York: MacMillan, 1976.

22. SCHEINMANN, F., ed., *An Introduction to Spectroscopic Methods for the Identification of Organic Compounds*. Oxford: Pergamon Press, 1973.

6

*Nuclear
Magnetic Resonance*

6.1
INTRODUCTION[1]

Nuclear magnetic resonance (NMR) spectroscopy involves transitions between nuclear spin states with the resultant absorption of energy. Nuclei possess mechanical spins, which, in conjunction with the charge of the nuclei, produce magnetic fields whose axes are directed along the spin axes of the nuclei. When a nucleus is placed in a magnetic field of strength H_0, the nucleus will assume $(2I + 1)$ spin orientations (spin states) with respect to the direction of the applied magnetic field, I being the nuclear spin quantum number of the nucleus. The possible spin states of a nucleus are labeled $-I, -I + 1, \ldots, +I$; for example, with I of $\frac{1}{2}$, spin states of $-\frac{1}{2}$ and $+\frac{1}{2}$ are possible; with I of 1, spin states of -1, 0, and $+1$ are possible. The spin quantum number I of any nucleus is related to the atomic and mass numbers as follows:

Atomic Number	Mass Number	Spin Quantum Number, I
Even or odd	Odd	$\frac{1}{2}, \frac{3}{2}, \frac{5}{2}, \ldots$
Even	Even	0
Odd	Even	$1, 2, 3, \ldots$

---

[1] Most modern introductory organic textbooks provide a simplified discussion on the subject of nuclear magnetic resonance spectroscopy. The beginning student may find it helpful to review this material before studying this chapter.

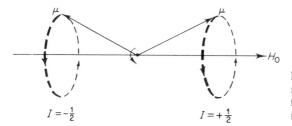

Fig. 6.1. Precession of the axes of the nuclear spin states about the lines of force of the applied magnetic field of a nucleus with $I = \frac{1}{2}$.

The magnetic and spin axis of a nucleus in a given spin state precesses about the direction of the lines of force of the applied field H_0 (Fig. 6.1). The precessional frequency (the Larmor frequency) of the nucleus is exactly equal to the frequency of the electromagnetic radiation required to induce transitions *between adjacent spin states*, for example, the $-\frac{1}{2} \rightarrow +\frac{1}{2}$ spin transition for a nucleus with I of $\frac{1}{2}$, or $-1 \rightarrow 0$ and $0 \rightarrow +1$ transitions for a nucleus with I of 1. Nuclei with I of 0 can exist in only one spin state; hence, they cannot undergo spin transitions and are not detected by NMR techniques. Figure 6.1 illustrates the two spin-state orientations in an applied magnetic field H_0 for a nucleus with I of $\frac{1}{2}$, for example, the hydrogen nucleus.

In the absence of a magnetic field at room temperature, the energies of the spin states of a nucleus are degenerate. As a magnetic field is impressed on the nucleus, the spin states lose their degeneracy and become separated by an energy difference (ΔE), which is directly proportional to the strength of the applied field (Fig. 6.2). The relationship between ΔE and the field strength (H_0) for a nucleus is given by Eq. (6.1),

$$\Delta E = h\nu = \frac{\mu \beta_N H_0}{I} \tag{6.1}$$

where h is Planck's constant, ν is the frequency of the exciting radiation, μ is the magnetic moment of the nucleus, and β_N is a constant called the nuclear magneton constant. Because various nuclei differ in their values of μ and I, they undergo nuclear spin transitions at different frequencies in the same applied magnetic field. Table 6.1 lists the more common nuclei encountered in organic and biochemical applications of NMR spectroscopy, along with their nuclear spin

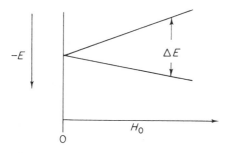

Fig. 6.2. Spin energy level separation for a nucleus with I of $\frac{1}{2}$ as a function of the applied field H_0.

Table 6.1.   Resonance Frequencies at 10,000 Gauss and Related Nuclear Properties

Nucleus	Spin Quantum Number, I	Magnetic Moment, μm	Resonance Frequency, MHz	Natural Abundance, %
^{1}H	$\frac{1}{2}$	2.79268	42.5759	99.9844
^{2}H	1	0.857386	6.53566	0.0156
^{13}C	$\frac{1}{2}$	0.70220	10.705	1.108
^{14}N	1	0.40358	3.076	99.635
^{15}N	$\frac{1}{2}$	-0.28304	4.315	0.365
^{17}O	$\frac{5}{2}$	-1.8930	5.772	0.037
^{19}F	$\frac{1}{2}$	2.6273	40.055	100.0
^{31}P	$\frac{1}{2}$	1.1305	17.236	100.0

quantum number I, μ, resonance frequency in a 10,000-gauss field, and natural abundance. Since the frequency and field strength are directly proportional, the data given in Table 6.1 can be scaled up or down to correspond to other applied field strengths.

Inspection of the data in Table 6.1 reveals that certain nuclei are present in very low natural abundance, for example, ^{2}H, ^{13}C, ^{15}N, and ^{17}O. Furthermore, the sensitivities, or signal intensities, of equal numbers of nuclei of different kinds are not the same, most of the low-abundance nuclei having sensitivities much lower than that of ^{1}H. In the past it was necessary to increase the atom concentration of low-abundance nuclei by chemical synthesis in order to record their NMR spectra. Such procedures were very tedious and costly. Even then, it was often necessary to employ computer averaging of transients (CAT) techniques, a process in which the spectrometer scans the resonance region many times, summing the resonance signals of each scan to enhance the signal-to-noise ratio. More recently, use of fast Fourier transform techniques (see later paragraphs of this section) has made it possible to record NMR spectra of low abundance nuclei such as ^{13}C in considerably less time than is required using the CAT technique. Until only recently the organic chemist utilized essentially only ^{1}H NMR in structural studies; now ^{13}C NMR spectroscopy has risen in prominence and complements the data available from ^{1}H NMR spectroscopy.

Continuous wave spectroscopy. The recording of an NMR spectrum of a sufficient quantity of sample (e.g., 10 mg or greater of sample) of a high-abundance nucleus requires different experimental techniques than are required for samples containing low-abundance nuclei. With the former, a single pass (sweep) through the appropriate electromagnetic region will suffice (termed continuous wave, or CW, spectroscopy), whereas with the latter, fast Fourier transform (FT) techniques are required. In the following paragraphs we will outline the general features of both techniques.

Absorption of energy results in the excitation of the nucleus to a higher-energy spin state. The probability of spontaneous emission from the higher spin state is negligible, although induced emission is just as probable as radiation-induced excitation. Thus a net initial absorption of energy is observed until equal populations are attained, whereupon further absorption will cease (*saturation*). The rate at which saturation is attained depends on the intensity (*amplitude*) of the exciting radiation and on the rate at which spin excited nuclei return to the spin ground state (*relaxation*). In the nuclear magnetic resonance experiment, it is most desirable to maintain sustained absorption of the radio-frequency energy and to avoid saturation. It is important, then, that we inquire as to how a nucleus in a higher spin state can return to a lower spin state (*relaxation*), enabling a sustained absorption of the radio-frequency energy.

Relaxation processes can be divided into two categories: spin-spin and spin-lattice relaxation. Spin-spin relaxation involves a transfer of spin energy of one nucleus to a neighboring nucleus, the rate of transfer being denoted as the relaxation time T_2 (the mean half-life required for a perturbed system to return to equilibrium). Spin-lattice relaxation (lattice being defined as the aggregate of atoms or molecules under study) involves the conversion of the spin energy into thermal energy. The mean half-life of spin-lattice relaxation (T_1) is affected by small fluctuations of the magnetic field throughout the sample. These fluctuations may arise from (1) individual aggregates of molecules, (2) by the presence of paramagnetic substances (oxygen and magnetic materials), or (3) by the presence of an atom possessing an electric quadrupole moment (atoms with $I > \frac{1}{2}$). (The reader is referred to more complete texts for more detailed descriptions of these processes.) The first effect is greatly influenced, in a complex manner, by the viscosity of the system. An increase in the viscosity generally decreases T_1. Similarly, the presence of paramagnetic materials or atoms possessing an electric quadrupole moment causes T_1 to decrease. Both T_1 and T_2 affect the width of the absorption line, the width increasing as T_1 and T_2 decrease.

From the foregoing discussion, it is obvious that certain precautions must be exercised in the preparation of the sample and in the recording of the spectrum. The solvent must be free of magnetic impurities, the sample must be free from paramagnetic species, and, for very careful measurement, the sample should be freeze-degassed to remove dissolved oxygen.

Adequate spin-lattice relaxation is required to maintain the unequal population of spin states for continuous radio-frequency absorption at a given excitation amplitude. Should this condition not exist, saturation will occur. Saturation interferes with peak shapes and, most importantly, with the integrated peak areas; it should be avoided if at all possible. Reducing the amplitude of the radio-frequency power input until a constant absorption response is obtained will usually avoid saturation.

Fast Fourier transform spectroscopy. When dealing with low-abundance and/or low-sensitivity nuclei, a single sweep through the NMR frequency region will not produce a sufficient signal at any given frequency to be detected above the electronic noise level of the spectrometer. Repeated sweeps through the resonance region with the summing of the signal intensity at each frequency (the CAT technique) requires considerable instrument time: up to 4 to 8 min/sweep when recording ^{1}H spectra! With low-abundance and low-sensitivity nuclei, such as ^{13}C, it may require up to several thousand sweeps to achieve a summed spectrum in which the signal-to-noise level is sufficient to produce a good spectrum. However, since most nuclei have spin relaxation half-lives on the order of 1 sec or less, if it were possible to excite all nuclei in the sample with a pulse of energy and then record the summed energy emitted by the various nuclei as they *relaxed* from their excited spin states to their ground spin states, and interpret this data in terms of the rates of relaxation of the nuclei, a considerable saving of time could be accomplished. This, in fact, is the basis of *fast Fourier transform (FT)* spectroscopy. As the spin excited nucleus relaxes, the amplitude of the precession frequency of a given nucleus about the lines of force of the applied magnetic field decreases in a sinusoidal manner (Fig. 6.3). The recorded signal is a decaying sine wave, more commonly called a free-induction decay (FID) signal. Since the various nuclei in a sample are precessing at their respective Larmor frequencies, each nucleus contributes its FID signal to the total of the sample. The summed FID of the nuclei in the sample is then converted to the normal CW-type spectrum by a mathematical process known as *Fourier transformation* (Fig. 6.3). The fast Fourier transform technique allows one to pulse, or "scan," the entire resonance region of a particular nucleus every second or less, allowing for a much faster accumulation of data. Since the nuclei in different regions of a molecule *relax* at different rates, their individual contributions to the FID will not be equivalent, and, as we shall see later, difficulties arise in the interpretation of the relative intensities of the resonance signals corresponding to each of the distinct nuclei in the sample.

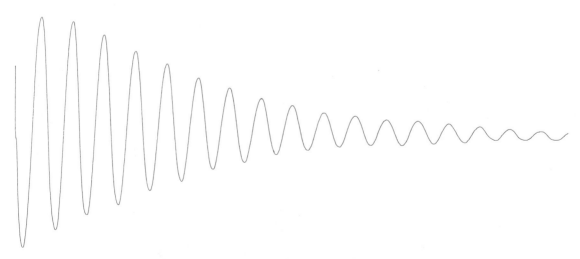

Fig. 6.3. (*a*) Expanded free induction decay signal of tetramethylsilane.

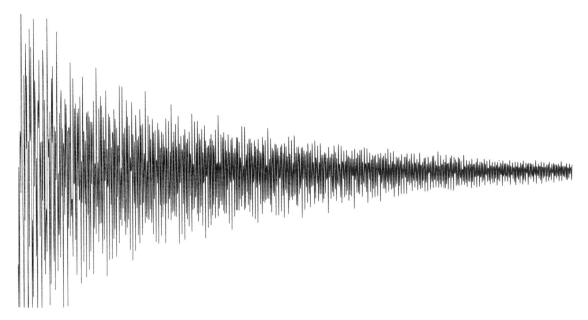

Fig. 6.3. (*b*) Free induction decay signal of benzyl acetate.

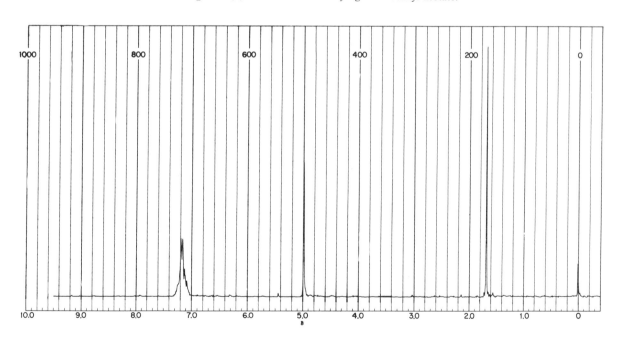

Fig. 6.3. (*c*) Spectrum of benzyl acetate derived from the Fourier transformation of the free induction decay signal in (*b*).

6.2
ORIGIN OF THE CHEMICAL SHIFT

Let us now consider a nucleus contained in a molecule that is in the vicinity of various other atoms, or groups of atoms. The magnetic field "felt" at the nucleus of our atom will not be equal to the applied field H_0, but may be less than or greater than the applied field H_0, and will be represented by H_{net} (net magnetic field). H_{net} is a function of the applied field strength H_0 and can be represented by Eq. (6.2), in which σ_A is the shielding constant for nucleus A, which we are considering.

$$H_{net} = H_0(1 - \sigma_A) \tag{6.2}$$

The shielding constant σ_A is composed of contributions as illustrated by Eq. (6.3).

$$\sigma_A = \sigma_{AA}^{dia} + \sigma_{AA}^{para} + \sum_{B \neq A} \sigma_{AB} + \sigma_A^{deloc} \tag{6.3}$$

The term σ_{AA}^{dia} is the contribution to the shielding constant of atom A by induced diamagnetic currents on atom A under the influence of the applied field H_0. The magnitude of σ_{AA}^{dia} is a function of the electron density about the nucleus of atom A; hence, it is an indirect function of the electronegativity of the atom or functional groups bonded to A. The term σ_{AA}^{para} is the contribution to σ_A by induced paramagnetic currents on atom A, and it arises as a consequence of the mixing of ground and excited electronic states (p or d-orbitals only; hence σ_{AA}^{para} is zero when the electrons of A are localized in pure s-orbitals, for example, the hydrogen atom) under the influence of the applied field. The diamagnetic term is generally the dominant contributor to the shielding of the hydrogen nucleus, whereas with ^{13}C and higher atomic number nuclei that have occupied p and d orbitals, the paramagnetic term is usually dominant. An important feature of this difference to the organic chemist is that the σ_A for ^1H and that for ^{13}C are not always similar. In many instances this can be an advantage to the organic chemist in that different kinds of structural information can be derived from the different NMR spectra.

The term σ_{AB} is the contribution to σ_A due to local induced currents, either diamagnetic (positive contribution) or paramagnetic (negative contribution), on the atom or functional group B, and it is sometimes referred to as the neighbor anisotropy effect. The magnitude of σ_{AB} is a function of the atom or functional group B, the distance between A and B, and the spatial orientation of B with respect to A. If the functional group B is axially symmetric, for example, a single atom or a methyl group, the magnitude of σ_{AB} can be calculated by use of Eq. (6.4), generally referred to as the McConnell equation,

$$\sigma_{AB} = \frac{1}{3N_0R_{AB}^3} \Delta\chi_B(1 - 3\cos^2\theta) \tag{6.4}$$

where N_0 is Avogadro's number, R_{AB} is the internuclear distance, $\Delta\chi_B = \chi_B^{\parallel} - \chi_B^{\perp}$ (difference in the longitudinal or parallel and transverse or perpendicular magnetic susceptibilities), and θ is the angle between the symmetry axis of B and the $A—B$ internuclear vector. The final term of Eq. (6.3), σ_A^{deloc}, arises from induced currents involving delocalized electrons, for example, the π-electron system of aromatics. The effect of neighboring groups on σ_A and σ_A^{deloc} is most apparent in ^1H NMR spectroscopy, and specific examples will be discussed later.

Substitution of the net field (H_{net}) felt at the nucleus of atom A in Eq. (6.1) gives Eq. (6.5).

$$\Delta E_A = h\nu = \frac{\mu \beta_N H_{net}}{I} \tag{6.5}$$

If one considers other nuclei in the same molecule, the remainder of the molecule will display different σ values with respect to the various nuclei unless the nuclei are identical with respect to the remainder of the molecule. The different shieldings of the other nuclei in the molecule result in different frequencies to induce spin transitions with these nuclei. The shielding constant σ_A may result in an increase (*shielding*) or a decrease (*deshielding*) in the resultant field felt by the nucleus. The change in frequency of resonance with respect to an arbitrarily chosen position is termed *chemical shift*. With nuclear magnetic resonance, we have an extremely sensitive probe to investigate the environment of various nuclei in a single molecule.

6.3
DETERMINATION OF CHEMICAL SHIFT

Equation (6.1) contains two potential variables, the frequency ν and the magnetic field strength H_0. Experimentally we might keep one of these variables constant, while varying the other. The frequencies required for resonance in a magnetic field of 10,000 gauss are in the region of 10^6 Hz. With proton magnetic resonance, differences in chemical shift are of the order of 1 to several hundred Hz, demanding an extremely fine control and measurement of frequency if the frequency is to be used as the measurable variable. Electronically, it is easier to maintain a constant radio frequency and vary the field strength. This is accomplished by summing a high-intensity, constant magnetic field with a weak, varying field controlled by a small flow of current through the sweep coils (Fig. 6.4). The sample is placed in the magnetic field, and a constant-energy radio frequency is impressed on the sample. The sample is spun at ~ 20 cycles/sec to time-average inhomogeneities in the sample and/or sample tube (see later discussion on sample preparation). As the flow of current is changed in the sweep coils, the magnetic field changes until the correct conditions for resonance occur with the absorption of radio-frequency energy. The reorientation of the spin angular momenta (relaxation) of the nuclei undergoing spin excitation induces a signal in the receiver coils (the plane of the receiver coil is maintained at a right angle with respect to the plane of the radio-frequency coil), which is amplified and sent to a recorder giving rise to an induction spectrum. Such a setup is referred to as a

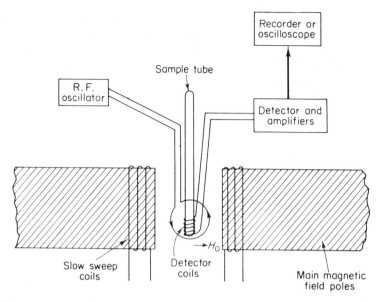

Fig. 6.4. Nuclear induction system of a nuclear magnetic resonance spectrometer.

double-coil system and is used only for high-resolution nuclear magnetic resonance spectroscopy. Another experimental arrangement involves the use of only a single coil for the radio-frequency coil and receiver coil; this is referred to as a *single-coil* system. Single-coil systems give rise to true absorption spectra and are used for both low- and high-resolution nuclear magnetic resonance spectroscopy.

The difficulty in specifying a resonance frequency to the degree of precision required to correlate NMR spectral data (~ 1 part in 10^6), such as is done in ultraviolet and infrared spectroscopy, is that it requires the use of a standard, which is added to the sample. It is assumed that the standard that is added undergoes resonance at a constant frequency regardless of the nature of the solvent and the sample. The standard should be chosen such that it has a single resonance frequency that is well displaced from the resonances of the sample, does not chemically or physically interact with the sample, and is easily removed so that the sample can be recovered. Tetramethylsilane [$(CH_3)_4Si$, TMS, bp 26°C] is ideal in all respects and is the most commonly used *internal* standard in ^1H and ^{13}C NMR spectroscopy. In some cases it may not be possible to use an internal standard that is added directly to the sample because of solubility or chemical reactivity problems. In such instances a standard is placed inside a fine capillary, and the capillary is then placed inside the regular NMR sample tube. Such standards are referred to as *external* standards. In both cases the standard is assigned a resonance position, or *chemical shift*, of 0. When standards other than TMS are used, corrections can be applied to change the chemical shifts relative to TMS.

Several chemical shift designation conventions have been employed to indicate the position of resonance of a nucleus. Although all of these conventions will be described, the use of a single convention is recommended, that being the δ system. Individual peak positions, reserved for use in cases in which a complete analysis of the NMR spectrum is not possible (Secs. 6.5 and 6.8), are reported in Hertz (Hz) relative to the standard.[2] In ^{1}H and ^{13}C NMR, peaks appearing at lower field than the standard are assigned a positive sign, whereas peaks appearing at higher field are assigned a negative sign.[3] As the resonance frequency of a nucleus is proportional to the strength of the applied field, when indicating resonance positions in Hz the spectrometer frequency and the standard must also be specified.

As NMR spectrometers of many different radio frequencies are currently in use (e.g., 40, 60, 100, and 220 MHz to cite a few), it is more convenient to use a chemical shift designation that is independent of the frequency of the spectrometer. Such is the δ system that is defined in Eq. (6.6),

$$\delta = \frac{(\text{chemical shift in Hz}) \cdot 10^6}{\text{spectrometer frequency}} \tag{6.6}$$

in which the chemical shift in Hz is the difference between the resonance frequency of a particular nucleus and TMS. For example, a proton resonance appearing at +72 Hz on a spectrometer with a frequency of 60 MHz has a δ value of 1.20. Note that the sign of δ is the same as that of the chemical shift in Hz, and that the spectrometer frequency and standard need not be specified.

Whereas the δ shift is used predominantly by American scientists, the *tau* (τ) scale is still often used. The τ scale implies the use of TMS as the standard, with τ being calculated according to Eq. (6.7). The τ scale can be used *only* for ^{1}H chemical shifts.

$$\tau = 10 - \frac{(\text{chemical shift relative to TMS}) \cdot 10^6}{\text{spectrometer frequency}} \tag{6.7}$$

(Note that $\tau = 10.0 - \delta$)

6.4
HYDROGEN MAGNETIC RESONANCE

In the following sections are discussed trends in chemical shifts in ^{1}H and ^{13}C NMR spectroscopy and the effect of neighboring nuclei with $I = \frac{1}{2}$ on the structure of the resonance peaks of the nuclei.

The factors contributing to the shielding of a nucleus in a complex molecule were discussed in Sec. 6.2; for ^{1}H, the σ_{AA}^{dia} term is generally dominant. The chemical shift of the proton is highly dependent on the electron density at the proton being observed, or associated with the atom to which it is bonded. As the

---

[2] Prior to 1970 such chemical shifts were reported in cycles per second (cps). Since that time cps has been replaced by Hz (1 cps = 1 Hz).

[3] It is not uncommon in the earlier literature to encounter an opposite sign assignment, and one must be careful in interpreting the data.

Table 6.2. Average Chemical Shifts (δ) of α-Hydrogens in Substituted Alkanes*

Functional Group X	CH$_3$X	—CH$_2$X	$>$CHX
H	0.233	0.9	1.25
CH$_3$ or CH$_2$	0.9	1.25	1.5
F	4.26	4.4	—
Cl	3.05	3.4	4.0
Br	2.68	3.3	4.1
I	2.16	3.2	4.2
OH	3.47	3.6	3.6
O—Alkyl	3.3	3.4	—
O—Aryl	3.7	3.9	—
OCO—Alkyl	3.6	4.1	5.0
OCO—Aryl	3.8	4.2	5.1
SH	2.44	2.7	—
SR†	2.1	2.5	—
SOR	2.5	—	2.8
SO$_2$R	2.8	2.9	3.1
NR$_2$	2.2	2.6	2.9
NR—Aryl	2.9	—	—
NCOR	2.8	—	3.2
NO$_2$	4.28	4.4	4.7
CHO	2.20	2.3	2.4
CO—Alkyl	2.1	2.4	2.5
CO—Aryl	2.6	3.0	3.4
COOH	2.07	2.3	2.6
CO$_2$R	2.1	2.3	2.6
CONH$_2$‡	2.02	2.2	—
CR≡CR^{1}CR^2	2.0–1.6	2.3	2.6
Phenyl	2.3	2.7	2.9
Aryl§	3.0–2.5	—	—
C≡CR	2.0	—	—
C≡N	2.0	2.3	2.7

*The data appearing in this table were taken from various published compilations and original literature references. The tabulated values are average values for compounds that do not contain another functional group within two carbon atoms from the indicated hydrogens. For methylene and methine hydrogens most values fall within ±0.15 ppm of the tabulated values.

† S-aryl derivatives absorb at somewhat lower fields.

‡ Replacement of NH$_2$ by N-alkyl$_2$ results in a slight shift upfield.

§ Includes polycyclic and many heterocyclic aromatics. The values for —CH$_2$—X and $>$CH—X probably appear at lower fields also.

electron density decreases, the proton experiences a greater deshielding and absorbs at lower field positions. The position of resonance shifts to lower field as one proceeds from left to right across the periodic table and from bottom to top. In fact, the chemical shifts of the methyl hydrogens of CH$_3$—X-type compounds show a nearly linear correspondence with the electronegativity of the X atom from lithium to oxygen. Chemical shifts for a variety of acyclic compounds are tabulated in Table 6.2. The shielding effects decrease by about 90% as one goes from the α- to the β-position, and from the β- to the γ-position, the γ-position being relatively unshifted compared to that in a hydrocarbon.

Cyclic hydrocarbons absorb at only slightly different frequencies relative to a CH_2 group in an acyclic hydrocarbon, except for cyclopropane, which absorbs at very high field (δ 0.222). Substituted cyclopropanes absorb in the δ 0.5 to 2.7 region[4]. Cyclobutane appears at δ 1.96, cyclopentane at 1.51, cyclohexane at 1.44, and cycloheptane at 1.53.

The presence of a second substituent Y on a carbon to give X—CH_2—Y-type compounds results in further deshielding, or shielding, of the hydrogens. Shoolery has developed an empirical method of calculating chemical shifts of hydrogens in CH_3X and XCH_2Y compounds by averaging the shifts caused by successive substitutions of hydrogens in methane, and then designating these shifts as effective shielding constants, $\sigma_{i_{eff}}$. Table 6.3 lists the $\sigma_{i_{eff}}$'s of a number of functional groups. These constants are used in Eq. (6.8) to calculate the chemi-

$$\delta = 0.233 + \sum \sigma_{i_{eff}} \tag{6.8}$$

cal shift of a hydrogen in a complex molecule. Calculation of the chemical shifts of hydrogens *beta* to a function can be done by multiplying $\sigma_{i_{eff}}$ in Eq. (6.8) by 0.1. The $\sigma_{i_{eff}}$'s are not always strictly additive, and severe deviations do occur, particularly in the calculation of the chemical shifts of methine hydrogens (e.g., in CHXYZ). Several comparisons of calculated with experimentally observed chemical shifts are included in Table 6.4.

Extension of the inductive effect explanation from aliphatic to aromatic, olefinic, and acetylenic compounds leads to a prediction of chemical shift trends in opposition to that experimentally observed. Before discussing the chemical shifts of protons attached to unsaturated centers, it is appropriate to discuss the shielding properties of various systems and the importance of these effects on

Table 6.3.   Shoolery's Effective Shielding Constants

Functional Group	$\sigma_{i_{eff}}$	Functional Group	$\sigma_{i_{eff}}$
—Cl	2.53	—CR^1=CR^2R^3	1.32
—Br	2.33	—C_6H_5	1.85
—I	1.82	—C≡C—R	1.44
—OH	2.56	—RC=O	1.70
—O—Alkyl	2.36	O	1.55
—O—Aryl	3.23	$\parallel$	
O	3.13	—C—OR	
$\parallel$		O	1.59
—OCR		$\parallel$	
—SR	1.64	—C—NR_2	
—N(Alkyl)$_2$	1.57	—CF_3	1.14
—CH_3	0.47	—C≡N	1.70

---

[4] For a tabulation of chemical shifts of monosubstituted cyclopropanes, see K. B. Wiberg, D. E. Barth and P. H. Schertler, *J. Org. Chem.*, **38**, 378 (1973).

Table 6.4.   Comparison of Observed and Predicted Chemical Shifts Using Shoolery's Rules

Compound	Calculated δ	Observed δ
BrC$\underline{H}_2$Cl	5.09	5.16
IC$\underline{H}_2$I	3.87	4.09
C$_6$H$_5$C$\underline{H}_2$OR	4.44	4.41
C$_6$H$_5$C$\underline{H}_2$CH$_3$	2.52	2.55
C$_6$H$_5$C$\underline{H}_2$C$_6$H$_5$	3.91	3.92
C≡C—C$\underline{H}_2$OH	3.92	3.91
C≡C—C$\underline{H}_2$—C≡C	2.87	2.91
CH$_3$C$\underline{H}_2$$\overset{\displaystyle O}{\overset{\displaystyle |}{C}}$—R	2.40	2.47
(C$_2$H$_5$O)$_3$C$\underline{H}$	7.31	4.96
(CH$_3$)$_2$C$\underline{H}$I	2.99	4.24

nuclei bonded to these systems and on nuclei in other portions of the molecule.

Doubly and triply bonded systems possess a more polarizable and delocalized electronic structure than singly bonded systems, and thus they produce significant induced fields when placed in a magnetic field. The magnitude and sign of the induced fields are highly directional with respect to the axes of the unsaturated center.

Hydrogens bonded to C=C and C=O absorb at much lower field positions than one might expect if the deshielding were due to local inductive effects of the doubly bonded system. The additional deshielding, approximately 4 to 5 ppm, of these hydrogens must be due to a long-range deshielding by the C=C and C=O. Theoretical and experimental evidence has indicated that the space about the C=C and the C=O can be divided into two regions, one producing paramagnetic shifts resulting in deshielding (indicated by a minus sign), and the other producing diamagnetic shifts resulting in shielding (indicated by a plus sign). The regions of paramagnetic and diamagnetic shielding about the C=C, C≡C, and C=O

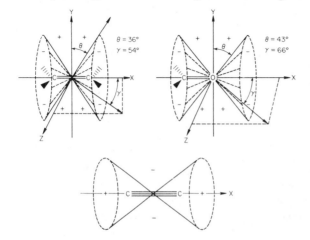

Fig. 6.5. Regions of diamagnetic ($+$) and paramagnetic ($-$) shielding about the C=C, C≡C, and C=O functions. [Representation of the shielding regions about the C=C and C=O are taken from J. W. ApSimon, P. V. Demarco, D. W. Mathieson, W. G. Craig, A. Karim, L. Saunders, and W. B. Whalley, *Tetrahedron*, **26**, 133 (1970).]

Table 6.5.   Chemical Shifts of Hydrogens Bonded to Unsaturated Centers

Type	Unconjugated, δ	Conjugated, δ
$\diagdown\!\!\!\diagup C\!=\!CH_2$	4.6–5.0	5.4–7.0*
$\diagdown\!\!\!\diagup C\!=\!C\diagup^{H}_{\diagdown}$	5.0–5.7	5.7–7.3*
Aromatic	6.5–8.3†	—
Nonbenzenoid aromatic	6.2–9.0	—
Acetylenic	2.3–2.7	2.7–3.2
Aldehydic	9.5–9.8	9.5–10.1*
$H\!-\!\overset{\overset{\textstyle O}{\|}}{C}\!-\!N\diagup^{}_{\diagdown}$	7.9–8.1	—
$H\!-\!\overset{\overset{\textstyle O}{\|}}{C}\!-\!O\!-\!$	8.0–8.2	—

* The position depends on the type of functional group in conjugation with the unsaturated group.

† The position of aromatic hydrogen absorption depends on the type of substituent attached to the aromatic ring (Table 6.6).

functions are illustrated in Fig. 6.5. Note that the shielding regions about the $C\!=\!C$ and $C\!\equiv\!C$ are opposite, resulting in the appearance of the acetylenic hydrogen at higher field than the vinyl hydrogen (Table 6.5).

The regions of paramagnetic and diamagnetic shielding produced by an aromatic ring are illustrated in Fig. 6.6. The effects are generally much greater than with the $C\!=\!C$, $C\!\equiv\!C$, or $C\!=\!O$, owing to the presence of the strong ring current of the aromatic system, and they can be observed at much greater distances from the ring than from the $C\!=\!C$, $C\!\equiv\!C$, or $C\!=\!O$ functions. The

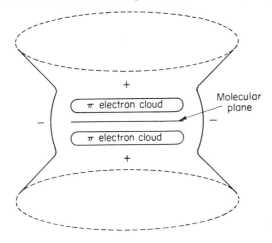

Fig. 6.6. Regions of diamagnetic (+) and paramagnetic (−) shielding of the benzene ring.

maximum diamagnetic shielding is observed directly over the π-electron system and not over the center of the aromatic ring. The very low-field chemical shifts of hydrogen attached to aromatic systems are due to the strong paramagnetic shielding in the plane of the ring.

Hydrogens in other parts of molecules may experience *long-range* shielding and deshielding by functions if the geometry of a molecule is such that the hydrogen resides in the shielding or deshielding regions of the group. In non-rigid molecules one or more molecular conformations may place the absorbing nucleus in these regions, resulting in diamagnetic or paramagnetic shifts. The magnitude of these shifts decreases as the distance between the group and the hydrogen increases. An example of the long-range shielding effect of the benzene ring is illustrated with structures **1** and **2**.

CH₃ absorption at δ 1.77   CH₃ absorption at δ 2.31

1   **2**

Long-range shielding effects may also be caused by electron currents in saturated bonds. In particular, in six-membered ring compounds, axial hydrogens appear at higher field (~ 0.6 ppm) than do their equatorial counterparts. This has been attributed to an anisotropic shielding of the axial hydrogen by the electrons of the β,γ-C—C bonds.

From the preceding paragraphs it is obvious that a certain degree of caution must be exercised in assigning hydrogen resonances based on the general regions of absorption. One must keep in mind that the geometry of the molecule, contributions by various molecular conformations, and the presence of functional groups can have a profound effect on the chemical shift of a given hydrogen. In many cases chemical evidence must be integrated with the spectral evidence in order to derive the correct structure.

Within the chemical shift regions given in Table 6.5, the chemical shift of a hydrogen attached to a C=C or a benzene ring is dependent on the type of functions attached to the C=C or the ring. By considering resonance and inductive interactions of functions with the C=C or aromatic ring, one can qualitatively predict the shielding effect exerted by the group. For example, the terminal methylene hydrogens of **3** appear at lower field than in **4** due to the contributions of resonance structures (b) to **3**, which results in deshielding of the

CH₂=CH—C—OR ⟷ CH₂—CH=COR

(a)   (b)

3

terminal hydrogens (δ 6.20 and 6.38), and to **4**, which results in shielding of the similar hydrogens (δ 4.55 and 4.85).

$$CH_2\!\!=\!\!CH\!\!-\!\!O\!\!-\!\!\overset{\overset{\displaystyle O}{\|}}{C}\!\!-\!\!R \quad\longleftrightarrow\quad \bar{C}H_2\!\!-\!\!CH\!\!=\!\!\overset{+}{O}\!\!-\!\!\overset{\overset{\displaystyle O}{\|}}{C}\!\!-\!\!R$$

(a)   (b)

4

In an approach similar to that of Shoolery, an empirical correlation has been developed to calculate the chemical shifts of vinyl hydrogens. The chemical shift of ethylenic hydrogens in carbon tetrachloride solution is calculated by use of Eq. (6.9),

$$\delta = \delta_{\text{ethylene}} + \sum_i Z_i \qquad (6.9)$$

where δ_{ethylene} is the chemical shift of the hydrogens of ethylene and is taken as $\delta = 5.28$, and Z_i is the $\Delta\delta$ that results on substitution of an ith functional group for hydrogen in ethylene. The values of Z_i depend on the stereochemical relationship of the substituent with respect to the hydrogen in question in the substituted ethylene.

$$\underset{R_{trans}}{\overset{R_{cis}}{\diagdown}}C\!\!=\!\!C\underset{R_{gem}}{\overset{H}{\diagup}}$$

The values for Z_i are given in Table 6.6; they give values of δ within 0.15 of the experimental values in 73.3% of the 1070 compounds correlated. Serious deviations sometimes occur in the more highly complex systems.[5] An example of the use of Eq. (6.9) is illustrated with *cis*- and *trans*-stilbene using the Z_i values given in Table 6.6.

$$\delta_{cis} = 5.28 + Z_{i_{gem}} + Z_{i_{trans}} = 5.28 + 1.35 + (-0.10) = 6.53$$

$$\text{Observed } \delta = 6.55$$

$$\delta_{trans} = 5.28 + Z_{i_{gem}} + Z_{i_{cis}} = 5.28 + 1.35 + 0.37 = 7.00$$

$$\text{Observed } \delta = 6.99$$

The chemical shift of hydrogens attached to a benzene ring is similarly quite sensitive to the type of substituent attached to the ring. Electron-withdrawing groups result in deshielding of the ring hydrogens, particularly the *ortho* hydrogens, while electron-donating functions result in shielding of the ring hydrogens, again primarily the *ortho* hydrogens. Table 6.7 lists the chemical shifts of *ortho*, *meta*, and *para* hydrogens in a number of substituted benzenes.

---

[5] U. E. Matter, C. Pascual, E. Pretsch, A. Pross, W. Simon and S. Sternhill, *Tetrahedron*, **25**, 2023 (1969).

Table 6.6. Substituent Constants for Substituted Ethylenes*

Substituent	gem	cis	trans	Substituent	gem	cis	trans
—H	0	0	0	—CHO	1.03	0.97	1.21
—Alkyl	0.44	−0.26	−0.29	—C(=O)—N<	1.37	0.93	0.35
—Cycloalkyl	0.71	−0.33	−0.30				
—CH₂O,—CH₂I	0.67	−0.02	−0.07	—C(=O)—Cl	1.10	1.41	0.99
—CH₂S	0.53	−0.15	−0.15	—OR†	1.18	−1.06	−1.28
—CH₂Cl,—CH₂Br	0.72	0.12	0.07	—OR‡	1.14	−0.65	−1.05
—CH₂N	0.66	−0.05	−0.23	—OCOR	2.09	−0.04	−0.67
—C≡C	0.50	0.35	0.10	—Aryl	1.35	0.37	−0.10
—C≡N	0.23	0.78	0.58	—Cl	1.00	0.19	0.03
—C=C§	0.98	−0.04	−0.21	—Br	1.04	0.40	0.55
—C=C‖	1.26	0.08	−0.01				
—C=O§	1.10	1.13	0.81	—N(R†)(R)	0.69	−1.19	−1.31
—C=O‖	1.06	1.01	0.95				
—COOH§	1.00	1.35	0.74	—N(R‡)(R)	2.30	−0.73	−0.81
—COOH‖	0.69	0.97	0.39				
—COOR§	0.84	1.15	0.56	—SR	1.00	−0.24	−0.04
—COOR‖	0.68	1.02	0.33	—SO₂R	1.58	1.15	0.95

* For carbon tetrachloride solutions. Taken from C. Pascual, J. Meier, and W. Simon, *Helv. Chim. Acta*, **49,** 164 (1966).

† R = aliphatic.

‡ R = unsaturated group.

§ For a single functional group in conjugation with the first C=C.

‖ For a functional group that is also conjugated to a further substituent (e.g., 1,3,5-hexatriene).

Table 6.7. Chemical Shifts of Hydrogens in Monosubstituted Benzenes

Functional Group	Chemical Shifts (δ) Relative to Benzene (δ 7.27)		
	ortho	*meta*	*para*
—NO$_2$	−0.97	−0.30	−0.42
—CO$_2$R	−0.93	−0.20	−0.27
—COCH$_3$	−0.63	−0.27	−0.27
—COOH	−0.63	−0.10	−0.17
—CCl$_3$	−0.80	−0.17	−0.23
—CH$_3$	0.10	0.10	0.10
—Cl	0.00	0.00	0.00
—OCH$_3$	0.23	0.23	0.23
—OH	0.37	0.37	0.37
—NH$_2$	0.77	0.13	0.40
—N(CH$_3$)$_2$	0.50	0.20	0.50

6.4.1 Chemical Shifts of Hydrogen Attached to Atoms Other than Carbon

6.4.1a Hydrogen Bonded to Oxygen

The chemical shifts of hydrogens bonded to oxygen are extremely sensitive to structural and environmental changes. The hydroxyl hydrogens of alcohols generally absorb in the δ 0.5 to 5.0 region (Table 6.8), the position being dependent on the concentration. This is due, in general, to rapid intermolecular hydrogen bonding and exchange, and results in a deshielding of the hydroxyl hydrogen. Distinction between intermolecular and intramolecular hydrogen bonding can be made by use of dilution studies. On dilution, a plot of the chemical shift *vs.* concentration results in a line with a significant slope for intermolecularly bonded hydrogen and a very small slope for intramolecularly bonded hydrogen (theoretically, one would predict a zero slope). The resonances of monomeric hydroxyl hydrogens are near δ 0.5 in carbon tetrachloride. Phenolic hydroxyl hydrogens absorb at lower fields, approximately δ 4.5 for monomeric species. Their chemical shift is also a function of the concentration.

The acidic hydrogens of enols and carboxylic acids appear at very low fields. Enolic hydroxyl hydrogens absorb near δ 15.5, the enhanced deshielding being due to the positive nature of the hydrogen and strong intramolecular hydrogen bonding. Carboxylic acid hydrogens absorb in the δ 9 to 12 region and are not affected by concentration changes. This is due to the fact that carboxylic acids normally exist in dimeric form in nonpolar solvents.

The resonance signal of hydroxyl hydrogens is further complicated by the possibility of chemical exchange of hydrogens between different molecules. The peak shape, both in broadness and splitting by spin-spin interaction with the carbinol proton (see discussions on spin-spin coupling, Sec. 6.5), of the hydrogen

Table 6.8.   Chemical Shifts of Hydrogen Bonded to Oxygen, Nitrogen, and Sulfur

	Functional Group		Chemical Shift, δ
OH	Aliphatic alcohols	0.5	(Monomeric)
		0.5–5	(Associated)
	Phenols	4.5	(Monomeric)
		4.5–8	(Associated)
	Enols	15.5	
	Carboxylic acids	9–12	(Dimeric)
	Hydrogen bonded to carbonyl systems	13–16	
NH_2	Alkylamine	0.6–1.6	
	Arylamine	2.7–4.0	
	Amide	7.8	
NH	Alkylamine	0.3–0.5	
	Arylamine	2.7–2.8	
$R_3\overset{+}{N}H$	Ammonium salts	7.1–7.7	(In trifluoroacetic acid)
SH	Aliphatic	1.3–1.7	
	Aromatic	2.5–4	

resonance peaks depends on the extent of chemical exchange and will be discussed in greater detail under time-averaging of chemical shifts (Sec. 6.6). Since the hydrogen exchange is acid catalyzed, it can be substantially reduced by careful purification of the compound and solvent, or by recording the spectrum in dimethyl sulfoxide solution.[6]

An alternative method of characterizing the hydroxyl hydrogen resonance is to remove the hydrogen and replace it with deuterium. This is accomplished by dissolving a small portion of the sample in ether or chloroform and shaking the solution several times with deuterium oxide. The organic phase is dried over a drying agent, and the organic compound is recovered by evaporation. Quite often, addition of one drop of deuterium oxide to the NMR tube followed by vigorous shaking is sufficient to exchange the hydroxyl protons.

---

[6] The exchange process is slowed considerably owing to hydrogen bonding between the hydroxyl hydrogen and the sulfoxide group. The hydroxyl hydrogen of a primary alcohol appears as a triplet (see Sec. 6.5 on spin-spin coupling), of a secondary alcohol as a doublet, and of a tertiary alcohol as a singlet [O. L. Chapman and R. W. King, *J. Am. Chem. Soc.*, **86**, 1257 (1964)].

6.4.1b Hydrogen Bonded to Nitrogen

The chemical shifts of hydrogens bonded to nitrogen are also quite sensitive to structural and environmental changes. The resonance signals are generally very broad, owing to an interaction with the electric quadrupole moment of the nitrogen, and they may not be readily apparent. Hydrogen exchange also occurs. Table 6.8 lists the resonance ranges for various types of N—H containing compounds.

6.4.1c Hydrogen Bonded to Sulfur

The chemical shifts of hydrogens bonded to sulfur in aliphatic thiols appear in the δ 1.2 to 1.6 region. Aromatic thiols absorb at a somewhat lower field: near δ 3.0 to 3.5 (Table 6.8.)

6.5
SPIN-SPIN COUPLING

The correlation of chemical shifts with structures provides very useful information for the determination of the structures of molecules. However, the value of nuclear magnetic resonance is greatly increased by the presence of another phenomenon, that of nuclear spin-spin interactions that result in the splitting of the resonance lines. In order to illustrate this let us consider the molecular fragment

$$
\begin{array}{ccc}
 & Z & A \\
 & | & | \\
Y\!-\!C\!&\!-\!C\!&\!-\!D \\
 & | & | \\
 & H_X & H_A
\end{array}
$$

in which the two hydrogens have different chemical shifts. If we were to ignore the fact that H_A is present, we would expect H_X to give rise to a single resonance line. In the real situation, H_X does sense the presence of H_A, and since H_A can be in either the $-\frac{1}{2}$ or $+\frac{1}{2}$ spin state, the resonance of H_X can occur while H_A is in either spin state. As the two spin states of H_A differ in magnetic orientation, the resonance of H_X will occur at two different frequencies, thus producing two resonance lines. The effect of the spin states of a neighboring nucleus is transmitted *via* the bonding electrons by a process termed spin polarization, i.e., the spin of the neighboring nucleus prefers spin pairing with one of the electrons of the bond most closely associated with that nucleus. This effect is then passed on through the bonding electrons to the nucleus under observation. The magnitude of the *spin-spin coupling*, i.e., the separation in energy of the individual resonance lines for a nucleus, generally decreases with the increase in the number of bonds between the interacting nuclei, and is specified as the *coupling constant J. J* has

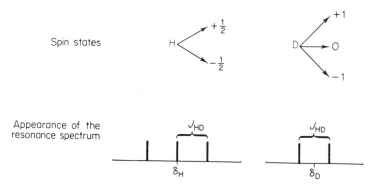

Fig. 6.7. Spin-spin splitting of the hydrogen and deuterium resonance lines in the HD molecule.

the dimensions of Hz and is *independent of the strength of the applied field H_0*. This is in contrast to the chemical shift, which is field dependent. The number of bonds through which the spin interaction is transmitted is sometimes indicated as a superscript before J, while identities of nuclei involved are indicated as subscripts after J; for example, in the above fragment it would be designated as $^3J_{HH}$.

Let us consider in more detail how spin-spin coupling affects the appearance of the resonances of nuclei. As a simple example we will start with the molecule HD, in which I for H is $\frac{1}{2}$ and for D is 1. The hydrogen possesses two spin states, $-\frac{1}{2}$ and $+\frac{1}{2}$, and the deuterium three spin states, -1, 0, and $+1$ (Fig. 6.7). The spin states of hydrogen have equal probability and thus will be equally populated, as will the three spin states of deuterium. Interaction of the spin states of deuterium with the hydrogen nucleus will result in a splitting of the hydrogen resonance into three distinct lines, one line being effectively deshielded, one remaining unchanged (interaction with the zero spin state of deuterium), and one line being effectively shielded. The intensities of the three lines will be equal, owing to the equal population of the three spin states of the deuterium (line intensity will be discussed later). The separation between the adjacent lines is the same and is the coupling constant J_{HD} given in Hz. The deuterium resonance is split into two lines, one deshielded and one shielded relative to a noncoupled deuterium resonance. The resonance lines of the deuterium are again of equal intensity. The separation of the deuterium resonance lines is the same as the separation of the hydrogen resonance lines (J_{HD}). When spin-spin coupling is present, the chemical shift of the absorbing nucleus is taken as the center of gravity of the resonance lines resulting from that nucleus (Fig. 6.7).

Let us now consider the effect of spin-spin coupling between hydrogens of differing chemical shift in a more complex organic molecule. If we consider the appearance of the H_A resonance in the fragment **5** in Fig. 6.8, the hydrogen H_A will interact with the two spin states of the adjacent hydrogen, giving a two-line spectrum similar to the deuterium resonance lines in HD. Hydrogen H_A in fragment **6** in Fig. 6.8 is coupled to the two adjacent methylene hydrogens (assumed to have identical chemical shifts, and widely separated from H_A in

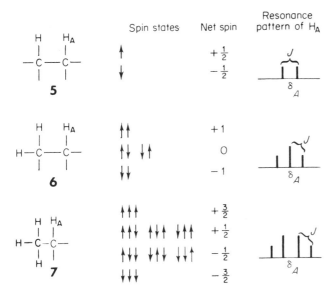

Fig. 6.8. Diagram of spin-spin interactions between adjacent hydrogens.

chemical shift). The net spin interaction with the two methylene hydrogens may be such that the methylene hydrogens give one spin state in which both hydrogens have $+\frac{1}{2}$ spins, two spin states of one hydrogen $+\frac{1}{2}$ with the other $-\frac{1}{2}$, and one spin state with both spins $-\frac{1}{2}$ (illustrated by means of the arrows in Fig. 6.8). All four net spin states are of equal probability and will thus produce a three-line pattern for H_A, in which the intensities of the lines will be $1:2:1$, the distance between two adjacent lines being the coupling constant J. Finally, if we consider the fragment **7** in Fig. 6.8, the three chemically equivalent methyl hydrogens produce the net spin states indicated to the right of **7**, in which one net spin state is $+1\frac{1}{2}$, three states have net spin of $+\frac{1}{2}$ and $-\frac{1}{2}$, and one net spin state is $-1\frac{1}{2}$. Again, since all net spin states are of equal probability, the absorption pattern for H_A will appear as four lines of relative intensity $1:3:3:1$ with a spacing of J.

In general, the multiplicity of a resonance pattern can be represented by Eq. (6.10), in which I is the spin quantum number of the coupled nucleus and n is the number of chemically equivalent coupled nuclei.

$$\text{no. of peaks} = 2In + 1 \tag{6.10}$$

The relative intensity of the peaks produced by spin-spin coupling between hydrogens that have widely differing chemical shifts approaches the coefficients of r in the expanded form of Eq. (6.11), in which n is the number of adjacent, chemically equivalent hydrogens.

$$(r + 1)^n \tag{6.11}$$

For example, the use of Eqs. (6.10) and (6.11) predicts a four-line pattern with relative intensities $1:3:3:1$ [coefficients of r in the expanded form of $(r+1)^3 = r^3 + 3r^2 + 3r + r^0$] for an interaction with a methyl group. The multiplicity of hydrogen absorption patterns can be referred to as doublets, triplets, quartets, etc. if the relative intensities of the peaks follow the coefficients of the expanded form of Eq. (6.11).

In most complex molecules, a particular hydrogen may be coupled to two or more sets of hydrogens of different chemical shift, as in the fragments **8** and **9**. In

$$-CH_n-\overset{|}{\underset{\underset{H_A}{|}}{C}}-CH_m \qquad\qquad -CH_n-\overset{\overset{\overset{|}{CH_p}}{|}}{\underset{\underset{H_A}{|}}{C}}-CH_m$$

8   **9**

8, the number of resonance lines for H_A will be $(2In+1)(2Im+1)$, and in **9** it will be $(2In+1)(2Im+1)(2Ip+1)$. In addition, the assumption that the hydrogens have greatly different chemical shifts (implying $\Delta\delta \gg J$), as employed in the previous paragraphs, is hardly ever applicable when dealing with typical organic molecules. As the difference in chemical shifts of the coupled nuclei decreases with respect to their coupling constants, the resonance patterns described in the previous paragraphs become highly distorted and a simple interpretation of the patterns is not always possible. The handling of such systems will be discussed later (Sec. 6.8.2).

In addition to the effect of the number of intervening bonds on the magnitude of J, the type of intervening bonds, the type of functions present in the molecule, and the geometrical relationship between the bonds also effect the magnitude of the coupling constant. Table 6.9 lists typical ranges for proton-proton coupling constants. (It will be noted that signs are associated with most of the constants given in Table 6.9. Although the mathematical signs are important in the detailed theory of spin-spin coupling, differences in sign do not affect the appearance of NMR spectra. The interested reader is referred to the more comprehensive NMR texts cited at the end of the chapter.)

The geminal coupling constant, $^2J_{HH}$, varies greatly: from approximately -17 to $+42$ Hz. Pople and Bothner-By[7] have summarized the effects of bonding features on the trends in $^2J_{HH}$. The rules are summarized as follows:

1. The geminal coupling constant becomes more positive as the hybridization on the C atom becomes more s-like.
2. Withdrawal of electron density from orbitals symmetric between hydrogen atoms (*sigma* orbital inductive effect) by groups directly bonded to the C

---

[7] J. A. Pople and A. Bothner-By, *J. Chem. Phys.*, **42**, 1339 (1965).

Table 6.9. Hydrogen-Hydrogen Coupling Constants

Structure	J_{HH} (Hz)
	$-9 - -20*$
	$0 - +9^\dagger$
	J_{12} $+6 - +8.5^\ddagger$ $\quad J_{25}$ $+5.5 - +8.3$ J_{14} $+2.9 - +5.2$ $\quad J_{23}$ $+8.8 - +10.9$ J_{24} $-3.3 - -6.3$
	$J_{12} \sim +5.5$ $\quad J_{23}$ $3-5$ J_{13} $2-3.5$
	$J_{12} \sim +2.0$
	J_{12} 6.9§ $\quad J_{23}$ 7.9 $\quad J_{45}$ 9.6 J_{14} 8.1 $\quad J_{25}$ 2.3 $\quad J_{43}$ 10.4
	J_{13}, J_{34} (*trans*-diaxial) $10-13$ J_{12}, J_{24} (*cis*-axial-equatorial) $4-7$ J_{25} (*trans*-diequatorial) $2-5$ J_{23}, J_{45} $11-13$

Table 6.9. (Continued)

Structure 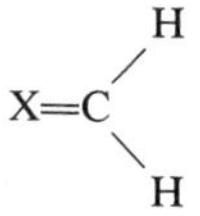	J_{HH} (Hz)

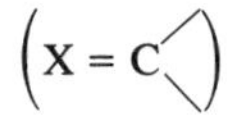

$$X=C \begin{matrix} H \\ H \end{matrix}$$

$$\left(X = C \triangleleft \right)$$ $-4.5 - +3.5^{\parallel}$

(X = N—R) ~ 16

(X = NOH) $\sim 8.5^{\P}$

(X = C=O, ketene) -15.8

(X = O) $+40 - +42^{\P}$

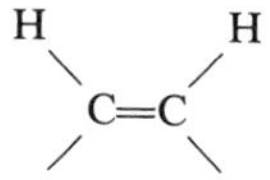

$$\begin{matrix} H \\ \diagdown \\ C=C \\ \diagup \end{matrix} \begin{matrix} H \\ \diagup \\ \diagdown \end{matrix}$$ $6\text{--}14^{\parallel}$

$$\begin{matrix} H \\ \diagdown \\ C=C \\ \diagup \end{matrix} \begin{matrix} \diagup \\ \diagdown \\ H \end{matrix}$$ $11\text{--}18^{\parallel}$

$$\triangle \begin{matrix} H \\ \\ H \end{matrix}$$ 1.4

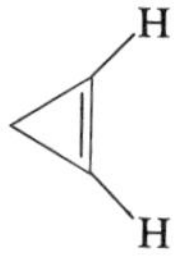

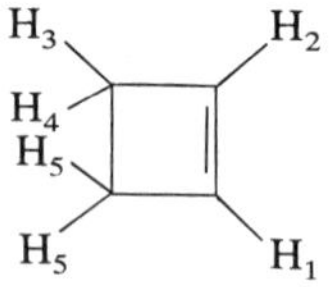

J_{23} +1.00 J_{34} −12.0 J_{12} 2.85
J_{35} +4.65 J_{36} +1.75

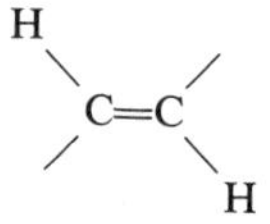

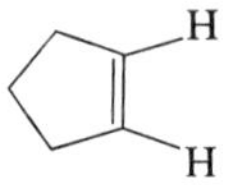

(H$_3$, H$_2$, H$_4$, H$_5$, H$_5$, H$_1$ cyclobutene structure)

$$\begin{matrix} H \\ \\ H \end{matrix}$$ (cyclopentene) ~ 5.5

$$(CH_2)_n \bigcirc \begin{matrix} H \\ \\ H \end{matrix} \quad (n = 4\text{--}6)$$ $10\text{--}11$

$$=C \begin{matrix} H \\ \\ C= \end{matrix} \begin{matrix} \\ H \end{matrix}$$ $10\text{--}11$

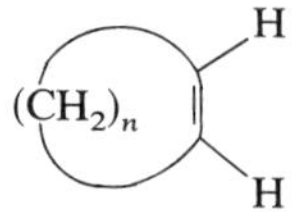

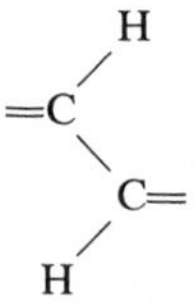

Table 6.9. (Continued)

Structure	J_{HH} (Hz)
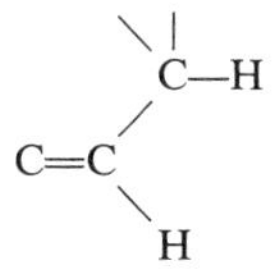	1–7†
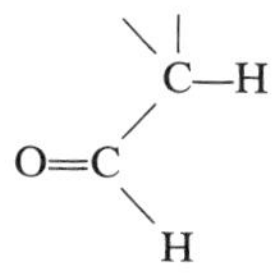	1–3
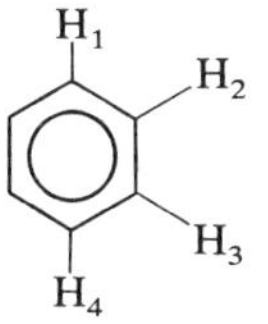	J_{12} 7–10 J_{13} 1–3 $J_{14} \sim 1$
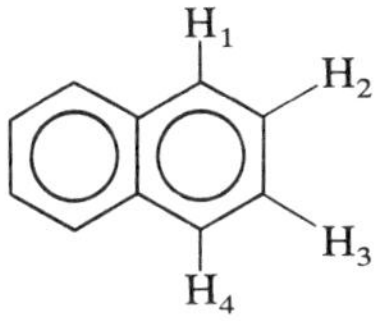	J_{12} 8.30 J_{14} 0.74 J_{13} 1.20 J_{23} 6.83
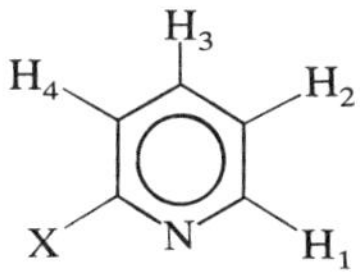	J_{12} 4.6–5.1 J_{23} 7.1–7.9 J_{13} 1.7–2.1 J_{24} 0.9–1.3 J_{14} 0.8–1.0 J_{34} 7.7–8.4
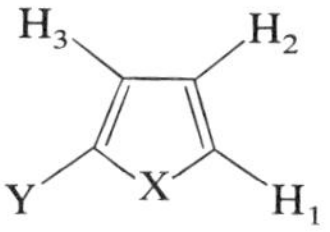	
(X = O) (X = NH) (X = S)	J_{12} 1–2 $J_{23} \sim 3.6$ J_{13} 1–2 $J_{12} \sim 2.6$ $J_{23} \sim 3.7$ J_{13} 1–2 $J_{12} \sim 5.5$ $J_{23} \sim 3.5$ J_{13} 1–2
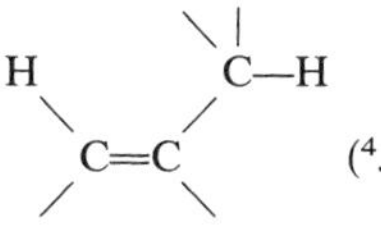	$-1.0 - -1.9†$

205

Table 6.9. (Continued)

Structure	J_{HH} (Hz)
H\C=C/ , \C—H (allylic, with H on C=C)	$-1.5 --2.5\dagger$
H\C=C=C—H	5–6
H—C≡C—C/ , H on C	2–4
H\C=C=C—C/ H $(^5J_{HH})^{**}$	2–3
H—C/ \C=C∿C/ H	0–3.0\dagger
H\C—C≡C—C/ H	2–3\dagger
H—C≡C—C≡C—C/ H $(^6J_{HH})$	1–2
H\C—(C≡C)$_2$—C/ H $(^7J_{HH})$	~1
H\C—(C≡C)$_3$—C/ H $(^9J_{HH})$	~0.4

* See Table 6.10 for individual values.
† J is a function of the dihedral angle and substituents (see discussion).
‡ K. B. Wiberg, D. E. Barth and P. H. Schertler, *J. Org. Chem.*, **38**, 378 (1973).
§ K. B. Wiberg and D. E. Barth, *J. Am. Chem. Soc.*, **91**, 5128 (1969).
‖ J is a function of the H—C—H bond angle and substituents (see discussion).
¶ Value of J depends on the solvent.
** For a review of long-range coupling constants see S. Sternhall, *Rev. Pure Appl. Chem.*, **14**, 15 (1964).

atom leads to a positive change in the coupling constant, and similar groups on the *beta* C atom lead to a negative change in the coupling constant.

3. Withdrawal of electron density from orbitals antisymmetric between hydrogen atoms (hyperconjugative effects) should lead to a negative change in the coupling constant.

4. Opposite changes occur in rules 2 and 3 when electron donation occurs.

Comparison of the values of the geminal coupling constant for alkanes, cyclobutanol, cyclopropane derivatives, and terminal alkenes illustrates the effect of changing the *s*-character of the orbital on carbon of the C—H bond on the magnitude of the coupling constant, the $^2J_{HH}$ becoming more positive as the *s*-character increases (in the order given).

Table 6.10 contains an extensive list of $^2J_{HH}$'s in substituted methanes (coupling between identical H's is not observable in NMR spectra; the determination of such coupling constants is discussed in Sec. 6.5.2). Comparison of the $^2J_{HH}$'s for X = Li, H, CH_3, NR_2, OCH_3 and F illustrates the change to more positive values with increasing electronegativity of the attached group, i.e., with increasing *sigma* orbital electron-withdrawing effect. (Comparison of the chemical shifts in the same series also illustrates the effect of electronegativity on chemical shifts, c.f., Table 6.10).

The effect of rule 3 is illustrated by the $^2J_{HH}$'s of acetone (-14.9 Hz) and acetonitrile (-16.9), which are indicative of extensive hyperconjugative electron withdrawal. In many cases more than a single effect is operative, for example, the $^2J_{HH}$ in formaldehyde ranges between $+40$ and $+42.2$ Hz (depending on the solvent).

Table 6.10. $^2J_{HH}$'s and Chemical Shifts in Substituted Methanes, CH_3—X

X	$^2J_{HH}$	δ
Li	−12.89	−1.74
$Si(CH_3)_3$	−14.05	0.0
$Ge(CH_3)_3$	−12.96	0.13
$Sn(CH_3)_3$	−12.37	0.06
$Pb(CH_3)_3$	−10.94	0.71
H	−12.4	0.23
CH_3	−12.56	0.90
I	−9.36	2.16
SCH_3	−11.87	2.08
Br	−10.1	2.68
$N(CH_3)_2$	−11.73	2.12
Cl	−10.7	3.05
OCH_3	−10.60	3.24
F	−9.5	4.26
$\overset{+}{N}H(CH_3)_2$	−11.72	3.05

The magnitude of $^3J_{HH}$ varies as a function of the dihedral angle between the C—H bonds on adjacent carbon atoms.

For ethane the relationship of $^3J_{HH}$ with ϕ is given by Eq. (6.12),

$$^3J_{HH} = 4.22 - 0.5 \cos \phi + 4.2 \cos^2 \phi \qquad (6.12)$$

a plot of which is illustrated in Fig. 6.9. Experimental values of $^3J_{HH}$ as a function of ϕ have been measured in conformationally rigid systems in which the bond angles are reasonably well known, as in conformationally biased cyclohexane derivatives (see entries in Table 6.9). These values are in reasonable agreement with the theoretically calculated values.

The magnitudes of $^3J_{HH}$'s are also affected by the substituents attached to the spin system. This effect has been evaluated by several groups,[8] and is apparent in the data given in Table 6.11.

The magnitude of $^3J_{HH}$ between vinyl hydrogens depends on their stereochemical relationship, electronegativity of functions attached to the C=C, and bond

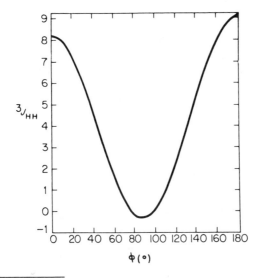

Fig. 6.9. Variation of $^3J_{HH}$ with dihedral angle, ϕ.

[8] M. Karplus, *J. Am. Chem. Soc.*, **85,** 2870 (1963); D. H. Williams and N. S. Bhacca, *J. Am. Chem. Soc.*, **86,** 2742 (1964); H. Booth, *Tetrahedron Lett.*, 411 (1965); H. R. Buys, *Recl. Trav. Chim. Pays-Bas*, **88,** 1003 (1969).

Table 6.11.   Hydrogen Coupling Constants (Hz) as a Function of the Substituent X in 10

X	J_{AX}	J_{BX}	J_{AB}
CN	4.6	9.3	−12.6
COOH	4.4	8.5	−12.6
C_6H_5	4.2	8.9	−12.7
Cl	3.2	8.0	−13.2
OH	2.4	7.4	−12.6
$OCOCH_3$	2.5	7.6	−13.3

$$
\begin{array}{cc}
H_A & H_X \\
| & | \\
-C- & -C- \\
| & | \\
H_B & X
\end{array}
$$

10

angle strain in the C=C. As might be suggested by Eq. (6.12), it is observed that *trans* $^3J_{HH}$ is greater than *cis* $^3J_{HH}$ (see entries in Table 6.9). The effect of the electronegativities of groups attached to the C=C on $^3J_{HH}$, as well as on $^2J_{HH}$, is given by Eqs. (6.13) to (6.15), in which E_X and E_Y are the Pauling electronegativities of the functions X and Y, or the attachment atoms of the functions (such as C of CH_3) in CHX=CHY (either X or Y can be H, with $E_H = 2.1$).[9]

$$^3J_{HH}\,(cis) = \frac{91.2}{E_X + E_Y} - 9.7 \qquad (6.13)$$

$$^3J_{HH}\,(trans) = \frac{75.0}{E_X + E_Y} \qquad (6.14)$$

$$^2J_{HH}\,(geminal) = \frac{61.6}{E_X + E_Y} - 12.9 \qquad (6.15)$$

The effect of bond angle strain in the C=C on $^3J_{HH}$ is reflected in the values of these coupling constants for cyclopropene, cyclobutene, and the larger ring cycloalkenes given in Table 6.9. As the ring strain decreases, the coupling constant increases. Comparisons between other cyclic and acyclic alkenes are given in Table 6.12. Note that in the acyclic alkenes, as the strain increases the coupling constant increases, the *cis* coupling constant increasing more rapidly than the *trans* coupling constant.

Long-range coupling, a term used to indicate coupling that is transmitted through more than three bonds, is generally more readily observed in unsaturated systems such as aromatic compounds, alkenes, and alkynes. Typical examples of long-range coupling are given in Table 6.9. Long-range coupling through saturated bonds is less frequently observed, and is highly dependent on the conformation of the system. When the bonded system exists in the **W**, or **M**, conformation,

---

[9] T. Schaefer and H. M. Hutton, *Can. J. Chem.*, **45**, 3153 (1967).

i.e.,

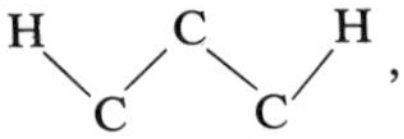

appreciable long-range coupling can be observed, as is illustrated in the structures **11** and **12**.

$^4J = 1$ Hz $^4J = 7$ Hz

11   **12**

Table 6.12. Comparison of $^3J_{HH}$'s in Strained and Unstrained Systems*

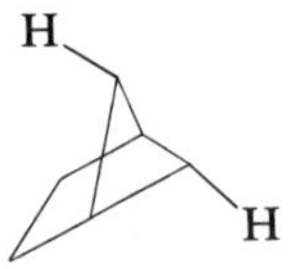

Strained Compound	$^3J_{HH}$	Less Strained Compound	$^3J_{HH}$
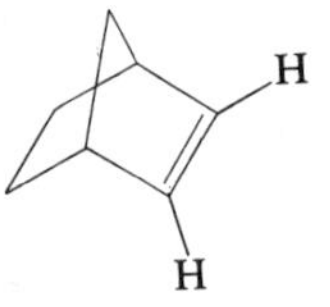	5.68	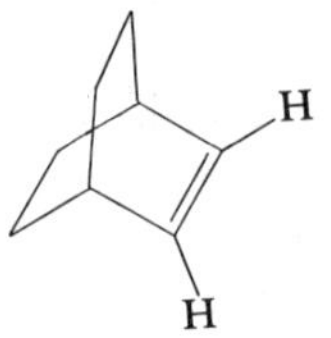	8.24
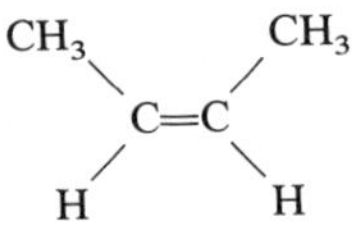 CH_3, CH_3, $C=C$, H, H	10.88 (11.0)†	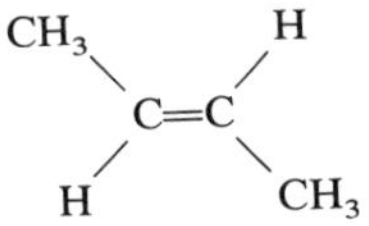 CH_3, H, $C=C$, H, CH_3	15.09 (17.0)
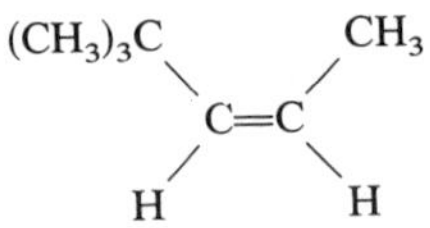 $(CH_3)_3C$, CH_3, $C=C$, H, H	12.02	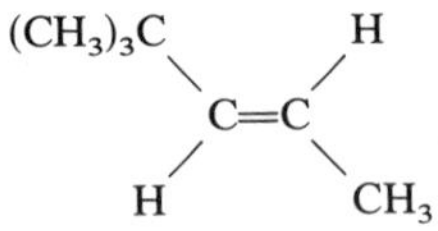 $(CH_3)_3C$, H, $C=C$, H, CH_3	15.50
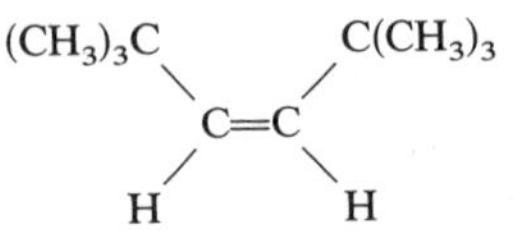 $(CH_3)_3C$, $C(CH_3)_3$, $C=C$, H, H	14.2	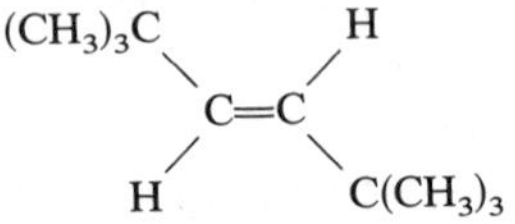 $(CH_3)_3C$, H, $C=C$, H, $C(CH_3)_3$	16.10

* Taken from M. A. Cooper and S. L. Manatt, *Org. Magnetic Res.*, **2**, 511 (1970).
† Parenthesized values are the calculated values, using 2.2 for E_C of the attached methyl and *t*-butyl groups.

6.5.1 Spin-Spin Coupling
Between Other Nuclear Combinations

As indicated in the opening paragraphs of this chapter, spin-spin coupling can exist between nuclei having I not equal to zero. It should now be obvious that the organic chemist is extremely fortunate that the predominant isotopes of carbon and oxygen, ^{12}C and ^{16}O, possess I values of zero and do not spin-spin couple with other nuclei. If this were not true, the complexity of hydrogen magnetic resonance spectra would be formidable. However, there are many other nuclei encountered in organic molecules that possess I values not equal to zero. Table 6.13 lists typical ranges of coupling constants for a variety of different nuclear combinations. These coupling constants also vary with bond angle, dihedral angle, the number and type of intervening bonds, and the electronegativity of the attached atoms or functions in the same manner as described for hydrogen-hydrogen coupling. Spin-spin coupling interaction of ^{13}C will be discussed in more detail in a later section.

Table 6.13. Coupling Constants for Various Nuclei Combinations

Coupling Constant	System	Coupling Constant Range (Hz)
J_{HD}	General	$0.154 J_{HH}$
$J_{^{10}BH}$	$^{10}B—H$	$\sim$30–50
$J_{^{11}BH}$	$^{11}B—H$	75–182
	$^{11}B———H———^{11}B$(bridged)	30–50
	$^{11}B—C—H$	$\sim$0
$J_{^{13}CH}$*	$^{13}CH_3X$	110–155
	$=\!^{13}C—H$	150–170
	$\equiv\!^{13}C—H$	$\sim$250
	$^{13}C—C—H$	4–6
	$^{13}C—C—C—H$	3–4
$J_{^{14}NH}$	NH_4^+	52.6
$J_{^{15}NH}$	NH_4^+	73.7

Table 6.13. (Continued)

Coupling Constant	System	Coupling Constant Range (Hz)
	(ortho) (meta) (para)	8–10 5–8 ~2
$J_{^{13}C^{19}F}$	$^{13}C-^{19}F$ $^{13}C-C-^{19}F$	250–300 30–40
$J_{^{19}F^{19}F}$		155–225
	CF—CF (trans) (gauche)	≈16 ≈18
		28–87
	C=C (cis)	20–58
	C=C (trans)	95–120
	(ortho) (meta) (para)	~20 2–4 11–15
$J_{^{29}SiH}$	$^{29}Si-H$	190–380
$J_{^{31}PH}$	$^{31}P-H$ $^{31}P-C-H$ $^{31}P-C-C-H$	179–700 ~4 ~12–14
$J_{^{199}HgH}$	$^{199}Hg-C-H$ $^{199}Hg-C-C-H$	80–235 115–200

* Trends in the magnitude of ^{13}C coupling constants with structure are discussed in Sec. 6.14.

The natural occurrence of more than one isotope of a particular atom, of which one, or both, can spin couple with other nuclei, leads to the superpositioning of two coupling patterns in the sample spectrum. The spectrum of chloroform, Fig. 6.10, in which ^{12}C is accompanied by a small natural abundance of ^{13}C of spin $\frac{1}{2}$ (Table 6.1), displays an intense singlet for the uncoupled absorption of the hydrogen in $^{12}CHCl_3$, and two weak outer bands (*satellites*) for the coupled absorption of the hydrogen in $^{13}CHCl_3$. The relative intensity of these ^{13}CH coupled peaks to that of the peak for $^{12}CHCl_3$ is directly equal to the $^{13}C/^{12}C$ ratio.

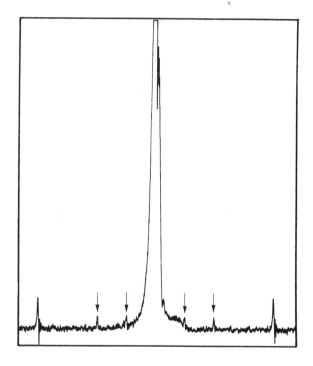

Fig. 6.10. Hydrogen resonance spectrum of chloroform displaying an intense, un-coupled central peak corresponding to $^{12}CHCl_3$, with the coupled peaks of $^{13}CHCl_3$ appearing as the weak outer peaks. The four very weak peaks indicated by the arrows are artifacts called spinning side bands. These often appear with very intense bands when the magnetic field is not perfectly adjusted. Their position depends on the rate at which the sample tube is spinning and is readily detected by changing the rate of the spinning of the sample tube.

6.5.2 Spin-Spin Coupling Between Chemically Identical Nuclei

In the earlier portions of this section we discussed spin-spin coupling between hydrogen atoms having different chemical shifts. Spin-spin coupling also occurs between nuclei having identical chemical shifts, although the splitting of reso-nance peaks due to this coupling is not observable and indirect methods are required to observe such coupling. Isotopic substitution, followed by determina-tion of the coupling constant between the isotopic atoms, allows one to calculate the coupling constant between nuclei of identical chemical shift by multiplying the

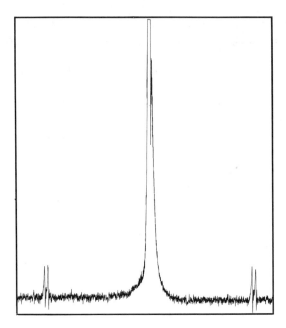

Fig. 6.11. NMR spectrum of 1,1,2,2-tetrachloroethane showing ^{13}C satellite bands with $J_{^{13}C-H} = 177.2$ Hz and $J_{HH} = 3.0$ Hz.

coupling constant observed between the isotopic atoms by the ratio of their gyromagnetic ratios. For example, the $^2J_{HH}$'s given in Table 6.10 for the substituted methanes could be determined by measuring the $^2J_{HD}$'s in the monodeuterio derivatives and multiplying that constant by 6.49 (the ratio of the gyromagnetic ratios of hydrogen and deuterium).

Another technique for determining the coupling constant between hydrogens of identical chemical shift on different carbon atoms involves analysis of the ^{13}C satellite patterns. The hydrogen on ^{13}C in the fragment $H-^{12}C-^{13}C-H$ is only very slightly different in chemical shift from the hydrogen on ^{12}C, but the peak(s) due to the hydrogen on ^{13}C are well displaced from the peak(s) arising from the hydrogen on ^{12}C due to the large ^{13}CH coupling constant. The hydrogen resonance is split by coupling with the ^{13}C and the hydrogen on the adjacent ^{12}C. This coupling constant is then taken as the coupling constant between the hydrogens of identical shift in the $H-^{12}C-^{12}C-H$ system. Figure 6.11 shows the splitting of the ^{13}C satellite bands in the spectrum of 1,1,2,2-tetrachloroethane.

6.5.3 Spin Decoupling

In spectra of relatively complex molecules it may be quite difficult to determine which nuclei are spin-coupled leading to the observed patterns. This is particularly true if several of the coupling constants are of similar size, if one or more nuclei are buried in a complex absorption multiplet, or if long-range coupling is present. In such cases it would be desirable to destroy the spin-spin coupling interaction between various nuclei in the system. Spin-spin splitting is observed when the nucleus that is responsible for the splitting remains in a given spin state for a period of time, which is long in comparison with the reciprocal of the difference in chemical shifts, in Hz, of the coupled nuclei. If we can cause the lifetime of a spin state to decrease sufficiently so that the absorbing nucleus "sees" only a time average of the various spin states possible, a single resonance line will result. Experimentally, it has been found possible to do this by irradiating the nucleus responsible for the spin-spin interaction at its resonance frequency. This is referred to as *double resonance*. When the nucleus being decoupled is the same as that being observed, the process is termed *homonuclear double resonance*; when the nucleus being decoupled is not the same as that being observed, it is termed *heteronuclear double resonance*. An example of homonuclear double resonance is shown in Fig. 6.12, Fig. 6.12a being the NMR spectrum of ethyl iodide showing a typical quartet and triplet for the CH_3CH_2— system, and Fig. 6.12b showing the resulting spectrum when the methylene hydrogens are irradiated at their resonance frequency destroying their coupling to the methyl hydrogens. An example of heteronuclear decoupling will be shown later in which the coupling of ^1H to ^{13}C is removed. Irradiation of two different sets of nuclei can also be accomplished and is referred to as *triple resonance*.

6.6
EFFECTS OF THE EXCHANGE
OF THE CHEMICAL ENVIRONMENT OF NUCLEI

Our discussions of chemical shift and spin-spin coupling phenomena have thus far been quite general. We have not stopped to consider in detail both the dynamic processes that may be occurring within or between molecules and the effect of these dynamic processes on the appearance of the nuclear magnetic resonance spectrum. Three dynamic processes are of great interest to the spectroscopist. These are (1) rotations about bond axes, (2) inversions at an atom (for example, nitrogen in amines), (3) intramolecular and intermolecular exchange of nuclei between functions (for example, hydroxyl hydrogen exchange). Each of these dynamic processes results in changes in the chemical environment of a given nucleus. Exchanges of chemical environments involving bond rotations are illustrated in structures **13** and **14**. In **13**, rotation about the central carbon-carbon

a   b   c

13

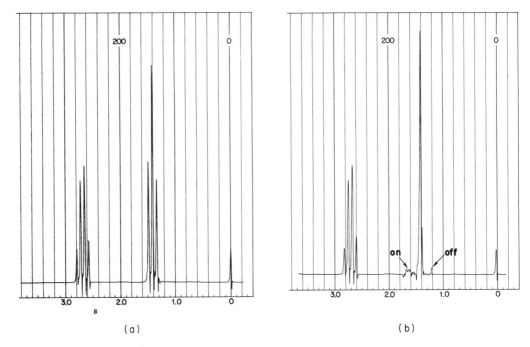

Fig. 6.12. Normal (a) and double resonance (b) spectra of ethyl iodide. The arrows signifying on and off in (b) indicate the positions where the double resonance frequency of the methylene group was turned on and off.

14

bond produces three rotational conformations in which the chemical environment of H_A, and also of H_B, is different in each conformation. Because the dihedral angles H_X—C—C—H_A and H_X—C—C—H_B also change in the three conformations, the coupling constant between H_A and H_X, and H_B and H_X, is also different in **13a**, **b**, and **c**.

The appearance of the nuclear magnetic resonance spectrum of a system such as **13** is a function of the rate of interconversion of the various conformations of the molecule, the difference in chemical shifts of a given nucleus in the available conformations (for example, H_A in **13a**, **b**, and **c**), and the population of the various conformations. If the rate, or frequency, of interconversion of conformations is much greater than the difference in chemical shifts in Hz of H_A in conformers **13a**, **b**, or **c**, a sharp, time-averaged resonance will result. If the frequency of interconversion approximates the difference in chemical shifts, a broad resonance line will result, and if the frequency of interconversion is much slower than the difference in chemical shifts, sharp, individual resonances will be obtained for the nuclei in each of the available conformations.[10] This phenomenon is illustrated in Fig. 6.13 for a nucleus equilibrating between two different sites in a molecule. P is the exchange frequency given in exchanges per second when $(\delta_A - \delta_{A'})$ is expressed in Hz.

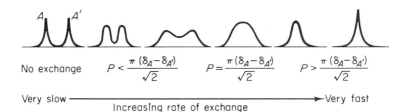

No exchange $P < \dfrac{\pi\,(\delta_A - \delta_{A'})}{\sqrt{2}}$ $P = \dfrac{\pi\,(\delta_A - \delta_{A'})}{\sqrt{2}}$ $P > \dfrac{\pi\,(\delta_A - \delta_{A'})}{\sqrt{2}}$

Very slow ──────────────────────────────────► Very fast
Increasing rate of exchange

Fig. 6.13. Line shapes of resonance peaks of an equilibrating system. (P is the exchange rate in sec^{-1}.)

The chemical shift of a nucleus in a rapidly interconverting system is given by Eq. (6.16),

$$\delta_{A_{av}} = \sum_i N_i \delta_{A_i} \tag{6.16}$$

in which N_i is the mole fraction of the ith conformer and δ_{i_A} is the chemical shift of H_A in the ith conformation. Similarly, the observed coupling constant is given by Eq. (6.17),

$$J_{AX_{av}} = \sum_i N_i J_{AX_i} \tag{6.17}$$

in which J_{AX_i} is the coupling constant between H_A and H_X in the ith conformation.

It is instructive to consider in greater detail the chemical shifts of H_A and H_B in structure **13**. Applying Eq. (6.16) to H_A and H_B gives

$$\delta_A = N_a \delta_{A_a} + N_b \delta_{A_b} + N_c \delta_{A_c}$$

and

$$\delta_B = N_a \delta_{B_a} + N_b \delta_{B_b} + N_c \delta_{B_c}$$

H_A and H_B are diastereotopic in all three of the rotamers **13a**, **b**, and **c**. (A *diastereotopic* relationship exists between two atoms or groups when replacement of each atom or group separately produces a pair of diastereomers.) In structure **13**, the diastereomeric environments are provided by the asymmetric carbon —CXYH$_X$. H_A and H_B are *not* chemically equivalent despite the fact that both hydrogens are bonded to the same carbon! The chemical shifts of H_A and H_B are

---

[10] The NMR spectrometer can be compared to a camera equipped with a relatively slow shutter. A clear picture will be obtained of a nearly stationary object, whereas a blurred picture will be obtained of a rapidly moving object.

consequently different in each of the three rotamers although the difference may be quite small, i.e., $\delta_A \neq \delta_B$ under any conditions of rapid rotation and rotamer populations. Similarly, $J_{AX} \neq J_{BX}$. Occasionally it is useful to modify a function to increase the difference between the chemical shifts of diastereotopic hydrogens. For example, the difference in the chemical shifts of the diastereotopic hydrogens of the 1,2-diphenylethyl system is greatly enhanced on converting the alcohol to the benzoate (Fig. 6.14), in which the diastereotopic hydrogens appear at δ 3.17 and δ 3.35. When R = H in **13**, the back carbon becomes a methyl carbon. The mole fractions of the three possible conformations will be equal, and all three methyl hydrogens will have exactly the same chemical shift and the same coupling constant with H_X.

Just as a pair of diastereotopic hydrogens or groups in the same molecule may have different chemical shifts, corresponding hydrogens or groups in two diastereomers, in principle, will have different chemical shifts. When the chemical shifts of hydrogens in diastereomers are of sufficient difference, nuclear magnetic resonance provides a very accurate method for the analysis of mixtures of diastereomers. For example, the NMR spectra in Fig. 6.15 are those of *dl*- and *meso*-2,3-dibromobutane, which are considerably different from each other.

The nuclear magnetic resonance spectra of equilibrating systems can provide a great deal of kinetic and thermodynamic information about the system. The equilibrium constant for a given system can be calculated by using Eq. (6.18), providing the chemical shifts of nucleus A in the two equilibrating species (δ_A and δ_A) can be independently determined.

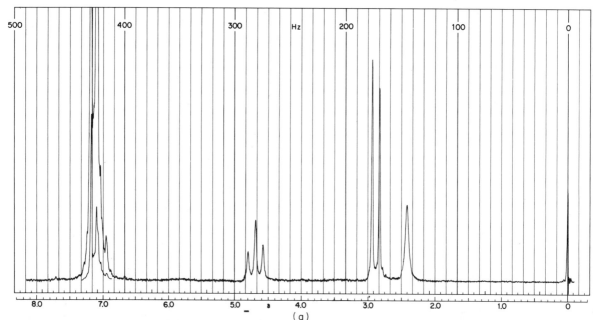

Fig. 6.14. NMR spectra of 1,2-diphenylethanol (*a*) and the corresponding benzoate ester (*b*).

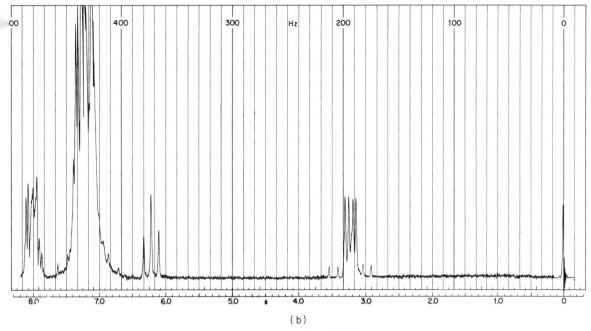

(b)

Fig. 6.14 (b)

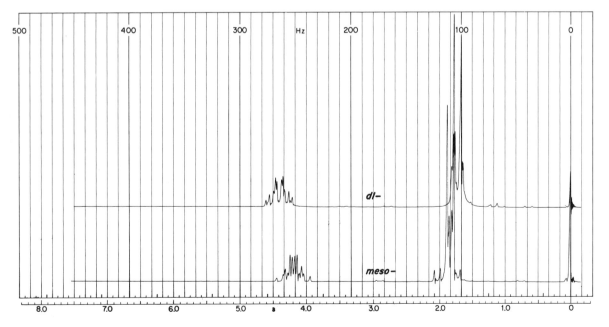

Fig. 6.15. NMR spectra of *dl-* and *meso*-2,3-dibromobutane.

$$K_{eq} = \frac{\delta_A - \delta_{A\,av}}{\delta_{A_{av}} - \delta_{A'}} \qquad (6.18)$$

The equilibrium constant is readily converted into ΔG for the reaction, which in turn, when determined as a function of temperature, allows calculation of ΔH and ΔS.

Determination of the values of δ_A and $\delta_{A'}$ can be carried out in two ways: the temperature of the sample can be lowered until the frequency of the interconversion of A to A' is less than $\Delta\delta_{AA'}$ (see the following paragraph), or an additional functional group can be incorporated in the molecule that restricts the nucleus as A or A'. The latter method demands that the new functional group does not contribute to the shielding or deshielding of the nucleus of interest. It has been demonstrated that the latter condition is very difficult to achieve and that δ_A and $\delta_{A'}$ should be determined by low-temperature nuclear magnetic resonance measurements.[11]

Kinetic information is obtained by analysis of the line shapes of the spectrum when $P \approx \pi(\delta_A - \delta_{A'})/\sqrt{2}$; however, this requires being able to record the resonance spectrum under these conditions. Associated with each transformation within a molecule is an activation energy required to affect the transformation. For most transformations, the thermal energy at room temperature is sufficient to overcome the activation energy, and one generally observes only time-averaged spectra. Lowering the temperature of the sample will reduce the available thermal energy, and, if the activation energy lies in the accessible thermal region, the rate of the exchange may be reduced until the peaks broaden and eventually separate into distinct resonance lines (transversing right to left in Fig. 6.13 as a result of lowering the temperature). The temperature at which the individual resonance lines merge into a broad resonance line is referred to as the *coalescence temperature*.

An excellent example of the effect of lowering the temperature of a sample is illustrated with compound **15**, which can exist either in conformation **15a** or **15b**. The resonance spectra of **15** recorded at various temperatures are shown in Fig. 6.16. At room temperature, only broad, time-averaged peaks are observed, whereas at approximately $-16°C$ individual peaks for both **15a** and **15b** are discernible.

15a   15b

[11] S. Wolfe and J. R. Campbell, *Chem. Comm.*, 872 (1967); E. L. Eliel and R. J. L. Martin, *J. Am. Chem. Soc.*, **90**, 682 (1968); F. R. Jensen and B. H. Beck, *J. Am. Chem. Soc.*, **90**, 3251 (1968).

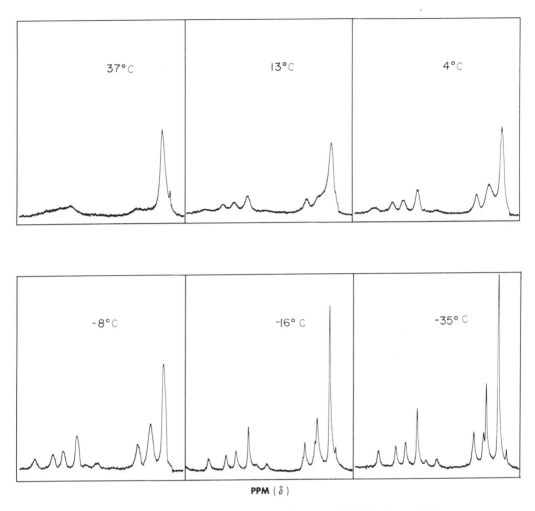

Fig. 6.16. Hydrogen resonance spectrum of compound **15** showing only the resonance peaks of the indicated aliphatic hydrogens in the δ 3.0 to 6.5 region as a function of the temperature of the sample. At −35°C the predominant isomer present is **15b**.

The complete analysis of NMR spectra at various temperatures to derive kinetic and thermodynamic information is very complex and requires computer techniques.

The exchange of nuclei between functions, either inter- or intramolecular, is illustrated in Fig. 6.17. The top spectrum is that of pure ethanol, in which the coupling between the hydroxyl proton and the methylene protons is apparent; the bottom spectrum is that of acidified ethanol, in which the exchange process occurs at a rate faster than that required for the methylene protons to "see" the hydroxyl proton.

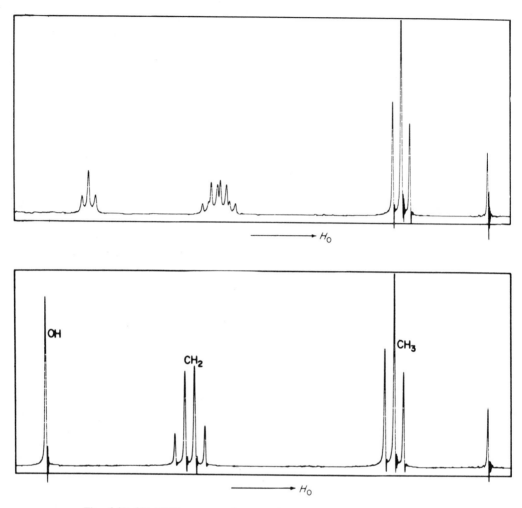

Fig. 6.17. ^1H NMR spectra of pure ethanol (top) and acidified ethanol (bottom) illustrating the loss of coupling under the rapid exchange conditions.

6.7
DESIGNATION OF SPIN SYSTEMS

In a spectrum containing the absorption peaks of a variety of differing nuclei, a system of designating the types of spin systems present is desirable. By a spin system, we mean a set of nuclei that are spin-spin coupled giving rise to recognizable resonance patterns. A single molecule may contain several such distinguishable spin systems.

The designation of the various nuclei in a spin system is based on the relative chemical shifts and the size of the chemical-shift difference with respect to the coupling constant J. The symbols used to designate individual nuclei and spin systems are included in Table 6.14. Examples of the use of these symbols appear in the following paragraphs.

Most of the entries in Table 6.14 are self-explanatory and will be illustrated by the use of actual resonance spectra. In general, spectra are presented in the literature with field strength increasing from left to right. The symbols designating the spin systems are employed such that the first letters of the alphabet (A, B, C, etc.) are used to represent nuclei appearing at lowest fields, while the latter letters (M, X, Y, etc.) represent nuclei appearing at progressively higher fields. For example, in an ABX system, the A nucleus would appear at the lowest field strength, the X nucleus appearing at the highest field strength.

The use of primed symbols, for example, in AA', also requires further explanation. Up to now we have considered only the chemical identity, or nonidentity, of nuclei in the resonance region. The term *magnetic identity* implies an equal spin-spin interaction, that is, an equal coupling constant between each nucleus of a given set of coupled nuclei and the absorbing nucleus. This condition is not always met. In many instances, two chemically identical nuclei interact with different coupling constants with another nucleus and, thus, are not magnetically identical with respect to that nucleus. The prime is used to indicate this magnetic nonidentity. For example, in the cyclopropane derivative **16**, we have three chemically different types of hydrogens, H_A, H_B, and H_X. The hydrogens on C_2 and C_3 *cis* to the X function are chemically identical and are designated as H_A's; similarly, the two hydrogens on C_2 and C_3 *trans* to the X function are chemically

Table 6.14.   Designation of Individual Nuclei and Spin Systems

Symbol	Description*
A	H_A with δ_A.
A_n	nH_A of same δ_A and magnetically identical.
AA'	Hydrogens H_A and $H_{A'}$ with same chemical shift, but magnetically different with respect to other nuclei.
AB	Hydrogens H_A and H_B with δ_A and δ_B, where $J_{AB} \geq (\delta_A - \delta_B)$.
AX	Hydrogens H_A and H_X with δ_A and δ_X, where $J_{AX} < (\delta_A - \delta_X)$.
AMX	Hydrogens H_A, H_M, and H_X with δ_A, δ_M, and δ_X, where $J_{AM} < (\delta_A - \delta_M)$ and $J_{XM} < (\delta_M - \delta_X)$.
ABX	Hydrogens H_A, H_B, and H_X with δ_A, δ_B, and δ_X, where $J_{AB} \geq (\delta_A - \delta_B)$, $J_{AX} < (\delta_A - \delta_X)$, and $J_{BX} < (\delta_B - \delta_X)$.
ABC	Hydrogens H_A, H_B, and H_C with δ_A, δ_B, and δ_C, where all $\Delta\delta$'s $< J$'s.

* Although all entries in this table utilize the hydrogen atom, these systems can be used with any other nuclear system or mixed nuclear system.

identical (H_B's), but are not chemically identical with the H_A's. Although the H_A's are chemically identical, they are not magnetically identical. H_A on C_2 is coupled to H_B on C_2 with a geminal coupling constant (^2J), but is coupled to the H_B on C_3 with a *trans* vicinal coupling constant (^3J). On the other hand, the H_A on C_3 is coupled to the H_B on C_2 with a ^3J and to the H_B on C_3 with a ^2J. Thus, although both H_A's are chemically identical, they are not equally coupled to the two H_B's on C_2 and C_3; they are therefore magnetically nonequivalent and are designated as H_A and $H_{A'}$. A similar analysis shows that the two H_B's are not magnetically identical with respect to their interactions with the two H_A's. It should be pointed out, however, that H_A and $H_{A'}$, as well as H_B and $H_{B'}$, are magnetically identical in their spin coupling to H_X, i.e., $J_{AX} = J_{A'X}$ and $J_{BX} = J_{B'X}$.

A similar situation exists in 1,3-butadiene (**17**). It is obvious that hydrogens H_A and $H_{A'}$ are chemically identical, as are H_B and $H_{B'}$, and H_C and $H_{C'}$; however, the spin-spin interactions between the C' and A', and C' and B' hydrogens are different from those between the C' and A, and C' and B hydrogens. Again magnetic nonidentity is indicated. Magnetic nonidentity may be obvious in some spectra, whereas in others, the complexity of the absorption patterns may not allow a straightforward analysis.

16   **17**

Certain ambiguities in the designation of spin systems may still arise when applying the nomenclature rules outlined in the foregoing paragraphs. For example, if we were to designate a system as $AA'BB'CC'$ (butadiene), we would not be clearly indicating the sequence of nuclei with respect to the carbon-atom framework of the molecule. To avoid such confusion, the symbols designating the types of nuclei bonded to one carbon atom can be separated by a hyphen from the symbols designating the types of nuclei on the adjacent carbon atom. Thus the preferred designation for butadiene would be $BC—A—A'—B'C'$.

6.8
EXAMPLES OF SIMPLE SPIN SYSTEMS

6.8.1 The Two-Spin System

We shall begin our discussion of simple spin systems by considering the two-spin system, the one-spin system being trivial and giving rise to a single resonance line. The simplest two-spin system is the A_2 system, which also gives

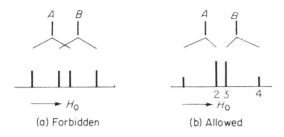

Fig. 6.18. Diagrammatic representation of the spin-spin interaction in the AB system.

rise to a single resonance line. If we allow the chemical shifts of the two nuclei to become slightly different, and then introduce coupling between the nuclei such that $J_{AB} > \Delta\delta_{AB}$, a number of lines appear in the resonance spectrum. If we were to split the resonance lines of nuclei A and B in a symmetrical fashion about their respective chemical shifts, the lower field line of the B nucleus doublet would appear at lower field than the higher field line of the A nucleus (crossing of energy states). However, the energy states of one nucleus produced by spin-spin interaction with a neighboring nucleus can not overlap (cross) the energy states of that nucleus. To avoid the crossing of states in the AB system, the inner lines of the A and B doublets become more intense than the outer lines so that the centers of gravity of the A and B doublets are at their respective chemical shifts. The spin-spin coupling interaction is diagrammed in Fig. 6.18. As the chemical shift difference ($\Delta\delta$) between the nuclei A and B increases relative to J_{AB}, the outer lines increase in intensity at the expense of the inner lines. Finally, as the system graduates to the AX system, the lines become more equal in intensity.

The relative intensities of the outer-to-inner lines in an AB or AX system and the difference in chemical shifts ($\Delta\delta$) are calculated using Eqs. (6.19) and (6.20), respectively.

$$\text{Relative intensity}\left(\frac{\text{line 1}}{\text{line 2}}\right) = \frac{Q - J}{Q + J} \qquad (6.19)$$

$$\Delta\delta = (Q^2 - J^2)^{1/2} \qquad (6.20)$$

The coupling constant J is the distance between lines 1 and 2, or 3 and 4, and the quantity Q is the distance between lines 1 and 3, or 2 and 4 expressed in Hz.

Examples of spectra containing the AB and AX systems are illustrated in Fig. 6.19. The hydrogen resonance spectrum of cis-β-phenylmercaptostyrene, Fig. 6.19a, shows an AB pattern centered at δ 6.48 for the two ethylenic hydrogens. The coupling constant J_{AB} is 10.7 Hz, and $\Delta\delta_{AB}$ is 0.095 ppm. The observed intensity ratio is 0.056; compared to 0.061 as calculated from Eq. (6.19). The complex resonance pattern at lower field is due to the aromatic hydrogens.

Figure 6.19b shows the hydrogen resonance spectrum of cis-β-ethoxystyrene, in which the ethylenic hydrogens appear as an AX system, with δ_A 5.18 and δ_X 6.06 and $J = 7.2$ Hz. The observed intensity ratio is 0.77, with a calculated

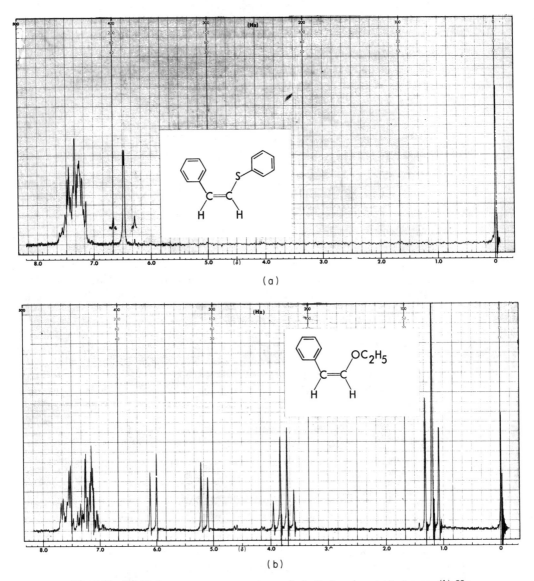

Fig. 6.19. (*a*) Hydrogen resonance spectrum of *cis-β*-phenylmercaptostyrene. (*b*) Hydrogen resonance spectrum of *cis-β*-ethoxystyrene.

value of 0.77. The triplet at δ 1.21 and the quartet at δ 3.77 represent the A_2X_3 pattern of the CH_3CH_2— group, while the aromatic hydrogens give rise to the complex multiplet centered at δ 7.3.

6.8.2 Three-Spin Systems

The simplest three-spin system is the A_3 system, for example, an isolated methyl group, which gives rise to a single resonance line. If one of the nuclei becomes slightly different in chemical shift relative to the other two nuclei such that $\Delta\delta$ is less than the coupling constant, we have an A_2B or AB_2 system. The spectra of such systems do not possess any symmetry and cannot be readily analyzed as described for the AB and AX systems. A similar situation is encountered with the ABC system, in which all three nuclei are slightly different in chemical shift. In spectra where there appears to be a three-spin system of the A_2B or ABC type, the student is referred to Wiberg and Nist's *The Interpretation of NMR Spectra* (ref. 14) in order to determine the chemical shifts and coupling constants. Wiberg and Nist have calculated the absorption patterns for various spin systems, maintaining a constant chemical shift difference (6 Hz) while varying the coupling constant. The student selects the calculated spectrum that most closely resembles the portion of the spectrum of his sample in peak shape, relative positions, and intensities. The difference in chemical shifts and coupling constants can then be scaled up to correspond to the observed pattern. Numerous computer programs are available for the calculation of spectra and for comparison with observed spectra.

The A_2X or AX_2 system, in which one nucleus is of a greatly different chemical shift, can be represented in a straightforward fashion as indicated in Fig. 6.20. The resonance of the A hydrogens is split into a 1:1 doublet by the single X hydrogen, while the X resonance is split into a 1:2:1 triplet by the two A hydrogens. The diagram in Fig. 6.20 is a splitting diagram that illustrates how the multiplicity of the resonance patterns is generated. The theoretical intensities of the lines in the A and X portions of the spectrum, as calculated from Eq. (6.11), apply only when $\Delta\delta_{AX}/J_{AX} \to \infty$. As the ratio of $\Delta\delta_{AX}$ to J_{AX} decreases, the absorption patterns become distorted, with the inner lines increasing in intensity at the expense of the outer lines, as in the $AX \to AB$ transformation.

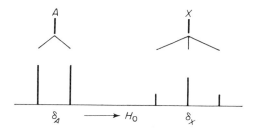

Fig. 6.20. Representation of the spin-spin splitting in the A_2X system.

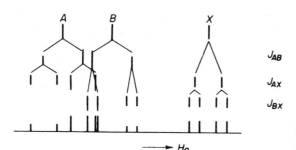

Fig. 6.21. Representation of the spin-spin splitting in the *ABX* system.

Another three-spin system is the *ABX* system where $\Delta\delta_{AB} < J_{AB}$, and $\Delta\delta_{AX}$ and $\Delta\delta_{BX} > J_{AX}$ and J_{BX}. The spin-spin interaction is diagrammed in Fig. 6.21. If we first couple the *A* and *B* nuclei independently of the *X* nucleus, we derive the typical *AB* pattern. Spin-spin interaction between the *A* and *X* nuclei doubles the *A* and *X* resonance lines, and interaction between the *B* and *X* nuclei doubles the *B* and again the *X* lines. It should be pointed out that coupling of the *A* and *B* nuclei with the *X* nucleus leads to an overlapping of the *A* and *B* resonance lines. (This is permissible except in cases when direct coupling between two nuclei might lead to crossing.)

Occasionally the *X* portion of the *ABX* spectrum may appear as an apparent triplet if $J_{AX} \approx J_{BX}$, which results in an overlap of the two central lines of the *X* portion. In such cases the term triplet should not be used; the pattern should be described as "the *X* portion of an *ABX* system." A typical example of a spectrum containing an *ABX* system is shown in Fig. 6.14 for 1,2-diphenylethyl benzoate. The *AB* portion arises from the dissimilarity of the diastereotopic methylene hydrogens, as described in Sec. 6.6.

The final three-spin system is the *AMX* system. The resonance pattern is predicted in the straightforward manner as illustrated in Fig. 6.22. The relative sizes of the coupling constants will vary, depending on whether we are dealing

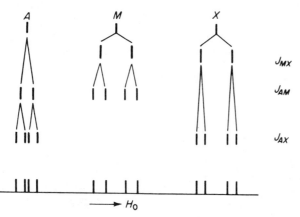

Fig. 6.22. Representation of the spin-spin coupling in the *AMX* system.

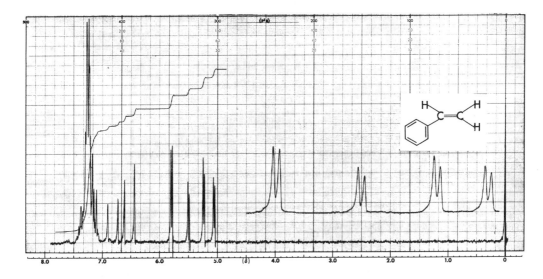

Fig. 6.23. Hydrogen resonance spectrum of styrene showing an *AMX* spin system. H_A appears as a double doublet at δ 6.62 with $J_{AM} = 17.5$ Hz and $J_{AX} = 11.7$ Hz, H_M at δ 5.68 and H_X at δ 5.16 with $J_{MX} = 1.7$ Hz. The tracing at the upper right is that of the H_X and H_M resonance region, which has been expanded to 100 Hz total sweep width with an offset of 296 Hz (to keep the resonance region on the chart paper). The vertical-stepped tracing is the integral line (see discussion in Sec. 6.9).

with an *AM-X* or an *A-M-X* system. A typical example of an *AMX* pattern is illustrated in the hydrogen resonance spectrum of styrene (Fig. 6.23).

6.8.3 Higher-Spin Systems

Except for the $A_n X_m$ spin systems, as the number of nuclei in a spin system increases the complexity of the NMR spectra rapidly increases, and first-order analyses of the spectra are not possible. Even in such complex cases, it may, however, be possible to interpret portions of a spin system. For example, in the NMR spectrum of 1-chlorobutane (Fig. 6.24), we can readily recognize a low-field, slightly distorted triplet representing the —CH_2Cl hydrogens, which are split by the adjacent methylene hydrogens. At higher field we can also make out a highly distorted triplet representing the methyl group [the distortion arises from the fact that the difference between the chemical shifts of the methyl and the adjacent methylene hydrogens is relatively small and only slightly larger than the coupling constant (~ 6.7 Hz)]. The resonance pattern centered at δ 2.6 represents the two methylene groups, which cannot be resolved and analyzed. The assignment that the multiplet at δ 2.6 represents two methylene groups, i.e., four hydrogens, can be made on the basis of the spectrum integral (Sec. 6.9). Therefore, had the spectrum in Fig. 6.24 been that of an unknown, we would

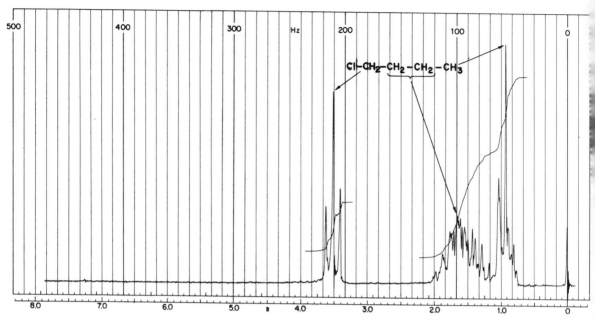

Fig. 6.24. NMR spectrum of 1-chlorobutane.

have been able to assign a unique structure to the unknown even though we could not make a complete analysis of the spectrum.

Even with much larger and more complex molecules than 1-chlorobutane, partial structures of a molecule can be assigned. By integrating chemical and other spectral data for the molecule, the complete assignment of structures of such molecules can usually be made.

6.9
QUANTITATIVE APPLICATIONS
OF ¹H NMR SPECTROSCOPY

In continuous wave (CW) NMR spectroscopy, the response (*sensitivity*) of equal numbers of the same type of nuclei is the same. Different nuclear species, however, possess different sensitivities and will not lead to equal integral absorptions for an equal number of nuclei. This is of little consequence because the resonances of two different nuclei do not occur within the same spectral region and, thus, never appear in the same spectral region.

The equal sensitivity, regardless of the chemical environment, provides the chemist with a very powerful quantitative tool. Measurement of the areas of individual resonance patterns allows one to derive the relative ratio of nuclei of different chemical environments within the molecule. If a definite structural assignment can be made for any one resonance pattern, and if the number of such functions within the molecule is known, the relative ratios of hydrogens can be transformed into an absolute ratio.

The analysis of mixtures can also be accomplished, provided that the resonance of one nucleus, or set of nuclei, appearing in each molecule present in the mixture can be distinguished and resolved from the resonance patterns of other nuclei in the other molecules.

The integration of an NMR spectrum is usually done electronically. Electronic integration is achieved by accumulating the voltage input to the recorder and plotting this accumulated voltage as a function of time or chart distance. Figure 6.23 illustrates the use of this method. The line appearing above the resonance peaks is the integration line, and the height of the vertical steps is proportional to the number of nuclei appearing under that resonance peak. Digital readout devices are also available to print out the accumulated voltages at desired intervals.

6.10
THE NUCLEAR OVERHAUSER EFFECT (NOE)

If intermolecular interactions are minimized when a compound is dissolved in a magnetically inert solvent,[12] most of the contributions to the spin-lattice relaxation of a nucleus will come from one or more of the other nuclei within the same molecule. Often the spin of a proton H_A and that of a spatially contiguous proton(s) H_B are coupled (not the same as spin-spin coupling) such that the relaxation of one spin affects the relaxation of the other. As a result, the populations of the nuclear energy levels of H_A and H_B are interdependent. With this kind of cross-relaxation, any change in the populations of the spin states of H_A will cause a redistribution in the populations of the corresponding levels of spin states of H_B. Nuclear Overhauser Effect (NOE) experiments take advantage of these changes, e.g., the transition of H_A is saturated (as in a double-resonance experiment), thereby equalizing the populations of the two spin states of that proton. The integrated intensity of H_B recorded during the saturation of H_A is compared with the intensity of H_B in the absence of double resonance. If protons H_A and H_B are spatially contiguous, the result will be seen in a substantial increase or, in some cases, decrease in the integrated NMR signal of H_B.

18   19   20

---

[12] In a solvent not containing nuclei with high magnetic moments, e.g., fluorine or hydrogen. Paramagnetic species such as molecular oxygen must also be absent.

On irradiation of $(CH_3)_{(a)}$ in **18**, $H_{(c)}$ shows a 17% increase in integrated intensity, showing that $H_{(c)}$ is relaxed by saturation of $(CH_3)_{(a)}$. Irradiation of the more distant $(CH_3)_{(b)}$ causes no change in the intensity of $H_{(c)}$. A similar effect is observed in dimethylformamide (**19**). In compound **20** irradiation of $H_{(b)}$ results in a 45% increase in the intensity of the very close $H_{(a)}$.[13] By acting as a probe for the establishment of proximity of protons that are not necessarily spin-spin coupled, the NOE provides a valuable aid for the assignment of NMR resonances and an important method for the study of conformational and other stereochemical problems.

NOE is also responsible for the enhanced intensity of ^{13}C resonance lines under proton decoupling (see the following section).

6.11
^{13}C NMR SPECTROSCOPY

The development of the application of Fourier transform (FT) techniques to NMR spectroscopy and advances in instrumentation during the late 1960's and early 1970's have made it possible to record natural abundance ^{13}C NMR spectra routinely. ^{13}C chemical shifts span slightly over 200 ppm in contrast to the typical 8 to 9 ppm range in ^1H NMR; thus considerably more structural information is generally available from ^{13}C NMR chemical shift data. A second, very important difference between ^1H and ^{13}C NMR spectroscopy is that whereas the diamagnetic term is dominant in the shielding of the hydrogen nucleus, the paramagnetic term is the dominant contributor to the shielding of the ^{13}C nucleus. Long-range shielding effects that were important in ^1H NMR are less important in ^{13}C NMR. As a result, ^1H chemical shifts generally do not parallel ^{13}C chemical shifts.

In ^1H continuous wave (CW) NMR spectroscopy, the responses of different nuclei are directly proportional to the numbers of the nuclei present. In ^{13}C FT NMR spectroscopy this is not so. The computer sums the free-induction decay (FID) signal for a short period of time (on the order of a few tenths of a second up to 1 to 2 sec), during which time varying percentages of the different nuclei have undergone relaxation. If the FID is recorded very shortly after the initial pulse, the nuclei undergoing rapid relaxation contribute more to the FID than do the more slowly relaxing nuclei. Delaying the recording of the FID results in proportionately greater contributions from the more slowly relaxing nuclei, the population of the rapidly relaxing nuclei having approached the equilibrium distribution; however, incorporating a substantial delay between the initial pulse and the recording of the FID greatly increases the instrument use time, due not only to the delay period but also to the lower intensity of the FID signal per scan, and a reasonable compromise must be achieved. In general, the rate of relaxation of a ^{13}C nucleus is directly proportional to the number of hydrogens directly attached to that carbon atom, i.e., methyl carbons undergo relaxation more

---

[13] F. A. L. Anet and A. J. R. R. Bourn, *J. Am. Chem. Soc.*, **87**, 5250 (1965).

rapidly than do methylene carbons, which in turn relax more rapidly than do methine carbons. Quaternary and carbonyl carbons generally possess long relaxation times. Examples of the variation in response is evident in the various ¹³C spectra presented later.

¹³C NMR spectra are generally recorded under double resonance conditions in which the coupling of ¹H to ¹³C is destroyed. (Coupling of ¹³C with ¹³C is generally not observed because of the low probability of having adjacent ¹³C's.) Complete ¹H decoupling is accomplished by irradiating the ¹H resonance region with a broad band width of radiofrequency radiation, termed "noise", sufficient to cover the entire ¹H resonance region. The ¹³C NMR spectra thus obtained contain only singlet resonances. Although this greatly simplifies the appearance of the spectra, it does result in the loss of information, i.e., the number of hydrogens attached to each carbon atom cannot be determined from the multiplicity of each ¹³C resonance signal according to Eq. (6.10). Two benefits accrue, however, by using the double resonance technique. As all lines in each multiplet are summed to give a single line in the decoupled spectra, the signal-to-noise ratio (noise here is the electronic background level variations in the spectrometer system) is considerably greater than in the nondecoupled recordings involving the same number of scans. The second benefit leading to increased signal-to-noise ratios arises from the operation of the Nuclear Overhauser Effect (NOE) (Sec. 6.10) under the double resonance conditions. Saturation of the hydrogens attached to the ¹³C nuclei results in enhanced populations of the excited spin states of ¹³C and induces more rapid relaxation. This results in a proportionately greater contribution by such a nucleus to the FID signal. The NOE enhancements are directly related to the number of attached hydrogens. Carbon atoms not bearing hydrogens, for example, quaternary and carbonyl carbons, are unaffected and are always of weaker intensity. In order to observe carbonyl carbon resonances, it is often necessary to incorporate a much longer delay period after the initial pulse before recording the FID signal. The emissions from the less rapidly relaxing nuclei are thus proportionately greater. The need for a greatly increased number of scans, with a resultant increase in instrument use time in recording a non-decoupled spectrum, is illustrated later in Fig. 6.26. Figure 6.26b is the ¹H decoupled spectrum of 4-methyl-1-pentene, which required 128 scans. Figure 6.26a is the proton coupled spectrum, which required 1244 scans! When sufficient sample is available, on the order of 1 g of a solid or 3 mL of a pure liquid, the non-decoupled spectrum can be recorded to aid in assignment of the individual resonances, as is illustrated in Fig. 6.26.

6.12
¹³C CHEMICAL SHIFTS

As in ¹H NMR, a standard must be used to correlate ¹³C chemical shifts. The ¹³C resonance of tetramethylsilane (TMS) is currently the most widely used standard, its chemical shift, δ, being assigned a value of 0.00, and in general, being at higher field than most ^{13}C resonances in organic compounds. Early studies used the ^{13}C resonance of benzene or carbon disulfide as standards. Corrections to the δ scale (implied use of TMS as standard) can be made by adding 128.7 to the chemical shifts when benzene was used as standard, or 193.7 with carbon disulfide.

Figure 6.25 illustrates ranges of chemical shifts for various types of carbon atoms. For several classes of compounds, specific correlations have been derived for calculating chemical shifts. These will be discussed in the following sections.

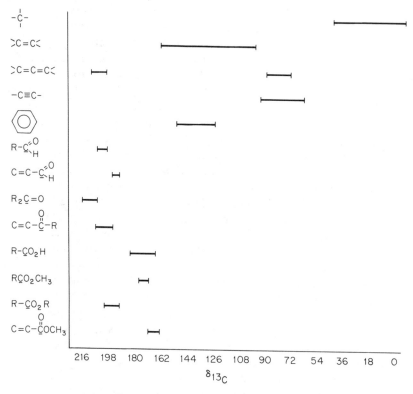

Fig. 6.25. Typical ranges of ^{13}C chemical shifts.

6.12.1 Alkanes and Cycloalkanes

Table 6.15 lists the ^{13}C chemical shifts for a number of alkanes and cycloalkanes. From such data, Grant and Paul[14] developed an empirical relationship [Eq. (6.21)] to calculate ^{13}C chemical shifts in which $B = -2.5$ for alkanes, A_j is the

$$\delta_{^{13}C} = B + \sum_j A_j n_{ij} \tag{6.21}$$

---

[14] D. M. Grant and E. G. Paul, *J. Am. Chem. Soc.*, **86**, 2984 (1964).

Table 6.15. ^{13}C **Chemical Shifts (δ) in Selected Alkanes and Cycloalkanes**

Compound	C$_1$	C$_2$	C$_3$	C$_4$	C$_5$
CH$_4$	−2.1				
CH$_3$CH$_3$	5.9	5.9			
CH$_3$CH$_2$CH$_3$	15.6	16.1	15.6		
CH$_3$CH$_2$CH$_2$CH$_3$	13.2	25.0	25.0	13.2	
CH$_3$CH$_2$CH$_2$CH$_2$CH$_3$	13.7	22.6	34.5	22.6	13.7
CH$_3$CH$_2$CH$_2$CH$_2$CH$_2$CH$_3$	13.9	22.9	32.0	32.0	22.9
CH$_3$CHCH$_3$ | CH$_3$	24.3	25.2	24.3		
CH$_3$CH$_3$ | CH$_3$CCH$_3$ | CH$_3$	31.5	27.9	31.5		
CH$_3$ | CH$_3$CH$_2$CCH$_3$ | CH$_3$	28.9	30.4	36.7	8.7	
Cyclopropane	−2.6				
Cyclobutane	23.3				
Cyclopentane	26.5				
Cyclohexane	27.8				
Cycloheptane	29.4				
Cyclooctane	27.8				

value for the appropriate shift parameter in Table 6.16, and n_{ij} is the number of each group present. The shift parameters are defined in the following manner. The α parameter, or more commonly termed the α "effect," is for the structural change —**C**—H → —**C**—CH$_3$, in which the bold face **C** is the ^{13}C under consideration. The β effect is for the structural change —**C**—C—H → —**C**—C—CH$_3$, and similarly, the γ, δ, and ε effects are for the substitution of a methyl carbon for

Table 6.16. Sheilding Parameters for Calculation of ^{13}C **Chemical Shifts in Acyclic Hydrocarbons**[*]

Shift Parameter	A$_j$ (Δδ)	Shift Parameter	A$_j$ (Δδ)
α	+9.43	2°(3°)	−2.5
β	+8.81	2°(4°)	−7.2
γ	−2.5	3°(2°)	−3.7
δ	+0.3	3°(3°)	−9.5
ε	+0.1	4°(1°)	−1.5
1°(3°)	−1.1	4°(2°)	−8.4
1°(4°)	−3.4		

[*] Taken from ref. 13, p. 58, except the α and β shift parameters, which are taken from D. K. Dalling and D. M. Grant, *J. Am. Chem. Soc.*, **96**, 1827 (1974).

hydrogen at the γ, δ, and ε positions from the ^{13}C under consideration. The reader should note that the γ effect is opposite that of the α, β, δ, and ε effects. This has been attributed to a steric effect in a synclinal, or *gauche*, conformation (this steric effect will be discussed in greater detail later). The last eight shift parameters in Table 6.16 are corrections that consider the degree of substitution at the α carbon. For example, 1°(3°) is the correction factor for a methyl, or primary carbon, when it is attached to a tertiary carbon, and 2°(4°) is the correction factor for a methylene, or secondary carbon, when it is attached to a quaternary carbon.

To illustrate the use of Eq. (6.21), we shall calculate the ^{13}C chemical shifts for 3-methylpentane: $CH_3CH_2CH(CH_3)CH_2CH_3$. For C_1, there are one α (+9.43), one β (+8.81), two γ (−2.5), and one δ (+0.3) contributions resulting in a value of $\sum_j A_j n_{ij} = 13.5$ and $\delta_{C_1} = 11.0$, compared to the observed value of δ 11.3. The correction for the degree of substitution at C_1, a 2°(1°) effect, does not appear in Table 6.16 and is assigned a value of zero. Similar calculations for C_2, C_3, and the methyl carbon attached to C_3 are detailed as follows:

C_2:			C_3:			C_3—$\underline{C}H_3$:		
	2α	+18.9		3α	+28.3		1α	+9.4
	2β	+17.6		2β	+17.6		2β	+17.6
	1γ	−2.5		2 3°(2°)	−7.4		2γ	−5.0
	2°(3°)	−2.5		$\sum_j A_j n_{ij}$	+38.5		1°(3°)	−1.1
	$\sum_j A_j n_{ij}$	+31.5					$\sum_j A_j n_{ij}$	+20.9
	Base	−2.5			−2.5		Base	−2.5
	δ_{C_2}	+29.0		δ_{C_3}	+36.0		$\delta_{\underline{C}H_3}$	+18.4
Observed		+29.3			+36.7			+18.6

Equation (6.21) cannot be used to calculate the ^{13}C chemical shifts for cyclic compounds due to the limited number of conformations possible with cyclic compounds compared to the large number of molecular conformations possible with acyclic structures which contribute to the A_j's. Separate empirical correlations are required for the cycloalkanes (see ref. in footnote to Table 6.16).

Another contribution to the chemical shifts of ^{13}C nuclei is the *steric compression shift*.[15] This occurs in sterically congested molecules (see also later discussion under alkenes). This effect is proportional to the force component along the ^{13}C—H bond arising from H—H nonbonded interactions, and is calculated according to Eq. (6.22), in which θ is the angle between the ^{13}C—H and H—H vectors and r is the H—H nonbonded distance in the partial structure.

$$\Delta\delta_{^{13}C}(\text{steric shift, ppm}) = +1680 \cos\theta \exp(-2.671r) \qquad (6.22)$$

---

[15] D. M. Grant and B. V. Cheney, *J. Am. Chem. Soc.*, **89**, 5315 (1967).

Table 6.17.   ¹³C Chemical Shifts (δ) in Alkenes

Compound	C_1	C_2	C_3	C_4
$H_2C{=}CH_2$	122.8	122.8		
$CH_3CH{=}CH_2$	115.4	135.7		
$CH_3CH_2CH{=}CH_2$	112.8	140.2		
$CH_3CH_2CH_2CH{=}CH_2$	113.5	137.6		
$CH_3CH_2\overset{|}{\underset{CH_3}{C}}{=}CH_2$	107.7	146.4		
$(CH_3CH_2)_2C{=}CH_2$	105.5	151.7		
$CH_3CH{=}CHCH_3$		123.3	123.3	
$CH_3CH_2CH{=}CHCH_3$		122.8	132.4	
$CH_3(CH_2)_3CH{=}CHCH_2CH_3$			130.5	129.3
Cyclopentene	130.6	130.6		
Cyclohexene	127.2	127.2		

Alkenes. The ¹³C chemical shifts for a number of acyclic and cyclic alkenes appear in Table 6.17. A number of trends are obvious even in the limited data presented in Table 6.17, which apply in general. In straight-chain alkenes, the difference in chemical shift $(\Delta\delta_C)$ between the two vinyl carbon atoms in a terminal alkene is ~ 24 ppm, that in Δ^2 alkenes is 7 to 10 ppm, and in Δ^3 alkenes is 1 to 2 ppm. Introduction of a methyl group onto one of the vinyl carbon atoms, i.e.,

$$C_\beta{=}C_\alpha\diagup{}^{H} \longrightarrow C_\beta{=}C_\alpha\diagup{}^{CH_3}$$

causes a deshielding of C_α by 5 to 7 ppm, but the C_β experiences a *shielding* of 5 to 7 ppm. This β effect is in the opposite direction of that observed in alkanes and is indicated as a β^π effect. When the methyl group is not attached directly to one of the vinyl carbon atoms, the β^σ effect is similar to that observed in alkanes, a β^σ effect being one in which the β carbon is attached to a saturated carbon atom. A correlation similar to that for alkanes has been derived in which $B = 122.1$, and the A_j's have the values given in Table 6.18 (definitions of the shielding parameter are given in the footnotes to the table). Correspondence between calculated and observed ¹³C chemical shifts is not as good as with the alkanes. Ring and steric strain cause a deshielding of the vinyl carbons, as is illustrated by the examples given in Table 6.18 and in the following structures.

In contrast to ¹H NMR, in which the C=C has a profound effect on the chemical shift of allylic hydrogens, the C=C has only a minor effect on the chemical shifts of carbon atoms separated from the C=C by only one or two bonds, with no effect observed on carbon atoms further removed. This is illustrated in the following comparison.

$$\begin{array}{c} 25.0 \quad 13.2 \\ \searrow \quad \downarrow \\ CH_3CH_2CH_2CH_3 \end{array}$$

$$\begin{array}{c} CH_3 \qquad\qquad CH_3 \\ \diagdown \qquad\qquad \diagup \\ \qquad C{=}C \qquad\qquad \diagup \qquad\qquad \diagdown \\ H \qquad\qquad H \end{array}$$ 10.6

$$\begin{array}{c} CH_3 \qquad\qquad H \\ \diagdown \qquad\qquad \diagup \\ \qquad C{=}C \\ \diagup \qquad\qquad \diagdown \\ H \qquad\qquad CH_3 \end{array}$$ 17.3

An interesting feature of the above data is the fact that the carbon atoms attached to the C=C in the *cis* alkenes are *shielded* by 6.0 ± 0.6 ppm relative to the *trans* isomer (a steric shift). This difference is often useful in distinguishing between conformational isomers of alkenes, particularly with trisubstituted alkenes.

Another contrast in the trends of ^1H and ^{13}C chemical shifts is that the introduction of a second double bond has relatively little effect on the ^{13}C chemical shifts of the vinyl carbon atoms, as illustrated by the comparison of the ^{13}C chemical shifts of C_1 and C_2 in 1-butene and 1,3-butadiene.

$$\begin{array}{c} CH_3CH_2CH{=}CH_2 \\ \diagup \qquad\qquad \diagdown \\ 140.2 \qquad\quad 112.8 \end{array} \qquad\qquad \begin{array}{c} H_2C{=}CH{-}CH{=}CH_2 \\ \diagup \qquad\qquad\qquad \diagdown \\ 137.2 \qquad\qquad 116.6 \end{array}$$

Alkynes. Acetylenic carbon atoms undergo resonance in a region intermediate between alkane and vinyl carbons (Table 6.19). The effect of substituents

Table 6.18. Shielding Parameters for Calculation of Vinyl ^{13}C Chemical Shifts in Alkenes*

Shift Parameter	$A_j\ (\Delta\delta)$
α	$+11.0 \pm 0.4$
β^π	-7.1 ± 0.4
β^σ	$+6.0 \pm 0.3$
γ^π	-1.9 ± 0.3
γ^σ	-1.0 ± 0.3
δ^π	$+1.1 \pm 0.3$
δ^σ	$+0.7 \pm 0.4$
ε	$+0.2 \pm 0.3$
cis	-1.2 ± 0.3
gem$^\alpha$	-4.9 ± 0.3
gem$^\beta$	$+1.2 \pm 0.3$
mult$^\sigma$	$+1.3 \pm 0.3$
mult$^\pi$	-0.7 ± 0.3

*Taken from p. 75 of ref. 13 at the end of this chapter. Superscript π is for when the substituent is attached to the C=C, and σ is for when it is attached to a saturated carbon of the alkene. *Cis* is the shielding effect on the vinyl carbon atoms for *cis*, tri-, or tetrasubstituted alkenes. *Gem*$^\alpha$ is the shift effect on the vinyl carbon bearing two groups, and *gem*$^\beta$ is the effect on the other vinyl carbon atom. *Mult*$^\sigma$ and *mult*$^\pi$ are effects caused by branching at a carbon directly attached to the C=C.

Table 6.19. ^{13}C Chemical Shifts (δ) in Alkynes

Compound	C_1	C_2	C_3
CH$_3$CH$_2$C≡CH	67.3	85.0	
CH$_3$C≡CCH$_3$		73.9	73.9
CH$_3$(CH$_2$)$_3$C≡CH	68.6	84.0	
CH$_3$(CH$_2$)$_2$C≡CCH$_3$		74.9	78.1
CH$_3$CH$_2$C≡CCH$_2$CH$_3$			81.1

on the chemical shifts of the acetylenic carbons is very similar to that observed with the alkenes; replacement of the acetylenic hydrogen by a methyl deshields that carbon by ~ 6.5 ppm, but shields the β carbon by ~ 5.0 ppm (a β^π effect). The β^σ effect is $\sim +7$ ppm, while the γ^σ effect is ~ -1 to 2 ppm, and that of γ^π is $+0.2$ to $+3.0$ ppm. The C≡C appreciably affects only the shielding of the attached carbon atoms (~ -9 to -13 ppm), the further removed carbon atoms having chemical shifts comparable to those in the corresponding alkane.

Aromatic hydrocarbons. Aromatic ring carbon atoms undergo resonance in the δ 125 to 150 region (Table 6.20). The introduction of a methyl group causes a deshielding of the *ipso* carbon (the ring carbon bearing the function) by $+9.2$ ppm, a slight deshielding of the *ortho* carbon, essentially no effect on the *meta* carbon, and a shielding of the *para* carbon. Again an empirical correlation has been developed for calculation of aromatic carbon chemical shifts using Eq. (6.21), in which $B = 128.3$ and the A_j's are those given in Table 6.21.

The effect of an aromatic ring on the ^{13}C chemical shifts of an attached alkyl group is to deshield the α carbon (the carbon attached to the ring) by ~ 23 ppm and the β carbon by ~ 9.5 ppm, and to shield the γ carbon by ~ 2 ppm. Carbon atoms further removed are unaffected. No appreciable long-range shielding by the phenyl group as was observed in ^{1}H NMR is observed in ^{13}C NMR.

Tables 6.22 and 6.23 list ^{13}C chemical shifts of ring carbon atoms in monosubstituted benzenes and in several selected monocyclic and heterocyclic aromatic compounds.

Table 6.20. ^{13}C Chemical Shifts (δ) of Aromatic Ring Carbon Atoms

Compound	C_1	C_2	C_3	C_4	C_5	C_6
Benzene	128.7	128.7	128.7	128.7	128.7	128.7
Methylbenzene	137.8	129.3	128.5	125.6	126.1	129.3
1,2-Dimethylbenzene	136.4	136.4	129.9	126.1	126.1	129.9
1,3-Dimethylbenzene	137.5	130.1	137.5	126.4	128.3	126.4
1,4-Dimethylbenzene	134.5	129.1	129.1	134.5	129.1	129.1
Ethylbenzene	144.1	128.1	128.5	125.9	128.5	128.1
n-Propylbenzene	142.5	128.7	128.4	125.9	128.4	128.7
t-Butylbenzene	149.2	125.4	128.4	125.9	128.4	125.4

Table 6.21. Chemical Shift Parameters for Calculation of ^{13}C Chemical Shifts of Aromatic Ring Carbon Atoms

Shift Parameter	$A_j\ (\Delta\delta)$
σ_α	$+9.2 \pm 0.2$
σ_o	$+0.8 \pm 0.2$
σ_p	-2.5 ± 0.2
σ_q	-2.3 ± 0.2
σ_t	$+1.2 \pm 0.2$

* σ_α, σ_o, and σ_p are the shielding parameters for the *ipso*, *ortho*, and *para* carbon atoms when a hydrogen on the *ipso* carbon atom is substituted by a methyl. σ_q is a shielding contribution to the carbon bearing a hindered methyl group (as in 1,2-dimethylbenzene), and σ_t is the contribution to the adjacent *tertiary* ring carbon atom (not the adjacent carbon atom bearing the other methyl group!).

Carbonyl compounds. The ranges of ^{13}C chemical shifts of the carbonyl carbon atoms in various carbonyl compounds are illustrated in Fig. 6.25. Changes in substituents have little effect on their chemical shifts (β effect of $\sim +2$, γ effect of ~ -1). α,β-Unsaturated compounds undergo resonance at ~ 7 to $10\ \text{ppm}$ higher field than do their saturated counterparts. Thioesters,

$$R\overset{\displaystyle O}{\overset{\|}{-C}}-SR,$$

undergo resonance at 15–20 ppm downfield relative to the corresponding ester analogs.

Table 6.22. ^{13}C Chemical Shifts (δ) of Aryl Carbon Atoms in Monosubstituted Benzenes, C_6H_5—X*

X	C_1	C_2	C_3	C_4
O^-	168.3	120.5	130.6	115.1
OH	155.6	116.1	130.5	120.8
OCH_3	158.9	113.2	128.7	119.8
$OCOCH_3$	151.7	122.3	130.0	126.4
NH_2	147.9	116.3	130.0	119.2
$N(CH_3)_2$	151.3	113.1	129.7	117.2
$NHCOCH_3$	139.8	118.8	128.9	123.1
F	163.9	114.6	130.3	124.3
Cl	135.1	128.9	129.7	126.7
Br	123.3	132.0	130.9	127.7
I	96.7	138.9	131.6	129.7
$CH{=}CH_2$	138.2	126.7	128.9	128.2
CO_2CH_3	130.0	128.2	128.2	132.2
COCl	134.5	131.3	129.9	136.1
CHO	137.7	129.9	129.9	134.7
$COCH_3$	136.6	128.4	128.4	131.6
CN	109.7	130.1	127.2	130.1
NO_2	148.3	123.4	129.5	134.7

* Taken from Table 5.58 on p. 197 of ref. 13 at the end of this chapter.

Table 6.23. ^{13}C Chemical Shifts (δ) in Heterocyclic Compounds

6.13
EFFECT OF HETEROFUNCTIONS
ON ^{13}C CHEMICAL SHIFTS

Table 6.24 lists the ^{13}C chemical shifts of a series of 1-substituted butanes. The change in chemical shift caused by the functions in other systems is fairly consistent with those observed in the butyl system. Note that in the case of all polar groups, the γ effect exerted by the functions is greater than that for a methyl group.

Empirical correlations to calculate ^{13}C chemical shifts in alcohols[16] and amines[17] have been developed.

^{13}C chemical shifts in alcohols are calculated according to Eqs. (6.23) to (6.26), in which the δ's are relative to carbon disulfide (193.7 ppm), and the δ_{C_α}, δ_{C_β}, etc. are defined in the conversion

$$C_\delta-C_\gamma-C_\beta-C_\alpha-CH_3 \quad \longrightarrow \quad C_\delta-C_\gamma-C_\beta-C_\alpha-OH$$

$$(\text{i.e., } R-CH_3 \quad \longrightarrow \quad R-OH)$$

---

[16] J. D. Roberts, F. J. Weigert, J. I. Kroschwitz, and H. J. Reich, *J. Am. Chem. Soc.*, **92,** 1338 (1970).

[17] H. Eggert and C. Djerassi, *J. Am. Chem. Soc.*, **95,** 3710 (1973).

$$\delta_{C_\alpha}^{ROH} = 0.83\delta_{C_\alpha}^{RCH_3} - 10.5 \tag{6.23}$$

$$\delta_{C_\beta}^{ROH} = \delta_{C_\beta}^{RCH_3} - 0.5 \tag{6.24}$$

$$\delta_{C_\gamma}^{ROH} = \delta_{C_\gamma}^{RCH_3} + 1.7 \tag{6.25}$$

$$\delta_{C_\delta}^{ROH} = \delta_{C_\delta}^{RCH_3} \tag{6.26}$$

The $\delta_C^{RCH_3}$'s are the calculated $\delta_{^{13}C}$'s for the corresponding alkane. (It must be noted here that the values of the $\delta_{^{13}C}$'s appear at higher field than does carbon disulfide, and they are assigned positive values.)

To illustrate the use of Eqs. (6.23) to (6.26), we shall calculate the $\delta_{^{13}C}$'s for 1-butanol. The calculated $\delta_{^{13}C}$'s for pentane (i.e., RCH_3 corresponding to 1-butanol) relative to TMS are C_α 22.7, C_β 34.0, C_γ 22.7, and C_δ 13.5, which, when converted to values relative to carbon disulfide, give $\delta_{C_\alpha}^{RCH_3} = 171.0$, $\delta_{C_\beta}^{RCH_3} = 159.7$, $\delta_{C_\gamma}^{RCH_3} = 171.0$, and $\delta_{C_\delta}^{RCH_3} = 180.2$. Using Eqs. (6.23) to (6.26) gives $\delta_{C_\alpha}^{ROH} = 131.4$, $\delta_{C_\beta}^{ROH} = 159.2$, $\delta_{C_\gamma}^{ROH} = 172.7$, and $\delta_{C_\delta}^{ROH} = 180.2$. Converting finally to δ's relative to TMS gives δ_α 62.3, δ_β 34.5, δ_γ 21.0, and δ_δ 13.5, which compare well with the values given in Table 6.23 for 1-butanol.

^{13}C chemical shifts for alkylamines are calculated according to Eq. (6.27) for the conversion

$$C_\gamma\!-\!C_\beta\!-\!C_\alpha\!-\!CH_3 \;\longrightarrow\; C_\gamma\!-\!C_\beta\!-\!C_\alpha\!-\!N\!\!\prec \quad (\text{i.e., } RCH_3 \;\longrightarrow\; R\!-\!N\!\!\prec)$$

$$\delta_C^{RN\prec} = (\delta_C^{RCH_3} \cdot A) + B \tag{6.27}$$

Table 6.24. ^{13}C Chemical Shifts (δ) in 1-Substituted Butanes

Compound	C_1	C_2	C_3	C_4
Butane	13.2	25.0	25.0	13.2
1-Chlorobutane	44.6	35.2	20.4	13.4
1-Bromobutane	33.2	35.4	21.7	13.5
1-Iodobutane	7.1	36.1	24.2	13.7
1-Butanol	61.7	35.3	19.4	13.9
Dibutyl ether	71.2	33.1	20.3	14.6
Butyl acetate	63.8	31.2	19.7	14.1
Butylamine	42.3	36.7	20.4	14.0
2-Hexanone*	43.5	31.9	23.8	14.0
Pentanoic acid†	34.1	27.0	22.0	13.5

* Considered to be 1-acetylbutane, i.e., $CH_3^4CH_2^3CH_2^2CH_2^1COCH_3$.
† Considered to be 1-carboxybutane, i.e., $CH_3^4CH_2^3CH_2^2CH_2^1CO_2H$. ^{13}C chemical shifts in esters are very similar to those in the corresponding carboxylic acids.

in which $\delta_C^{RCH_3}$'s are the calculated values of the alkane relative to TMS. The parameters for primary, secondary, and tertiary amines for use in Eq. (6.27) are contained in Table 6.25. As an example of the use of Eq. (6.27), we shall calculate the ^{13}C chemical shifts for butylamine. Using the calculated values for the ^{13}C chemical shifts of pentane given earlier, values for butylamine are calculated to be $\delta_{C_\alpha}^{RNH_2} = 42.3$, $\delta_{C_\beta}^{RNH_2} = 35.5$, $\delta_{C_\gamma}^{RNH_2} = 21.2$, and $\delta_{C_\delta}^{RNH_2} = 13.5$; these values agree well with those given in Table 6.23 for butylamine. (The value of $\delta_{C_\delta}^{RNH_2}$ is taken to be the same as that in RCH_3.)

Undoubtedly, correlations for the calculation of chemical shifts in other classes of compounds will ultimately be developed.

Table 6.25. Parameters for Calculation of ^{13}C Chemical Shifts in Amines*

Primary amines	A	B
C_α	0.846	23.09
C_β	0.955	3.00
C_γ	0.941	−0.07
Secondary amines		
C_α	0.900	22.88
C_β	0.942	2.07
C_γ	0.951	−0.68
Tertiary amines		
C_α	0.914	22.62
C_β	0.999	0.45
C_γ	0.934	−0.43

* J. D. Roberts, F. J. Weigert, J. I. Kroschwitz, and H. J. Reich, *J. Am. Chem. Soc.*, **92**, 1338 (1970).

Table 6.26. Values of $^1J_{^{13}C-H}$ in Substituted Methanes, CH_3-X*

X	$^1J_{^{13}C-H}$ (Hz)	X	$^1J_{^{13}C-H}$ (Hz)
Li	98	CN	133
$MgCH_3$	105.5	SCH_3	138
$C(CH_3)_3$	124	OCH_3	140
H	125	OH	142
CH_3	125	NO_2	147
C_6H_5	126	F	149
$COCH_3$	127	Cl	150
$C{\equiv}CH$	131	Br	151
NH_2	133	I	151

* Taken from Chapter 10 of ref. 13 at the end of this chapter.

6.14
SPIN COUPLING INVOLVING ^{13}C

The most extensively studied coupling interaction of ^{13}C is that with ^{1}H. The magnitude of $^1J_{^{13}C-H}$ has been found to be very sensitive to the electronegativity of the function(s) attached to the carbon atom under consideration. Table 6.26 lists $^1J_{^{13}C-H}$'s for a number of substituted methanes in which, in general, the coupling constant is observed to increase with increasing electronegativity of the attached group. This effect is rather large, as exemplified in the series CH_3Cl (148.6 Hz), CH_2Cl_2 (176.5 Hz), and $CHCl_3$ (208.1 Hz).

In addition to the effect of substituent electronegativity, $^1J_{^{13}C-H}$ is also sensitive to the degree of hybridization of the orbitals on carbon involved in the C—H bond, increasing markedly as the s-character of the orbital on the carbon increases. For example, with sp³-hybrid orbitals (25% s-character), $^1J_{^{13}C-H}$ is ~ 125 Hz, with sp²-hybrid orbitals (33% s-character) ~ 160 Hz, and for sp-hybrid orbitals (50% s-character) ~ 250 Hz; a nearly linear relationship of percent s-character with the magnitude of $^1J_{^{13}C-H}$! The values of $^1J_{^{13}C-H}$ in strained ring systems have been used to calculate the degree of hybridization in such systems. Table 6.27 lists values of $^1J_{^{13}C-H}$'s for a variety of systems.[18]

Values for $^2J_{^{13}C-H}$ (−2 to −6 Hz) and $^3J_{^{13}C-H}$ (4 to 6 Hz) have been measured and trends evaluated in terms of electronegativity, bond angles, and dihedral angles. Remarkably, the trends observed parallel those observed with $^2J_{HH}$'s and $^3J_{HH}$'s.

Values for ^{13}C-^{13}C coupling constants have also been determined. Because of the low abundance of ^{13}C, $^{13}C-^{13}C$ coupling constants are difficult to measure even using FT techniques. Such determinations generally require the preparation of ^{13}C enriched compounds. Values of the various $^{13}C-^{13}C$ coupling constants are ^1J, 35 to 80 Hz; ^2J, 0 to 2 Hz; ^3J, 0 to 5.2 Hz, these values depending on the dihedral angle between the C—C bonds as in $^3J_{HH}$, and ^4J, 0 to 0.6 Hz. In general, $J_{^{13}C^{13}C} = 0.27J_{^{13}C-H}$.

Table 6.27.   Values of $^1J_{^{13}C-H}$ in Various Systems*

Compound	J (Hz)	Compound	J (Hz)
CH_3—CH_3	125	—H	159
	160.5	H_4 H_3 H_2	(2) 180 (3) 157 (4) 160
	136	H_3 H_2	(2) 184 (3) 170

---

[18] For a more detailed discussion of ^{13}C coupling constants, see Chapter 10 in ref. 13 at the end of this chapter.

Table 6.27.   (continued)

Compound	J (Hz)	Compound	J (Hz)
(cyclopentane)	128	(furan with H_3, H_2)	(2) 201
			(3) 175
(cyclohexane)	125		
		$HC\equiv CH$	248.7
$H_2C{=}CH_2$	156.2	$CH_3C{\equiv}\underline{CH}^*$	247.6
$H_2C{=}CH{-}CH{=}\underline{CH_2^*}$	158	$CH_3C\overset{O}{\underset{H}{\diagup\!\!\diagdown}}$	172.4
$H_2C{=}C{=}CH_2$	168	(oxirane)	176
(cyclobutene, H)	170	(aziridine, H, N)	171
(cyclopentene, H)	160	(H_3, H_2, H_1)	(1) 169
			(2) 153
			(3) 205
(cyclohexene, $\underline{H}$)	157	($\underline{H}^*$)	212

*Values are for the underlined $\underline{C}\underline{H}$ bond.

6.15
EXAMPLES OF ^{13}C NMR SPECTRA

Let us begin by considering the proton nondecoupled and decoupled spectra of 4-methyl-1-pentene shown in Fig. 6.26. The proton decoupled spectrum shows three peaks in the saturated carbon region at δ 22.5, 28.7, and 43.9, and two peaks in the vinyl carbon region at δ 115.6 and 137.8 (Fig. 6.26b), representing the five chemically different carbon atoms in 4-methyl-1-pentene. Although we could reasonably assign each of the resonances in the decoupled spectrum to specific carbon atoms by the use of the shift parameters given earlier, the multiplicities of the resonances in the nondecoupled spectrum allow for immediate assignment. [In FT NMR spectroscopy, the relative intensities of the components of a multiplet do not always conform to those calculated by Eq. (6.11).] In the nondecoupled spectrum, the resonance at δ 22.5 appears as a quartet ($^1J_{^{13}C-H}$ = 123 Hz), and thus must represent the two methyl carbon atoms (C_5). The resonances at δ 28.7 and 43.9 appear as a doublet (126 Hz) and triplet (128 Hz) and must represent C_4 and C_3 respectively. The relative magnitudes of the chemical shifts of the saturated carbon atoms are in the expected sequence, C_5 having one α and two β carbons, C_4 having three α and one β, and C_3 having two α and three β. The vinyl carbon resonances at δ 115.6 and 137.8 appear as a triplet (151 Hz) and doublet (148 Hz), representing C_1 and C_2 respectively. The observed $\delta_{^{13}C}$'s of C_1 and C_2 correspond closely to the calculated values of 115.4 (one β^π, one γ^σ and two δ^σ) and 137.1 (one α, one β^σ, and two γ^σ), respectively.

It should be noted that of the many C_6H_{12} isomeric structures, only 4-methyl-1-pentene can give rise to the ^{13}C NMR spectra shown in Fig. 6.26. It is left as an exercise for the reader to outline the appearance of the ^{13}C NMR spectra of the other C_6H_{12} isomeric alkenes.

Figure 6.27 shows the proton nondecoupled and decoupled spectra of dipropyl ether. In the decoupled spectrum, each singlet represents two chemically identical carbon atoms. In the nondecoupled spectrum, the resonance at δ 10.7 appears as a quartet (128 Hz), while those at δ 23.3 and 72.5 appear as triplets (127 and 141 Hz respectively). As in the case of 4-methyl-1-pentene, the NMR spectra in Fig. 6.27 uniquely represent dipropyl ether, only one of the many possible $C_6H_{14}O$ structures.

Figure 6.28 shows the proton decoupled spectrum of methyl benzoate. The carbonyl carbon resonance appears at δ 167.0, characteristic of an unsaturated ester carbon. The four resonances characteristic of a monosubstituted benzene appear at δ 129.0, 130.2, 131.0, and 133.3, while that of the methyl carbon appears at δ 52.0 typical of a methyl attached to an oxygen atom.

Figure 6.29 shows the decoupled spectrum of 3,5-dimethylphenol. Assignment of the resonances in the aromatic region to specific carbons can be made by comparison of the chemical shifts with those of phenol (Table 6.22). The low field peaks at δ 113.5, 122.7, and 155.2 represent C_2, C_4, and C_1, respectively, all of which are reasonably close to the values for phenol. The peak at δ 139.7 represents C_3, which is deshielded, relative to phenol, by the attached methyl group. Of the dimethyl phenols, only 2,6- and 3,5-dimethylphenol have chemically identical methyl groups and four chemically nonidentical aromatic ring carbon atoms.

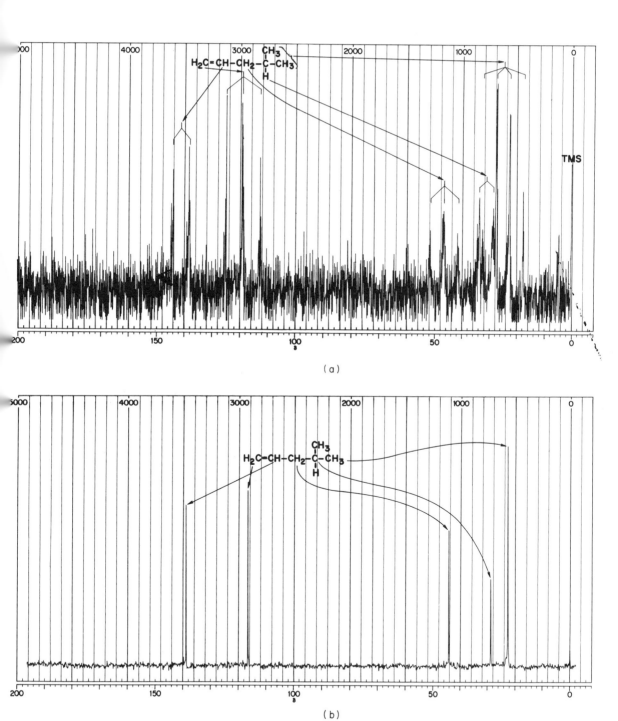

Fig. 6.26. Proton coupled (*a*) and decoupled (*b*) ^{13}C NMR spectra of 4-methyl-1-pentene.

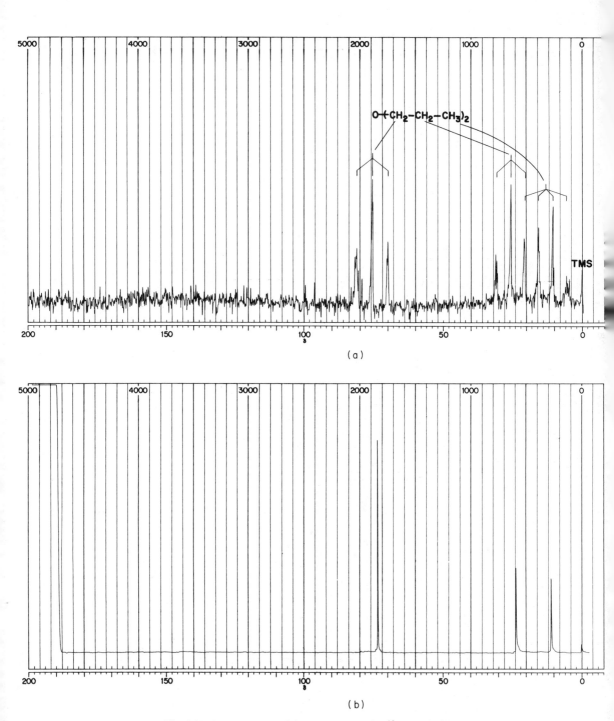

O—(CH₂—CH₂—CH₃)₂

TMS

(a)

(b)

248   **Fig. 6.27.** Proton coupled (*a*) and decoupled (*b*) ¹³C spectra of dipropyl ether.

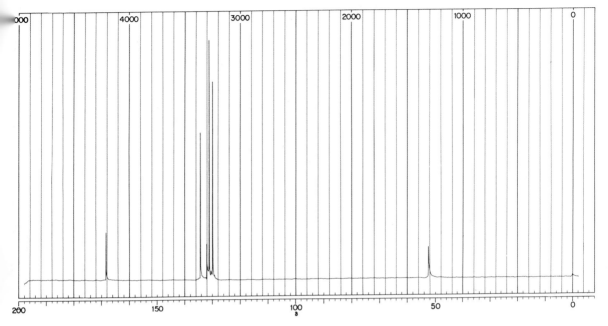

Fig. 6.28. Proton decoupled ^{13}C NMR spectrum of methyl benzoate.

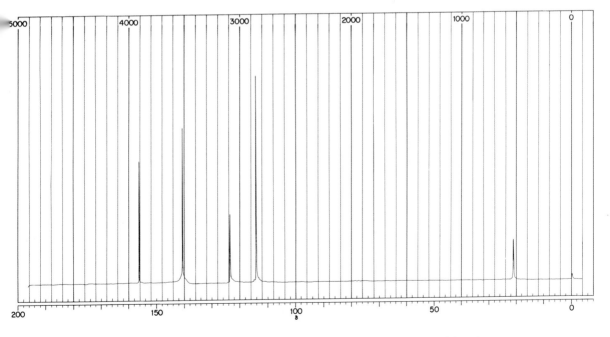

Fig. 6.29. Proton decoupled ^{13}C NMR spectrum of 3,5-dimethylphenol.

6.16
CHEMICAL SHIFT REAGENTS

Certain paramagnetic lanthanide metal complexes produce changes in chemical shifts when added to solutions of organic compounds that contain Lewis basic functions. The magnitude of the induced shift depends on the nature of the metal and the ligands attached to the metal and the type of function and structural features in the organic molecule.

The most commonly used complexes are the *tris*-octahedral complexes of europium, praesidynium, and ytterbium with dipivaloylmethane (**21**) and 2,2-dimethyl-6,6,7,7,8,8,8-heptafluorooctan-3,5-dione (**22**), generally abbreviated M(DPM)$_3$ or M(thd)$_3$ for the former, and M(fod)$_3$ for the latter.

Addition of the lanthanide complex to a solution of an organic compound containing a Lewis basic function establishes the equilibrium

$$R—X: + M(L)_3 \rightleftharpoons R—X—M(L)_3$$

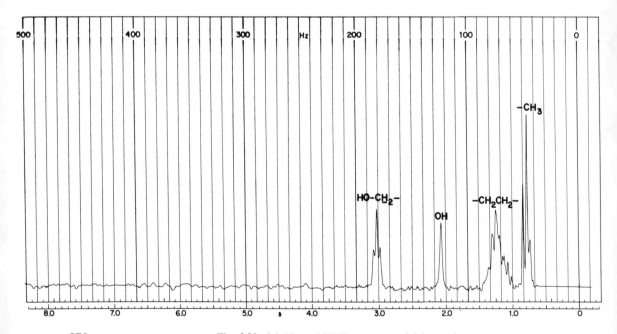

Fig. 6.30. (*a*) Normal NMR spectrum of 1-butanol.

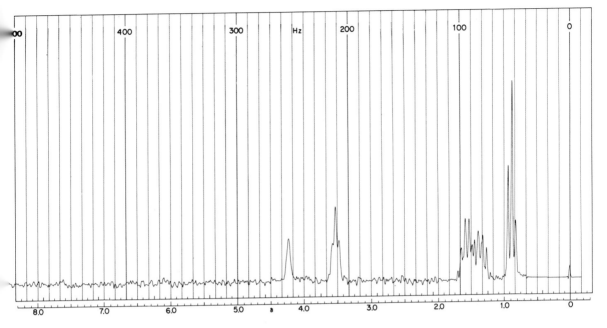

Fig. 6.30 (b) 1:0.25 molar ratio of 1-butanol to Eu (fod)$_3$.

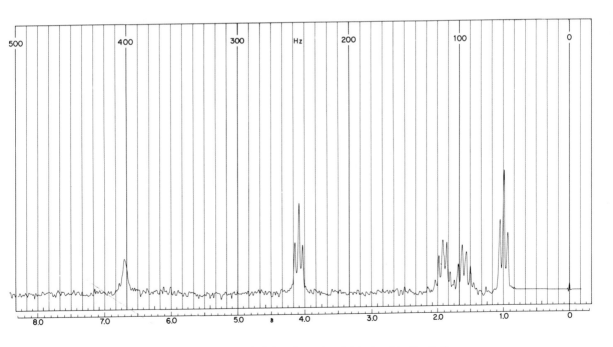

Fig. 6.30 (c) 1:0.5 molar ratio of 1-butanol to Eu (fod)$_3$.

The induced shift, $\Delta\nu$ (ppm), in the $1:1$ complex is determined by measuring the induced shifts at several low ratios of lanthanide complex to organic substrate, and extrapolating to the $1:1$ molar ratio.[19] $\Delta\nu$ in the $1:1$ complex is given by Eq. (6.28), in which ε and $f(g)$ are constants characteristic of the metal complex, θ is the angle between the symmetry axis of the metal complex (generally taken to be the X—M bond axis) and the nucleus being observed (H in the diagram), and R is the distance between the nucleus and the metal atom.

$$\Delta\nu \text{ (ppm)} = -\frac{3}{5}\,\varepsilon f(g)\left[\frac{(3\cos^2\theta - 1)}{R^3}\right] \tag{6.28}$$

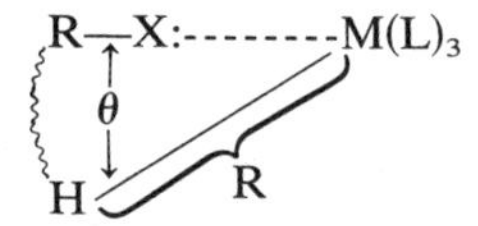

The largest induced shifts are observed with amines and alcohols, moderate shifts with carbonyl compounds, and very small shifts with ethers and sulfur-containing compounds. In conformationally mobile acyclic compounds, the induced shifts decrease with increasing distance from the X function, generally resulting in a considerable simplification of the NMR spectrum. For example, in the NMR spectrum of 1-butanol [Fig. 6.30(a)], the methylene hydrogens on C_2 and C_3 possess very similar chemical shifts and appear as a very complex multiplet. The addition of increments of Eu(fod)$_3$ results in a dramatic deshielding of the —OH, with less extensive deshielding of the hydrogens at greater distances from the metal atom. In Fig. 6.30(c), all of the resonances are well separated, and the spectrum can be easily interpreted. With conformationally biased cyclic compounds, the use of shift reagents can often allow assignment of the stereochemistry of isomers. For example, in the cis- and trans-4-t-butylcyclohexanols (**23** and **24**, respectively), the axial hydrogen attached to C_2 experiences induced shifts of -8.2 and -14.7 ppm in the cis- and trans-isomers, respectively, while the axial hydrogens at C_3 undergo induced shifts of -13.6 and -5.4 ppm in the cis- and trans-isomers, respectively.

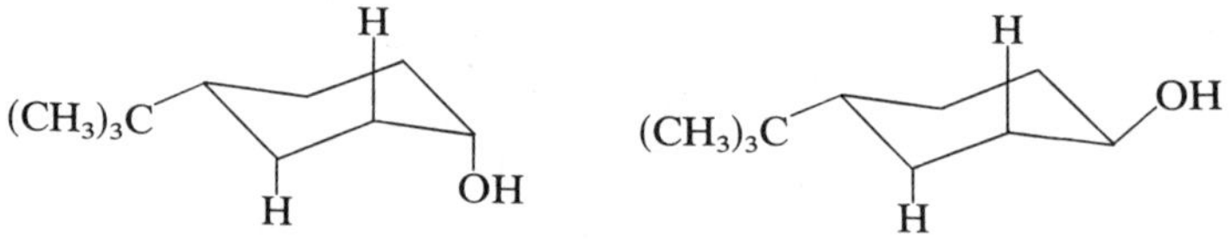

23   **24**

[19] The direct measurement of the chemical shifts in the $1:1$ complex is generally not possible owing to the limited solubility of the lanthanide complex and to extensive line width broadening caused by the paramagnetic metal atom.

Chiral shift reagents are also available and are useful for the analysis of the composition of mixtures of enantiomers. Enantiomers possess identical NMR spectra; however, in the presence of a chiral environment, such as a chiral shift reagent, the complexes that are formed are diastereomeric and possess different NMR spectra. Integration of the diastereomerically related sets of hydrogens provides the most accurate method for determining optical purities. The structures of two chiral shift reagents are given below.

Extensive line broadening occurs when any of the shift reagents are used, placing a practical limit on the concentration of the shift reagent that can be used. This line broadening is due to interaction of the paramagnetic metal atom with the hydrogen nuclei.

Chemical shift reagents are also effective in ^{13}C NMR spectroscopy; however, the ^{13}C resonances in the ligands attached to the metal atom complicate the ^{13}C NMR spectra.

6.17
SAMPLE PREPARATION

One of the most important factors in the production of good spectra is care in the preparation of the sample. The sample should be of high purity and free from traces of lint, dust, or other foreign material, particularly impurities having paramagnetic properties. The presence of such materials causes extensive line broadening.

Nonviscous liquids can be run neat (without solvent) with approximately 3% by weight tetramethylsilane added as an internal standard for hydrogen magnetic resonance. Other suitable standards are available for other nuclear species. Viscous liquids and solids must be dissolved in a suitable solvent, generally 10 to 25% by weight, to give a solution of low viscosity. The choice of the solvent will depend on the solubility characteristics of the sample. The solvent should not absorb in regions that will obscure the absorption patterns arising from the sample. Solvent absorption interferences can be avoided by the use of fully deuterated or halogenated molecules, for example, $CDCl_3$ and CCl_4. A great many deuterated solvents suitable for nuclear magnetic resonance studies are commercially available. With certain types of spectrometers, deuterium-containing solvents are required to provide a frequency "lock." The instructor or NMR operator should be consulted as to the solvent requirements, if any, of the spectrometer.

The neat liquid sample, or solution, is placed in a carefully cleaned sample tube. Special tubing is available specifically for use in nuclear magnetic resonance spectrometers. The wall diameter should be relatively thin and kept within close thickness tolerances. This provides for as great a volume of sample within the magnetic field as is possible and a minimum of field distortion due to variances in wall thickness. The sample tube should be sealed or capped. For very careful work the tube should be sealed under reduced pressure, using degassing techniques to remove dissolved oxygen, which may lead to line broadening.

6.18
NMR SPECTRAL PROBLEMS

From the data given and the NMR spectrum, assign structures to the compounds in each of the following problems.

1. C_6H_{12}. NMR spectrum is shown in Fig. 6.31.

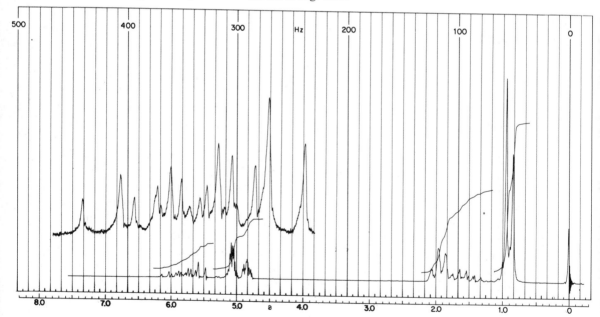

Fig. 6.31. NMR spectrum of unknown in problem 1.

2. $C_{10}H_{12}O_2$. Absorbs in the IR at 1740 cm^{-1} and in the UV at 255 nm. NMR spectrum is shown in Fig. 6.32.

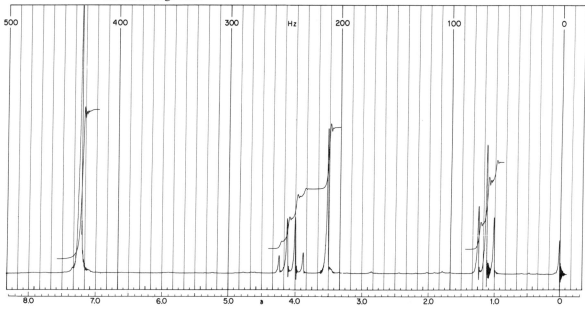

Fig. 6.32. NMR spectrum of unknown in problem 2.

3. C_5H_8O. Transparent in the 3600 and 2000 to 1600 cm^{-1} regions. NMR spectrum is shown in Fig. 6.33.

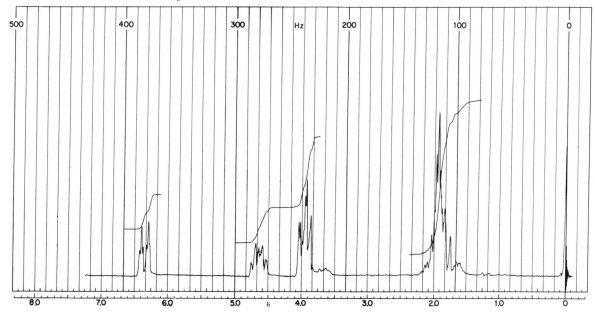

Fig. 6.33. NMR spectrum of unknown in problem 3.

4. C_9H_8O. Absorbs in the IR at 1685 cm^{-1}. NMR spectrum is shown in Fig. 6.34.

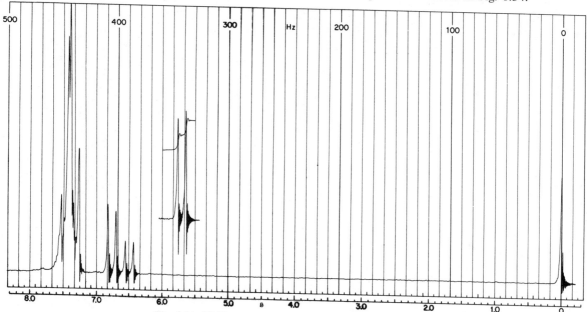

Fig. 6.34. NMR spectrum of unknown in problem 4. Insert is offset 250 Hz.

5. C_4H_8O. Absorbs in the IR at 3500 cm^{-1} (broad). NMR spectrum is shown in Fig. 6.35.

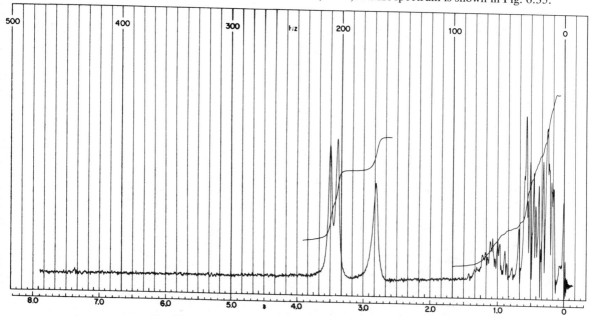

Fig. 6.35. NMR spectrum of unknown in problem 5 recorded at 250 Hz width.

6. $C_9H_{14}O$. Absorbs in the IR at 1680 cm^{-1} and at 239 nm in the UV. NMR spectrum is shown in Fig. 6.36.

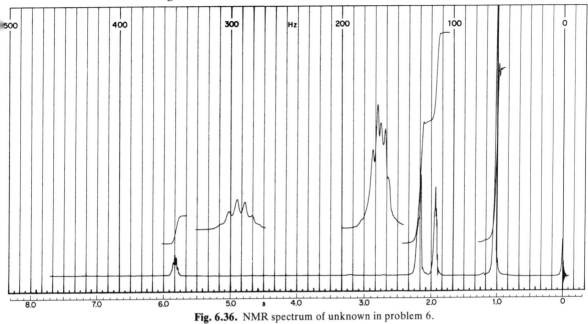

Fig. 6.36. NMR spectrum of unknown in problem 6.

7. $C_8H_{11}N$. Absorbs in the IR at 3450 cm^{-1}. NMR spectrum is shown in Fig. 6.37.

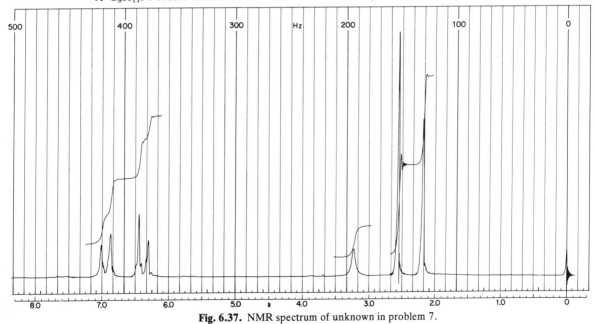

Fig. 6.37. NMR spectrum of unknown in problem 7.

8. $C_6H_{10}O$. Absorbs in the IR at 3600 and 3300 cm^{-1}. NMR spectrum is shown in Fig. 6.38.

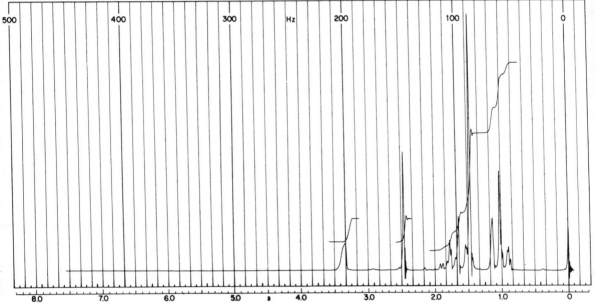

Fig. 6.38. NMR spectrum of unknown in problem 8.

9. $C_8H_{11}N$. Absorbs in the IR at 3500 and 3400 cm^{-1}. NMR spectrum is shown in Fig. 6.39.

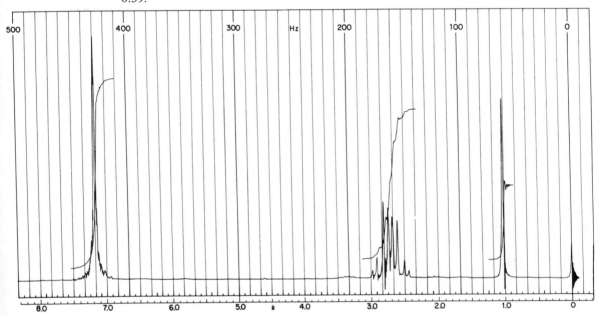

Fig. 6.39. NMR spectrum of unknown in problem 9.

10. C_3H_6O. Absorbs in the IR at 3600 cm^{-1}. ^{13}C NMR spectrum is shown in Fig. 6.40.

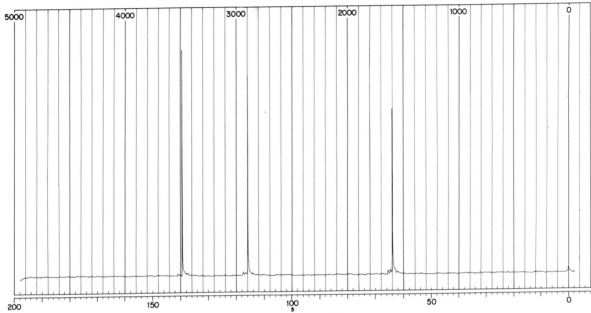

Fig. 6.40. ^{13}C NMR spectrum of unknown in problem 10.

11. $C_5H_{10}O$. Absorbs in the IR at 1710 cm^{-1}. ^{13}C NMR spectrum is shown in Fig. 6.41.

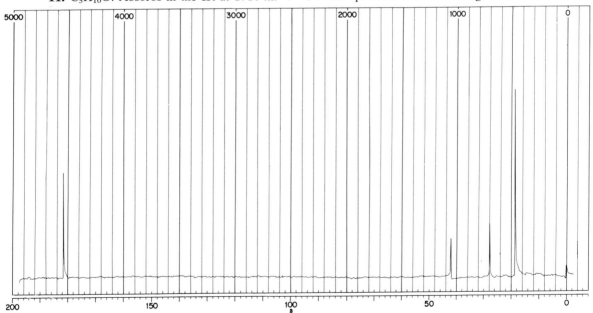

Fig. 6.41. ^{13}C NMR spectrum of unknown in problem 11.

12. $C_7H_{12}O$. Absorbs in the IR at 3600 and 3300 cm^{-1}. ^{13}C NMR spectrum is shown in Fig. 6.42. Calculate the $\delta_{^{13}C}$'s for your answer using the correlations discussed earlier and compare with the observed values in the spectrum.

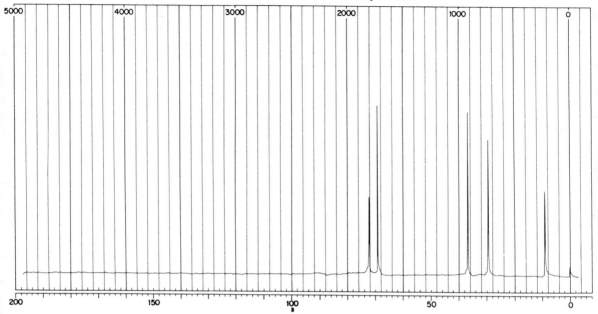

Fig. 6.42. ^{13}C NMR spectrum of unknown in problem 12.

13. $C_{10}H_{12}O$. Absorbs in the IR at 1685 cm^{-1}. ^{13}C NMR spectrum is shown in Fig. 6.43.

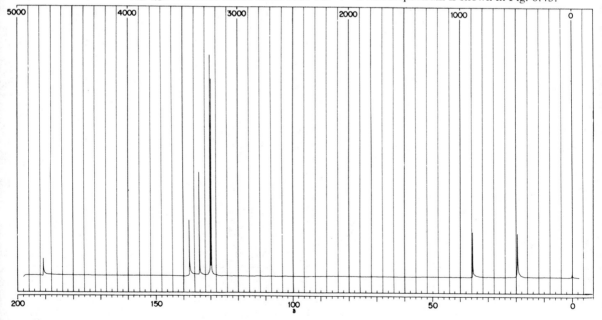

Fig. 6.43. ^{13}C NMR spectrum of unknown in problem 13.

14. $C_6H_{14}O$. Absorbs in the IR at 3600 cm^{-1}. ^{13}C NMR spectrum is shown in Fig. 6.44. Calculate the $\delta_{^{13}C}$'s for your answer using the correlations discussed earlier and compare with the observed values in the spectrum.

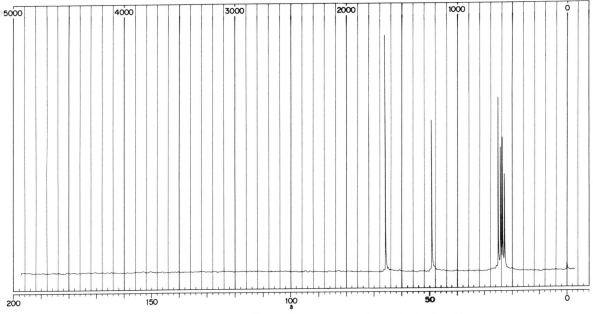

Fig. 6.44. ^{13}C NMR spectrum of unknown in problem 14.

15. $C_{10}H_{12}O_2$. Absorbs in the IR at 1740 cm^{-1}. ^{13}C NMR spectrum is shown in Fig. 6.45.

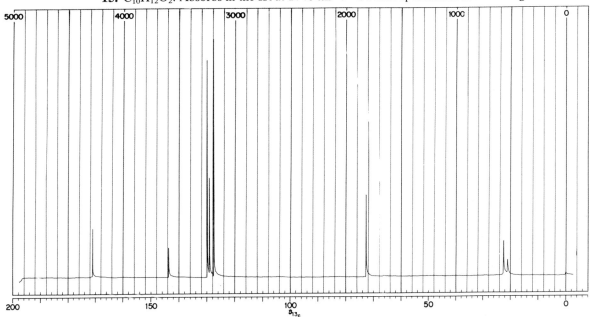

Fig. 6.45. ^{13}C NMR spectrum of unknown in problem 15.

16. $C_8H_{19}N$. ^{13}C NMR spectrum is shown in Fig. 6.46.

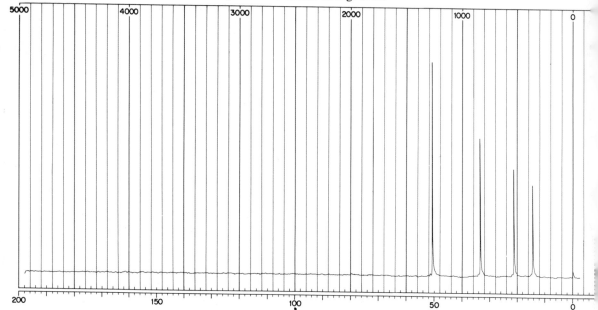

Fig. 6.46. ^{13}C NMR spectrum of unknown in problem 16.

17. $C_6H_{10}O$. ^{13}C proton nondecoupled spectrum is shown in Fig. 6.47a, and the proton decoupled spectrum is shown in Fig. 6.47b.

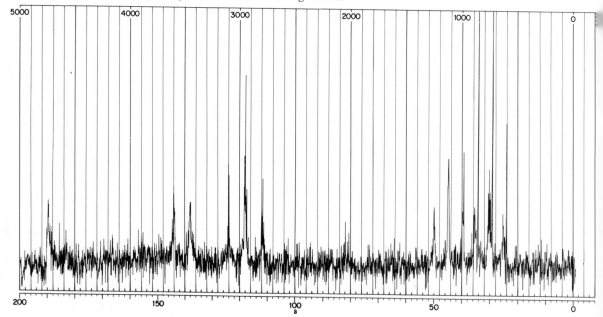

Fig. 6.47. ^{13}C NMR spectra of unknown in problem 17. (*a*) Proton non-decoupled spectrum.

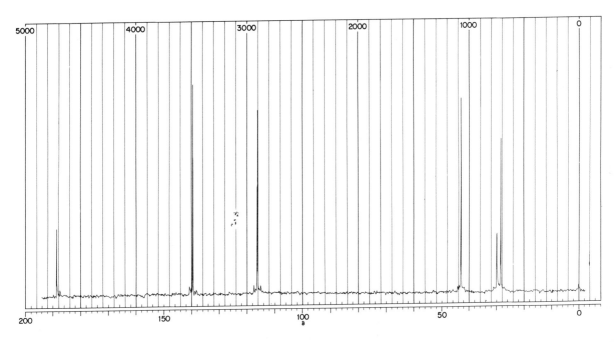

Fig. 6.47. (*b*) Proton decoupled spectrum.

6.19
NMR LITERATURE

The most recent comprehensive text on nuclear magnetic resonance is that by Emsley, Feeney, and Sutcliffe (ref. 5). Volume 1 discusses the theoretical aspects of nuclear magnetic resonance and the analysis of high-resolution spectra. Volume 2 contains discussions of correlations of resonance spectral parameters with molecular structure for various nuclear species. Both volumes contain a great many references to the original literature. Another comprehensive text is that by Pople, Schneider, and Bernstein (ref. 10), although this text is outdated by the recent advances in NMR spectroscopy with respect to the structure-spectra correlations. The texts by Levy and Nelson (ref. 8) and Stothers (ref. 13) deal specifically with ^{13}C NMR spectroscopy.

The analysis of resonance spectra (the determination of chemical shifts and coupling constants) in complex spin systems is greatly facilitated by Wiberg and Nist's *The Interpretation of NMR Spectra* (ref. 14), in which line positions and intensities have been calculated as functions of the coupling constant and the difference in chemical shifts.

Yearly reviews and reports on various aspects of NMR spectroscopy appear in refs. 15 and 16.

^{1}H NMR data or spectra for specific compounds appear in refs. 17, 18, 19, 22, 25, and 26, while ref. 20 gives similar data for ^{19}F NMR, and ref. 23 for ^{13}C data.

NMR, EPR, and NQR Current Literature Service, published periodically by Butterworth (ref. 24), provides a continuing survey and compilation of literature citations in various areas of nuclear magnetic resonance.

6.20
REFERENCES

GENERAL

1. BECKER, E. D., *High Resolution NMR: Theory and Chemical Applications.* New York: Academic Press, 1969.
2. BHACCA, N. S., and WILLIAMS, D. H., *Applications of NMR Spectroscopy in Organic Chemistry.* San Francisco: Holden-Day, Inc., 1964.
3. BIBLE, R. H., *Interpretation of NMR Spectra: An Empirical Approach.* New York: Plenum Press, 1965.
4. BOVEY, F. A., *Nuclear Magnetic Resonance Spectroscopy.* New York: Academic Press, 1969.
5. EMSLEY, J. W., FEENEY, J., and SUTCLIFFE, L. H., *High Resolution Nuclear Magnetic Resonance Spectroscopy,* Vols. 1 and 2. New York: Pergamon Press, 1965.
6. FARRAR, T. C., and BECKER, E. D., *Pulse and Fourier Transform NMR: An Introduction to Theory and Methods.* New York: Academic Press, 1971.
7. JACKMAN, L. M., and STERNHELL, S., *Applications of Nuclear Magnetic Resonance Spectroscopy in Organic Chemistry.* New York: Pergamon Press, 1969.
8. LEVY, G. C., and NELSON, G. L., *Carbon-13 NMR in Organic Chemistry.* New York: Wiley-Interscience, 1972.
9. *Nuclear Magnetic Resonance for Organic Chemists,* D. W. Mathieson, ed. New York: Academic Press, 1967.
10. POPLE, J. A., SCHNEIDER, W. G., and BERNSTEIN, H. J., *High Resolution Nuclear Magnetic Resonance.* New York: McGraw-Hill Book Company, 1959.
11. ROBERTS, J. D., *An Introduction to the Analysis of Spin-Spin Splitting in High-Resolution Nuclear Magnetic Resonance Spectra.* New York: W. A. Benjamin, Inc., 1962.
12. ROBERTS, J. D., *Nuclear Magnetic Resonance. Application to Organic Chemistry.* New York: McGraw-Hill Book Company, 1959.
13. STOTHERS, J. B., *Carbon-13 NMR Spectroscopy.* New York: Academic Press, 1972.
14. WIBERG, K. B., and NIST, B. J., *The Interpretation of NMR Spectra.* New York: W. A. Benjamin, Inc., 1962.

YEARLY REVIEWS

15. *Annual Review (Reports) on NMR Spectroscopy,* E. F. Mooney, ed. New York: Academic Press, beginning 1968, continuing to present.
16. *Progress in Nuclear Magnetic Resonance,* J. W. Emsley, J. Feeney, and L. H. Sutcliffe, eds. New York: Pergamon Press, beginning 1966, continuing to present.

Compilations of NMR Spectra and Literature References

17. *A Catalogue of the Nuclear Magnetic Resonance Spectra of Hydrogen in Hydrocarbons and Their Derivatives*, Humble Oil and Refining Co., Baytown, Texas.

18. Bovey, F. A., *NMR Data Tables for Organic Compounds*, Vol. 1. New York: Interscience Publishers, Inc., 1967.

19. Brugel, W., *Nuclear Magnetic Resonance Spectra and Chemical Structures*. New York: Academic Press, 1967.

20. Dugan, C. H., and Van Wazer, J. R., *Compilation of Reported F^{19} NMR Chemical Shifts (1951–1967)*. New York: Wiley-Interscience, 1970.

21. *Formula Index to NMR Literature Data*, M. G. Howell, A. S. Kende, and J. S. Webb, eds. New York: Plenum Press, 1965.

22. *High Resolution NMR Spectra Catalogue*, Vol. 1 (1962) and Vol. 2 (1963), Varian Associates, Palo Alto, California.

23. Johnson, L. F., and Jankowski, W. C., *Carbon-13 NMR Spectra; A Collection of Assigned, Coded and Indexed Spectra*. New York: Wiley-Interscience, 1972.

24. *NMR, EPR and NQR Current Literature Service*, Butterworth Inc., Washington, D.C.

25. *Nuclear Magnetic Resonance Spectral Data*. American Petroleum Institute, Project No. 44, Chemical Thermodynamic Properties Center, Agricultural and Mechanical College of Texas, College Station, Texas.

26. Pouchert, C. J., and Campbell, J. R., *The Aldrich Library of NMR Spectra*, Vols. I–XI. Aldrich Chemical Company, Inc., Milwaukee, Wisconsin, 1974–1975.

27. *Sadtler NMR Spectra*, Sadtler Research Laboratories, Philadelphia, Pennsylvania.

7

Mass Spectrometry

7.1
INTRODUCTION

Most of the spectroscopic and physical methods employed by the chemist in structure determination are concerned only with the physics of molecules; mass spectrometry deals with both the chemistry and the physics of molecules, particularly with gaseous ions. In conventional mass spectrometry, the ions of interest are positively charged ions. The mass spectrometer has three functions:

1. To produce ions from the molecules under investigation.
2. To separate these ions according to their mass-to-charge ratio.
3. To measure the relative abundances of each ion.

Mass spectrometry is the latest method to join the array of spectroscopic methods at the disposal of the organic chemist. Interestingly, the demonstration of the basic principles of mass spectrometry preceded that of most of the other physical methods currently used for structure determination. As early as 1898, Wein showed that positive rays could be deflected by means of electric and magnetic fields; in 1912, J. J. Thompson recorded the first mass spectra of simple low-molecular-weight molecules. The earliest of the prototypes of today's instruments were the mass spectrometer of Dempster (1918) and the mass spectrograph of Aston (1918). By the early 1940's, instruments suitable for the examination of the spectra of organic molecules of moderate molecular weight were commercially available. The earliest extensive studies of organic systems were made by the

266

petroleum industry on hydrocarbons. Although mass spectrometry was used for the quantitative analysis of hydrocarbon mixtures, there seemed to be little, if any, recognizable relationship between structure and spectra. This situation led to a dormant period in the utility of mass spectrometry for studies of organic molecules. In the late 1950's, Beynon, Biemann, and McLafferty clearly demonstrated the role of functional groups in directing fragmentation, and the power of mass spectrometry for organic structure determination began to develop. At that time the technology employed was essentially that available since 1940. Today mass spectrometry has achieved status as one of the primary spectroscopic methods to which a chemist faced with a structural problem turns; great advantage is found in the extensive structural information that can be obtained from submilligram or even submicrogram quantities of material.

7.2 INSTRUMENTATION

A rather large variety of commercial mass spectrometers is currently available. There is considerable variation in the mechanics by which the various mass spectrometers accomplish their tasks of production, separation, and measurement of ions. It is entirely beyond the scope and purpose of this book to provide the reader with a detailed discussion of instrumentation. However, so that the reader can at least have general ideas of the principles by which most mass spectrometers operate, representative examples will be discussed briefly.

Throughout this text the term *low resolution* will refer to instruments capable of distinguishing only between ions of differing nominal mass, i.e., ions whose mass-to-charge ratios differ by one mass unit. *High resolution* will refer to spectrometers capable of distinguishing between ions whose masses differ in the third decimal place or less, for example, $C_2H_6^+$, CH_2O^+, and CH_4N^+, with masses (based on carbon 12.00000) of 30.0469, 30.0105, and 30.0344, respectively.

7.2.1 Magnetic Focusing Instruments

A schematic diagram of a representative single-focusing, low-resolution mass spectrometer is shown in Fig. 7.1. Sufficient sample is introduced into the ionization chamber to produce a pressure of 10^{-5} mm Hg. The vapor in the ionization chamber is bombarded with an electron beam of variable energy [usually 50 to 70 electron volts (eV), the latter value being most commonly employed]. A small percentage of the molecules are ionized, by electron impact, into positively charged ions that subsequently form fragment ions. Negative ions are also formed to a small extent. The small repeller potential between the back wall of the ionization chamber and the first accelerator plate attracts the negative ions to the back wall and discharges them. At the same time, this potential pushes

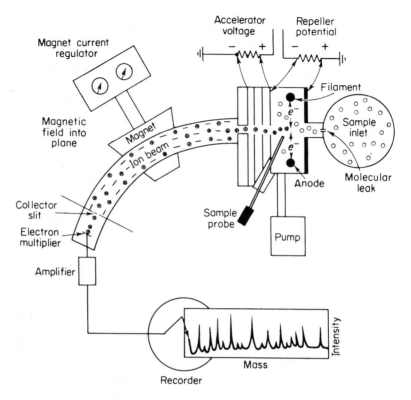

Fig. 7.1. Schematic diagram of a magnetic single-focusing mass spectrometer with a 60° sector magnet.

the positive ions toward the accelerating region. There they are accelerated by a potential of approximately 2000 V between plates that have a slit in them. The ions are focused on an exit slit by means of subsidiary accelerating plates and slits. The positive ions are accelerated by this potential according to Eq. (7.1),

$$\tfrac{1}{2}mv^2 = eV \tag{7.1}$$

where m is the mass of the ion, e the electronic charge, V the potential of the ion accelerating plate, and v the velocity of the particle. The accelerated ions then pass into the magnetic field H generated between the two poles of an electromagnet. In the magnetic field, the ions are deflected along a circular path according to Eq. (7.2),

$$r = \frac{mv}{eH} \tag{7.2}$$

where r is the radius of the path. Elimination of v from the preceding two equations yields Eq. (7.3).

$$\frac{m}{e} = \frac{H^2 r^2}{2V} \tag{7.3}$$

From this equation one can conclude that, at given values of H and V, only particles with a particular mass-to-charge ratio will arrive at the collector slit placed along the fixed path r. It can be further seen from the equation that particles with mass-to-charge ratios of $m/1$, $2m/2$, and $3m/3$ would all follow the same path and be detected and recorded simultaneously. Fortunately, under the conditions usually employed in recording mass spectra of organic compounds, most of the particles are singly charged, ions of higher charge generally being produced in insignificant quantities. Equation (7.3) also indicates that a spectrum of ions could be obtained at the collector slit by variation of either the magnetic field strength or the accelerating voltage. In instruments with a 180° magnetic arc path, the scanning is accomplished by decreasing the accelerating voltage at a suitable rate. In the 60° and 90° arc path instruments, the accelerating voltage is fixed and the magnetic field is varied.

The collector assembly consists of a series of collimating slits and an ion current detector. Modern electron multipliers are capable of detecting single ions. The current that is amplified and measured is directly related to the abundance of the ions at the mass being examined.

Two necessary features of an adequate recording system for mass spectrometry are (1) a fast response time and (2) an ability to accurately record peaks of widely varying intensity on the same spectrum. Although fast-response pen-and-ink recorders with manual or automatic attenuation are occasionally employed, the most practical solution is afforded by a multi-trace recording oscillograph. Ultraviolet light beams from an oscilloscope are played upon light-sensitive paper, simultaneously producing four or five tracings, depending on the type of the recorder, of intensity ratios $1:\frac{1}{3}:\frac{1}{10}:\frac{1}{30}$, or $1:\frac{1}{3}:\frac{1}{10}:\frac{1}{30}:\frac{1}{100}$, permitting intensity variations of up to 50,000 times in ion signal strength. This type of recording system is very fast; typically, a complete spectrum can be recorded in a few seconds or less. Many of the more modern instruments are equipped with computerized data acquisition systems that provide both a printout of the spectrum and a tabulation of m/e's and intensities. Using the data acquisition systems, spectra can be scanned within a few tenths of a second or less, with the ability to repeatedly scan the complete m/e region every few tenths of a second. This type of data collection system is extremely useful in recording the mass spectra of peaks coming out of a gas chromatograph, which are directly introduced into the mass spectrometer.

The high-resolution, double-focusing instruments (Figs. 7.2 and 7.3) incorporate most of the principles and methods, with additional refinements, outlined in the foregoing discussion of low-resolution instruments. It will be seen in Fig. 7.3 that, at a constant magnetic field H, any spread in the magnitude of velocity

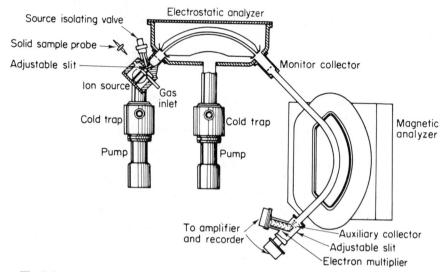

Fig. 7.2. Schematic diagram of a Nier-Johnson double-focusing, high-resolution mass spectrometer showing the radial electric analyzer and the magnetic analyzer. (Courtesy of Associated Electrical Industries, Ltd.)

(equivalent to the spread of *V*) will result in a spread in *r* for any given value of *m/e*. The peaks obtained in low-resolution mass spectrometry are indeed broadened by such a spread, which is caused by the contributions of initial kinetic or thermal energy to the kinetic energy gained by the particle during acceleration in the ionization chamber.

A several-thousand-fold increase in the resolving power of an instrument can be achieved by elimination of this energy spread in the ion beam before it enters the magnetic field. In the double-focusing instrument, ions are passed through a

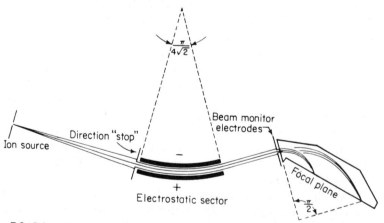

Fig. 7.3. Schematic diagram of a double-focusing mass spectrometer showing Mattauch-Herzog geometry. (Courtesy of Consolidated Electrodynamics Corporation.)

radial electrostatic field that sorts them into monoenergetic paths (velocity focusing) before they are passed into the magnetic field where they are refocused according to their mass-to-charge ratio. According to the design of the instrument, the spectrum can be scanned by the conventional sweeping of the magnetic field or the accelerating potential, the ions being recorded, as in the case of the low-resolution instrument, as they come into focus on the collector (Fig. 7.2), or all ions can be simultaneously recorded on a photographic plate placed in the focal plane of the instrument (Fig. 7.3). As will be discussed later, this latter method is especially valuable for obtaining the complete high-resolution spectra of microsamples whose concentrations in the ion source are rapidly changing and whose residence times in the instrument are short, as in the continuous monitoring of gas chromatography.

7.3
SAMPLE HANDLING

The sample must be introduced into the ionization chamber in the vapor state in sufficient quantity so that the vapor pressure exerted by the sample is in the range of 10^{-5} to 10^{-9} mm Hg. Most spectrometers have several sample introduction systems available; the method of choice is governed by the volatility of the sample.

For most organic compounds, direct insertion of the sample into the highly evacuated ionization chamber is advantageous. In a commonly employed system, the sample is placed in a small crucible on the end of a probe that is inserted into the ionization chamber through a vacuum lock system; if necessary to produce sufficient vapor pressure, the sample can be heated either by heating the entire ionization chamber or by heating elements associated with the probe. With probe heating, temperatures as high as 1000°C can be achieved; the method is limited only by the thermal stability of the compound.

For gases and liquids or solids of high volatility, the *inlet system* is often used for the introduction of the sample into the ion source. The inlet system consists of a series of vacuum locks, a reservoir, a sensitive pressure-measuring device, and a molecular leak into the ion source (Fig. 7.4). The volume of the reservoir on commercial instruments varies from 1 to 5 L; these relatively large volumes require proportionately larger samples compared with direct insertion. Sufficient sample is introduced into the reservoir to produce a pressure of 10^{-2} mm Hg. The entire system is in an oven that can be heated if necessary to produce sufficient pressure. The sample in the reservoir, under a pressure higher than that required in the ion source, is allowed to leak through a very small hole, producing a constant stream of the sample into the ionization region. The large volume of the reservoir minimizes depletion of the sample during the run; this is necessary if reproducible spectra are to be obtained at slow scan speeds.

An extremely useful and versatile method for the direct introduction of

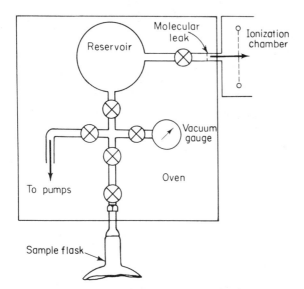

Fig. 7.4. Schematic diagram of an inlet system.

samples into the ionization chamber involves the direct hookup of a mass spectrometer and a gas-liquid chromatograph. Fortunately, the sensitivities of mass spectrometer and gas chromatographic detectors are comparable; the quantities of material usually eluted from a gas-liquid chromatography (glc) column are much too small for analysis by infrared spectroscopy and most other identification techniques. The eluent from a glc column can be monitored by direct introduction into the ionization chamber. The dilution of the sample by the carrier gas can be reduced by means of special separators that pump off the carrier gas preferentially. Since the concentration of any eluent is constantly and rapidly changing, this direct monitoring can be achieved only with spectrometers having rapid recording systems such as those spectrometers equipped with computer data acquisition systems or photographic plate detectors.

Even with the variety of sample-handling techniques, mass spectrometry cannot be applied routinely to all compounds; the limiting factor is often thermal stability of the sample at the temperature necessary to produce the minimum pressure. It is often possible to convert thermally unstable compounds to more stable and/or more volatile derivatives, for example, acids to esters, and alcohols to acetates or trimethylsilyl ethers.

7.4
PRODUCTION AND REACTIONS OF IONS
IN THE MASS SPECTROMETER

As indicated in the introduction, mass spectrometry involves the formation of and reactions of ions in the gas phase. In the ionization chamber of the mass spectrometer, high-energy electrons emitted from a heated filament collide with the molecules of the sample. If the energy of the electrons is greater than the ionization potential of the molecule, an electron will be ejected from the molecule, producing a molecular radical cation (see the reaction scheme below). The ionization potentials of most molecules fall in the range of 8 to 11 electron volts (eV); however, most mass spectrometers generally induce ionization with electrons having an average energy of 70 eV. When such an energetic electron collides with a molecule, the resulting molecular ion that is formed contains considerable excess internal energy, which is dissipated through subsequent fragmentation and/or rearrangement reactions.

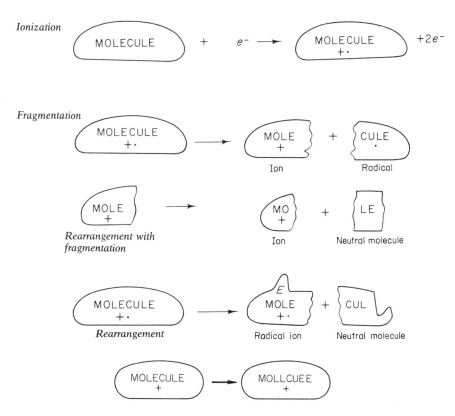

Due to this great excess of energy imparted to the parent molecular ion, in many cases the intensity of the parent ion is very weak, and in some cases is not detectable. To circumvent this problem, the energy of the ionizing electrons can be lowered; however, this also results in a general lowering in intensity of all of the peaks in the mass spectrum. An alternative technique that has proved to be extremely useful for the formation of molecular ions with little excess internal energy, thus resulting in relatively little fragmentation, involves ionization induced by interaction with an ion whose neutral parent has an ionization potential only slightly greater than that of the molecule to be ionized. In this case, very little excess energy is imparted to the molecular ion, and fragmentation occurs to a far less degree, with the molecular ion being more intense. This form of mass spectrometry is called *chemical ionization* mass spectrometry. Methane and argon are commonly used as the sources of the primary ions, which then cause electron transfer from the sample molecules, resulting in formation of the molecular ions. The methane and argon are used in great excess ($>10^3$ times) over the sample molecules so that initial ionization by the 70 eV electron beam results in essentially only ionization of the methane or argon. Chemical ionization mass spectrometry utilizes much higher pressures ($\sim 10^{-2}$ mm Hg) than does conventional mass spectrometry. Figure 7.5 illustrates the differences in relative intensities between electron impact and chemical ionization mass spectrometry.

In the following paragraphs the characteristics of parent ions and of the fragmentation and rearrangement reactions are described.

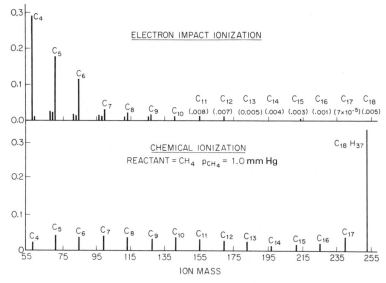

Fig. 7.5. Electron impact and chemical ionization spectra of $C_{18}H_{38}$.

7.4.1 Determination of Molecular Formulas

The mass spectrum of an organic compound indicates that the compound is composed of several different molecular species of different combinations of naturally occurring isotopes. The masses that we should calculate for mass spectrometry should not be based on the chemical scale, but should be the sum of the masses of the isotopes occurring in the species under consideration. For example, the molecular weight of methyl bromide is 94.95 on the chemical scale. This chemical scale weight is based on the weighted averages of the naturally occurring isotopes

C	12.011
H	1.008
Br	79.909

However, in the mass spectrum, we, of course, do not see the average molecular species but rather the molecular species corresponding to each possible isotopic combination according to the following naturally occurring abundances:

^{12}C 98.89% ^{1}H 99.985% ^{79}Br 50.54%
^{13}C 1.11% ^{2}H 0.015% ^{81}Br 49.46%

The low-resolution mass spectrum of methyl bromide in the region of the molecular ion is shown in Fig. 7.6.

In this text we shall refer to the radical cation produced by the removal of a single electron from a molecule composed of the lightest naturally occurring isotopes of the elements present in the compound as the *molecular ion*, symbolized by M. The terms M + 1, M + 2, etc. signify peaks at 1, 2, etc. nominal mass units higher than M.

If a molecular ion has sufficient stability to accord it a lifetime of approximately 10^{-5} seconds, it will be fully accelerated and recorded at its corresponding m/e value. Thus, for the large majority of compounds, mass spectrometry provides an exact and unambiguous method for the determination of the molecular weight of a molecule. The molecular ion and related isotopic species correspond to the peaks of highest mass in the spectrum in the absence of collision processes (see below). However, in a number of cases, compounds give molecular ions of very low or negligible intensity, and care must be taken not to confuse peaks due to impurities or fragments with molecular ions. Reduction of the ionizing voltage leads to higher molecular ion intensities.

The stabilization of the positive charge in the molecular ion and the tendency toward fragmentation influence the intensity of the molecular ion peak. Biemann

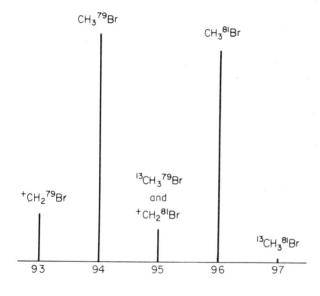

Fig. 7.6. Mass spectrum of methyl bromide in the region of the molecular ion. Small contributions by low-abundance ^{2}H are not indicated. Contributions by species containing both ^{13}C and ^{2}H are negligible.

has suggested the following approximate order of decreasing stability of the molecular ion:

Aromatic compounds—conjugated olefins—alicyclic compounds— sulfides—straight chain hydrocarbons—mercaptans—ketones—amines —esters—ethers—carboxylic acids—branched hydrocarbons—alcohols

Cyclic compounds naturally tend to give rise to correspondingly more intense molecular ion peaks than their acyclic analogs do because in the cyclic structure, cleavage of one bond does not lead to the splitting off of a fragment of lower mass. It should be emphasized that one should always check to see that the fragmentation pattern exhibited in the rest of the spectrum can be accommodated by the suggested molecular ion and that the intensity of that peak is of the magnitude expected for the proposed structure. In some cases, even though a compound does not exhibit a molecular ion, it is possible to arrive at the molecular weight of the compound by the partial interpretation of the mass spectrum. Another approach often used for determining molecular weights when compounds do not exhibit recognizable molecular ions is the formation of a derivative chosen to give a predicted mode of cleavage or a more stable molecular ion.

As an aid in determining whether a peak is due to a molecular ion or a fragment, the so-called *nitrogen rule*, which holds for all compounds containing carbon, hydrogen, oxygen, nitrogen, sulfur, and/or the halogens, as well as phosphorus, arsenic, and silicone, can be employed. The rule states that a neutral molecule of even-numbered molecular weight must contain either no nitrogens or an even number of nitrogen atoms; an odd-numbered molecular weight compound must contain an odd number of nitrogen atoms. Simple fragments formed from even-numbered compounds will always have odd mass numbers, with the exception of rearrangement peaks which are usually formed by the elimination of a neutral molecule with the formation of a fragment of even mass number.

Since only a very small fraction of the molecules in the ionization chamber are actually ionized, those ions that are formed are present in a high population of un-ionized molecules. Ion-molecule collisions, which result in the abstraction of an atom or a group of atoms from the neutral molecule by the positive ion, may occur, resulting in a particle of mass higher than the expected molecular ion. At the pressures usually employed in mass spectrometry, the only significant reaction of this type is the abstraction of a hydrogen atom by the molecular ion resulting in a peak at $M + 1$. This $M + 1$ peak becomes particularly important in those cases (ethers, esters, amines, aminoesters, nitriles) in which the molecular ion is relatively unstable, whereas the corresponding protonated species is quite stable. Since this $M + 1$ peak is formed in a bimolecular process, it can be recognized as such by the proportionality of its intensity to the square of the sample pressure. As the sample pressure is increased, the intensity of the $M + 1$ peaks will increase relative to the intensity of the other peaks in the spectrum. An $M + 1$ peak can also be recognized by its sensitivity to the repeller potential; an increase in the

Table 7.1. Masses and Natural Abundances of Nuclides of Importance in High-Resolution Mass Spectrometry of Organic Compounds

Nuclide	Mass	Natural Abundance (%)
^{1}H	1.007825	99.985
^{2}H	2.01400	0.015
^{12}C	12.00000	98.9
^{13}C	13.00335	1.11
^{14}N	14.00307	99.63
^{16}O	15.994914	99.80
^{19}F	18.998405	100.00
^{28}Si	27.976927	
^{31}P	30.973763	
^{32}S	31.972074	94.62
^{34}S	33.96786	4.2
^{35}Cl	34.968855	24.5
^{37}Cl	36.9658	75.5
^{79}Br	78.918348	49.5
^{81}Br	80.983	50.5
^{127}I	126.904352	100.00

repeller potential will decrease the residence time of ions in the ionization chamber and thus decrease the likelihood of collision processes.

As indicated earlier in this chapter, with high-resolution mass spectrometry it is possible to measure masses to four decimal places. Table 7.1 gives the exact masses of the nuclides frequently found in organic compounds.

It can be readily seen that each formulation of elements will have a unique mass associated with it. Measurement of the masses with sufficient accuracy allows one to determine the molecular formula. An excellent example of this approach is found in the work of Djerassi on the alkaloid vobtusine. For this alkaloid, earlier work based on microanalysis and other classical data had suggested possible molecular formulas of $C_{45}H_{54}N_4O_8$ (778), $C_{42}H_{48}N_4O_6$ (704), and $C_{42}H_{50}N_4O_7$ (722). Low-resolution mass spectrometry showed the molecular weight to be 718, which is inconsistent with all of the foregoing, but is consistent with the formulations of $C_{43}H_{50}N_4O_6$ (718.3730) and $C_{42}H_{46}N_4O_7$ (718.3366). A high-resolution mass spectrum indicated a molecular weight of 718.3743, clearly showing that the formula $C_{43}H_{50}N_4O_6$ was the correct one.

An extensive table of elemental composition *vs.* mass for use in high-resolution mass spectroscopy has been constructed by McLafferty (ref. 15). A sampling from this table is shown in Table 7.2. From high-resolution spectra, one obtains the elemental composition not only of the molecular ion, but of each ion that is examined. The spectrum is then interpreted directly according to the elemental formula of the molecule and its fragments.

Table 7.2. Mass *vs.* Elemental Composition at Nominal Mass 43

Mass	Elemental Composition
43.0058	CHNO
43.0184	C_2H_3O
43.0269	CH_3N_2
43.0421	C_2H_5N
43.0547	C_3H_7

From low-resolution spectra, it is sometimes possible to determine a unique molecular formula, or at least to severely limit the number of possibilities, by consideration of the relative intensities of the various isotope peaks. In this discussion we shall restrict our considerations to compounds containing carbon, hydrogen, oxygen, nitrogen, sulfur, fluorine, chlorine, bromine, and iodine. The principal stable heavier isotopes of each are listed in Table 7.1. To determine the molecular formula, one measures the intensity of the molecular ion $M^{\ddagger}$ and expresses the intensity of the $M^+ + 1$ peak and the $M^+ + 2$ peak as percentages of the molecular ion peak intensity. For any possible molecular formula, the observed values of the $M^{\ddagger} + 1$ and $M^{\ddagger} + 2$ peaks can be compared with the theoretical relative intensities calculated by the equation

$$I_n = \sum_{\substack{\text{all isotopic} \\ \text{combinations}}} x^{(f)^a y(i-f)^b}$$

where n is the number of higher-mass isotopic species present ($I = 1.00$ for the molecular species containing the lowest-mass isotopes of all atoms contained in the species), x is the statistical factor for the number of possible positional combinations for the lower-mass isotope, f is the fractional abundance of that lower-mass isotope, a is the number of lower-mass isotopes present, y is the statistical factor for the number of possible positional combinations for the higher-mass isotopic species, $(i - f)$ is the natural abundance of the higher-mass isotopic species, and b is the number of atoms of the higher-mass isotope; these are summed over all possible isotopic variations. Tables of such calculations for various combinations of C, H, N, and O have been published and appear in ref. 2 by Beynon, and, in part, in the book by Silverstein, Bassler, and Morrill (ref. 21).

An example of the use of this method follows. From a mass spectrum we obtain the following data:

m/e	Percent
100(M)	(100)
101(M + 1)	5.64
102(M + 2)	0.60

We can readily eliminate from further consideration formulas containing sulfur, chlorine, or bromine from the relatively small contribution of the M + 2 peak (these elements are usually easy to detect because of their high contribution to the M + 2 peak). Iodine need not be considered because its mass alone is 127. The relatively large contribution of the M + 1 peak suggests that fluorine is also not present. If we look in the Beynon table of C, H, N, O compound formulas of mass 100, we find the following data:

100	M + 1	M + 2
$C_2H_2N_3O_2$	3.42	0.45
$C_2H_4N_4O$	3.79	0.26
$C_3H_2NO_3$	3.77	0.65
$C_3H_4N_2O_2$	4.15	0.47
$C_3H_6N_3O$	4.52	0.28
$C_3H_8N_4$	4.90	0.10
$C_4H_4O_3$	4.50	0.68
$C_4H_6NO_2$	4.88	0.50
$C_4H_8N_2O$	5.25	0.31
$C_4H_{10}N_3$	5.63	0.13
$C_5H_8O_2$	5.61	0.53
$C_5H_{10}NO$	5.98	0.35
$C_5H_{12}N_2$	6.36	0.17
$C_6H_{12}O$	6.72	0.39
$C_6H_{14}N$	7.09	0.22
C_7H_2N	7.98	0.28
C_7H_{16}	7.82	0.26
C_8H_4	8.71	0.33

On the basis of the nitrogen rule, we need not consider further any formulas containing an odd number of nitrogens. From the M + 1 and M + 2 values, the formula $C_5H_8O_2$ is found to be the best fit. The measured isotope peaks are usually slightly higher than the calculated ones because of small contributions from bimolecular collisions, impurities, M − 1 peaks, background, etc. Moreover, it should be emphasized that the method is limited to compounds that produce relatively intense molecular ions and to compounds of low-to-moderate molecular weight. Intense molecular ion peaks are necessary if the isotope peaks are to be of measurable intensity. In compounds that contain large numbers of atoms, the isotopic distribution functions are too complex for accurate analysis, and the distinction between possible formulas is not possible owing to experimental uncertainties in the peak intensities.

In the formation of a molecular ion by electron bombardment, removal of an electron from a nonbonded electron pair on a heteroatom appears easier than removal of a pi electron, which in turn, is easier to remove than a sigma electron. It is therefore possible and often desirable to depict fragmentation processes with the positive charge and free valence localized on a specific atom, e.g., $CH_3CH_2\overset{+\cdot}{-}O-CH_2CH_3$. In diagrams in this text, we shall occasionally choose to show charges and radicals localized, but more often we shall represent ions and radical-ions by enclosing the structural formulas in brackets, e.g., $[CH_3CH_2—O—CH_2CH_3]^{\ddot{+}}$.

7.4.2 Fragmentation

The mass spectrum of a compound provides the structural chemist with two types of information: the molecular weight of the compound and the mass of the various fragments produced from the molecular ion. To obtain structural information, the chemist attempts on paper to reassemble these fragments in a way not unlike the assembling of a jigsaw puzzle. Hypothetical molecules containing these fragments are constructed, and the fragmentation patterns predicted for these models are checked against the spectrum obtained.

Fragmentation is a chemical process that results in bond breaking; the energetic considerations that are applicable to classical chemical reactions are also applicable to these fragmentation processes. There are two major differences between conventional ion chemistry and chemistry in the mass spectrometer. First, we are dealing with particles in excited states, and the energies involved are considerably higher than those from typical chemical reactions. Second, at the very low operating pressures ($\approx$ low concentration) of the mass spectrometer, we are dealing with unimolecular reactions, and energy is not dissipated to any appreciable extent by collisions with other molecules of the compound or solvent as might occur in solution chemistry; thus, further fragmentation of the initially formed ions occurs extensively. It must be pointed out, therefore, that the intensity of a given ion peak depends not only on its rate of formation, but also on the rate of its subsequent fragmentation.

Fragmentations are best interpreted based on the known stability of carbenium ions and free radicals in solution. In the two possible modes of fragmentation shown in the following equation, the dominant process in general will be to

$$[A—B]^{\ddot{+}} \begin{array}{c} \diagup A^+ + B\cdot \\ \diagdown A\cdot + B^+ \end{array}$$

generate the most stable cation, although the stability of the neutral radical product can be very important. In cases where both A^+ and B^+ cations are of similar stability, the positive charge will reside mostly with the highest-molecular-weight cationic fragment. Thus in the mass spectrum of propiophenone, the molecular ion fragments to produce the ions with relative intensities are as indicated.

$$
\begin{array}{ccc}
 & C_6H_5\overset{+}{C}{=}O & (100) \\
\overset{\displaystyle O}{\underset{\displaystyle \|}{[C_6H_5CCH_2CH_3]^{\ddot{+}}}} \longrightarrow & CH_3CH_2\overset{+}{C}{=}O & 0\cdot 64 \\
 & C_6H_5^+ & 43.4 \\
 & CH_3CH_2^+ & 1.85
\end{array}
$$

Often in the description of mass spectra it is convenient to refer to a particular fragment ion in terms of the molecular ion less the mass or elemental composition of the lost neutral fragment(s). The $C_6H_5CO^+$ ion at mass 105 in the spectrum of propiophenone can be characterized as the $M - 29$ or the $M - C_2H_5$ ion.

As an aid in the visualization of fragmentation and rearrangement processes, arrows are often used to symbolize bond breaking and bond making [Eq. (7.4)].

$$\left[\begin{array}{c} \overset{H}{\underset{\displaystyle CH_3-\overset{\displaystyle O}{\overset{\|}{C}}\diagdown \atop \displaystyle CH_2}{} \end{array} \right]^{\ddagger} \longrightarrow \left[\begin{array}{c} OH \\ CH_3-C \\ \diagdown CH_2 \end{array} \right]^{\ddagger} + CH_2 = CH_2 \qquad (7.4)$$

Since in conventional chemical notation such arrows usually indicate two-electron shifts, other authors have suggested different representations, e.g.,

$$CH_3-\overset{..+}{\overset{\|}{C}}\quad CH_2 \qquad CH_3-\overset{+\,.}{\overset{\|}{C}}\quad CH_2$$

($\frown$ = one electron shift)

In the absence of a well-established convention, this text will continue to employ the method illustrated in Eq. (7.4). The student is cautioned to be well aware that the arrows in such a representation do not have physical significance; they indicate neither the direction of electron flow nor the numbers of electrons moving. It is simply a method of "electron bookkeeping."

Most of the important types of fragmentation are summarized in general form below.

1. Simple carbon-carbon bond cleavages

$$a. \left[R-\overset{|}{\underset{|}{C}}- \right]^{\ddagger} \longrightarrow R^+ + \cdot \overset{|}{\underset{|}{C}}-$$

$$b. \; R-\overset{|}{\underset{|}{C}}-\overset{|}{C}{}^+ \longrightarrow R^+ + \;\diagdown\!\!C\!\!=\!\!C\diagup$$

The relative importance of R^+ increases in the order

$$CH_3 < CH_2R' < CHR_2' < CR_3' < CH_2-CH=CH_2 < CH_2-\!\!\left\langle\!\!\bigcirc\!\!\right\rangle$$

$$c. \left[R-\overset{O}{\overset{\|}{C}}-R \right]^{\ddagger} \longrightarrow R\cdot + R\overset{+}{C}\!\!=\!\!O$$

$$d. \; R-\overset{+}{C}\!\!=\!\!O \longrightarrow R^+ + C\!\!=\!\!O$$

2. Cleavages involving heteroatoms

$$X = halogen, OR', SR', NR'_2 \ (R' = H, Alkyl, Aryl)$$

a. $\left[-\overset{|}{\underset{|}{C}}-X \right]^{\ddagger} \longrightarrow -\overset{|}{\underset{|}{C}}{}^+ + X\cdot$

b. $\left[R-\overset{|}{\underset{|}{C}}-X \right]^{\ddagger} \longrightarrow R\cdot + \overset{\diagdown}{\underset{\diagup}{C}}{}^+-X$

3. Concerted cleavages

a.

b.

7.4.3 Rearrangements

In the mass spectra of almost all compounds, there are fragment peaks whose presence cannot be accounted for by the simple cleavage of bonds in the parent compound. These fragments are the result of rearrangement processes. Some rearrangements occur during a fragmentation process and are classified as specific rearrangements. Other rearrangements occur, leading to hydrogen and carbon atom scrambling within the cationic species, and are termed random rearrangements. Subsequent fragmentation produces ions not directly related to the parent ion.

The specific rearrangement processes involve the migration of a hydrogen atom from one part of the parent ion to another during the fragmentation process. Such rearrangements typically involve the migration of a specific hydrogen atom that is sterically accessible to a carbenium or radical cation site. (Recall that in the ionization, electrons are most easily removed from π electron systems and nonbonded pairs of electrons on heteroatoms.) Perhaps the most common rearrangement of this type is the McLafferty rearrangement involving the migration of a hydrogen atom to a π electron system *via* a six-membered ring transition state. This type of rearrangement is illustrated in general terms as follows:

X, Y, and Z can be O, N, C, S, and any combinations thereof.

Other common rearrangements involving hydrogen atom migration include elimination reactions in which 1,4-elimination *via* a six-centered transition state is generally the dominant process.

$$X = OH, OR, NR_2, SR, Halogen$$

Carbenium ions undergo 1,3 migrations with elimination of an alkene.

Another type of rearrangement process is that involving 1,2-shifts. As in ordinary chemical reactions, the driving force for this type of rearrangement is the formation of a more stable species.

One can readily envision how 1,2-shifts might ultimately lead to fragments that seemingly have little to do with the initial structure of the compound, e.g., aryl sulfones lose carbon monoxide during fragmentation. Perhaps the most extensively studied rearrangement of this type is the formation of tropylium ion from toluene. In this case it appears that ring expansion occurs in the parent ion before loss of a hydrogen atom. It has been demonstrated that in the intermediate $C_7H_8^+$ all hydrogens are equivalent. The spectra of thirteen nonaromatic isomers of toluene have been studied. In all cases, the spectra are remarkably similar and, in every example, the most abundant ion is mass 91 ($C_7H_7^+$), apparently formed from a common $C_7H_8^+$ intermediate.

Mass spectral studies with ^{13}C and ^2H labeled molecules have revealed that in most cases the carbon and hydrogen atoms become essentially randomly distributed in the carbenium ion species formed after fragmentation of the parent radical cation, and in the succeeding carbenium ion species. Two processes appear to be responsible; one involving the well-known typical 1,2-hydride and alkyl shifts

$$-\overset{\overset{\displaystyle H}{|}}{C}-\overset{+}{C}\diagdown \longrightarrow \diagup\overset{+}{C}-\overset{\overset{\displaystyle H}{|}}{C}-$$

in carbenium ions (outlined above), and the other involving formation of protonated cyclopropanes by carbenium ion insertion into a γ-C—H bond. Protonated cyclopropanes are well known to undergo rapid hydrogen migration around the ring. Opening of the protonated cyclopropanes in any of the three possible manners results in both hydrogen and carbon atom scrambled species.

Obvious examples in which such processes must be occurring are illustrated in the following equations.

7.4.4 Metastable Peaks

If the rate of decomposition of an ion m_1 formed in the ionization chamber is very fast, almost all such ions will decompose before reaching the acceleration region, and only the fragments m_2 will be accelerated, deflected, and detected in any

$$m_1^+ \longrightarrow m_2^+ + \text{neutral fragment}$$

significant quantity. If the original ion (m_1) is very stable, it will be observed as an intense peak relative to any daughter ions (m_2). Ions that decompose at an intermediate rate should show considerable intensity for both the primary (m_1) and the daughter (m_2) ions. With this intermediate rate of decomposition, some of the primary ions will decompose into fragments while traveling through the accelerating region of the instrument. Such ions will at first be accelerated as mass m_1, decompose with loss of some kinetic energy to the neutral fragment, and then continue to be accelerated and deflected as mass m_2. Such an ion will be recorded as a broad peak of low intensity (generally 1% or less) at mass m^*, as follows:

$$m^* = \left[\frac{(m_2)^2}{m_1} \right] \tag{7.5}$$

The metastable peaks (m^*) are usually found at 0.1 to 0.4 mass units higher than calculated from this equation. An example of a characteristic metastable peak is shown in Fig. 7.7.

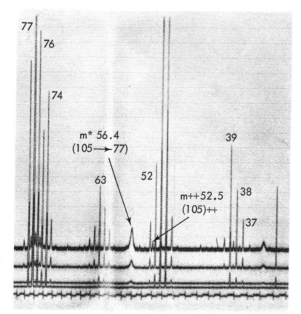

Fig. 7.7. A portion of a mass spectrum traced by a four-element galvanometer illustrating normal peaks, metastable peaks, and peaks due to doubly charged ions.

Metastable peaks are very important in the deduction of fragmentation mechanisms, for they indicate that the fragment of mass m_2 is formed in a one-step process from mass m_1. For example, in the mass spectrum of phthalic anhydride the following principal fragmentation pattern is indicated:

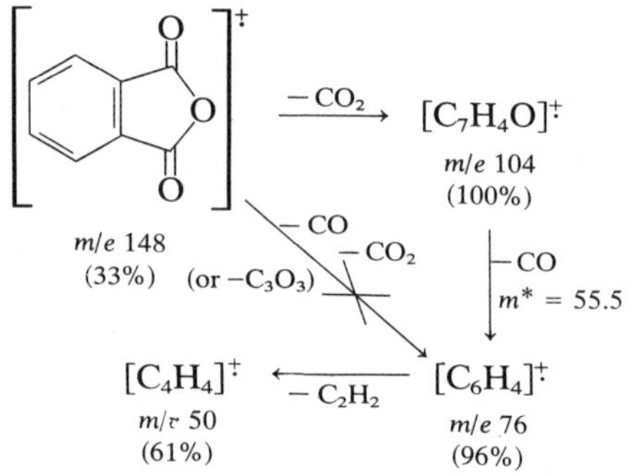

The metastable peak at 55.5 (calculated. 55.5) tells us that the ion at m/e 76 (m_2) (benzyne?) is formed from the m/e 104 (m_1) ion and not by concerted loss of carbon monoxide and carbon dioxide from the molecular ion.

7.5
ISOTOPIC LABELING

Using isotopic labeling at strategically located points in a molecule and then following this label through to the products is a well-established method of studying chemical and biological reaction mechanisms. For the detection of radioisotopes, various counting techniques are employed; for the detection of deuterium and sometimes carbon-13, NMR techniques can be used (deuterium can also be measured by combustion of the sample to water and measurement of the density of the water). However, mass spectrometry is the only generally applicable method for the detection of stable isotopes. The mass spectrometric technique has the advantages of examining the intact molecule and often pinpointing the location of the label.

Isotopic labeling is also a very useful method of establishing reaction pathways that occur in the mass spectrometer. The method has special pertinence to low-resolution spectrometry, which cannot distinguish between ions of the same nominal mass as illustrated by the following: The mass spectrum of cyclopentanone exhibits a prominent peak at mass 28 that could be attributed to either carbon monoxide or ethylene. Since this fragment retains two deuterium atoms (producing a peak at m/e 30) in the spectrum of the tetradeuterio derivative, the peak is due to ethylene.

7.6
HANDLING AND REPORTING OF DATA

Because of space limitations in journals and books, just as in other forms of spectroscopy, it is impractical to present reproductions of the original mass spectra. Likewise, as is characteristic of other forms of spectroscopy, the methods of reporting data are considerably varied.

Data are frequently presented in tabular form, in which the mass numbers and the intensities of the peaks are listed. Two methods are commonly used to express intensity. The simplest and most frequently used method, though perhaps not the best, expresses the intensities in relation to the most intense peak in the spectrum. This peak, known as the *base peak*, is arbitrarily assigned a value of 100. The second method relates the intensity of any given peak with the total intensity of the spectrum, i.e., the sum of the intensities of all the peaks, or the *total ionization* (Σ). The intensity of a peak is then expressed as percent of Σ, and it indicates the extent to which the molecular ions decompose into this fragment. It is frequently not practical to measure the peaks from $m/e = 1$ to the molecular weight; instead, they are measured over a more restricted range. In this case the designation Σ_m is used, where the value of m indicates the lowest mass in the range, e.g., Σ_{30} signifies that the intensities have been summed from mass 30 to the highest mass in the spectrum.

Chemists, especially organic chemists, immediately associate the term spectrum with a picture with lines on it. Mass spectral data are often presented in the form of a bar graph, the ordinate indicating the relative abundances (intensity) either as percent of the base peak or percent Σ, or both, and the abscissa indicating the mass-to-charge ratios. This method of presentation has the advantage that the overall characteristics of the spectrum are available at a glance.

The most extensive use of the large amount of data made available on a complete high-resolution mass spectrum involves the *element mapping* technique initiated by Biemann and co-workers. With this method, exact mass measurements are made on all, or almost all, of the ions produced, and possible elemental formulas are calculated for each ion. With large molecules this can involve hundreds of ions. The spectrum is then interpreted directly according to the elemental composition of ions responsible for each peak in the spectrum. The handling of the vast quantities of data necessary for the complete analysis of the spectrum is made practical only by the use of computer techniques. For high-resolution instruments employing the Nier–Johnson geometry with scanning (Fig. 7.2), the data can be digitized directly on magnetic tape or punched cards, which can then be analyzed by means of a computer. The Mattauch–Herzog instruments (Fig. 7.3) record the entire spectrum simultaneously on a photographic plate. The line intensity and position data from the plate are then digitized by means of an automatic or semiautomatic comparator. The calculated line positions are converted to exact masses by comparison with reference lines of a standard spectrum (usually perfluorokerosene) recorded simultaneously. Although the former method has the advantage of not requiring the intervention of an expensive and somewhat time-consuming comparator, it has the disadvantage of requiring a relatively long time to record the complete spectrum. Thus it is not as suitable for use with direct hookup with a gas-liquid chromatograph.

From the digitized data, the computer calculates possible elemental compositions for each ion, sorts these into columns according to the content of heteroatoms, and prints out in the form of an element map. Figure 7.8 is the element map obtained from 3-octanone. In the first column, the nominal masses of the various ions are given. The second column indicates ions containing only carbon and hydrogen; the number of each element present is listed, followed by the difference in found and calculated mass expressed in millimass units, and then followed by 1 to 10 asterisks indicating on a log scale the approximate intensity of the particular ion. The third column gives the same information for ions containing carbon, hydrogen, oxygen, and so on. In the element map, the heteroatom composition increases from left to right, and the mass increases from top to bottom. The molecular ion is found in the entry at the bottom right, in this case, mass 128 and composition $C_8H_{16}O$. For an example of the interpretation of the information, we shall examine the information given at nominal mass 56. We note that there are two entries on this line, indicating two ions of the same nominal mass: the $C_4H_8^+$ ion, which appears in the spectrum 1 millimass unit higher than calculated and with a relative intensity of 3 on a log scale, and the $C_3H_4O^+$ ion, which appears in the spectrum 1 millimass unit lower than calculated with a relative intensity of 4.

The element map indicates that the largest saturated alkyl fragment (C_nH_{2n+1}) in the molecule is a pentyl group. The oxygen-containing fragments indicate acyl ions ($C_nH_{2n-1}O$) corresponding to $n = 2$, 3, and 6 predominate. Acyl ions of $n = 3$ and 6 are indicative of a ketone with 8 carbons with the carbonyl oxygen at C_3. Furthermore, the rearrangement ion corresponding to C_4H_8O implies that the chain is unbranched at the 4 position. At this point we can write the following structure:

$$C_4H_9—CH_2—CO—CH_2CH_3$$

A somewhat unexpected peak appears at mass 81 with composition C_6H_9, which probably arises from loss of water from the $C_6H_{11}O$ acyl ion. This example

		CH		CHO	CHO$_2$	CHO$_2$
40	3/4	0**				
41	3/5	0*****				
42	3/6	0****				
43	3/7	0******	2/3	0*****		
44			2/4	0**		
45			2/5	0**		
50	4/2	0**				
51	4/3	0***				
52	4/4	0*				
53	4/5	0***				
54	4/6	0**				
55	4/7	1*****	3/3	1****		
56	4/8	1***	3/4	−1****		
57	4/9	1****	3/5	1******		
58			3/6	0***		
59			3/7	0**		
62	5/2	1*				
63	5/3	0*				
65	5/5	0*				
67	5/7	0**				
68	5/8	0**				
69	5/9	0****	4/5	0****		
70	5/10	0***	4/6	0***		
71	5/11	0******	4/7	0****		
72			4/8	0******		
73			4/9	−2*****		
75	6/3	1*				
79	6/7	0*				
80	6/8	0*				
81	6/9	0***				
83			5/7	0**		
84			5/8	0*		
85			5/9	1****		
86			5/10	0***		
97			6/9	0**		
98			6/10	0*		
99			6/11	0*****		
114			7/14	0*		
128			8/16	0*****		
		CH		CHO	CHO$_2$	CHO$_3$

Fig. 7.8. Element map of 3-octanone.

illustrates the utility of the element map for the interpretation of structural information and, therefore, for uncovering unexpected fragmentation processes.

Figure 7.8 is but one of several different ways of presenting computer-interpreted data; however, all of the methods present basically the same type of data and comparisons of observed with calculated formula weights.

7.7
MASS SPECTRA OF
SOME REPRESENTATIVE COMPOUNDS

The important fragmentation patterns for a number of compounds representative of the most common chemical classes are described in this section. For a more comprehensive treatment, the general references cited at the end of this chapter should be consulted. *Mass Spectrometry of Organic Compounds* by Budzikiewicz, Djerassi, and Williams is a very thorough and suitable treatment for chemists interested in organic structure determination. In many cases, the structural formulas shown in this section for mass spectral fragments should be considered only as probable representations of the observed elemental compositions of the fragments.

7.7.1 Hydrocarbons

7.7.1a Saturated Hydrocarbons

The utility of mass spectra of hydrocarbons is greatly enhanced by the availability of a large number of reference spectra. Straight-chain hydrocarbons exhibit clusters of peaks, 14 mass units (CH_2) apart, of decreasing intensity with increasing fragment weight (Fig. 7.9). The molecular ion peak is almost always present, albeit of low intensity. Branching causes a preferential cleavage at the point of the

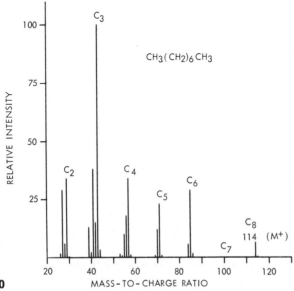

Fig. 7.9. Mass spectrum of a straight-chain hydrocarbon.

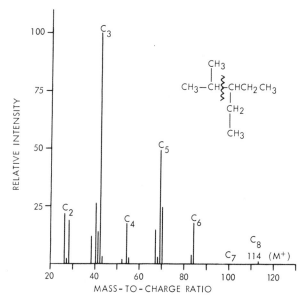

Fig. 7.10. Mass spectrum of a branched hydrocarbon.

branch, resulting in the formation of a secondary carbenium ion (Fig. 7.10). Low-intensity ions from random rearrangements are common.

Saturated cyclic hydrocarbons give rise to complex spectra. The molecular ion peak is usually rather intense. Characteristic peaks are usually formed from the loss of ethylene and side chains.

7.7.1b Unsaturated Hydrocarbons

The molecular ion peak, apparently formed by removal of a π electron, is usually distinct. The prominent peaks from monoalkenes have the general formula C_nH_{2n-1} (allyl carbenium ions), but fragments of the composition C_nH_{2n} formed in McLafferty rearrangements are also common (Fig. 7.11).

7.7.1c Aromatic Hydrocarbons

Hydrocarbons containing an aromatic ring usually give readily interpretable mass spectra, which exhibit strong molecular ion peaks. The molecular ion of benzene

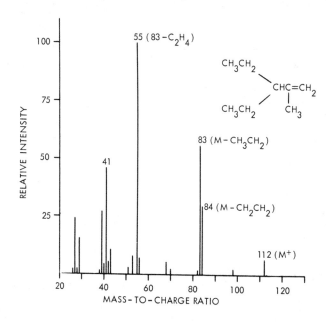

Fig. 7.11. Mass spectrum of a typical alkene.

$[C_6H_6]^+$ accounts for 43% of the total ion current at 70 eV in the mass spectrum of benzene. The most characteristic cleavage of the alkyl benzenes occurs at the bonds beta to the aromatic ring. In the usual case, substituents are preferentially lost resulting in the formation of tropylium or substituted tropylium ions.

If side chains of propyl or larger are present, a McLafferty rearrangement may also occur (Fig. 7.12).

7.7.2 Halides

Iodine and fluorine are monoisotopic, whereas chlorine and bromine occur naturally as mixtures of two principal isotopes; hence a molecular ion or fragment ion containing a chlorine will show a M + 2 iostope peak amounting to about a third of the intensity, whereas a bromine-containing ion will be accompanied by a

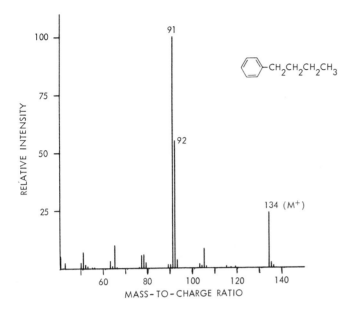

Fig. 7.12. Mass spectrum of an aromatic hydrocarbon.

M + 2 peak of almost equal intensity. For organic halides, the abundance of the molecular ion in a given series of compounds increases in the order F < Cl < Br < I. The intensity of the molecular ion peak decreases with increasing size of the alkyl group and α-branching. The following fragmentation processes are listed in approximately decreasing order of importance.

1. $R—X^+ \rightarrow R^+ + X\cdot$. Most important for X = I or Br.
2. $R—CH_2CH_2X^+ \rightarrow [RCH{=}CH_2]^{\ddagger} + HX$. More important for X = Cl and F than for I or Br.
3. $R—CH_2X^+ \rightarrow R\cdot + CH_2{=}X^+$. Heaviest group preferentially lost.

4.

$$R\overset{\frown}{—}CH_2 \quad \overset{\nearrow}{X^+} \qquad \longrightarrow \qquad \begin{matrix} CH_2{—}X^+ \\ | \quad\quad | \\ CH_2 \quad CH_2 \\ \diagdown \quad \diagup \\ CH_2 \end{matrix} \quad + R\cdot$$

X = Br or Cl

The molecular ion peak of aromatic halides is usually abundant. An M − X peak is usually observed (Fig. 7.13c).

7.7.3 Alcohols and Phenols

The molecular ion peak of primary and secondary alcohols is usually weak; that of tertiary alcohols is usually not detectable. The most important general

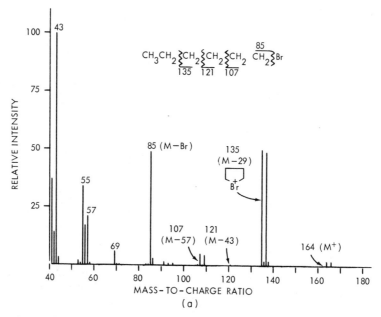

Fig. 7.13. Mass spectra of organic halides.

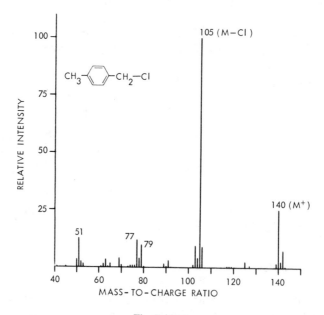

Fig. 7.13(*b*)

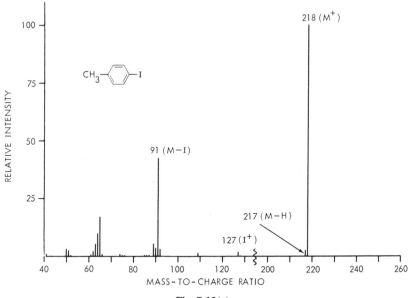

Fig. 7.13(*c*)

fragmentation process involves cleavage of the bond *beta* to the oxygen atom. The largest group is lost most readily.

$$R_2-\underset{\underset{R_3}{|}}{\overset{\overset{R_1}{|}}{C}}-\overset{+}{\overset{\cdot\cdot}{O}}H \longrightarrow R_1^{\cdot} + \underset{R_3}{\overset{R_2}{>}}C=\overset{+}{\overset{\cdot\cdot}{O}}H$$

Loss of water generally occurs by a cyclic mechanism where $n = 2$ is the dominant process.

$$\left[\overset{(CH_2)_n}{RCH\underset{H\cdot OH}{\diagdown}CHR}\right]^{\ddagger} \xrightarrow[n=1\ or\ 2]{} \left[\overset{(CH_2)_n}{RCH\underset{}{\underline{\qquad}}CHR}\right]^{\ddagger} + H_2O$$

A concurrent elimination of water and an alkene is usually observed with alcohols of four or longer carbon chains.

$$\underset{\underset{\underset{CH_2}{\diagdown}}{\overset{CH_2}{\diagup}}\underset{CHR}{\overset{\cdot O^+}{\overset{|}{\underset{}{}}}} }{ } \xrightarrow[-C_2H_4]{-H_2O} [CH_2=CHR]^{\ddagger}$$

Examples of mass spectra are shown in Fig. 7.14.

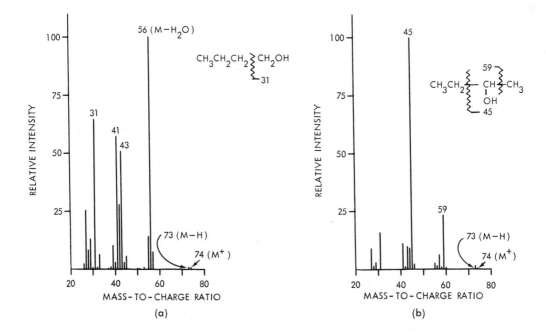

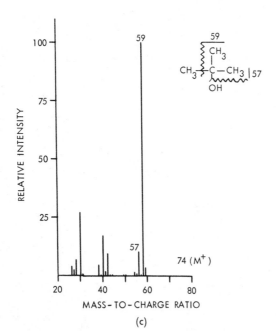

(c)

Fig. 7.14. Mass spectra of isomeric butanols.

The fragmentation of a typical cyclic alcohol is illustrated below.

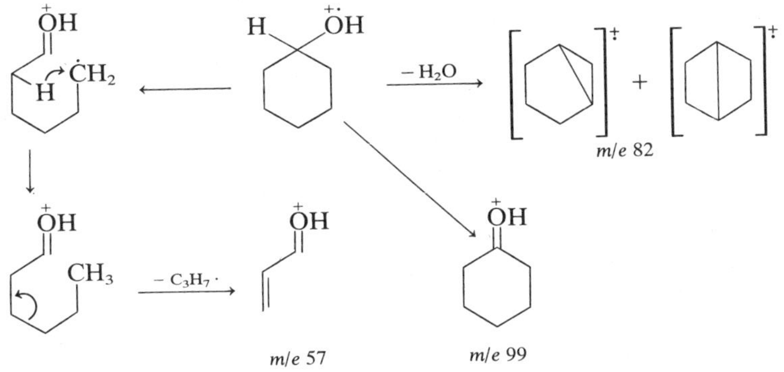

m/e 82

m/e 57   *m/e* 99

Benzylic alcohols typically exhibit intense molecular ion peaks, as well as M − 17 (loss of ·OH) peaks and rearrangement fragment peaks.

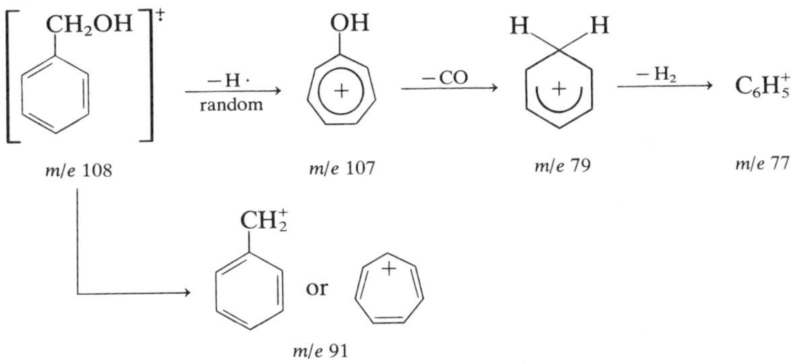

m/e 108   *m/e* 107   *m/e* 79   *m/e* 77

or

m/e 91

Phenols also exhibit intense molecular ion peaks as illustrated in Fig. 7.15.

Trimethylsilyl ethers of alcohols are widely employed in mass spectrometry and gas liquid chromatography because of their higher volatility compared with the parent alcohol. Although the molecular ion peak of these ethers is often of low intensity, the molecular weight can usually be inferred from the strong M − CH_3 peak.

$$CH_2=\overset{+}{O}Si(CH_3)_3 \xleftarrow{-R\cdot} RCH_2\overset{+\cdot}{O}Si(CH_3)_3 \xrightarrow{-CH_3\cdot} RCH_2\overset{+}{O}Si(CH_3)_2$$

m/e 93   M − 15

$(CH_3)_3Si^+$ $(CH_3)_3SiO^+$ $HO-\overset{+}{Si}(CH_3)_2$

m/e 73   *m/e* 89   *m/e* 75

7.7.4 Ethers, Acetals, and Ketals

The molecular ion peak of ethers is weak but can usually be observed. M + 1 peaks frequently occur at higher pressures. The ethers undergo fragmentations

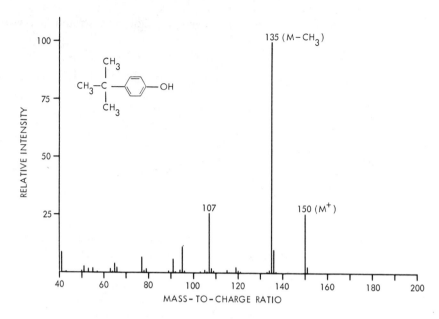

Fig. 7.15

similar to those of alcohols. The two most important pathways are:

1. Cleavage of a bond *beta* to the oxygen.

$$R-CH_2\overset{+\cdot}{O}R \longrightarrow CH_2=\overset{+}{O}R + R\cdot$$

In the case of branching at the α-carbon, the largest R group is lost preferentially.

2. Cleavage of a C—O bond.

$$R-\overset{+\cdot}{O}-R \longrightarrow RO\cdot + R^+$$

Acetals and ketals behave in a similar fashion.

Important fragmentations of aromatic ethers are illustrated below and in Fig. 7.17.

$$\left[\bigcirc\right]^{\ddagger} \xleftarrow{-CH_2O} \left[\bigcirc\!\!-OCH_3\right]^{\ddagger} \xrightarrow{-CH_3^{\cdot}} \left(\bigcirc\!\!\overset{O}{+}\right) \xrightarrow{-CO} C_5H_5^+$$

$$\left[\overset{OCH_2CH_2R}{\bigcirc}\right]^{\ddagger} \longrightarrow \left[\overset{OH}{\bigcirc}\right]^{\ddagger} + CH_2=CHR$$

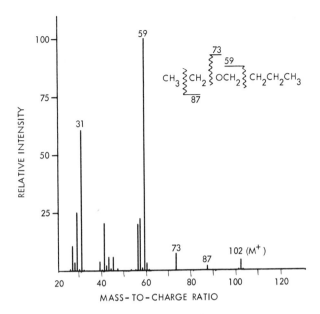

Fig. 7.16

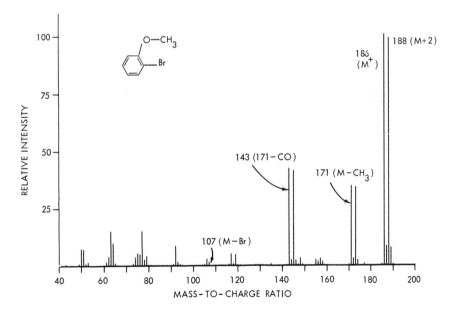

Fig. 7.17

7.7.5 Aldehydes

Aliphatic aldehydes generally exhibit weak molecular ion peaks. α-Cleavage occurs extensively to produce M − 1 and M − 29 peaks.

$$\alpha\text{-Cleavage} \quad [R\text{—CHO}]^{+} \longrightarrow R\text{—}\overset{+}{C}\text{=}O + H\cdot$$

$$[R\text{—CH}_2\text{—CHO}]^{+} \longrightarrow R\text{—CH}_2^{+} + H\dot{C}O$$

$$\beta\text{-Cleavage} \quad [R\text{—CH}_2\text{—CHO}]^{+} \longrightarrow R^{+} + H_2C\text{=}CH\text{—}O\cdot$$

McLafferty rearrangements generally result in the formation of the dominant fragment peaks.

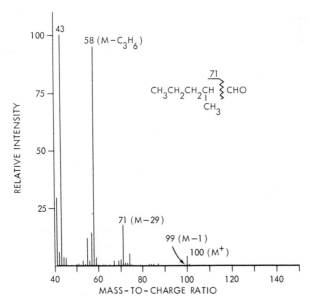

Aromatic aldehydes typically exhibit intense molecular ion and M − 1 peaks.

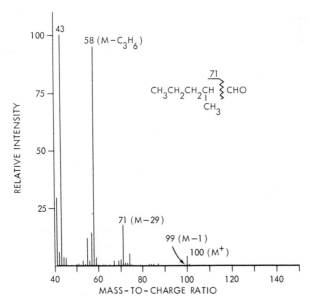

Fig. 7.18. Mass spectrum of an aliphatic aldehyde.

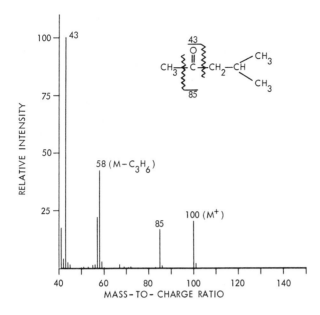

Fig. 7.19. Mass spectrum of an aliphatic ketone.

7.7.6 Ketones

The molecular ion of aliphatic ketones is generally intense. Fragmentation pathways are similar to those of aldehydes. The peak arising from α-cleavage with loss of the larger alkyl group is more intense than that from loss of the smaller one. Thus the base peak for most methyl ketones is m/e 43(CH_3CO^+) (intense

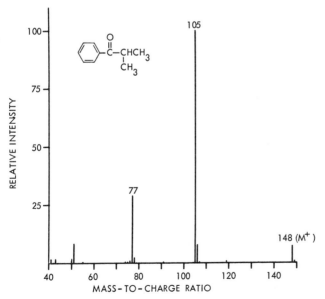

Fig. 7.20. Mass spectrum of an aromatic ketone.

peaks may also occur at *m/e* 43 with propyl or isopropyl ketones representing loss of $C_3H_7\cdot$ from the molecule ion).

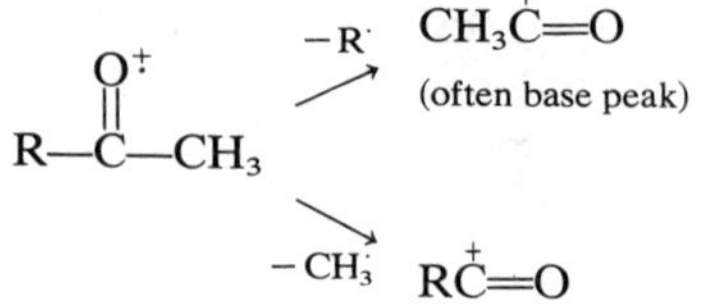

$$R-\overset{\overset{O^{+}}{\|}}{C}-CH_3 \quad \overset{-R\cdot}{\nearrow} \quad CH_3\overset{+}{C}=O \quad \text{(often base peak)}$$

$$\overset{-CH_3\cdot}{\searrow} \quad R\overset{+}{C}=O$$

With ketones containing γ-hydrogens, McLafferty rearrangements are prevalent.

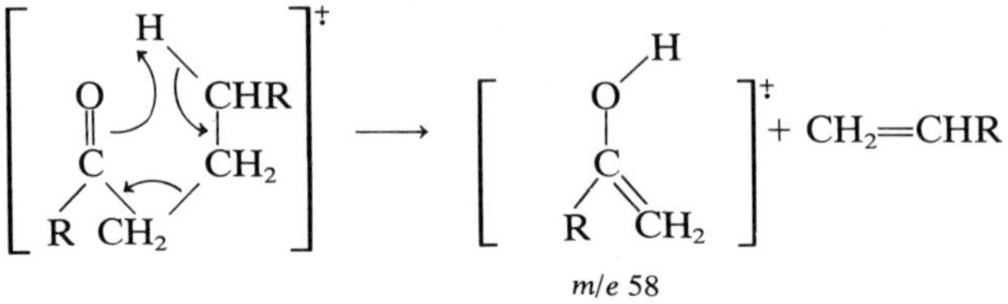

m/e 58

Important modes of fragmentation of cyclic and aromatic ketones are illustrated below.

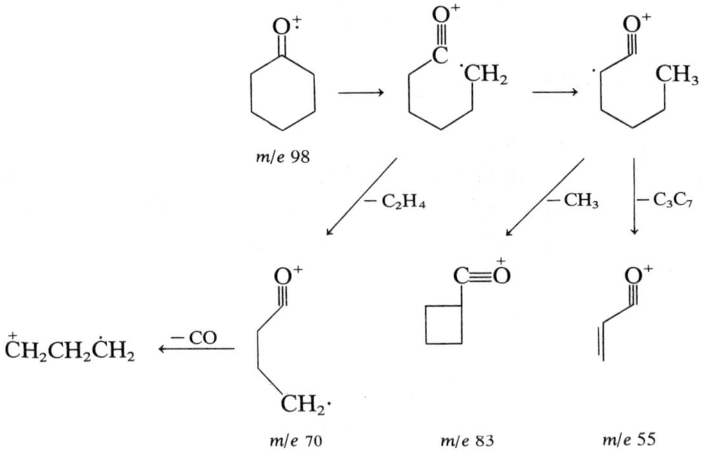

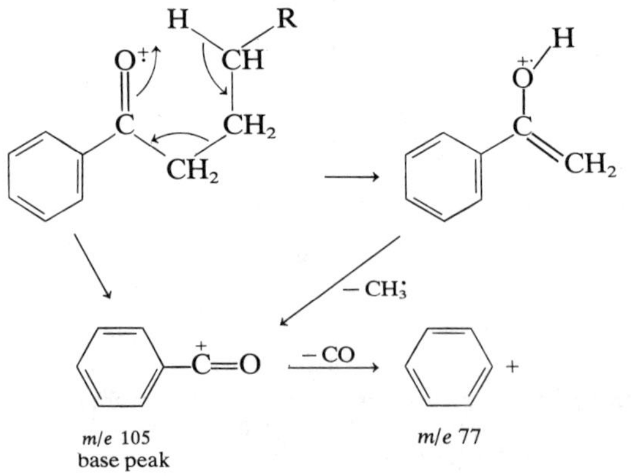

m/e 98

$-C_2H_4$ $-CH_3$ $-C_3C_7$

$\overset{+}{C}H_2CH_2\dot{C}H_2 \quad \overset{-CO}{\longleftarrow}$

m/e 70   *m/e* 83   *m/e* 55

$-CH_3^{\cdot}$

$\overset{-CO}{\longrightarrow}$

m/e 105
base peak

m/e 77

7.7.7 Carboxylic Acids

Acids are more frequently and often more conveniently examined as their methyl esters. The free acids exhibit weak but discernible molecular ion peaks. Normal aliphatic acids in which a γ-hydrogen is available for transfer have their base peak at m/e 60.

$$\text{R—C} \quad \overset{H}{\diagup} \quad \text{O}^+ \quad \longrightarrow \quad \left[\begin{array}{c} \text{HO} \\ \text{C} \\ \text{CH}_2 \quad \text{OH} \end{array} \right]^{\ddagger} + \text{RCH}=\text{CH}_2$$

$$m/e\ 60$$

Both aliphatic and aromatic acids exhibit a peak due to [COOH] at m/e 45.

$$[\text{R—COOH}]^+ \xrightarrow{-\text{R}\cdot} \ ^+\text{COOH}$$

$$m/e\ 45$$

In the spectra of aromatic acids, the M − OH peak is prominent.

$$\left[\text{C}_6\text{H}_5\text{—C} \overset{O}{\underset{OH}{}} \right]^{\ddagger} \xrightarrow{-\text{OH}\cdot} \text{C}_6\text{H}_5\text{—C}\equiv\text{O}^+ \xrightarrow{-\text{CO}} \text{C}_6\text{H}_5 +$$

$$m/e\ 105 \qquad\qquad m/e\ 77$$

7.7.8 Esters and Lactones

7.7.8a Methyl esters

The molecular ion is usually discernible. Fission of bonds adjacent to the carbonyl group may occur to yield four ions.

$$\left[\text{R}\!-\!\overset{O}{\overset{\|}{C}}\!-\!\text{OCH}_3 \right]^{\ddagger} \longrightarrow \text{RC}\equiv\text{O}^+ > \ ^+\text{O}\equiv\text{C—OCH}_3 > \ ^+\text{OCH}_3 > \text{R}^+$$

The McLafferty rearrangement is a most important process in the longer chain methyl esters. The base peak in the mass spectrum of methyl esters of C_6 to C_{26} fatty acids occurs at m/e 74.

$$[\text{RCH}_2\text{CH}_2\text{CH}_2\text{COOCH}_3]^{\ddagger} \longrightarrow \left[\text{CH}_2\!=\!\text{C} \overset{OH}{\underset{OCH_3}{}} \right]^{\ddagger} + \text{RCH}=\text{CH}_2$$

$$m/e\ 74$$

7.7.8b Higher esters

The higher esters undergo fragmentations similar to those of the methyl esters, but the overall spectrum is complicated by additional fragmentation involving the alkoxy group. Such esters usually exhibit a peak corresponding to a protonated acid that may arise by the following pathway.

Rearrangement ions occur at m/e 60 and 61, corresponding to acetic acid and protonated acetic acid in the spectrum of ethyl or higher esters of butyric or higher acids.

Benzyl esters fragment with the loss of ketene.

The loss of the alkoxy group is a very important fragmentation process for alkyl benzoates.

It is typical that *ortho* groups exert a marked influence on fragmentation processes.

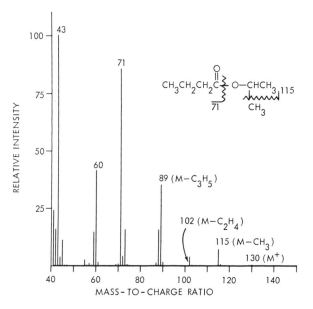

Fig. 7.21

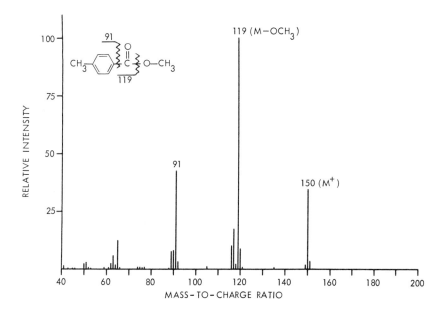

Fig. 7.22

Some of the important ions recorded in the spectrum of γ-valerolactone are illustrated below.

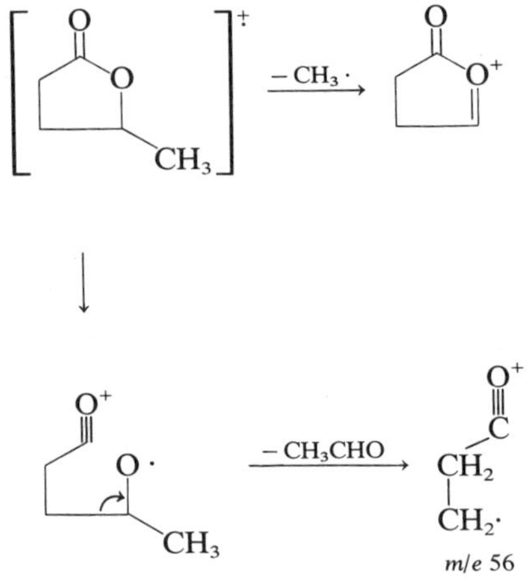

7.7.9 Amines

Amines and other nitrogen compounds containing an odd number of nitrogen atoms have an odd-numbered molecular weight.

7.7.9a Aliphatic amines

The molecular ion peak is weak to absent. The most intense peak results from β-cleavage.

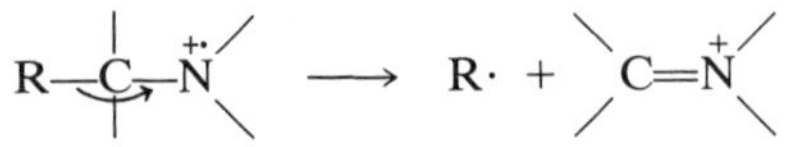

Again the largest alkyl group is preferentially lost. The base peak for all primary amines unbranched at the α-carbon is at m/e 30 [$CH_2=\overset{+}{N}H_2$]. The presence of this peak is good but not conclusive evidence for a primary amine since this fragment may arise from sequential fragmentation from secondary and tertiary amines. Cyclic fragments apparently occur during the fragmentation of longer chain amines.

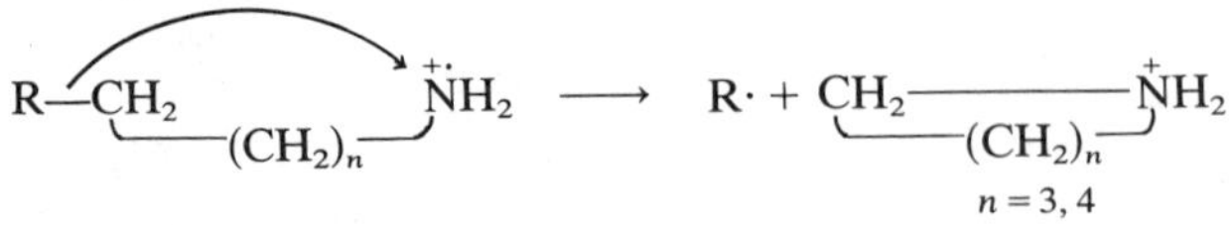

Cyclic amines give rise to strong molecular ion peaks; primary fragmentation occurs at α-carbons.

Aromatic amines give very intense molecular ion peaks accompanied by a moderate M − 1 peak.

7.7.10 Amides

The molecular ion peak is usually observed. McLafferty rearrangements are important. A strong peak at m/e 44 is indicative of a primary amide.

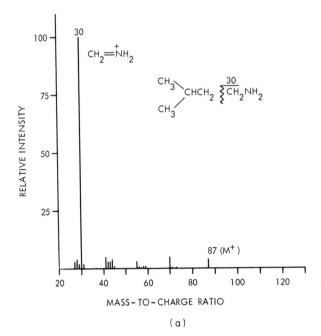

Fig. 7.23. Mass spectra of aliphatic and aromatic amines.

(a)

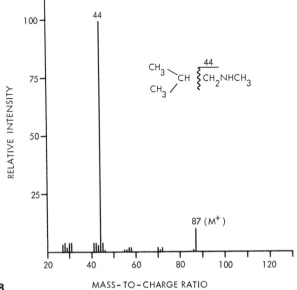

Fig. 7.23(*b*)

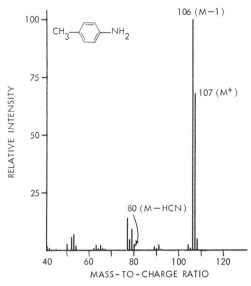

Fig. 7.23(c)

7.7.11 Nitriles

The molecular ion of aliphatic nitriles is usually not observed. At higher pressures an M + 1 ion may appear. A relatively weak but useful M − 1 ion due to R—CH=C=N⁺ is often found. The base peak generally results from a McLafferty rearrangement and is found at *m/e* 41 in the spectrum of C₄ to C₉ straight chain nitriles.

7.7.12 Nitro Compounds

The molecular ion peak of aliphatic nitro compounds is seldom observed. The spectra of such compounds are composed mainly of hydrocarbon ions, with observable peaks containing nitrogen occurring at *m/e* 30 (NO) and 46 (NO₂). On the other hand, nitrobenzene exhibits an intense molecular ion. A rearrangement reaction also occurs to give $C_6H_5O^+$.

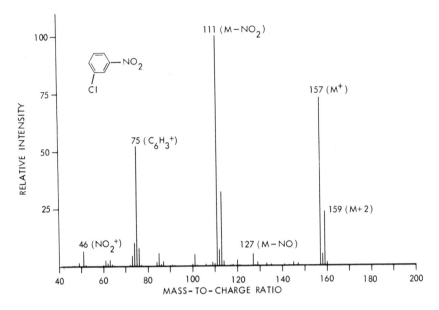

Fig. 7.24

7.7.13 Sulfur Compounds

The fragmentation of thiols and sulfides parallels those of alcohols and ethers. In each case, the molecular ion peak of the sulfur compounds is more intense. Cyclic ions of the type

$$
\begin{pmatrix} \overset{\displaystyle H}{\underset{(CH_2)_n}{\overset{|}{S^+}}} \end{pmatrix}
$$

have been postulated in a number of cases.

The principal fragments from disulfides are produced by elimination of alkenes.

$$[CH_3CH_2—S—S—CH_2CH_3]^+ \xrightarrow{-\ CH_2=CH_2} [CH_3CH_2SSH]^+$$

m/e 122   *m/e* 94

$$\downarrow -\ CH_2=CH_2$$

$$[HSSH]^+$$

m/e 66

The principal fragmentation of a typical aliphatic sulfoxide is shown below.

$$[CH_3CH_2CH_2\overset{\overset{O}{\|}}{S}CH_2CH_2CH_3]^+ \xrightarrow{-\ C_3H_6} [CH_3CH_2CH_2SOH]^+$$

$$\downarrow -\ OH\cdot \qquad\qquad\qquad\qquad \downarrow -\ CH_3CH_2\cdot$$

$$CH_3CH_2CH_2—\overset{+}{S}=CHCH_2CH_3 \qquad CH_2=\overset{+}{S}—OH$$

m/e 117   *m/e* 63

Aromatic sulfoxides are interesting in that apparently an electron impact-induced aryl migration from sulfur to oxygen occurs.

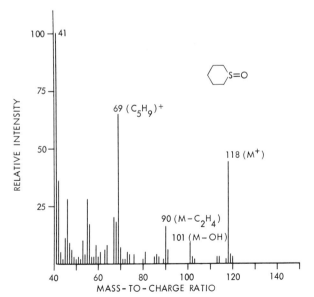

Fig. 7.25. Mass spectrum of a typical cyclic sulfoxide.

Similar migrations occur with aromatic sulfones.

7.8
MASS SPECTRAL PROBLEMS

Identify each of the following unknowns on the basis of the mass spectral and other structural data given. Suggest plausible explanations for the mass spectral patterns observed.

1. Compound **A** exhibits an intense infrared band at $1715\ cm^{-1}$. The important ions appearing in the mass spectrum of **A** are listed below:

m/e	% base peak	m/e	% base peak
41	44	85	66
42	8	100	12
43	10	113	6
55	10	142	12
57	100		
58	58	metastable peak at 38.3	
71	6		

Compound **A** reacts with semicarbazide to form a single derivative, mp 89 to 90°C.

2. The infrared spectrum of compound **B** displays strong bands at 1690 and $826\ cm^{-1}$. The NMR spectrum consists of a pair of **AB** doublets near δ 7.60 (4) and a singlet at δ 2.45 (3). The peaks at highest m/e in the mass spectrum of **B** are found at 198 (26), 199 (4), 200 (25), 201 (3), and the base peak is found at 183 (100).

3. Compound **C** was isolated after treatment of a methylene chloride extract of a bacterial culture with ethanol and dry hydrogen chloride. Oxidation of **C** with iodine and sodium hydroxide produced hexadecanoic acid. Selected high resolution mass spectral data on compound **C** is at the top of the following page.[1]

---

[1] The high resolution mass spectral data for this problem were kindly supplied by Dr. Walter J. McMurray, Yale University.

Compound C

m/e	Intensity	Composition			
		CH	CHO	CHO$_2$	CHO$_3$
327	2				20/39
326	5				20/38
239	15		16/31		
196	2	14/28			
143	23				7/11
131	25				6/11
130	100				6/10
115	6				5/7
102	5				4/6
97	11	7/13	6/9	5/5	
88	20			4/9	
71	17	5/11	4/7		

4. Compound **D**, mp 115 to 116°C, gave the mass spectrum reproduced in Fig. 7.26.

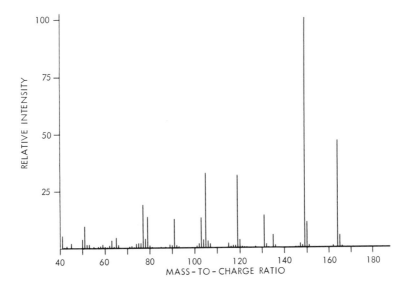

Fig. 7.26. Mass spectrum of unknown **D**, problem 4.

5. The mass spectrum of compound **E** is shown in Fig. 7.27. At low-operating pressures the following data are obtained:

m/e (relative intensity): 114 (12.0), 115 (0.81), 116 (0.07)

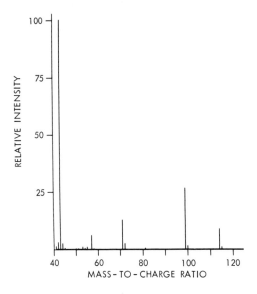

Fig. 7.27. Mass spectrum of unknown **E**, problem 5.

6. The infrared spectrum of compound **F** is devoid of absorption in the 1700 cm^{-1} region; mass spectrum $(70e)$ m/e (rel. intensity) 27 (26), 29 (39), 31 (18), 45 (100), 47 (16), 61 (6), 73 (44), 75 (8), 89 (3), 103 (15), 117 (2).

Anal. Found: C, 61.10; H, 11.9; N, 0.00.

Compound **F**, bp 102°C, upon treatment with 2,4-dinitrophenylhydrazine reagent, forms a yellow precipitate, mp 166 to 168°C.

7. The infrared spectrum of compound **G** contains a carbonyl band at 1830 cm^{-1}. The NMR spectrum consists of a singlet at $\delta 7.34$. Important ions in the mass spectrum of **G** are indicated below.

m/e	% base peak
63	48
64	100
92	56.5
136	75.5
137	6.1
138	0.8

Metastable peaks are found at 44.5 and 62.6.

8. Extraction of the leaves of *Nepeta cataria Linné*, otherwise known as catnip, with methylene chloride followed by chromatography of the crude extract over alumina, employing carbon tetrachloride-acetone as the eluent, resulted in the isolation of an oil, compound H, $\lambda_{\max}^{\text{EtOH}}$ 240 nm.

Mass Spectral Data for Compound H

m/e	% base peak	m/e	% base peak
27	43	53	12
28	7	55	100
29	46	70.3 (m^*)	—
36.8 (m^*)	—	83	96
39	40	84	5
41	12	98	51
43	92	99	3.5
51.1 (m^*)	—	100	0.2

9. Compound **J**, n_D^{20} 1.4800, bp 77°C/10 mm, gave the mass spectrum shown in Fig. 7.28.

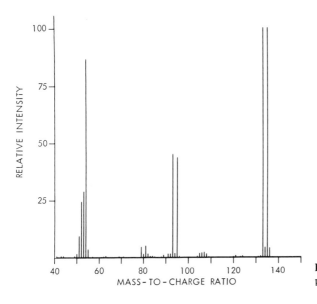

Fig. 7.28. Mass spectrum of unknown **J**, problem 9.

10. Compound **K**, which has an exceptionally disagreeable odor, yields the following mass spectral data:

m/e	% base peak	m/e	% base peak	m/e	% base peak
27	24	57	38	83	46
29	32	58	2.9	84	51
39	19	59	3.0	89	11
41	76	61	18	98	4
42	47	68	18	112	23
43	74	69	59	146	37
47	29	70	70	147	3.9
55	74	71	15	148	1.8
56	100	82	11		

7.9
GENERAL REFERENCES ON MASS SPECTROMETRY

1. BEYNON, J. H., *Mass Spectrometry and Its Application to Organic Chemistry*, Amsterdam: Elsevier Publishing Company, 1960.
2. BEYNON, J. H., and WILLIAMS, A. E., *Mass and Abundance Tables for Use in Mass Spectrometry*, Amsterdam: Elsevier Publishing Company, 1963.
3. BIEMANN, K., *Mass Spectrometry, Applications to Organic Chemistry*, New York: McGraw-Hill Book Company, 1962.
4. BUDZIKIEWICZ, H., DJERASSI, C., and WILLIAMS, D. H., *Interpretation of Mass Spectra of Organic Compounds*. San Francisco; Holden-Day, Inc., 1964.
5. BUDZIKIEWICZ, H., DJERASSI, C., and WILLIAMS, D. H., *Mass Spectrometry of Organic Compounds*. San Francisco: Holden-Day, Inc., 1967. This is the most complete general reference written for the organic chemist interested in employing mass spectrometry in structure determination.
6. BUDZIKIEWICZ, H., DJERASSI, C., and WILLIAMS, D. H., *Structure Elucidation of Natural Products by Mass Spectrometry*, Vols. I and II, San Francisco: Holden-Day, Inc., 1964.
7. "Catalog of Mass Spectra Data," American Petroleum Institute Res. Project No. 44, Carnegie Institute of Technology, Pittsburgh, Pennsylvania, and the Manufacturing Chemists Association Research Project, Agricultural and Mechanical College of Texas, College Station, Texas.
8. CORNU, A., and MASSOT, R., *Compilation of Mass Spectral Data*. London: Heyden and Sons Ltd., 1966.
9. ELLIOTT, R. M., ed., *Advances in Mass Spectrometry*. New York: Pergamon Press, 1963.
10. *Index of Mass Spectral Data*, A.S.T.M. Special Technical Publication No. 356, American Society for Testing and Materials, Philadelphia, Pennsylvania, 1963.
11. KISER, R. W., *Introduction to Mass Spectroscopy and Its Applications*, Chapter 5. Englewood Cliffs, New Jersey: Prentice-Hall, Inc., 1965.
12. LEDERBERG, J., *Tables and an Algorithm for Calculating Functional Groups of Organic Molecules in High Resolution Mass Spectrometry*, NASA Sci. Techn. Aerosp. Rep., N64-21426 (1964).
13. MCDOWELL, C. A., ed., *Mass Spectrometry*. New York: McGraw-Hill Book Company, 1963.
14. MCLAFFERTY, F. W., *Interpretation of Mass Spectra*. New York: W. A. Benjamin, Inc., 1966.
15. MCLAFFERTY, F. W., *Mass Spectral Correlations*, Advances in Chemistry Series No. 40. American Chemical Society, Washington, D.C., 1963.
16. MCLAFFERTY, F. W., "Mass Spectrometry," Chapter 2, Vol. II, in *Determination of Organic Structure by Physical Methods*, F. C. Nachod and W. D. Philips, eds. New York: Academic Press., 1962.
17. MCLAFFERTY, F. W., and PENZELIK, J., *Index and Bibliography of Mass Spectrometry*, 1963–1965. New York: Interscience Publishers, Inc., 1967.
18. MEAD, W. L., ed., *Advances in Mass Spectrometry*. Amsterdam: Elsevier Publishing Company, 1966.

19. REED, R. I., *Application of Mass Spectrometry to Organic Chemistry.* New York: Academic Press, Inc., 1966.

20. REED, R. I., *Ion Production by Electron Impact.* New York: Academic Press, 1962.

21. SILVERSTEIN, R. M., BASSLER, G. C., and MORRILL, T. C., *Spectrometric Identification of Organic Compounds,* 3rd ed. New York: John Wiley & Sons, Inc., 1974.

22. WALDRON, J. D., ed., *Advances in Mass Spectrometry.* New York: Pergamon Press, 1959.

CHEMICAL
AND
PHYSICAL TECHNIQUES
FOR IDENTIFICATION
OF
ORGANIC COMPOUNDS

8

Characterization of an Unknown Compound

In Part III we shall be concerned with the classification of a compound with respect to the functional groups present and the complete identification of the compound. Such a task is most efficiently accomplished through a systematic integration of physical and chemical methods. Just as the various forms of spectroscopy furnish us information in a synergistic manner, the combination of physical data with chemical observations also acts in a synergistic manner. Through the careful and systematic selection of appropriate physical and chemical methods, the task of reaching our goal—the final identification of an organic structure—can be greatly simplified, and the consumption of both time and material can be greatly reduced.

Of the various forms of spectroscopy, the state of the art, the cost and availability of the instruments, and the ease and rapidity of operation place infrared and nuclear magnetic resonance spectroscopy as the primary spectroscopic methods to which the organic chemist turns when faced with a structural problem. Typically, as soon as there is some reasonable assurance of the purity of a compound, IR and NMR spectra should be obtained. Chemists often find it convenient to examine the IR and/or NMR spectrum of a mixture, not only as an aid in deciding on purification methods, but also as a method of checking to see that no artifacts appear during the purification process, and to provide assurance that all components of a mixture are isolated during a separation process (Sec. 2.7).

Even upon cursory examination of the infrared spectrum of a compound, one can often classify the material with respect to functional groups present. Equally important, and often more so, one can readily establish the absence of many functional groupings, thus immediately eliminating many wet tests, which may be laborious and costly in materials. For example, the casual observation of the absence of infrared bands in the 3600 to 3100 cm^{-1} (2.8 to 3.2 μm) region and of strong bands in the 1850 to 1650 cm^{-1} (5.4 to 6.1 μm) region readily eliminates from further consideration all N—H compounds (primary and secondary amines and amides, etc.), all O—H compounds (alcohols, phenols, carboxylic acids, oximes, etc.). and all C=O compounds (acids, esters, ketones, aldehydes, etc.). The same information gained from this quick observation (information that should be accessible in not more than 20 min from the time the sample is obtained) might also be obtained in a considerably longer time by the indiscriminant use of the commonly used chemical classification tests.

Compounds will be classified according to functional group; the most useful spectroscopic and chemical methods for the identification of the functional group will be outlined. Finally, the crucial physical and chemical methods that allow structure assignment will be discussed, and the procedures will be outlined in some detail. The final chapter in this section provides instructions for making a thorough literature search for a compound, its reactions, and its properties.

8.1
STEPS FOR THE IDENTIFICATION OF AN UNKNOWN

Step 1. Gross Examination

A. Physical state. The physical state of the unknown should be indicated. Additional description, such as "amorphous powder," "short needles," and "viscous liquid," often proves useful.

B. Color. Many organic compounds possess a definite color owing to the presence of chromophoric groups. Among the most common of the simple chromophoric groups are nitro, nitroso, diazo, azo, and quinone. Compounds with extensive conjugation are likely to be colored. The color of many samples is due to impurities frequently produced by air oxidation. Freshly purified aromatic amines and phenols are usually colorless; on storage, small amounts of these compounds are oxidized to highly colored quinone impurities. Stable, colorless liquids or white crystalline solids are not likely to contain the usual chromophoric groups or functional groups that are easily oxidized.

C. Odor. Many organic compounds are exceedingly toxic and will produce at least temporary discomfort upon inhalation. The reader is cautioned about the indiscriminate "whiffing" of compounds or mixtures about which little is known. To note the odor of an unknown substance, hold the tube pointed away from the face in one hand and gently wave the vapors from the tube toward the nose with the other hand. With a good nose educated by experience, a chemist can often make a tentative identification of a common chemical, or he can make an intelligent guess as to functional groupings present. Alcohols, phenols, amines, aldehydes, ketones, and esters all have odors more or less characteristic of the general group. Mercaptans, low-molecular weight-amines, and isonitriles (particularly toxic) have characteristic odors that one seldom mistakes once experience has been gained. The student will soon learn to recognize the commonly used solvents by their odors.

Step 2. Determination of Purity and Physical Constants

One should be assured that the physical constants used in a structure identification are obtained with pure material. The necessity for purification or fractionation can be indicated by melting-point and/or boiling-point ranges, behavior on thin-layer or gas-liquid chromatography, or any inhomogeneity or discoloration.

Step 3. Classification by Functional Group

A. Determination of acidity or basicity and solubility behavior. The acidity or basicity of a compound, as determined with indicators or potentiometer or by solubility in acids or alkali, as well as general solubility behavior (Sec. 8.4), is useful, not only in the determination of possible chemical classes to which the compound may belong, but also to serve as a guide in the choice of solvents and procedures used for spectroscopic analyses.

B. Classification of functional groupings by spectroscopic and chemical means. As indicated earlier, in this text we shall use spectroscopy with significant emphasis on infrared as the primary method of determining the major functional groups present. Supplemental chemical tests will be used to confirm or clarify the assignments made by spectroscopic means.

C. Elemental analysis. Qualitative elemental analysis (Sec. 8.2) by the sodium fusion method is not a procedure that the chemist finds necessary to apply to every unknown compound with which he is faced. The presence or, more likely, the absence of many heteroatoms can be inferred from the source and/or history of the sample. The presence or absence of many heteroatoms can also be inferred from spectroscopic data. High-resolution mass spectrometry will provide exact elemental compositions. Evidence of the presence or absence of sulfur, halogens, and nitrogen can usually be obtained from low-resolution mass spectrometry. Infrared spectroscopy, by virtue of detection of various functional groups, can establish the presence or absence of certain elements. The presence of an amino, nitrile, or nitro group in the infrared spectrum provides obligatory evidence for the presence of a nitrogen containing compound. The basicity of a compound may be sufficient to establish the presence of nitrogen. The presence of sulfur can be inferred from mercaptan, sulfone, sulfoxide, etc. bands in the infrared spectrum.

A simple Beilstein test is usually reliable for establishing the presence or total absence of halogen in a compound. Finally, quantitative microanalytical data for C, H, and N may often be available. Such data, together with the known presence or absence of oxygen as inferred by chemical or spectroscopic means, may also be sufficient to eliminate or establish the presence of other elements within the molecule.

Step 4. Final identification

At this point, based on the foregoing chemical and physical data, classification of the unknown compound as belonging to a specific functional group class should be possible. A comparison between the physical data obtained for an unknown compound and information about known compounds listed in the literature should be made, and an initial list of possibilities that melt or boil within 5° C of the value observed for the unknown should be prepared. This initial list is usually obtained by consulting tables of compounds arranged in order of melting points or boiling points and according to the functional groups present. The most extensive of the tables—which also gives data on derivatives—is the *Handbook of Tables for Organic Compound Identification*, published by the Chemical Rubber Publishing Co. It should be apparent that no one compendium is available that provides complete coverage of the literature on physical constants useful in identification work (refer to the more extensive discussion of literature searching provided in Chapter 10). The tables provided in Chapter 9 provide only a sampling of the more commonly encountered compounds.

The initial list of possibilities can be immediately condensed to include only those compounds that conform to the data on elemental composition, solubility, chemical behavior, and other physical and spectral properties. In most cases, a tentative list of possibilities will not contain more than three to five compounds. A more extensive literature search is then made to ascertain additional properties of the compounds listed as possibilities. The final proof of structure of an unknown is accomplished through the preparation of selected derivatives and/or the collection, comparison, and evaluation of other physical and chemical data, such as neutralization equivalent, nuclear magnetic resonance, ultraviolet, and mass spectra, optical rotation or optical rotatory dispersion, and dipole moments.

The classical criteria for the conclusive identification of a compound are that (1) the compound has the expected chemical and spectral properties, and the physical constants match those given in the literature; and (2) the appropriate derivatives prepared from the unknown melt within 1 to 2°C of the melting points given in the literature. The actual criteria necessary to establish the identity of the compound beyond reasonable doubt are, of course, subject to considerable variation, depending on the kind and content of the information available to the experimenter and the complexity of the compound in question. In many cases, one derivative may be sufficient, and, in some cases, it may not be necessary to prepare derivatives at all. Indeed, more rigorous and detailed proof of structure can often be obtained from a combination of carefully obtained and interpreted spectral and other physical data (e.g., molecular weights) chemical reactivity (e.g., kinetics, pK_a's, etc.), and microanalytical data (e.g., found percentages of C, H, and other elements agreeing with the theoretical within 0.3%).

The task of the rigorous establishment of the identity of two compounds is considerably simplified if a known sample is available. In such a case, identity can be established by one or more of the following methods.

A. Mixture melting point. Two samples having the same melting point are identical if, upon admixture of the two samples, the melting point observed is not depressed relative to that of pure samples (Sec. 3.2.2). This procedure, wherever possible, should be applied not only to the original unknown but also to derivatives thereof.

B. Spectral comparison. In cases of molecules of moderate complexities, the superimposability of the infrared spectrum of the unknown with that of the known establishes the identity of the two. However, different molecules of simple structure, such as undecane and dodecane, may have remarkably similar spectra, whereas in exceedingly complicated molecules, such as steroidal glycosides, peptides, and many antibiotics, the resolution obtained in the infrared region is usually not sufficient to allow identification with certainty by means of spectral comparison. The method applies equally well to the use of spectra recorded in the literature; in either event, care must be taken to be sure that the spectra are recorded under identical conditions.

The comparison of spectra obtained by other spectroscopic methods also provides helpful criteria for the establishment of identity. Since NMR spectra run the gamut from the exceedingly simple to the extraordinarily complex, direct comparison may provide anything from very tenuous to very conclusive information. Mass spectral data can be expected to match very closely only when the spectra are obtained under exactly identical conditions (usually run one after the other on the same instrument). It should be noted that the mass spectral fragmentation pattern is often not sensitive to differences in spatial orientation of atoms or groups; the mass spectra of geometrical isomers are often almost indistinguishable. The superimposability of ultraviolet spectra only serves to indicate that compounds have the same chromophore.

C. Chromatographic comparison. Good but not necessarily rigorous criteria of the identity of two compounds can be established by comparative gas-liquid, thin-layer, and/or paper chromatography (Sec. 2.6). The standard procedure involves chromatographing the two compounds separately and as a mixture. Safer conclusions can be drawn if the comparisons are made under several conditions and by more than one method.

It should be apparent to the reader that it is impossible to set down an exacting set of conditions that must be met each time a compound is said to be conclusively identified. There are numerous methods by which identity can be established; some are more rigorous than others, and some are more appropriate in one case than in another. In any event, there is no substitute for common sense and the careful and thoughtful planning and execution of crucial experiments, and a thorough evaluation of data.

D. Selection and preparation of derivatives. Ideally, the derivatives should be easily and quickly prepared from readily available reagents and should be easily purified. The best derivatives meet the following standards:

1. The derivative should be crystalline solid melting between 50 and 250°C. Solids melting below 50°C are often difficult to crystallize and recrystallize. Accurate determinations of melting points above 250°C are difficult.
2. The derivative should have a melting point and other physical properties that are considerably different from the original compound.
3. Most importantly, the derivative chosen should be one whose melting point will single out one of the compounds from the list of possibilities. The melting points of the derivatives to be compared should differ from each other by a minimum of 5°C.

In this text, for each functional group class, detailed procedures will be found only for the most generally suitable derivatives for which extensive data are readily available. In addition to the tables of melting points of derivatives in Chapter 9, more extensive compilations of the melting points of the derivatives chosen will be found in *Handbook of Tables for Organic Compound Identification*, cited earlier.

8.2
QUALITATIVE ELEMENTAL ANALYSIS

Before proceeding with chemical and spectral functional group analysis, it is helpful to determine what elements are present in a molecule and in what ratio these elements are present. The former data are derived via qualitative tests that can be readily carried out by the student, whereas the latter data are usually obtained by trained analysts in laboratories specifically set up for such quantitative analysis. In addition to quantitative analysis of the elements present, these laboratories are usually capable of performing quantitative functional group analyses, for example, for methyl, methoxyl, acetoxyl, and deuterium. The following paragraphs of this section outline the handling and interpretation of such analytical data.

The presence of carbon in organic compounds will be assumed, as will the presence of hydrogen, except in perhalogenated compounds; no specific tests for the presence of carbon and hydrogen will be given. (The presence of hydrogen is

readily indicated by infrared and hydrogen nuclear magnetic resonance spectroscopy, as discussed in detail in Chapters 5 and 6.) Heteroatoms most commonly
encountered in organic compounds include the halogens, oxygen, sulfur, and
nitrogen. The detection of oxygen is relatively difficult by qualitative analytical
means, and we shall rely on infrared spectral analysis and on solubility data to
indicate its presence.

   The presence of metallic elements, for example, lithium, sodium, and potassium (usually as the salts of acidic materials), can be indicated by an ignition test.
A few milligrams of the substance are placed on a clean stainless steel spatula and
carefully burned in a colorless burner flame. The presence of a noncombustible
residue after complete ignition indicates the presence of a metallic element. The
color of the flame during ignition may give some indication of the type of metal
present. The ignition residue is dissolved in two drops of concentrated hydrochloric acid, and a platinum wire is dipped into the solution and placed in a colorless
burner flame. The following colors are produced on ignition: blue by lead, green
by copper and boron, carmine by lithium, scarlet by strontium, reddish-yellow by
calcium, violet by potassium, cesium, and rubidium, and yellow by sodium.
Precautions must be taken to avoid contamination of the sample by sodium
because the intense yellow flame produced by sodium generally masks all other
elements. If only trace amounts of sodium are present, the initial yellow flame will
soon disappear, leaving the final colored flame of the elements present in greater
amounts.

   The qualitative tests for most of the elements are based on reactions involving
an anion of the element, for example, the halide, sulfide, and cyanide; hence, a
reductive decomposition of the compound is required. This is generally accomplished by means of fusion with metallic sodium.

$$\text{C, H, O, N, S, X,} \xrightarrow[\Delta]{\text{Na}} \text{NaX, NaCN, Na}_2\text{S, NaSCN}$$

A few of the elements are more readily detected as the oxyanions, formed in an
oxidative decomposition of the compound by fusion with sodium peroxide. Such
examples include phosphorus, sulfur, silicon, and boron. The reductive and
oxidative procedures will be discussed separately.

8.2.1 Sodium Fusion

   To accomplish the decomposition of organic compounds with sodium, the
compound is fused with sodium-lead alloy[1] or metallic sodium. Since the sample
sizes employed are generally quite small, 1 to 10 mg, and the detection tests are
quite sensitive, the student must take certain precautions to avoid contamination
of the sample during and after fusion. All glassware must be thoroughly cleaned

---

[1] J. A. Vinson and W. T. Grabowski, *J. Chem. Ed.*, **54**, 187 (1977).

and rinsed with distilled water, or preferably deionized water, and the solvents and reagents must be of analytical quality. *Safety goggles should be worn during these tests*! Procedure 8.2.1a is recommended, as it avoids the use of pure metallic sodium.

8.2.1a Procedure: Sodium fusion

(a) In a small test tube place about 0.5 g of sodium-lead alloy (dri-Na® from J. T. Baker Chemical Company). Clamp the tube in a vertical position and heat with a flame until the alloy melts and fumes of sodium are seen up the walls of the tube. *Do not heat the alloy to redness.* Add a few drops of liquid sample or 10 mg of solid. During addition be careful not to get any sample on the sides of the tube. Heat gently, if necessary, until the reaction is initiated; remove the flame until the reaction subsides, and then heat to redness for 1 or 2 min and let cool. Add 3 mL of distilled water and heat gently for a few minutes to react the excess sodium with the water. Filter the solution if necessary. Wash the filter paper or dilute the decanted reaction mixture with about 2 mL of distilled water, and proceed with the elemental analysis. This final solution will be referred to as the fusion solution in discussions that follow.

(b) If the organic sample to be analyzed is quite volatile, a 1 to 10-mg sample of the material is placed in a 4-in. test tube and 30 to 50 mg of sodium, a piece about half the size of a pea, is cautiously added to the test tube. (The mouth of the test tube should be pointed away from the experimenter and other people in the laboratory to prevent their being splashed with material from the tube should a violent reaction ensue on the addition of the sodium.) The test tube is then gently heated until decomposition and charring of the sample occur. When it appears that all the volatile material has been decomposed, the test tube is strongly heated until the residue becomes red. The test tube is allowed to cool to room temperature, and a few drops of methanol are added to decompose the excess sodium. If no gas evolves on the addition of the methanol, an excess of sodium was not present, and there is a distinct possibility of incomplete conversion of the elements to their anions. The fusion should be repeated with a larger quantity of sodium.

The contents from the tube are boiled with 1.5 to 2.0 mL of distilled water, diluted to 10 mL with distilled water, and the mixture is then filtered or centrifuged. The decomposition of the organic material usually leads to the extensive formation of carbon, which may prove very difficult to remove. Occasionally filtration through a filtering aid, for example, Celite, which has been thoroughly washed with distilled water, will remove the finely divided carbon particles. The resulting solution should be clear and nearly colorless. If the solution is highly colored, the entire fusion process should be repeated because the color may interfere with the detection tests. This final solution will be referred to as the fusion solution in the discussions of the following elemental tests.

Although the sodium fusion procedure for the decomposition of organic compounds is the simplest and most useful, several other reductive decompositions have been employed. These procedures involve the use of magnesium, magnesium and an alkali carbonate, zinc and an alkali carbonate, or soda lime. In the decompositions utilizing magnesium and soda lime, the organic nitrogen is eventually converted into ammonia. Organic nitrogen is converted to cyanide in the zinc and alkali carbonate decomposition. The decomposition mixtures are dissolved in water, clarified, and used in the following elemental detection tests.

8.2.2 Detection of Halides

The presence of chlorine, bromine, and iodine can be readily detected by the precipitation of the corresponding silver halides on treatment with silver ion. Fluorine cannot be detected in this test because silver fluoride is soluble.

8.2.2a Procedure: Detection of halides

The presence of nitrogen or sulfur interferes in this test; sulfide and cyanide must be removed before treatment of the solution with silver ion. This is accomplished by acidification of the fusion solution followed by gentle heating to boil off the hydrogen cyanide and hydrogen sulfide formed on acidification. *This procedure must be carried out in a hood.* A great excess of sulfuric acid should be avoided since silver sulfate can precipitate from solutions containing a high concentration of sulfate ions. Take 0.5 mL of the fusion solution and carefully acidify by the dropwise addition of 10% sulfuric acid. (The pH of the solution can be checked readily by immersing a clean stirring rod into the acidified solution and then applying a piece of pHydrion paper to the liquid adhering to the stirring rod.) The solution is gently boiled over a microburner for about 3 to 5 min. One or two drops of aqueous silver nitrate solution are added; the formation of a substantial quantity of precipitate indicates the presence of chloride, bromide, or iodide. A white precipitate soluble in ammonium hydroxide indicates the presence of chloride, a pale yellow precipitate slightly soluble in ammonium hydroxide indicates bromide, and a yellow precipitate insoluble in ammonium hydroxide indicates iodide. Fluoride does not form a precipitate with silver ion. The formation of a slight cloudiness in the solution is not indicative of a positive test. It is recommended that a halide analysis on a known compound be carried out for comparison purposes.

Tests for all of the halogens must be carried out if the initial silver nitrate test is positive. This prevents missing one or more of the halogens if more than one halogen was originally present in the unknown. The chemical distinction between chloride, bromide and iodide is based on the ease of oxidation of iodide over bromide, and bromide over chloride, and the ability of bromine, but not iodine, to add to an active site of unsaturation.

8.2.2b Procedure: Distinction between chloride, bromide, and iodide

A 0.5-mL portion of the fusion solution is placed in a small test tube and acidified with 10% sulfuric acid. It is then gently boiled in the hood to remove any hydrogen cyanide and hydrogen sulfide, as described before. To this solution are added four to five drops of $0.1\,N$ potassium permanganate. The contents of the test tube are shaken for approximately 1 min. 10 to 20 mg oxalic acid is added to discharge the color of the excess permanganate, and 0.5 mL of carbon disulfide is added to the tube (carbon disulfide is extremely flammable and toxic, and due precautions should be taken). The contents of the tube are again shaken and the two phases allowed to separate. The presence of a color in the carbon disulfide layer indicates the presence of bromine or iodine; a purple color is formed if iodine is present, and a red-brown color if bromine or both bromine and iodine are present. The formation of no color in the carbon disulfide layer indicates the presence of chlorine, although further confirmation is required.

Distinction between bromine and/or iodine is readily made by the addition of one to two drops of allyl alcohol. If only iodine is present, the color of the carbon disulfide layer will remain, whereas with only bromine, the color will be immediately discharged. If both bromine and iodine are present, the color of the carbon disulfide layer will turn from red-brown to purple.

To check for the presence of chloride ion, the aqueous layer is removed from the test tube by means of a capillary pipette and placed in another clean test tube. Several drops of nitric acid $(6\,N)$ are added, and the contents of the tube are gently boiled for 2 to 3 min (the nitric acid oxidizes any bromide or iodide, but not chloride, to the corresponding free halogen, which is vaporized during the boiling process). The solution is then cooled, and two drops of $0.1\,M$ silver nitrate solution are added. The production of a white precipitate (but not a light cloudiness) indicates the presence of chloride ion.

Detection of fluorine: Owing to the solubility of silver fluoride in aqueous solutions, the presence of fluorine cannot be detected by the use of silver nitrate. The simplest procedure for the detection of fluoride involves the reaction of fluoride ion with the red zirconium alizarin complex to produce free alizarin (yellow in color) and the colorless zirconium hexafluoride complex.

8.2.2c Procedure: Determination of fluoride

A 0.5-mL portion of the fusion solution is acidified with three to four drops of concentrated hydrochloric acid. To this solution are added two to three drops of the zirconium alizarin complex solution, prepared by mixing equal volumes of a 1% ethanol solution of alizarin and a 2% solution of zirconium chloride, or nitrate, in 5% hydrochloric acid. The change in color from red to yellow indicates the presence of fluoride ion.

Beilstein test for halogens: The presence of chlorine, bromine, or iodine in organic compounds can be detected by the Beilstein test. The test depends on the production of a volatile copper halide when an organic halide is strongly heated with copper oxide. The test is extremely sensitive, and a positive test should always be confirmed by other methods.

8.2.2d Procedure: Beilstein test

A small loop in the end of a copper wire is heated to redness (an oxide film is formed) in a Bunsen flame until the flame is no longer colored. After the loop has cooled, it is dipped into a little of the compound to be tested and then reheated in the nonluminous Bunsen flame. A blue-green flame produced by volatile copper halides constitutes a positive test for chlorine, bromine, or iodine (copper fluoride is not volatile).

Very volatile compounds may evaporate before proper decomposition occurs, causing the test to fail. Certain compounds such as quinoline, urea, and pyridine derivatives give misleading blue-green flames owing to the formation of volatile copper cyanide.

Mass spectrometric detection of halides: The halogens Cl and Br exist as a mixture of isotopes. The mass spectrum of a Cl- or Br-containing compound will display several peaks in the parent mass region, which will be separated by the difference in mass of the halogen isotopes and will be in an intensity ratio equal to the natural abundance ratio (Sec. 7.7.2). Mass spectrometry is also valuable in indicating the number of halogen atoms that occur in a single molecule (Sec. 7.4.1).

8.2.3 Detection of Nitrogen

The detection of nitrogen is based on the conversion of the nitrogen present in a given molecule to the cyanide ion, which is usually detected as the cyano complex formed from ferrous ammonium sulfate (Prussian blue).

8.2.3a Procedure: Detection of nitrogen

A 0.5-mL portion of the fusion solution is adjusted to pH 13 (as determined by testing with pHydrion paper) by the addition of $6 N$ sodium hydroxide. Two drops each of a saturated ferrous ammonium sulfate solution and 30% potassium fluoride solution are added, and the resulting solution is boiled gently for 30 sec. The solution is immediately acidified by the addition of 30% sulfuric acid until the colloidal iron hydroxide dissolves. An excess of acid must be avoided as it may interfere with the test. The production of a blue color indicates the presence of nitrogen in the original compound.

The formation of a very faint blue color, or a pale green color, is indicative of a poor fusion, resulting in an incomplete conversion of the nitrogen in the original molecule to cyanide ion. This is particularly true of compounds in which the nitrogen is in a high state of oxidation, for example, nitro compounds. The fusion should be repeated with the use of a greater excess of sodium and more drastic heating. It is recommended that, in such cases, a control fusion be carried out on a known nitro derivative. It should be pointed out here that the functional groups containing nitrogen in a high oxidation state are quite readily detected by infrared spectroscopy (Chapter 5), and it is not necessary to be concerned about poor fusion results.

The presence of sulfide interferes in the foregoing test for nitrogen. If sulfur is present in the original molecule, the iron sulfide formed when the ferrous ammonium sulfate is added must be removed by centrifugation and decantation of the supernatant liquid. The test for nitrogen is then continued.

An alternate and more sensitive procedure for the detection of cyanide utilizes p-nitrobenzaldehyde in dimethyl sulfoxide.[2]

8.2.3b Procedure: Detection of nitrogen or sulfur

Put about ten drops of the fusion solution in a small test tube and saturate it with solid sodium bicarbonate. Shake to ensure saturation and check to see if excess solid is present. Then add one or two drops of the saturated solution (it is permissible to transfer some of the solid sodium bicarbonate) to a test tube containing about twenty drops of a 1% p-nitrobenzaldehyde solution in dimethyl sulfoxide (PNB reagent). This reagent should be discarded if the initial yellow darkens, and it should be stored in a brown bottle. A purple color indicates nitrogen is present. A green color indicates sulfur is present. If both sulfur and nitrogen are present, only a purple color will be visible. Therefore, if a positive test for nitrogen is observed with PNB, a test for sulfur by the lead acetate method (Sec. 8.2.4a) should be carried out. If only a positive test for sulfur is observed by the present method, then nitrogen is definitely absent.

8.2.4 Detection of Sulfur

8.2.4a Procedure: Detection of sulfide
(See also Sec. 8.2.3b)

A 1-mL portion of the fusion solution is acidified with acetic acid, and two to three drops of dilute lead acetate solution are added. The formation of a black precipitate indicates the presence of sulfur.

8.2.5 Detection of Other Metallic and Nonmetallic Elements

If the ignition test reveals the presence of metallic elements, decomposition of the material in the presence of ammonium nitrate is carried out, and the detection of the elements is carried out by the general inorganic qualitative analysis scheme.

8.2.5a Procedure: Fusion with ammonium nitrate

Approximately 200 to 250 mg of compound is mixed with 500 mg of ammonium nitrate in a clean crucible, and the mixture is strongly heated for 15 to 20 min. The

---

[2] J. A. Vinson and W. T. Grabowski, *J. Chem. Ed.*, **54,** 187 (1977).

residue is allowed to cool and is then dissolved in concentrated sulfuric acid or aqua regia. The excess acid is evaporated (*in a fume hood!*). and the residue is ignited again for a short period of time. If carbonaceous material still remains, a few drops of concentrated nitric acid are added and ignition is repeated. The resulting residue is analyzed by the usual inorganic qualitative analysis scheme. Mercury is not detected in this procedure since it distills off as free mercury. Mercury can be detected by boiling 10 to 20 mg of the compound with 10 mL of concentrated potassium chlorate solution until the reaction mixture is colorless. To test the solution for mercury, a piece of clean copper wire is immersed into the solution for 10 min. The formation of a mercury amalgam, as evidenced by the color of the wire, indicates the presence of mercury.

8.2.6 Oxidative Decomposition

Oxidative decomposition of organic compounds is generally accomplished by fusion of the compound with sodium peroxide. This procedure requires the use of a small bomb. The qualitative tests employed are those outlined earlier and found in inorganic qualitative analysis texts.

Swift and Nieman have developed procedures for the complete analysis of organic compounds based on sodium peroxide fusion.[3]

8.3
QUANTITATIVE ELEMENTAL ANALYSIS

Data on the quantitative elemental composition of an unknown are particularly useful in establishing structure. Traditionally, microanalytical data have been accepted as important criteria for proof of structure and purity. In most primary chemical journals, microanalytical data are given for all new compounds reported in a paper; the data are usually presented in the following form:

> *Anal.*   Calcd for $C_{10}H_{14}N_2O_4$: C, 53.09; H, 6.24; N, 12.38
>   Found: C, 53.20; H, 6.35; N, 12.44

(In writing a formula, after C and H, other elements are listed in alphabetical order.) Such quantitative elemental analyses are not generally performed by the individual chemist; they are usually made by trained technicians in laboratories (often outside independent laboratories) specifically designed for this purpose.

The most common, and hence least expensive, analytical data are those for carbon-hydrogen and nitrogen. In these analyses, the sample is quantitatively combusted to yield carbon dioxide, water, and nitrogen. In the classical technique,

---

[3] *A System for the Ultimate Analysis of Chemical Warfare Agents*, published by the Chemical Warfare Service, U.S.A., and briefly described in *Analytical Chemistry*, **26**, 538 (1954).

the water is collected by adsorption on calcium chloride, the carbon dioxide is collected on Ascarite (sodium hydroxide on asbestos), and the nitrogen is determined by volume. In this method, a sample of about 5 mg is required for a carbon-hydrogen determination, and an additional 5 mg for nitrogen and each other determination.

The simultaneous microdetermination of carbon, hydrogen, and nitrogen can be achieved by automatic gas chromatographic methods. With such instruments, a sample of about 1 mg is converted to water, carbon dioxide, and nitrogen, which are recorded as a three-peak chromatogram. The composition can be determined by peak height to an accuracy of about 0.3%.

To derive the empirical formula of a compound that contains n different elements, elemental analyses of $n - 1$ of the elements present must be determined. The percentage of the remaining element can be determined as the difference between the sum of the percentages of the other elements and 100%. (The percentage of oxygen is usually determined by difference.) Certain errors may be introduced by this method. The acceptable experimental error involved in the determination of the percentage of each element is $\pm 0.3\%$. If the analysis of a majority of the elements is high, or low, then the analysis of the element obtained by difference may be anomalously low, or high. With molecules of a high degree of complexity, such errors may result in an inability to distinguish between two or more empirical formulas unless auxiliary information is available. It is therefore recommended that, in cases when complex structures are thought to be involved, the quantitative analysis of each element contained in the compound be determined. Multiple analyses are also recommended in such cases, using an average value for each element in the calculations of the empirical formula.

For the conversion of microanalytical data to an empirical formula, the first step is to divide the percentage of each element by its atomic weight; and the second is to divide the resulting numbers by the smallest one to determine the atomic ratios.

Element	Percentage		Atomic weight						Atomic ratio
C	67.38	÷	12.01	=	5.61				10
H	7.92	÷	1.01	=	7.84	$\times$	$\dfrac{1}{0.56}$	=	14
N	15.72	÷	14.01	=	1.12				2
O	8.98	÷	16.00	=	0.56				1

The empirical formula for the above compound is $C_{10}H_{14}N_2O$. Conversion of an empirical formula to a molecular formula requires molecular weight data (see discussion on the determination of molecular weights, Sec. 3.8). Sometimes, it may be known from other data that an unknown contains only one chlorine, two nitrogens, etc.; in such cases, the molecular formula may be determined directly from analytical data.

A compilation of calculated analytical data for a great number of compounds containing C, H, N, O, and S in various combinations has been published in book form and is very useful in handling analytical data.[4]

It should be obvious from the foregoing paragraphs that great care must be exercised in the preparation of a sample for analysis. The sample must be of very high purity. Recrystallized solids should be heated at low temperatures, below the melting or decomposition points, and under vacuum to remove solvent molecules. Hygroscopic compounds must be protected from moisture. The presence of as little as 1 mol% of water, which is generally a very low weight percent, may be sufficient to produce a bad analysis. The analyst should always be informed about the hygroscopic nature of a compound, and should also be warned of compounds of explosive or very toxic nature.

8.3.1 Interpretation of Empirical Formula Data

Once the chemist has determined the molecular formula of a compound, valuable information can be derived from the molecular formula with respect to gross features of the molecule. In particular, the molecular formula data can be used to calculate the total number of rings and/or double and triple bonds (sites of unsaturation) in the unknown molecule.

The number of sites of unsaturation in a molecule is conveniently calculated by means of Eq. (8.1),

$$N = \frac{\sum_i n_i(\nu_i - 2) + 2}{2} \tag{8.1}$$

where N is the number of sites of unsaturation, n_i is the number of atoms of element i, and ν_i is the absolute value of the valence of element i. The following examples will illustrate the use of Eq. (8.1).

EXAMPLE 1.   Elemental analysis indicates an empirical formula of C_6H_6. Using ν_i of 4 for carbon and 1 for hydrogen, we have

$$N = \frac{6(4 - 2) + 6(1 - 2) + 2}{2} = \frac{8}{2} = 4$$

sites of unsaturation.

EXAMPLE 2.   Elemental analysis indicates an empirical formula of C_7H_6ClNO. Using ν_i of 4 for carbon, 1 for hydrogen, 1 for chlorine, 3 for nitrogen, and 2 for

---

[4] George H. Stout, *Composition Tables for Compounds Containing C, H, N, O, S*. W. A. Benjamin, Inc., New York (1963).

oxygen, we have

$$N = \frac{7(4-2) + 6(1-2) + 1(1-2) + 1(3-2) + 1(2-2) + 2}{2} = 5$$

sites of unsaturation.

EXAMPLE 3.   Elemental analysis indicates an empirical formula of $C_{19}H_{18}BrP$. The physical and chemical properties indicate that the bromide is ionic and that we must be dealing with a pentavalent phosphorus. Thus, using ν_i of 4 for carbon, 1 for hydrogen and bromine, and 5 for phosphorus, we have

$$N = \frac{19(4-2) + 18(1-2) + 1(1-2) + 1(5-2) + 2}{2} = 12$$

sites of unsaturation.

In these calculations, no distinction can be made among the various types of sites of unsaturation. The experimentalist must rely on spectral and chemical tests to indicate what types of functional groups, and hence their worth in sites of unsaturation, are present in any given molecule. A single ring, C=C, C=O, C=N, N=O, or any other doubly bonded system is considered as a single site of unsaturation; C≡C and C≡N are considered as two sites of unsaturation. The phosphorus-oxygen and sulfur-oxygen coordinate covalent bonds ($\overset{+}{>}P{-}O^-$ and $\overset{+}{>}S{-}O^-$) are incorporated as single bonds and thus are *not* sites of unsaturation. A benzene ring containing a combination of a ring and double bonds represents four sites of unsaturation.

8.3.2 Quantitative Functional Group Analysis

Most analytical laboratories are capable of carrying out quantitative functional group analyses. (See Table 8.1 for examples of typical available functional group analyses in addition to elemental analysis.) Since most of these analyses involve a chemical degradation, the yields of the fragments to be analyzed will not correspond to 100% of the functional group contained in the starting molecule. The net result is that less than integral functional group analytical results will be obtained, and a correction factor must be applied. This correction factor is generally 1/0.75 to 1/0.85 (to be multiplied times the observed results).

8.3.3 Active Hydrogen Determination

Active hydrogen can be classified in two categories: The first consists of the relatively acidic hydrogens bonded to oxygen or sulfur (e.g., alcohols, thiols, and acids), and the second type consists of the weakly acidic hydrogens bonded to a

Table 8.1.   Available Analysis for Elements and Functional Groups*

Elements	Functional Groups
Carbon-hydrogen	Acetyl
Nitrogen	Alkoxyl
Phosphorus	Benzoyl
Oxygen	C-Methyl (Kuhn-Roth)
Sulfur	N-Methyl
Fluorine	Amino, primary (van Slyke)
Chlorine	Hydroxyl
Bromine	Peroxide oxygen
Iodine	Bromine or iodine number
Metallic elements of	Microhydrogenation
groups Ia, IIa, IIIa,	Active hydrogen (Zeisel)
IVa, and a few selected	Ash
transition metals	Lactone titration
	Molecular weight (Rast and osmometric)
	Neutralization equivalent
	pK of acids and bases
	Saponification equivalent
	Volatile acid

* Selected from price lists of a variety of analytical companies.

carbon alpha to a strong electron withdrawing group (e.g., a carbonyl or cyano group). The first category of active hydrogen is determined readily by treatment of the compound with methyl magnesium iodide (Zerevitinov determination), the methyl Grignard reagent reacting to give 1 mol of methane for each mole of active hydrogen. A carefully weighed portion of the sample is dissolved in a high boiling anhydrous ether, for example, dimethoxyethane, and placed in a flask attached to a gas measuring apparatus (an inverted, filled burette whose opening is maintained below the surface of water, or other neutral liquid, in a beaker and connected to the flask), and equipped with a pressure equilibrating addition funnel. The quantity of material used will depend on the expected number of acidic hydrogens and on the volume of the gas measuring burette. The volume of the solvent should be kept to a minimum. The Grignard reagent is prepared from methyl iodide and excess magnesium in the solvent used, and it is put in the addition funnel. The funnel is stoppered, and the methyl Grignard reagent is slowly added to the compound. The volume of the generated gas is measured, corrected to standard pressure and temperature and for liquid vapor pressure, and is converted into moles of methane produced. The ratio of the number of moles of methane produced to the number of moles of unknown employed is the number of acidic hydrogens. In a similar test, the volume of hydrogen produced from the reaction of active hydrogens with lithium aluminum hydride is measured.

The second category of active hydrogen bonded to carbon is determined by deuterium exchange experiments. A small amount of compound (100 mg) is dissolved in deuterium oxide or an ethanol-O-*d* deuterium oxide mixture in the presence of a small amount of basic catalyst, usually sodium carbonate or occasionally very dilute sodium deuteroxide. The solution is allowed to stand for several hours, and the compound is recovered by extraction or evaporation. The process is repeated to ensure maximum isotope concentration in the sample. The carefully purified sample is analyzed for deuterium by combustion, nuclear magnetic resonance, or mass spectrometry (the sample should be kept away from sources of replaceable protium). Much smaller samples can be used when analysis is available by mass spectrometry. Most hydrogens *alpha* to an activating group capable of stabilizing anions will undergo exchange under the foregoing conditions.

$$\underset{H}{\overset{|}{-C}}-\overset{/}{\underset{X}{C}} \xrightarrow{\text{base}} \overset{\backslash}{\underset{\ominus}{C}}-\overset{/}{\underset{X}{C}} \longleftrightarrow \overset{\backslash}{C}=\overset{/}{C} \underset{X^{\ominus}}{} \xrightarrow{D^{\oplus}} \underset{D}{\overset{|}{-C}}-\overset{/}{\underset{X}{C}}$$

A further distinction among active hydrogens on methine, methylene, and methyl positions can be made. Activated methylene and methyl groups undergo condensation reactions with benzaldehyde to give benzylidene derivatives, as illustrated by the following equation:

$$\overset{X}{\underset{\parallel}{-C}}-CH_3 + C_6H_5CHO \xrightarrow{\text{base}} \overset{X}{\underset{\parallel}{-C}}-CH=C\overset{H}{\underset{C_6H_5}{\diagup}}$$

The sample is dissolved in absolute ethanol in the presence of benzaldehyde and sodium ethoxide (prepared by the cautious addition of sodium metal to absolute ethanol) or sodium hydroxide. The benzylidene derivative may precipitate and be collected by filtration, or it can be recovered by extraction with ether after dilution of the reaction mixture with water. The sample is purified and its molecular formula determined by combustion analysis or molecular weight determination. The introduction of each benzylidene moiety will result in an increase in the molecular formula by C_7H_4. Activated methine positions do not undergo condensation reactions to produce benzylidene-type derivatives.

8.4
CLASSIFICATION BY SOLUBILITY

The solubility characteristics of a compound may be very useful in providing structural information. Fairly elaborate solubility and indicator determination schemes have been presented in previous texts on this subject. The authors of this text contend that such elaborate schemes are not required because of the present availability of spectral methods of functional group detection. In addition, the solubility test results obtained with rather complex molecules may not lend themselves to unambiguous interpretation. In view of these comments, a rather limited series of solubility tests are recommended. The solvents recommended for use in these tests are water, 5% hydrochloric acid, 5% sodium hydroxide, and 5% sodium hydrogen carbonate. The effect of structure on the acidity or basicity of organic molecules will not be discussed in this text owing to a necessary limitation of space. For discussions concerning the inductive, resonance, and steric effects on the acidity and basicity of organic compounds, the student is referred to the modern organic chemistry texts.

The quantities of compound and solvent used in the solubility tests are critical if reliable data are to be derived. A compound will be considered "soluble" if the solubility of that compound is greater than 3 parts of compound per 100 parts of solvent at room temperature (25°C). The experimentalist should use as little of the unknown compound as is necessary, generally 10 mg in 0.33 mL of solvent. Employing such small quantities of material requires a fairly accurate determination of the weight of the material to be used; however, it is not necessary to weigh accurately the amount of compound each time. A 10-mg portion of the compound should be accurately weighed out. Subsequent 10-mg portions of the compound can then be estimated visually.

A summary of the solubility characteristics of some of the most important classes of compounds is found in Table 8.2.

8.4.1 Solubility in Water

Water is a highly polar solvent possessing a high dielectric constant and is capable of acting as a hydrogen bond donor or acceptor. As a result, molecules possessing highly polar functional groups capable of entering into hydrogen bonding with water (hydrophilic groups) display a greater solubility in water than do molecules without such functional groups. Irrespective of the presence of a highly polar functional group in a molecule, a limiting factor on its solubility in water is the amount of hydrocarbon structure (*lipophilic* portion of the molecule) associated with the functional group. For example, methanol, ethanol, 1-propanol, and 1-butanol all possess solubilities in water at room temperature, which would allow classification as "water-soluble"; however, 1-pentanol would be classified as "insoluble." The solubility within a series of compounds is also dependent on the extent and position of chain branching, the solubility in water increasing as the branching increases. For example, all of the isomeric pentanols except 3-methyl-1-butanol and 1-pentanol are soluble in water.

Similar solubility trends are observed with aldehydes, amides, amines, carboxylic acids, and ketones, although minor variations in the cutoff solubility point may occur in the various classes of compounds. The upper limit of water solubility for most compounds containing a single hydrophilic group occurs at about five carbon atoms. Most difunctional and polyfunctional derivatives are soluble in water, as are most salts of organic acids and bases.

Table 8.2.   General Solubility Guidelines

Soluble in Water and Ether

Monofunctional alcohols, aldehydes, ketones, acids, esters, amines, amides, and nitriles containing up to five carbons.

Soluble in Water: Insoluble in Ether

Amine salts, acid salts, polyfunctional compounds such as polyhydroxy alcohols, carbohydrates, polybasic acids, amino acids, etc.

Insoluble in Water: Soluble in NaOH and NaHCO₃

High-molecular-weight acids and negatively substituted phenols.

Insoluble in Water and NaHCO₃: Soluble in NaOH

Phenols, primary and secondary sulfonamides, primary and secondary aliphatic nitro compounds, imides, and thiophenols.

Insoluble in Water: Soluble in Dilute HCl

Amines except diaryl and triaryl amines, hydrazines, and some tertiary amides.

Insoluble in Water, NaOH, and Dilute HCl, but Contain Sulfur or Nitrogen

Tertiary nitro compounds, tertiary sulfonamides, amides, azo compounds, nitriles, nitrates, sulfates, sulfones, sulfides, etc.

Insoluble in Water, NaOH, and HCl; Soluble in H_2SO_4

Alcohols, aldehydes, ketones, esters, ethers (except diaryl ethers), alkenes, alkynes, and polyalkylbenzenes.

Insoluble in Water, NaOH, HCl, and H_2SO_4

Aromatic and aliphatic hydrocarbons and their halogen derivatives, diaryl ethers, perfluoro-alcohols, -esters, -ketones, etc.

Distinction between monofunctional and polyfunctional compounds and salts can usually be made by testing their solubilities in diethyl ether, a relatively nonpolar solvent. Glycols, diamines, diacids, or other similar di- and polyfunctional derivatives, as well as salts, are not soluble in diethyl ether.

Further solubility tests of water-soluble compounds in 5% hydrochloric acid, 5% sodium hydroxide, or 5% sodium hydrogen carbonate are meaningless in that their solubilities in these solvents are primarily dependent on the solubility in water and not on the pH of the solvent. It is still possible, however, to detect the presence of acidic and basic functional groups in a water-soluble compound. The pH of the aqueous solution of the compound, derived from the solubility test, is determined by placing a drop of the solution on a piece of pHydrion paper. If the compound contains an acidic functional group, the solution will be acidic (with phenols and enols, the solution will be only very weakly acidic, and a control test should be run on an aqueous solution of a phenol for comparison); for compounds containing a basic functional group, the solution will be basic. It is usually possible to distinguish between a strong and a weak acid by the addition of one drop of dilute sodium hydrogen carbonate solution. Carbon dioxide bubbles will appear if a strong acid is present (pK_a less than 7). The pK of the acidic or basic functional group can be determined readily if it is desired (Sec. 8.5).

8.4.2 Solubility in 5% Hydrochloric Acid

Compounds containing basic functional groups, for example, amines (except triarylamines), hydrazines, hydroxylamines, aromatic nitrogen heterocyclics, but *not* amides (although some N,N-dialkyl amides are soluble), are generally soluble in 5% hydrochloric acid. Occasionally an organic base will form an insoluble hydrochloride salt as rapidly as the free base dissolves, thus giving the appearance of an insoluble compound. One should observe the sample carefully as the solubility test is run.

8.4.3 Solubility in 5% Sodium Hydroxide

Compounds containing acidic functional groups with pK_a's of less than approximately 12 will dissolve in 5% sodium hydroxide. Compounds falling into this category include sulfonic acids (and other oxysulfur acids), carboxylic acids, β-dicarbonyl compounds, β-cyanocarbonyl compounds, β-dicyano compounds, nitroalkanes, sulfonamides, enols, phenols, and aromatic thiols. Certain precautions should be observed since some compounds may undergo reaction with sodium hydroxide, for example, reactive esters and acid halides, yielding reaction products that may be soluble or insoluble in the 5% sodium hydroxide solution. Certain active methylene compounds (β-dicarbonyl compounds, etc.) may undergo facile condensation reactions, producing insoluble products. If any reaction appears to occur during the course of the solubility test, it should be noted. Long chain carboxylic acids (C_{12} and longer) do not form readily soluble sodium salts, but tend to form a "soapy" foam.

8.4.4 Solubility in 5% Sodium Hydrogen Carbonate

The first ionization constant of carbonic acid (H_2CO_3) is approximately 10^{-7}, which is less than the ionization constant of strong acids (carboxylic acids) but is greater than the ionization constant of weak acids (phenols). Therefore, only acids with pK_a's less than 6 will be soluble in 5% sodium hydrogen carbonate. This category includes sulfonic (and other oxysulfur acids) and carboxylic acids and highly electronegatively substituted phenols (for example, 2,4-dinitrophenol and 2,4,6-trinitrophenol).

8.4.5 Additional Solubility Classifications

Two additional solubility tests can be used to further classify compounds, although these tests are not generally necessary in that more valuable and definitive structural information can be derived from the various spectra of the compound.

Compounds containing sulfur and/or nitrogen, but which are not soluble in water, 5% hydrochloric acid, or 5% sodium hydroxide, are almost always sufficiently strong bases to be protonated and dissolved in concentrated sulfuric acid. For such compounds, the sulfuric acid solubility test provides no additional information. Compounds falling into this solubility category include most amides, di- and triarylamines, tertiary nitro, nitroso, azo, azoxy and related compounds, sulfides, sulfones, disulfides, etc.

Compounds not soluble in water, 5% hydrochloric acid, or 5% sodium hydroxide and that do not contain nitrogen or sulfur can be further classified based on their solubilities in concentrated sulfuric acid and 85% phosphoric acid. Compounds insoluble in concentrated sulfuric acid include hydrocarbons, unreactive alkenes and aromatics, halides, diaryl ethers, and many perfluoro compounds.

Compounds soluble in concentrated sulfuric acid include compounds that contain very weakly basic functional groups. This solubility class includes alcohols, aldehydes, ketones, esters, aliphatic ethers, reactive alkenes and aromatics, and acetylenes. Dissolution in concentrated sulfuric acid is often accompanied by reaction, as indicated by color changes or decomposition, and should be noted. This solubility class can be subdivided further on the basis of solubility in 85% phosphoric acid. The foregoing compounds that are soluble in sulfuric acid and contain *5 to 9 carbon atoms* generally will be soluble in 85% phosphoric acid.

8.5
DETERMINATION OF IONIZATION CONSTANTS

The ionization constant of an acidic or basic functional group provides important information concerning the type of functional group present and the type of substituents near the functional group. Ionization constant data are particularly useful in the determination of structures of complex natural products, for example, alkaloids and proteins. The use of ionization constant data in structure work requires the availability of ionization constant data for simple model systems that contain all the essential structural features in the immediate vicinity of the acidic or basic functional group. The original literature abounds with ionization constant data for a great variety of systems, and no compilation of data will be presented in this text.

There are two relatively simple methods for the determination of ionization constants; one is a titrametric method requiring the use of a pH meter, and the other is spectrophotometric, requiring the use of an ultraviolet-visible spectrophotometer.

In the case of amines, especially liquid amines, it is usually more convenient and accurate to make these determinations on a nonhygroscopic crystalline salt of the amine by titration with base rather than attempting a direct titration of the organic base with an acid.

8.5.1 Titrametric Method

The titrametric method of ionization constant determination involves the titration of a known quantity of the acid or base with a base or acid, respectively, of known concentration. The pH, actually the hydrogen ion concentration, is determined at several degrees of neutralization, and the ionization constant is calculated by means of Eqs. (8.2) and (8.3) for each point (method *a*) where [H-anion] and [cation-OH] represent the concentrations of undissociated acid and base, respectively.

$$K_A = \frac{[H^+][anion]}{[H\text{-anion}]} \tag{8.2}$$

$$K_B = \frac{[cation][OH^-]}{[cation\text{-OH}]} \tag{8.3}$$

The ionization constant should remain constant throughout the neutralization curve. [Actually, the ionization constants calculated from Eqs. (8.2) and (8.3) will vary throughout the titration curve owing to changes in total ion concentration (ionic strength) and, in cases of relatively weak acids and bases, hydrolysis of the resulting ions. More elaborate calculations incorporating Debye–Hückel approximations and hydrolysis constants can be employed if very accurate ionization constants are desired.]

Another simple titrametric procedure involves the determination of the entire neutralization curve, pH *vs.* volume of titrant (method *b*). The pH at the half-neutralization point is taken as the value of the pK. This method should be used only with acids having pK_a's less than 4 and bases having pK_b's less than 4, to avoid errors introduced owing to hydrolysis of the ions in solution. This method requires the determination of the entire neutralization curve in order to calculate accurately the half-neutralization point.

8.5.1a Procedure: Determination of pK

A 50 to 100 mg accurately weighed portion of the acid or base is dissolved in 50 mL of distilled water. The electrodes of a pH meter, previously standardized carefully with a buffer having a pH in the expected pK region of the acid or base, are immersed in the solution, and quantities of carefully standardized 0.05 *N* acid or base are added from a burette. After each addition of titrant, the solution is allowed to mix thoroughly (the use of a magnetic stirrer is recommended) until equilibrium is attained, and the pH of the solution is recorded.

If one employs method *a* to calculate the ionization constant, the concentration of all species in solution must be calculated (the student should note that the total volume of the solution changes during the titration, and appropriate corrections in the calculations must be made). Utilization of method *b* does not require the exact calculation of the concentrations of the various species in solution.

If the quantity of sample is limited, as little as 5 mg of sample can be used. Corresponding reductions in the amount of solvent and in the concentration of the titrant may be required. Many times the solubility of the compound will not permit the determination of the ionization constant in water. Mixed or nonaqueous solvents can be used, for example, water-ethanol, water-dioxane, or pure ethanol or dioxane; however, pK values determined in these solvents will differ substantially from the values determined in pure water. Ionization constant comparisons should be made only when the same solvent system is employed.

8.5.2 Spectrophotometric Method

The electronic absorption spectra of acids and their conjugate bases, and, similarly, bases and their conjugate acids, generally differ in the wavelength of maximum absorption and/or in extinction coefficient. The extent of the difference depends on the type of chromophore present in the molecule. The greater this difference, the more accurately one can determine the ionization constant. The spectroscopic method is generally applicable only to conjugated systems that display absorption in the ultraviolet and visible regions. The principal advantage of the spectroscopic method is the small amount of sample required, the final solutions being 10^{-3} to 10^{-5} molar.

To determine the ionization constant by spectroscopic methods, the absorption spectra of the protonated and unprotonated species must be available. This is accomplished by recording the ultraviolet or visible spectrum of the compound in a sufficiently acidic solution such that the compound is completely protonated, and in a sufficiently basic solution such that only the unprotonated form is present. These spectra provide the extinction coefficients and wavelengths required in the calculations. The spectrum of the compound is then recorded in buffer solutions in which the compound will be present in both the protonated and unprotonated forms. The concentrations of the protonated and unprotonated forms are then calculated (see Sec. 4.2 for the details of calculating extinction coefficients), and the ionization constant is calculated using Eqs. (8.2) and (8.3), in which the hydrogen ion, or hydroxide ion, concentration is calculated from the pH of the buffer used as solvent. The ionization constant should be determined with the use of at least two different buffers. A plot of the concentration of the protonated, or unprotonated, form *vs.* buffer pH will resemble a typical neutralization curve.

8.6
NEUTRALIZATION AND SAPONIFICATION
EQUIVALENTS

The neutralization equivalent is defined as the equivalent weight of an acid, or base, as determined by titration with base, or acid. The neutralization equivalent can be used to determine the empirical formula of a molecule or the number of acidic or basic functional groups contained in the molecule, if the molecular weight is known.

The neutralization equivalent is calculated from Eq. (8.4). For procedural details, see Sec. 9.10.1c.

$$\text{N.E.} = \frac{\text{weight of sample}}{\text{volume of titrant} \cdot \text{normality}} \tag{8.4}$$

The saponification equivalent is the equivalent weight of an ester or amide based on the ester and amide functional groups. The saponification equivalent is used in the same way as the neutralization equivalent to determine the empirical formula of the ester or amide. For procedural details, see Secs. 9.13.1c and 9.13.1d.

9

Functional Classification and Characterization

This text gives procedures using semimicro quantities of materials, typically in the 50- to 500-mg range. The experienced and careful chemist will have no trouble manipulating quantities on a much smaller scale. Valuable material can be conserved by running solubility and classification tests in capillary tubes. In most cases, infrared spectra of compounds typical of the functional class under consideration are reproduced in this chapter.

Tables 9.1 to 9.19 provide identification data on representative compounds of various classes. In large part, the compounds listed are readily available from chemical supply houses. For the identification of an unknown, it may be necessary to consult a more extensive compilation of data, e.g., *Handbook of Tables for Identification of Organic Compounds*, 3rd ed., Chemical Rubber Co. Press, Cleveland, Ohio, 1973. The reader is also advised to see Chapter 10 on searching the literature.

Adequate precautions should always be taken to avoid unnecessary exposure to chemicals; see list on page xv for chemicals that are known or suspected to be carcinogenic.

9.1
HYDROCARBONS

9.1.1 Alkanes and Cycloalkanes

9.1.1a Classification

The following observations are indicative of saturated hydrocarbons:

1. *An exceptionally simple infrared spectrum.* In the infrared spectra of hydrocarbons, the strongest bands appear near 2900, 1460, and 1370 cm^{-1}

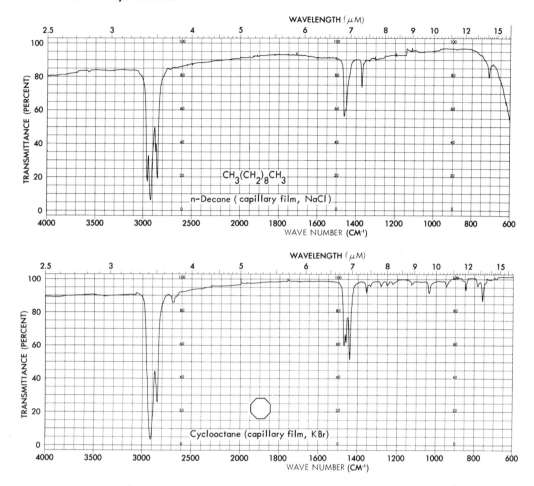

Fig. 9.1. Typical infrared spectra of alkanes. Note the absence of a 1370 cm⁻¹ band for methyl in the spectrum of cyclooctane.

(3.45, 6.85, and 7.30 μm) (Fig. 9.1). In simple cyclic hydrocarbons containing no methyl groups, the latter peak is absent. Other compounds that typically give rise to rather simple infrared spectra include aliphatic sulfides and disulfides, aliphatic halides, and symmetrically substituted alkenes and alkynes.

2. *Negative iodine charge-transfer test.* When iodine is dissolved in compounds containing π electrons or nonbonded electron pairs, a brown solution results. The brown color is due to the formation of a complex between the iodine and the π or nonbonded electrons. These complexes are called charge-transfer or π complexes. Solutions of iodine in nonparticipating solvents are violet.

$$\cdots I_2, \qquad R_2O \cdots I_2$$

9.1.1b Procedure: Iodine charge-transfer test

On a spot-test plate place a very small crystal of iodine; add one or more drops of the unknown liquid. Saturated hydrocarbons and their fluorinated and chlorinated derivatives, and arenes and halogenated derivatives give violet solutions; all other compounds give brown solutions. The test should be employed only with liquid unknowns. For color comparisons, it is strongly recommended that knowns be run concurrently with the unknowns.

3. *Insoluble in cold concentrated sulfuric acid.* Compounds that are unsaturated or possess a functional group containing nitrogen or oxygen are soluble in cold concentrated sulfuric acid. Saturated hydrocarbons, their halogen derivatives, simple aromatic hydrocarbons, and their halogen derivatives are insoluble.

4. *Negative test for halogens.* On the basis of the chemical, physical, and spectroscopic behavior, the most likely compounds to be confused with saturated hydrocarbons are the alkyl halides. If an unknown is suspected of being a hydrocarbon, the absence of halides should be verified by the Beilstein or sodium fusion test. The presence or absence of halogens can also be indicated by quantitative or qualitative measurements of the density of the unknown; hydrocarbons having densities less than one, halides greater than one.

5. *No resonance appears below $\delta 2$ in the NMR spectrum.* In most functionalized compounds, resonance of one or more hydrogens is usually found downfield from $\delta 2$; exceptions include compounds in which the functional group is located on a tertiary carbon and certain alkyl metal derivatives.

6. *No absorption above 185 nm in the ultraviolet spectrum.*

9.1.1c Characterization

There are no suitable general chemical methods for the conversion of saturated hydrocarbons to useful derivatives; they are typically inert or undergo nondiscriminate reactions that produce mixtures that are difficult to separate. Saturated hydrocarbons are, therefore, best identified through physical constants and spectroscopic methods.

The most useful physical constants are the boiling point and refractive index (Table 9.1). Specific gravity measurements are sometimes employed. Standard pycnometers or gravitometers require careful temperature control and large amounts of material; however, accurate determinations of specific gravity can be made on very small amounts of material by means of a capillary technique (Sec.

Table 9.1. Physical Properties of Alkanes, Alkenes, and Alkynes

Compound	bp (°C)*	D_4^{20}†	n_D^{20}†
	Alkanes		
2,2-Dimethylpropane	9.5	0.596	1.3513
2-Methylbutane	28	0.620	1.3580
Pentane	36	0.626	1.3577
Cyclopentane	49.3	0.746	1.4068
2,2-Dimethylbutane	49.7	0.649	1.3689
2,3-Dimethylbutane	58	0.662	1.3750
2-Methylpentane	60.3	0.653	1.3716
3-Methylpentane	63.3	0.664	1.3764
Hexane	68.3	0.659	1.3749
Cyclohexane (mp 6.5°C)	80.7	0.778	1.4264
3,3-Dimethylpentane	86	0.693	1.3911
2-Methylhexane	90	0.679	1.3851
3-Methylhexane	92	0.687	1.3887
3-Ethylpentane	93.5	0.698	1.3934
Heptane	98.4	0.684	1.3877
2,2,4-Trimethylpentane	99.2	0.692	1.3916
Methylcyclohexane	100.9	0.769	1.4231
2,2-Dimethylhexane	106.8	0.695	1.3930
2,5-Dimethylhexane	109.1	0.694	1.3930
2,4-Dimethylhexane	109.4	0.700	1.3953
3,3-Dimethylhexane	112	0.708	1.3992
Cycloheptane	118–20	0.8099	1.4440
Octane	125.7	0.703	1.3975
Ethylcyclohexane	131.7	0.788	1.4332
Nonane	150.8	0.718	1.4056
Cyclooctane (mp 14°C)	150/750 mm	0.8349	1.4586
Cyclononane (mp 9.7°C)	170–2	0.8534	1.4328[16]
Decane	174	0.730	1.4114
trans-Decahydronaphthalene	187.2	0.870	1.4695
cis-Decahydronaphthalene	195.7	0.896	1.4810
Undecane	196	0.702	1.4190
Dodecane	217	0.749	1.4216
Tridecane	235.5	0.7563	1.4256
Tetradecane	253.5	0.764	1.4289
Pentadecane	270.7	0.769	1.4310
Hexadecane	287	0.773	1.4352
Heptadecane	302.6	0.7767	1.4360[25]
Octadecane	317.4	0.7767	1.4367[28]
Nonadecane	331.6	0.7776	
Eicosane	345.1	0.7777	1.4307[50]
	Alkenes		
Methylpropene	−6.9	0.6266[−6.6]	1.3462
1-Butene	−6.3	0.6255[−6.5]	1.3465
cis-2-Butene	3.73	0.6303[1]	
3-Methyl-1-butene	20	0.6320	1.3643

Table 9.1   (Continued)

Compound	bp (°C)*	D_4^{20}†	n_D^{20}†
	Alkanes		
1,4-Pentadiene	26.1	0.6607	1.3887
1-Pentene	30.1	0.6410	1.3710
2-Methyl-1-butene	31	0.6504	1.3778
trans-2-Pentene	35.85	0.6486	1.3790
cis-2-Pentene	37.0	0.6562	1.3822
2-Methyl-2-butene	38.5	0.6620	1.3878
3,3-Dimethyl-1-butene	41	0.652	1.3766
Cyclopentene	44.2	0.7736	1.4225
1,5-Hexadiene	60	0.692	1.4045
1-Hexene	64	0.673	1.3875
trans-3-Hexene	67	0.677	1.3940
trans-2-Hexene	69	0.699	1.3929
Cyclohexene	83	0.8088	1.4465
1-Heptene	93.6	0.6971	1.3998
trans-2-Heptene	98.5	0.701	1.4035
2,4,4-Trimethyl-1-pentene	101.2	0.7151	1.4082
Cycloheptene	112	0.824	1.4585
2-Ethyl-1-hexene	120	0.7270	1.4157
1-Octene	121.3	0.7160	1.4088
Styrene	145.2	0.9056	1.5470
1-Nonene	147	0.7292	1.4154
Allylbenzene	157	0.8912	1.5042^{25}
α-Methylstyrene	165.4	0.9106	1.5386
1-Decene	170.5	0.7408	1.4220
Indene	181	0.9915	1.5764
trans-Stilbene (mp 180.5°C)	306	0.971	
	Alkynes		
1-Pentyne	39.7	0.6945	1.3847
2-Pentyne	55	0.710	1.4045
1-Hexyne	71	0.719	1.3990
3-Hexyne	81	0.724	1.4115
1-Heptyne	100	0.7338	1.4084
1-Octyne	131.6	0.7470	1.4172
Ethynylbenzene	141.7	0.9281	1.5485

  * BP at 760 mm Hg pressure unless otherwise noted.
  † Superscript numbers indicate the temperature (°C) at which the value was determined.

3.5). Retention times in gas-liquid chromatography are often very helpful in the tentative identification of a hydrocarbon.

Of the spectroscopic methods, mass spectroscopy is perhaps the most useful for the final identification of the saturated hydrocarbons. The utility of the method is greatly enhanced by the large number of reference spectra available. For the larger saturated hydrocarbons, NMR is of limited utility because of the small differences in chemical shift of the various types of hydrogens. In some cases, integration of the NMR spectrum may provide information on the ratio of methyl to other types of hydrogens in the molecule. In those cases in which the ring bears hydrogens, NMR is the best way to detect the presence of cyclopropanes because of their unique high-field resonance. Careful examination of the infrared spectrum can provide information about certain structural features of an unknown hydrocarbon, for example, geminal methyl groups, but typically cannot conclusively lead one to the assignment of structure. There is even some danger in assigning structures to hydrocarbons based upon comparison of infrared spectra because of the small differences in the relatively simple spectra of compounds with closely related structures, e.g., 4-methyl- and 5-methylnonane. Comparative gas-liquid chromatography is used in distinguishing hydrocarbons once some idea of the structure or molecular weight is available.

9.1.2 Alkenes

9.1.2a Classification

The infrared spectra of alkenes are generally considerably more complex than those of the saturated hydrocarbons (Figs. 9.1 and 9.2). The C=C absorption appears in the 1680 to 1620 cm^{-1} (5.95 to 6.17 μm) region; unfortunately, this band varies from very intense to entirely absent in completely symmetrical alkenes. However, unsaturation can also be detected by the vinyl hydrogen band just above 3000 cm^{-1} (3.33 μm) (if they are not hidden by the stronger, main band below 3000 cm^{-1}) or by the C—H out-of-plane bending bands in the 1000 to 700 cm^{-1} (10.0 to 14.3 μm) region. In many cases, the latter may be intense

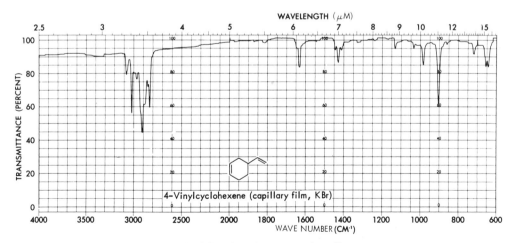

Fig. 9.2. Infrared spectrum of an alkene.

and very diagnostic. The presence of vinyl and allylic hydrogens can often be inferred from NMR spectra by the characteristic bands near $\delta 5$ to $\delta 7$ and near $\delta 2$, respectively. Unconjugated alkenes show only simple end absorption in the ultraviolet spectra recorded to 200 nm. The characteristic maxima and extinction coefficients at higher wavelengths for conjugated systems make ultraviolet spectroscopy quite suitable for dealing with conjugated systems.

As noted in Sec. 9.1.1, alkenes can be distinguished from saturated hydrocarbons by the positive iodine charge-transfer test and by their solubility in concentrated sulfuric acid.

Compounds suspected of being alkenes or alkynes should be tested with *bromine in carbon tetrachloride.* The majority of alkenes and alkynes add bromine quite rapidly; ethylenes or acetylenes substituted with electronegative groups add bromine slowly or not at all.

$$\text{Addition:} \quad \overset{\diagdown}{\underset{\diagup}{}}C{=}C\overset{\diagup}{\underset{\diagdown}{}} + Br_2 \quad \longrightarrow \quad \overset{\overset{\displaystyle Br}{|}}{\underset{\underset{\displaystyle Br}{|}}{-\underset{|}{C}-\underset{|}{C}-}}$$

$$\text{Substitution:} \quad -COCH_3 + Br_2 \quad \longrightarrow \quad -COCH_2Br + HBr \text{ (gas)}$$

Substitution reactions accompanied by the liberation of hydrogen bromide gas occur with phenols, amines, enols, aldehydes, ketones, and other compounds containing active methylene groups. With amines, the first mole of hydrogen bromide is not evolved but reacts with the amine to form the amine hydrobromide.

9.1.2b Procedure: Bromine in carbon tetrachloride

Dissolve 50 mg or two drops of the unknown in 1 mL of carbon tetrachloride, and add dropwise a 2% solution of bromine in the solvent until the bromine color persists. If more than two drops of the bromine solution are required to cause the color to remain for 1 min, an addition or substitution reaction has occurred. Substitution reactions are indicated by the evolution of hydrogen bromide, which is insoluble in carbon tetrachloride. The evolved hydrogen bromide can be detected by blowing across the top of the tube and noting the fog that is produced, or better, by denoting the acidic reaction on dampened pH paper held across the mouth of the tube.

A *permanganate test* can also be used to detect unsaturation. The positive test given by alkenes and alkynes is indicated by the disappearance of the purple permanganate ion and the appearance of the sparingly soluble brown manganese oxide. Positive tests are also given by phenols, aryl amines, aldehydes, primary and secondary alcohols (reaction often slow), and all organic sulfur compounds in which the sulfur is in a reduced state.

$$3 \overset{\diagdown}{\underset{\diagup}{}}C{=}C\overset{\diagup}{\underset{\diagdown}{}} + 2\,KMnO_4 + 4\,H_2O \quad \longrightarrow \quad 3 \underset{\underset{\displaystyle HO\ \ OH}{}}{-\underset{|}{C}-\underset{|}{C}-} + 2\,MnO_2 + 2\,KOH$$

9.1.2c Procedure: Permanganate test

Dissolve 25 mg or one drop of the unknown in 2 mL of water or reagent grade acetone in a small test tube. Add dropwise with vigorous shaking a 1% aqueous solution of potassium permanganate. If more than one drop of the permanganate is reduced, the test is positive.

Tetranitromethane forms colored charge-transfer or π complexes with alkenes and other unsaturated compounds. Many times, unsaturation can be detected with this reagent when other methods fail. The reagent is toxic, explosive, and expensive, and it should be employed in as small a quantity as practical. The substitution pattern of alkenes can be determined by measurement of the absorption maxima of the complexes in the visible region.[1]

$$\underset{\displaystyle \diagdown \diagup}{\overset{\displaystyle \diagup \diagdown}{\underset{C}{\overset{C}{\|}}}} + C(NO_2)_4 \longrightarrow \underset{\displaystyle \diagdown \diagup}{\overset{\displaystyle \diagup \diagdown}{\underset{C}{\overset{C}{\|}}}} \cdots C(NO_2)_4$$

Yellow-to-red complex

9.1.2d Procedure: Tetranitromethane test

Add a small amount of the compound to be tested to a 25% solution of *tetranitromethane* in dichloromethane. (The reagent solution can be made up in advance and stored in the refrigerator.) The color increases from yellow to dark red with increasing unsaturation. Simple alkenes and benzene give a yellow coloration; naphthalene gives an orange color. The test is positive with cyclopropane, feeble with α,β-unsaturated carbonyl compounds and other electrophilic alkenes, and negative with allylic alcohols and alkynes.

9.1.2e Characterization

Although alkenes undergo numerous reactions, mainly involving addition to or cleavage of the carbon-carbon double bond, relatively few of these reactions have wide general applicability for the preparation of suitable derivatives. Like saturated hydrocarbons, many of the simple alkenes can be identified solely by reference to physical and/or spectral properties (Table 9.1).

Upon oxidation, alkenes yield aldehydes, ketones, and/or acids. The product obtained depends upon the reagent and the conditions as well as on the degree of substitution about the double bond. *Ozonization* is the classical method for determining positions of double bonds in alkenes. The ozonides that are formed in these reactions are usually not isolated since they are often unstable and may decompose with explosive violence. Hydrolysis of ozonides with water and

---

[1] E. Heilbronner, *Helv. Chim. Acta*, **36**, 1121 (1953).

reduction with zinc yield aldehydes and ketones. When the ozonides are worked up in the presence of acidic hydrogen peroxide, the aldehydic components are oxidized to carboxylic acids. Ozonization reactions have the advantage that they are relatively simple to run and work up. An excellent technique for the ozonolysis of double bonds is to add the ozone to a solution of the alkene in dichloromethane: pyridine (2:1) at $-70°C$. The reaction mixture is then allowed to slowly warm to room temperature, is poured into water, and the aldehydes and/or ketones are isolated by ether extraction. The pyridine acts as a reducing agent. Ozone can be generated from oxygen by a commercially available apparatus[2] or can be purchased diluted with Freon-13 in stainless stell cylinders.[3]

$$R_2C{=}CHR' \xrightarrow{O_3} R_2C\underset{O}{\overset{O-O}{\diagdown\diagup}}CHR' \begin{array}{l} \xrightarrow{\text{pyridine}} R_2C{=}O + R'CHO \\ \\ \xrightarrow[H^+]{H_2O_2} R_2C{=}O + R'COOH \end{array}$$

9.1.2f Procedure: Ozonolysis

A solution of 0.01 mol of the alkene in 20 mL of dichloromethane and 10 mL of pyridine is chilled to $-70°C$ in a Dry Ice-acetone bath, and a stream of ozone is bubbled through the solution until the effluent gas causes a solution of potassium iodide and starch indicator to turn purple. The reaction mixture is poured into a 250-mL beaker and is allowed to warm slowly to room temperature with occasional stirring. The resulting brown solution is poured into a separatory funnel and is washed with 10% hydrochloric acid until free of pyridine. The organic layer is dried over magnesium sulfate, the solvent is then removed, and the carbonyl compound(s) is (are) purified by recrystallization or distillation.

An aqueous solution of excess *periodate ion with a catalytic amount of permanganate ion*[4] is capable of the oxidative cleavage of carbon-carbon double bonds. The best conditions for carrying out an oxidative cleavage of alkenes with the reagent are the use of a large excess of periodate and 5 to 10 mol % of permanganate in an aqueous solution at pH 7 to 10. If the alkene is insoluble in water, then *t*-butyl alcohol, dioxane, or pyridine can be used as a co-solvent up to 25%. The reaction is illustrated by the following equations:

$$\begin{array}{c} H \quad\quad R \\ \diagdown\quad\diagup \\ C{=}C \\ \diagup\quad\diagdown \\ R \quad\quad R \end{array} + MnO_4^- \xrightarrow[OH^-]{H_2O} \begin{array}{c} H \quad\quad R \\ \diagdown\quad\diagup \\ C{-}C \\ \diagup | \;\; | \diagdown \\ R \;\; OH \;\; OH \;\; R \end{array} + MnO_3^-$$

---

[2] "Organic Ozone Reactions and Techniques" and "Basic Manual of Applications and Laboratory Ozonization Techniques," The Welsbach Corp., Ozone Process Division, Westmoreland and Stokley Sts., Philadelphia, Pennsylvania, 19129.

[3] The Matheson Company, P.O. Box 85, East Rutherford, New Jersey.

[4] R. U. Lemieux and E. von Rudloff, *Can. J. Chem.*, **33**, 1701, 1710 (1955).

$$\underset{R}{\overset{H}{\diagdown}}\underset{OH}{\overset{|}{\underset{|}{C}}}-\underset{OH}{\overset{|}{\underset{|}{C}}}\overset{R}{\underset{R}{\diagup}} + IO_4^- \longrightarrow RCHO + R_2CO + IO_3^-$$

$$RCHO + MnO_4^- \xrightarrow[\text{OH}^-]{\text{H}_2\text{O}} RCOO^- + MnO_3^-$$

$$MnO_3^- + IO_4^- \longrightarrow MnO_4^- + IO_3^-$$

A useful modification of this reaction involves substituting osmium tetroxide for permanganate.[5] In this case, the oxidation is stopped at the aldehyde stage. The expensive and poisonous osmium tetroxide is used only in catalytic amounts, so the major objections to this reagent are minimized. Aqueous dioxane is used as a solvent, or the reaction is run in a two-phase system of water and ether.

Alkenes can be directly oxidized to carboxylic acids by shaking or refluxing with *alkaline permanganate*. Under these conditions, small amounts of other carboxylic acids often arise owing to over-oxidation. This can lead to difficulties in purification and in the interpretation of the results.

$$RCH{=}CHR \xrightarrow[\text{(2) H}_3\text{O}^+]{\text{(1) KMnO}_4,\ \text{OH}^-} 2RCOOH$$

9.1.2g Procedure: Alkaline permanganate oxidation

The compound (500 mg) is added to 50 mL of water containing 2 g of potassium permanganate; 1 mL of 5% potassium hydroxide solution is added, and the mixture is shaken at intervals for 10 to 15 min, or until the purple color of the permanganate disappears. For less reactive compounds, the mixture is heated under reflux for 1.5 to 3 hr, and then allowed to cool. The mixture is carefully acidified with sulfuric acid. Any excess manganese dioxide can be removed by the addition of small amounts of sodium hydrogen sulfite. The acid is collected by filtration or by extraction with ether or dichloromethane.

Unsaturated hydrocarbons can also be characterized by the addition of reagents to the double bond. Addition of bromine is a particularly facile reaction. In some cases, the dibromide formed is a suitable solid derivative and can be characterized and identified by means of physical constants and spectral data. The bromide, if a liquid, can be converted to a 2,4-dinitrophenylthioether by reaction with 2,4-dinitrothiophenol. The dibromide is a particularly useful derivative for the identification of gaseous or low-boiling alkenes that may distill out of reaction mixtures. In this case, the derivative can be prepared by passing the gas, aided by a nitrogen stream, directly into a carbon tetrachloride solution containing bromine. Any excess bromine can be removed from the carbon tetrachloride by shaking with aqueous sodium hydrogen sulfite.

---

[5] R. Pappo, D. S. Allen, Jr., R. U. Lemieux, and W. S. Johnson, *J. Org. Chem.*, **21**, 478 (1956).

The *Kharasch reagent* (*2,4-dinitrobenzenesulfenyl chloride*) readily adds to many alkenes to yield crystalline chlorosulfides. With symmetrical alkenes or styrene derivatives, a single adduct is formed. Other alkenes often yield mixtures of isomeric sulfides. The reagent will also add to alkynes and allenes. The reagent adds in a stereospecific *trans* manner; thus *cis*- and *trans*-2-butene give different products. Tetra-substituted alkenes do not react.

9.1.2h Procedure: Addition of 2,4-dinitrobenzenesulfenyl chloride

CAUTION: The reagent 2,4-dinitrobenzenesulfenyl chloride is known to be explosive. Do not heat above 100°C.

Mix 200 mg of 2,4-dinitrobenzenesulfenyl chloride with an equal amount of the alkene in 5 mL of glacial acetic acid. Heat on a steam bath for 15 min or until the reaction is complete. The presence of unreacted reagent can be detected by the liberation of iodine on starch iodide paper.

The mixture is cooled and the precipitate is collected by vacuum filtration. If crystals do not form, the reaction mixture is poured onto a small amount of crushed ice. The crude derivative can be recrystallized from ethanol.

9.1.2i Chemical conversion to other functional classes

When the saturated hydrocarbon is well characterized, identification of an alkene can be expedited by conversion to the saturated hydrocarbon. Addition of hydrogen across the double bond can be effected through catalytic hydrogenation, dissolving metal reductions, diimide reduction, or the hydroboration-protonolysis sequence. The catalytic hydrogenation of very small amounts of material has often been used as a means of determining the amount of unsaturation in a molecule. Sensitive gas burettes or pressure gauges are used to quantitatively measure the amount of hydrogen uptake. In most catalytic hydrogenations, the hydrogen is supplied to the reaction vessel at atmospheric or higher pressures from a storage tank. The Brown-II Hydrogenator[6] employs hydrogen generated *in situ* by the action of acid on sodium borohydride. The apparatus consists of a reaction flask, a generator flask, and a burette, which is used to add standard sodium borohydride solution to the generator through an automatic valve. The catalyst used with

---

[6] H. C. Brown and C. A. Brown, *J. Am. Chem. Soc.*, **85**, 1005 (1963) and previous papers.

the apparatus can be prepared by the reduction of nickel or platinum metal salts with borohydride immediately prior to use. Diimide reductions can be performed by the generation of the diimide *in situ* by the oxidation of hydrazine with hydrogen peroxide under the influence of copper catalyst.[7]

$$NH_2NH_2 + H_2O_2 \longrightarrow H-N=N-H + 2H_2O$$

In a reaction known as hydroboration,[8] diborane (B_2H_6) adds as BH_3 to alkenes to give organoboranes. Diborane and alkylboranes are very reactive, toxic substances (the lower-molecular-weight alkyl boranes are also highly flammable), but for most synthetic purposes, the diborane and the alkylboranes are generated and reacted *in situ*. Diborane is conveniently prepared by reaction of boron trifluoride etherate with sodium borohydride in glyme, or by reaction of the boron trifluoride with lithium borohydride in diethyl ether. Standardized solutions of borane in tetrahydrofuran are commercially available. The utility of the hydroboration reaction is illustrated by the following equations:

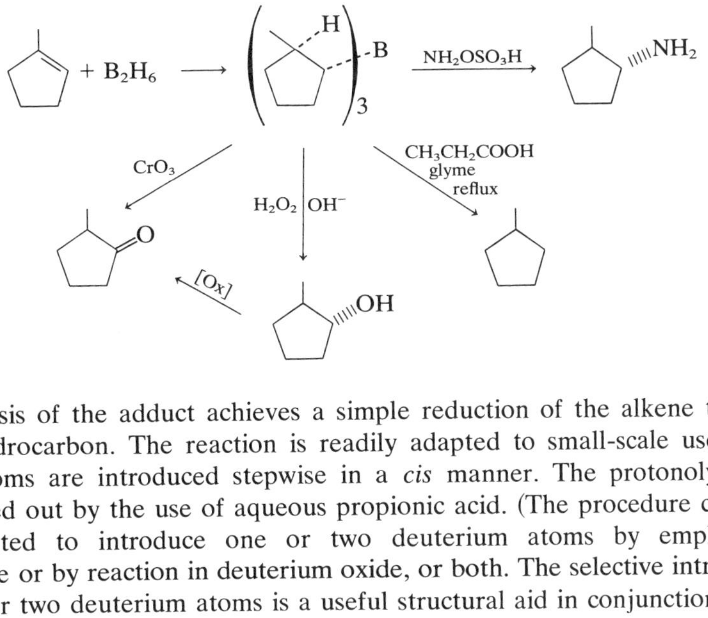

Protonolysis of the adduct achieves a simple reduction of the alkene to the saturated hydrocarbon. The reaction is readily adapted to small-scale use; the hydrogen atoms are introduced stepwise in a *cis* manner. The protonolysis is usually carried out by the use of aqueous propionic acid. (The procedure can be readily adapted to introduce one or two deuterium atoms by employing borodeuteride or by reaction in deuterium oxide, or both. The selective introduction of one or two deuterium atoms is a useful structural aid in conjunction with

---

[7] E. J. Corey, W. L. Mock, and D. J. Pasto, *Tetrahedron Letters*, 347 (1961).

[8] H. C. Brown, *Hydroboration*, W. A. Benjamin, Inc., New York, 1962; G. Zweifel and H. C. Brown, *Org. Reactions*, **13**, 1 (1963). For experimental techniques see H. C. Brown, *Organic Syntheses via Boranes*, Wiley-Interscience, New York, 1975.

mass spectrometry or NMR spectroscopy.) The peroxide reaction achieves the anti-Markonikoff addition of water to the carbon-carbon double bond. The alcohols thus obtained can be converted to solid derivatives or can be further oxidized to aldehydes or ketones for identification purposes. Alkyl boranes can be directly oxidized to ketones. In certain cases, the conditions necessary to effect this direct oxidation may induce rearrangements. It is often advantageous to first prepare the alcohol and then oxidize to the ketone. This procedure—conversion of an alkene to a ketone—has utility in that it may indicate the size of the ring in which the carbon-carbon double bond is located by the carbonyl frequency of the final product. Alkyl boranes also react with hydroxylamine-O-sulfonic acid to produce amines.

The Diels-Alder reaction may occasionally be found to be the most suitable method for preparing a derivative of an alkene or a diene.[9] Maleic anhydride and N-phenylmaleiimide are among the most suitable of the dienophiles. They are rather reactive and usually form solid derivatives that can be characterized by further chemical conversion if necessary. Diphenylisobenzofuran[10] is recommended as a highly reactive diene that can be used to trap reactive alkenes and alkynes.[11]

The additions of nitrosyl chloride, nitrogen trioxide, and nitrogen tetroxide have found special applications in the characterization of terpenes.[12]

9.1.3 Alkynes and Allenes

Alkynes and allenes, like alkenes, are soluble in concentrated sulfuric acid. In addition they add bromine in carbon tetrachloride and are oxidized by cold aqueous permanganate.

[9] M. C. Kwitzel, *Org. Reactions*, **4**, 1 (1948); H. L. Holmes, *Org. Reactions*, **4**, 60 (1948).

[10] M. S. Newman, *J. Org. Chem.*, **26**, 2630 (1961).

[11] G. Wittig and R. Pohlke, *Chem. Ber.*, **94**, 3276 (1961).

[12] J. L. Simonsen and L. N. Owen, *The Terpenes*, Vols. I and II, Cambridge University Press, Cambridge, 1947–1957.

9.1.3a Classification of alkynes

There are two general types of alkynes—terminal and nonterminal. The terminal alkynes are easily detected by their very characteristic infrared spectra and by their ease of formation of metal derivatives, e.g., mercury alkynides (see below). The terminal alkynes show a sharp characteristic $\equiv$C—H band at 3310 to 3200 cm^{-1} (3.02 to 3.12 μm) and a triple-bond stretch of medium intensity at 2140 to 2100 cm^{-1} (4.67 to 4.76 μm). In the NMR, the $\equiv$C—H occurs as a singlet near $\delta2.8$ to 3.0. The nonterminal alkynes display triple-bond absorption at 2250 to 2150 cm^{-1} (4.44 to 4.65 μm), which may be absent in symmetrically substituted cases (Fig. 9.3).

9.1.3b Characterization of alkynes

Like other hydrocarbons, the characterization of alkynes is heavily dependent on physical properties (Table 9.1). Only a few general methods have been developed for the preparation of solid derivatives.

Alkynes can be *hydrated to ketones* by the action of sulfuric acid and mercuric sulfate in dilute alcohol solution. Terminal alkynes give methyl ketones; nonterminal, unsymmetrically substituted alkynes often give mixtures of ketones.

$$RC\equiv CR + H_2O \xrightarrow[H_2SO_4]{HgSO_4} RCOCH_2R$$

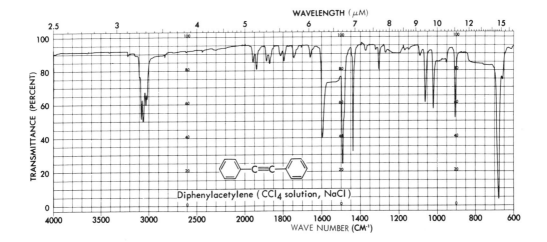

Fig. 9.3. Infrared spectrum of a symmetrically substituted acetylene. Compare with spectrum of phenylacetylene (Fig. 5.10).

9.1.3c Procedure: Hydration of alkynes

A solution of 0.2 g of mercuric sulfate and three to four drops of concentrated sulfuric acid in 5 mL of 70% aqueous methanol is warmed to 60°C. The alkyne (0.5 g) is added dropwise, and the solution is maintained at 60°C with stirring for 1 to 2 hr. The methanol is distilled off, and the residue is saturated with salt and extracted with ether. Appropriate spectral and physical properties of the ketone are determined, and a solid derivative is made.

Addition of 2,4-dinitrobenzenesulfenyl chloride proceeds readily to give colored, crystalline products.

$$ArSCl + RC \equiv CR' \longrightarrow \underset{Cl}{\overset{R}{\diagdown}} C = C \underset{R'}{\overset{SAr}{\diagup}}$$

Ar = 2,4-dinitrophenyl
R' = R or H

9.1.3d Procedure: Addition of 2,4-dinitrobenzenesulfenyl chloride

Dissolve the reagent in 1,1-dichloroethane, cool to 0°C, and add dropwise an excess of the alkyne. Keep the reaction mixture at 0°C for 2 hr or longer. Remove the solvent and the excess alkyne by vacuum evaporation. Keep the residue cold until crystallization occurs. Recrystallize from ethanol using activated carbon, if necessary, to remove any highly colored impurities.

When a terminal alkyne is added to the *alkaline mercuric iodide* reagent, an immediate white or gray-white precipitate forms. This reaction can be used as a convenient test for terminal alkynes, as well as a method for the preparation of a derivative. *Certain heavy metal acetylides are explosive when dry and therefore should not be isolated.*

$$2 \, RC \equiv CH \xrightarrow[\text{NaOH}]{\text{K}_2\text{HgI}_4} (RC \equiv C)_2 Hg$$

9.1.3e Allenes

Allenes are usually identified by spectral and physical characteristics. The best method of detection is by the presence of a strong stretching vibration, often observed as a doublet, in the 2200 to 1950 cm^{-1} (4.55 to 5.13 μm) region. Allenes absorb only below 200 nm in the ultraviolet. In the NMR, terminal hydrogens are near $\delta 4.4$ and allenic hydrogens near $\delta 4.8$ with ^4J values of up to 7 Hz. Allenes are capable of exhibiting optical isomerism.

Derivation can be made by the addition of 2,4-dinitrobenzenesulfenyl chloride (Sec. 9.1.3d); however, mixtures may be formed.

9.1.4 Aromatic Hydrocarbons

9.1.4a Classification

Benzene and a large number of its alkylated derivatives are liquids. Most compounds containing more than one aromatic ring, either fused or nonfused, are solids. Aromatic hydrocarbons burn with a characteristic sooty flame and give negative tests with bromine in carbon tetrachloride and alkaline permanganate (Sec. 9.1.2a). The simple aromatic hydrocarbons are insoluble in sulfuric acid but soluble in fuming sulfuric acid. With two or more alkyl substituents, the nucleus is sufficiently reactive to sulfonate, and such compounds dissolve in concentrated sulfuric acid.

The infrared spectra of aromatic hydrocarbons exhibit the expected aromatic (and, in the case of alkylated nuclei, aliphatic) C—H stretching and bending deformations, plus characteristic ring absorption bands (Sec. 5.2.1b). The spectra show only bands of moderate intensity above 1000 cm^{-1} (10 μm) and generally only moderate to weak bands between 1000 and 800 cm^{-1} (10.0 and 12.5 μm) (Fig. 9.4).

In the NMR spectra, the presence, position, and, especially, intensity (integration) of the aryl hydrogens, plus the chemical shift and coupling pattern of any aliphatic protons, provide important and often definitive structural information.

In all alkyl benzenes, the substituted tropylium ion is an important and diagnostic fragment in the mass spectra.

9.1.4b Characterization

The *nitro* and *polynitro derivatives* are generally useful in identification of aromatic hydrocarbons. In certain cases, it is advantageous to reduce mononitro compounds to primary amines from which solid derivatives are made.

The nitration of aromatic compounds should always be done with great caution and on a small scale, especially when dealing with unknowns. A number of different procedures are used for nitration. The conditions that should be employed depend on the reactivity of the substrate and on the degree of nitration desired. It is very helpful to know something about the structure and, hence, reactivity of the compound before nitration is attempted. The following methods in procedure 9.1.4c are listed in order of increasingly vigorous conditions. Method *b* usually leads to mononitro compounds, except for highly reactive substrates such as phenol and acetanilide. Method *c* usually gives di- or trinitro derivatives.

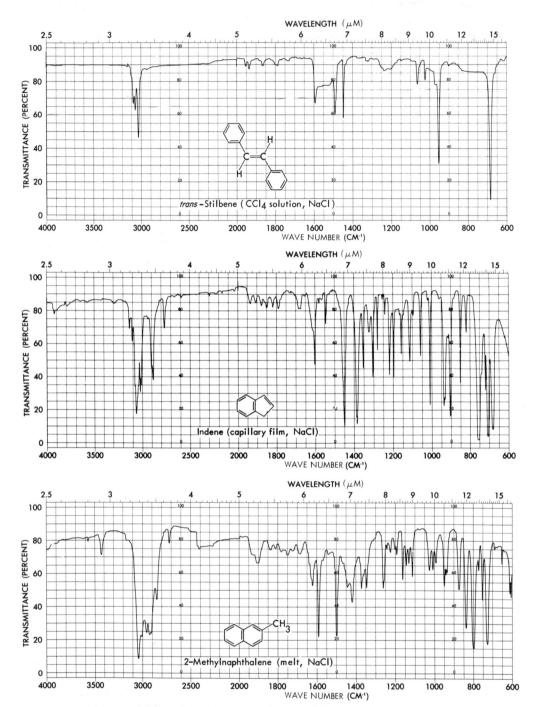

Fig. 9.4. Typical infrared spectra of aromatic hydrocarbons. Note the absence of a $\nu_{C=C}$ in the spectrum of *trans*-stilbene.

9.1.4c Procedure: Nitration

(a) Reflux a mixture of 0.5 g of the compound, 2 mL of glacial acetic acid, and 0.5 mL of fuming nitric acid for 10 min. Pour into ice water. Filter the precipitate, wash with cold water, and recrystallize from aqueous ethanol.

(b) Add 0.5 to 1 g of the compound to 2 to 4 mL of concentrated sulfuric acid. An equal volume of concentrated nitric acid is added drop by drop, with shaking after each addition. Heat on a water bath at 60°C for 5 to 10 min. Remove the tube from the bath and shake every few minutes. Cool and pour into ice water. Filter off the precipitate, wash with water, and recrystallize from aqueous ethanol.

(c) Follow the foregoing procedure, substituting fuming nitric acid instead of concentrated nitric acid. Heat the mixture for 10 min on a steam bath.

The clean and efficient nitration of aromatic hydrocarbons and derivatives can be achieved by use of nitronium fluoroborate in sulfolane or acetonitrile.[13] The method has the advantages that acid labile groups are not hydrolyzed during the nitration or work-up, and the stages of nitration can be easily controlled.

$$R\text{-ring}(G) + NO_2^+BF_4^- \longrightarrow R\text{-ring}(G)(NO_2) + HBF_4$$

Aromatic hydrocarbons (and halogen derivatives) react with phthalic anhydride under Friedel-Crafts conditions to yield aroylbenzoic acids, which can be characterized by their melting points and neutralization equivalents.

$$\text{phthalic anhydride} + ArH \xrightarrow{AlCl_3} \text{(o-COOH, COAr)}$$

9.1.4d Procedure: Aroylbenzoic acids

CAUTION: Carbon disulfide is highly inflammable.

In a small apparatus equipped with a reflux condenser and a hydrogen chloride trap, place 0.4 g of phthalic anhydride, 10 mL of carbon disulfide, 0.8 g of anhydrous aluminum chloride, and 0.4 g of the aromatic hydrocarbon. Heat the mixture on a water bath until no more hydrogen chloride is evolved, or about 30 min. Cool under tap water. Decant the carbon disulfide layer slowly. Add 5 mL of concentrated hydrochloric acid with cooling and then 5 mL of water to the residue. Shake or stir the mixture thoroughly. Cool if necessary to induce crystallization. Collect the solid,

---

[13] S. J. Kuhn and G. A. Olah, *J. Am. Chem. Soc.*, **83**, 4564 (1961); reagent available from Ozark-Mahoning Co., 310 West 6th St., Tulsa, Oklahoma 74119.

wash with cold water, and recrystallize from aqueous ethanol. If the product fails to crystallize, take the oil up in dilute ammonium hydroxide, treat with activated carbon, filter, cool, and neutralize with concentrated hydrochloric acid.

With chromic acid or alkaline permanganate, *alkyl benzenes are oxidized to aryl carboxylic acids.* This procedure is recommended when there is one side chain, in which case benzoic or a substituted benzoic acid is obtained, or when there are two *ortho* side chains, in which case *o*-phthalic acids are obtained (the melting points of a number of substituted *o*-phthalic acids and anhydrides have been recorded). Permanganate oxidation is usually the preferred procedure.

9.1.4e Procedure: Permanganate oxidation

In a small flask equipped for reflux, place 1.5 g of potassium permanganate, 25 mL of water, 6.5 mL of 6 N potassium hydroxide, 2 boiling chips, and 0.4 to 0.5g of the alkyl benzene. Reflux gently for 1 hr, or until the purple color of the permanganate has disappeared. Cool the reaction mixture and acidify with dilute sulfuric acid. Heat to boiling; add a pinch of sodium hydrogen sulfite if necessary to destroy any brown manganese dioxide. Cool and filter the acid. Purify by recrystallization from water or aqueous ethanol, or by sublimation.

Polynitro aromatic compounds form stable charge-transfer complexes[14] with many electron-rich aromatic systems. The picrates, formed from 2,4,6-trinitrophenol, of polynuclear aromatic hydrocarbons and aryl ethers are well-known examples (Sec. 9.4). The *complexes of 2,4,7-trinitro-9-fluorenone (TNF)* and aromatic hydrocarbons, or derivatives, are of considerable utility in identification work, especially with polynuclear systems.[15]

2,4,7-trinitro-9-fluorenone (TNF)

[14] L. N. Ferguson, *The Modern Structural Theory of Organic Chemistry,* Prentice-Hall, Inc., Englewood Cliffs, New Jersey, 1963.

[15] M. Orchin and E. O. Woolfolk, *J. Am. Chem. Soc.,* **68,** 1727 (1946); M. Orchin, L. Reggel, and E. O. Woolfolk, *J. Am. Chem. Soc.,* **69,** 1225 (1947); D. E. Laskowski and W. C. McCrone, *Anal. Chem.,* **30,** 1947 (1958).

9.1.4f Procedure: TNF complexes

Prepare separate, nearly saturated hot solutions of 2,4,7-trinitro-9-fluorenone (TNF) and the aromatic substance in absolute ethanol or glacial acetic acid. Mix the two hot solutions and heat for 1 min, cool and recrystallize the complex from any of the above solvent systems.

Table 9.2 lists the physical properties and derivatives of selected aromatic hydrocarbons.

Table 9.2. Physical Properties of Aromatic Hydrocarbons and Derivatives

	Liquids					mp (°C) of Derivatives	
						Aroyl-benzoic	
				Nitro		Acid	Picrate*
Compound	bp (°C)	D_4^{20}	n_D^{20}	Position	mp (°C)	mp (°C)	mp (°C)
Benzene	80.1	0.8790	1.5011	1,3	89	127	84u
Methylbenzene	110.6	0.8670	1.4969	2,4	70	137	88u
Ethylbenzene	136.2	0.8670	1.4959	2,4,6	37	122	96u
1,4-Dimethylbenzene (p-Xylene)	138.3	0.8611	1.4958	2,3,5	139	132	90u
1,3-Dimethylbenzene (m-Xylene)	139.1	0.8642	1.4972	2,4,6	183	126	91u
1,2-Dimethylbenzene (o-Xylene)	144.4	0.8802	1.5054	4,5	118	178	88u
Isopropylbenzene (cumene)	152.4	0.8618	1.4915	2,4,6	109	133	
Propylbenzene	159.2	0.8620	1.4920			125	103u
1,3,5-Trimethylbenzene	164.7	0.8652	1.4994	2,4	86	211	97u
tert-Butylbenzene	169.1	0.8665	1.4926	2,4	62		
1,2,4-Trimethylbenzene	169.2	0.8758	1.5049	3,5,6	185		97u
Isobutylbenzene	172.8	0.8532	1.4865				
sec-Butylbenzene	173.3	0.8621	1.4901				
1,2,3-Trimethylbenzene	176.2	0.8944	1.5139				
1-Isopropyl-4-methylbenzene	177.1	0.8573	1.4909	2,6	54	123	
Indene	182	0.857					98
Butylbenzene	183.3	0.8601	1.4898				
1-Methyl-4-propylbenzene	183.5	0.859	1.493				
1,2,3,4-Tetramethylbenzene	197.9	0.8899	1.5125	4,6	181	213	
Pentylbenzene	205.4	0.8585	1.4878				
Tetrahydronaphthalene	206	0.971		5,7	95	153	
p-Diisopropylbenzene	210.4	0.8568	1.4898				
1-Methylnaphthalene	244.8	1.0200	1.6174	4	71	68	142
2-Ethylnaphthalene	257.9	0.9922	1.5995				71
1-Ethylnaphthalene	258.3	1.0076	1.6052				98
1,6-Dimethylnaphthalene	262	1.003	1.607				114
1,7-Dimethylnaphthalene	263	1.012	1.607				121
1,3-Dimethylnaphthalene	263	1.002	1.6078				118
1,4-Dimethylnaphthalene	263	1.008	1.6127				140
1,1-Diphenylethane	270	1.003	1.5761				

Table 9.2 (Continued)

Compound	Solids mp (°C)	bp (°C)	n_D^{20}	mp (°C) of Derivatives Nitro Position	Nitro mp (°C)	Aroyl-benzoic Acid mp (°C)	Picrate* mp (°C)
Diphenylmethane	25.3			2,4,2',4'	172		
2-Methylnaphthalene	34.4	241		1	81		116
Pentamethylbenzene	51			6	154		131
1,2-Diphenylethane	53			4,4'	180		
				2,2',4,4'	169		
Biphenyl	69.2			4,4'	237	224	
				2,2',4,4'	150		
Naphthalene	80.2	218		1	61	172	149
Triphenylmethane	92			4,4',4"	206		
Acenaphthene	96.2			5	101	198	161
Phenanthrene	96.3						144
2,3-Dimethylnaphthalene	104						124
2,6-Dimethylnaphthalene	111						143
Fluorene	113.5	295		2	156	227	87
Pyrene	148	385					220
Hexamethylbenzene	165						170
Anthracene	216						138u

*u indicates an unstable complex that is difficult to purify.

9.2
HALIDES

9.2.1a General Classification

Halogen substituents can be found in combination with all other functional groupings. When another functional group is present, chemical transformations for the purpose of making derivatives are usually carried out on that functional group rather than the halogen, e.g., esters are made from chloroacids, carbonyl derivatives are made from chloroketones, and urethanes are made from bromoalcohols.

The presence or absence of halogen in an unknown can seldom be inferred from its infrared spectrum (Fig. 9.5). The carbon fluorine stretching absorptions occur in the 1350 to 960 cm^{-1} (7.41 to 10.42 μm) region, those for chlorine occur in the 850 to 500 cm^{-1} (11.7 to 20 μm) region (often obscured by chlorinated solvents or aromatic substitution bands), while those for bromine and iodine occur below 667 cm^{-1} (16 μm) and are not observed in the normal infrared region.

The presence or absence of halogen can be determined by the Beilstein test, analysis of the filtrate from a sodium fusion, or by application of the silver nitrate or sodium iodide tests (see below). The very characteristic cracking patterns and/or isotope peaks observed in the mass spectra of halogenated compounds

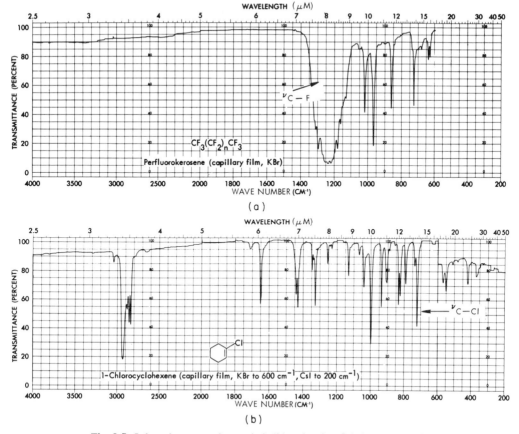

Fig. 9.5. Infrared spectra of organic halides showing C-halogen stretching bands.

provide definitive evidence for the presence and types of halogen(s). The presence of halogen is sometimes suggested by the characteristic chemical shift in the NMR of hydrogens on the carbon atom bearing the halogen. The presence of fluorine can be demonstrated by NMR spectroscopy. In hydrogen spectra, hydrogen-fluorine coupling can be observed; in fluorine spectra, the relative number and kinds of fluorines present can be determined through integration and measurement of characteristic fluorine chemical shifts.

In addition to alkyl and aryl halides, halogen occurs in the following types of compounds:

1. Salts (amine hydrochlorides, ammonium, phosphonium and sulfonium salts, etc.).
2. Acid halides (acyl halides, sulfonyl halides, etc.).
3. Haloamines and haloamides (N-bromosuccimide, Chloramine-T, etc.).
4. Miscellaneous compounds (iodobenzene dichlorides, alkyl hypochlorites, halosilanes, alkyl mercuric halides, and other organometallic halides etc.).

Alkyl and aryl halides (along with hydrocarbons, diaryl ethers, and most perfluoro compounds) are insoluble in concentrated sulfuric acid.

The following tests are useful in classifying the various types of alkyl and aryl halogen compounds. It is advisable to test the reactivity of any unknown halide toward each reagent.

9.2.1b Alcoholic silver nitrate

$$AgNO_3 + X^- \longrightarrow AgX + NO_3^-$$

The test depends on the rapid and quantitative reaction of silver nitrate with halide ion to produce insoluble silver halide (excepting the fluoride).

Silver nitrate in aqueous or ethanolic solution gives an immediate precipitate with compounds that contain ionic halide, or with compounds such as acid chlorides, which react immediately with water or ethanol to produce halide ion. Many other halogen-containing substances react with silver nitrate to produce insoluble silver halide. The rate of such reactions is a measure of the reactivity of the substrate in S_N1 reactions.

$$RX \longrightarrow R^+ + X^-$$

Silver and other heavy metal salts catalyze S_N1 reactions of alkyl halides by complexation with the unshared electrons of the halide, making the leaving group a metal halide rather than halide ion.

$$R\text{—}\ddot{X}\text{:} \xrightarrow{Ag^+} [R\text{—}\overset{\delta^+}{\ddot{X}}\text{:---}\overset{\delta^+}{Ag}] \xrightarrow{slow} R^+ + AgX \downarrow$$
$$R^+ \longrightarrow \text{alcohols, ethers, alkenes, etc.}$$

The rate of precipitation of silver halide depends on the leaving group, I > Br > Cl (silver fluoride is soluble), and upon the structure of the alkyl group. Any structural factors that stabilize the electron-deficient carbonium ion (R^+) will accelerate the reaction. We thus find the expected order of reactivity

benzyl ≈ allyl > tertiary > secondary > primary > methyl ≫ vinyl ≈ aryl.

The following equations illustrate other structural classes that are unusually reactive in S_N1 reactions.

G = amino, sulfide, iodide, aryl, etc.

G = amino, alkoxy, thioalkoxy, etc.

Factors that tend to destabilize the incipient carbonium ion decelerate the rate of S_N1 reactions. The following compounds are not reactive under S_N1 conditions:

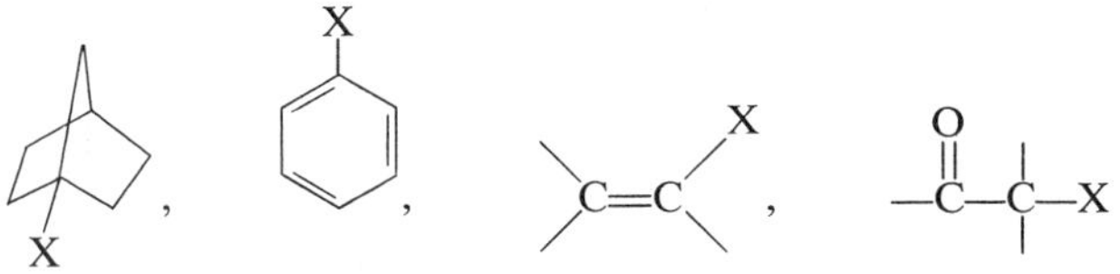

A summary of types of compounds in order of decreasing reactivity of silver nitrate with various organic compounds containing halogen is as follows (X=I, Br, or Cl):

1. Water-soluble compounds that give immediate precipitates with aqueous silver nitrate.
   a. salts
   b. halogen compounds that are hydrolyzed immediately with water, such as low-molecular-weight acyl or sulfonyl halides
2. Reactivity of water-insoluble compounds with alcoholic silver nitrate.
   a. immediate precipitation

$$RCOCl \qquad\qquad RSO_2Cl \qquad\qquad RO-\overset{|}{\underset{|}{C}}-X$$

$$R-CH=CH-CH_2-X \qquad\qquad R-CHBrCH_2Br$$

$$ArCH_2X \qquad\qquad R_3CX \qquad\qquad RI$$

   b. silver halide precipitates slowly or not at all at room temperature, but readily when warmed on the steam bath

$$RCH_2Cl \quad R_2CHClBr \quad RCHBr_2 \quad O_2N-\underset{NO_2}{\underset{|}{\bigcirc}}-X$$

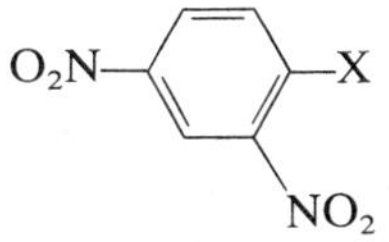

   c. inert toward hot silver nitrate solutions

$$ArX \quad RCH=CHX \quad HCCl_3 \quad -\overset{O}{\overset{||}{C}}-CH_2X$$

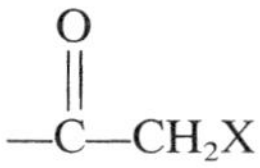

9.2.1c Procedure: Silver nitrate test

Add one drop (or several drops of an ethanol solution) of the halogen compound to 2 mL of 2% ethanolic silver nitrate. If no reaction is observed after 5 min, heat the solution to boiling for several minutes. Note the color of any precipitate formed. Add two drops of 5% nitric acid. Certain organic acids give insoluble silver salts. Silver halides are insoluble in dilute nitric acid; the silver salts of organic acids are soluble.

For water-soluble compounds, the test should be run using aqueous silver nitrate.

9.2.1d Sodium iodide in acetone

$$\text{NaI} + \text{RX} \xrightarrow[\text{acetone}]{} \text{RI} + \text{NaX}$$

$$\text{X} = \text{Br or Cl}$$

The rate of the reaction of sodium iodide in acetone with a compound containing covalent bromine or chlorine is a measure of the reactivity of the compound toward bimolecular nucleophilic substitution (S_N2 reaction). The test depends on the fact that sodium iodide is readily soluble in acetone, whereas sodium chloride and bromide are only slightly soluble. (Sodium fluoride is soluble in acetone; thus no precipitate formation is observed.)

S_N2 reactions are concerted and involve backside nucleophilic displacements that result in inversion of the configuration of the carbon atom at the reaction site.

As might be expected from the consideration of the transition state, such reactions are relatively insensitive to electronic effects of the R groups, but steric effects are of primary importance. In contrast to S_N1 reactions, the rates of S_N2 reactions of alkyl halides follow the order

$$\text{methyl} > \text{primary} > \text{secondary} > \text{tertiary.}$$

It should also be noted that α-halocarbonyl compounds and α-halonitriles are highly reactive under S_N2 conditions. This has been attributed to stabilization of the transition state by a partial distribution of the charge from the entering and leaving groups onto the carbonyl oxygen, etc. Also important may be the fact that the stereochemistry of the carbonyl and nitrile groups provides a favorable environment for nucleophilic attack.

A summary of the results to be expected in the sodium iodide in acetone test follows:

1. Precipitate within 3 min at room temperature.
   a. primary bromides
   b. acyl halides
   c. allyl halides
   d. α-haloketones, -esters, -amides, and -nitriles
   e. carbon tetrabromide
2. Precipitate only when heated up to 6 min at 50°C.
   a. primary and secondary chlorides
   b. secondary and tertiary bromides
   c. geminal polybromo compounds (bromoform)

3. Do not react in 6 min at 50°C.
   a. vinyl halides
   b. aryl halides
   c. geminal polychloro compounds (chloroform, carbon tetrachloride, trichloroacetic acid)
4. React to give precipitate and also liberate iodine.
   a. vicinal halides

$$-\overset{|}{\underset{Cl}{C}}-\overset{|}{\underset{Cl}{C}}- \xrightarrow{\ NaI\ } \ \underset{/}{\overset{\backslash}{C}}=\underset{\backslash}{\overset{/}{C}} + I_2 + 2NaCl$$

   b. sulfonyl halides

$$ArSO_2Cl + NaI \longrightarrow ArSO_2I + NaCl \quad \text{(immediate)}$$
$$\qquad\qquad\qquad\qquad NaI\Big\downarrow$$
$$\qquad\qquad\qquad ArSO_2Na + I_2$$

   c. triphenylmethyl halides

$$(C_6H_5)_3CX + NaI \longrightarrow NaX + (C_6H_5)_3CI \xrightarrow{\ NaI\ } Dimer + I_2$$

9.2.1e Procedure: Sodium iodide in acetone

Reagent: Dissolve 15 g of sodium iodide in 100 mL of pure acetone. The reagent should be stored in a dark bottle. On standing, it slowly discolors. It should be discarded when a definite red-brown color develops.

To 1 to 2 mL of the reagent in a small test tube, add 2 drops (or 0.1 g in a minute volume of acetone) of the compound. Mix well and allow the solution to stand at room temperature for 3 to 4 min. Note whether a precipitate forms and whether the solution acquires a red-brown color. If no change occurs at room temperature, warm the tube in a water bath at 50°C for 6 min. Cool to room temperature and record any changes.

9.2.2 Alkyl Halides

9.2.2a Characterization

A number of the commonly encountered alkyl halides can be identified by reference to physical and spectral properties or conversion to suitable derivatives (Table 9.3). A good number of alkyl halides are readily available in most organic laboratories. The utility of comparative glc and infrared spectroscopy should not be overlooked. Useful methods for preparing solid derivatives of polyhalogenated nonbenzenoid hydrocarbons are not available. Identification in such cases rests solely on spectral and physical data and chromatographic comparisons.

The most readily formed solid derivatives of primary and secondary alkyl halides are the *S-alkylthiuronium picrates*. The initial reaction involves the direct displacement of halide ion by the strongly nucleophilic thiourea; tertiary halides do not react.

$$RX + 2\,NH_2\overset{\overset{\displaystyle S}{\|}}{C}NH_2 \longrightarrow NH_2\overset{\overset{\displaystyle S-R}{|}}{C}{=}NH + NH_2{=}\overset{\overset{\displaystyle SH}{\|}}{\overset{+}{C}}{-}NH_2 \;\; Br^-$$

$$NH_2\overset{\overset{\displaystyle SR}{|}}{C}{=}NH + \text{(2,4,6-trinitrophenol)} \longrightarrow NH_2\overset{\displaystyle SR}{C}{=}NH \cdot \text{(picrate)}$$

S-Alkylthiuronium picrate

9.2.2b Procedure: S-alkylthiuronium picrates

CAUTION: *Thiourea may be a carcinogen.*

The alkyl halide (1 mmol) and thiourea (2 mmol) are dissolved in 5 mL of ethylene glycol contained in a small tube or flask fitted with a condenser. The mixture is heated in an oil bath for 30 min. Primary alkyl iodides react at 65°C, others require temperatures near 120°C. Add 1 mL of a saturated solution of picric acid in ethanol and continue to heat for an additional 15 min. Cool the reaction mixture and add 5 mL of cold water. Allow the mixture to stand in an ice bath. Collect the precipitate and recrystallize from methanol.

Alkyl 3,5-dinitrobenzoates can be prepared by reaction of alkyl iodides with the silver salt of 3,5-dinitrobenzoic acid.

$$RI + \text{(3,5-dinitrobenzoate)}{-}COOAg \longrightarrow \text{(3,5-dinitrobenzoate)}{-}COOR + AgI$$

9.2.2c Procedure: 3,5-Dinitrobenzoates

The alkyl iodide and a slight excess of finely powdered silver 3,5-dinitrobenzoate are refluxed in a small volume of alcohol until conversion to silver iodide appears complete. Evaporate the mixture to dryness, and extract the ester with ether. Recrystallize from alcohol.

9.2.2d Derivatives prepared from Grignard reagents

The alkyl halide can be converted to a Grignard reagent, which, in turn, can be treated with an isocyanate to form a substituted amide, with mercuric halide to form an alkylmercuric halide, or with carbon dioxide to form an acid. The latter should be attempted only in cases where solid acids are obtained.

$$RX \xrightarrow[\text{Et}_2\text{O}]{\text{Mg}} RMgX \begin{cases} \xrightarrow[\text{(2) H}_2\text{O}]{\text{(1) ArNCO}} RCONHAr \\ \xrightarrow{\text{HgX}_2} RHgX + MgX_2 \\ \xrightarrow[\text{(2) H}_3\text{O}^+]{\text{(1) CO}_2} RCOOH \end{cases}$$

Grignard reagents of the allyl or benzyl type are likely to give rearranged and/or dimeric products.

$$\underset{\text{CH}_3}{\overset{\text{CH}_2\text{Br}}{\bigcirc}} \xrightarrow[\text{(2) CO}_2 \;\; \text{(3) H}_3\text{O}^+]{\text{(1) Mg/(C}_2\text{H}_5)_2\text{O}} \underset{\text{CH}_3}{\overset{\text{CH}_3}{\overset{\text{COOH}}{\bigcirc}}} + \underset{\text{CH}_3}{\overset{\text{CH}_2\text{COOH}}{\bigcirc}}$$

$$\text{Minor}$$

$$\text{CH}_3\text{CH}{=}\text{CH}{-}\text{CH}_2\text{Br} \xrightarrow[\text{(2) CO}_2 \;\; \text{(3) H}_3\text{O}^+]{\text{(1) Mg/(C}_2\text{H}_5)_2\text{O}} \underset{\underset{\text{(in part)}}{\text{COOH}}}{\text{CH}_3\overset{|}{\text{CH}}\text{CH}{=}\text{CH}_2}$$

$$\text{CH}_2{=}\text{CH}{-}\text{CH}_2\text{Br} \xrightarrow[\text{(C}_2\text{H}_5)_2\text{O}]{\text{Mg}} \underset{\text{(in part)}}{\text{CH}_2{=}\text{CH}{-}\text{CH}_2{-}\text{CH}_2{-}\text{CH}{=}\text{CH}_2}$$

9.2.2e Procedure: Preparation of Grignard reagents

Grignard reagent: The choice of apparatus is governed by the steps involved in the reaction of the prepared Grignard reagent. The apparatus should consist of a flask, a tube or a small separatory funnel used as a reaction vessel, a condenser, and a drying tube. Into the carefully dried apparatus, place about 0.5 mmol (120 mg) of finely cut magnesium turnings, 0.55 mmol of the halide, 5 mL of absolute ether, and a small crystal of iodine. If the reaction is low in starting, warm it with a beaker of warm water. The reaction is normally complete in 5 to 10 min.

9.2.2f Procedure: Alkylmercuric halides

CAUTION: *These materials are highly toxic and should be handled with care.*

The Grignard reaction mixture (equivalent to 0.5 mmol Grignard) is filtered through a microporous disc or a little glass wool into a test tube containing 1 to 2 g of mercuric chloride, bromide, or iodide (corresponding to the halogen of the alkyl halide). Stopper the tube and shake the mixture vigorously. Remove the stopper, warm the tube on the steam bath for 1 or 2 min, and shake again. Evaporate to dryness, add 8 to 10 mL of ethanol, and heat on the steam bath until the alcohol boils. Filter and dilute with $\frac{1}{2}$ volume of water. Cool in an ice bath; when crystallization is complete collect the product and recrystallize from dilute ethanol.

9.2.2g Procedure: *N*-Aryl amides

Prepare the Grignard solution as above. Immerse the reaction vessel in cool water. Dissolve 0.5 mmol of α-naphthyl, phenyl, or *p*-tolyl isocyanate in 10 mL of absolute ether. By means of a dropper, add the isocyanate solution in small portions through the condenser into the Grignard solution. Shake the mixture and allow it to stand for 10 min. Pour the reaction mixture into a separatory funnel containing 20 mL of ice water and 1 mL of concentrated hydrochloric acid. Shake well and discard the lower aqueous layer. Dry the ether solution with anhydrous magnesium sulfate. Evaporate the ether and recrystallize the crude product from alcohol, aqueous alcohol, or petroleum ether.

If the apparatus or reagents are not kept dry, diphenylurea (mp 241°C), di-*p*-tolylurea (mp 268°C) or di-1-naphthylurea (mp 297°C) is obtained. These isocyanates are liquids and should not be used without purification if crystals (substituted urea formed by hydrolysis) are present in the reagent bottle.

9.2.2h Procedure: Carboxylic acids

(a) Filter the Grignard reagent into a mixture of finely crushed Dry Ice and ether. Pour the reaction mixture into 20 mL of water and 1 mL of concentrated hydrochloric acid contained in a separatory funnel. After the reaction subsides, separate the ether layer.

(b) In areas of high humidity, large amounts of water may condense in the Dry Ice/ether mixture described earlier. In such cases it is advantageous to bubble carbon dioxide into, or over, the Grignard solution with stirring. The carbon dioxide can be obtained from a tank or from Dry Ice.

9.2.3 Alkyl Fluorides and Fluorocarbons

9.2.3a Characterization

Simple alkyl and cycloalkyl fluorides are identified by reference to physical constants and spectral characteristics.

In recent years, the chemistry of highly fluorinated compounds ("fluorocarbons") has attracted considerable attention. Perfluoro compounds (all hydrogen atoms, except those in the functional group, are replaced by fluorine) and compounds containing the trifluoromethyl group are commonplace. The chemistry and physical properties of these materials differ considerably from their nonfluorinated analogues. The boiling points are usually lower than those of the hydrogen analogues, e.g., perfluoropentane (29°C) and pentane (36°C), acetophenone (202°C) and trifluoromethyl phenyl ketone (152°C).

Highly fluorinated olefins do not add bromine under the usual conditions. They do, however, react with potassium permanganate. Most perhalogenated compounds fail to burn.

Fluorination increases the acidity of acids and alcohols and reduces the basicity of amines. Perfluoroaldehydes and ketones exhibit abnormally high carbonyl frequencies in the infrared. They are readily cleaved into acids and highly volatile monohydrogen perfluoroalkanes by the action of hydroxide (haloform reaction). The usual carbonyl derivatives of these compounds can be made by the appropriate carbonyl reagents (semicarbazide hydrochloride, etc.) in alcohol, but require long reaction times.

9.2.4 Vinyl Halides

9.2.4a Characterization

Vinyl halides, like aryl halides, are much less reactive than saturated halides. Identification can usually be made by reference to spectral and physical data. The nature and stereochemistry of substitution at the double bond can be inferred from the C=C stretching and C—H bending bands in the infrared spectrum (Sec. 5.2.1b) and from the coupling constants in the NMR (Sec. 6.5). For vinyl fluorides, J_{FH} may be particularly useful.

Grignard reagents can be made from vinyl chlorides, iodides, and bromides, provided that dry tetrahydrofuran is substituted for diethyl ether as solvent (Sec. 9.2.2d).

9.2.5 Aryl Halides

9.2.5a Characterization

As in the case of many other types of aromatic compounds, the best procedures for making solid derivatives of aryl halides involve additional substitution of the aromatic nucleus. Nitration by the procedures outlined for aromatic hydrocarbons (Sec. 9.1.4) is generally useful.

The aryl halides react readily with the chlorosulfonic acid to produce sulfonyl chlorides, which in turn can be treated with ammonia to yield sulfonamides.

The procedure outlined under aromatic ethers (Sec. 9.4.1f) can be employed. Sometimes the intermediate sulfonyl chloride will serve as a suitable derivative. With polyhalo aromatics, more vigorous conditions may be necessary to produce the sulfonyl chloride; the chloroform solution can be warmed or the reactants warmed neat to 100°C. Sometimes side reactions become predominant, products resulting from nuclear chlorination are obtained, or the reaction can provide diaryl sulfones.

Aryl bromides and chlorides having aliphatic side chains can be oxidized to the corresponding acids. Aryl iodides and bromides can be converted to Grignard reagents and subsequently carbonated to yield acids or reacted with isocyanates to form amides (Sec. 9.2.2g). Aryl chlorides form Grignard reagents only in tetrahydrofuran; in ether, bromochlorobenzenes form the Grignard reagent exclusively at the bromo position.

Occasionally aromatic polyvalent iodine compounds are encountered; representative examples are as follows:

$$ArICl_2$$
Iodoarene dichloride

$$ArI(OAc)_2$$
Iodoarene diacetate

$$Ar_2I^+X^-$$
Diaryliodonium salts

$$ArIO$$
Iodosoarene

$$ArIO_2$$
Iodoxyarene

The iodoarene dichloride acts as a ready source of chlorine. It will add chlorine to alkenes. The oxygen derivatives all act as oxidants. With the exception of the diaryliodonium salts, all will liberate iodine from acidified starch iodide paper. The diaryliodonium salts are arylating agents; they can be pyrolyzed to yield ArI and ArX.

Table 9.3 lists the physical properties of organic halides and their derivatives.

Table 9.3. Physical Properties of Alkyl and Aryl Halides and Derivatives

	bp(°C)			mp(°C) of Derivatives		
	Chloride	Bromide	Iodide	Anilide	S-Alkyl-thiuronium Picrate	Alkyl-mercuric Chloride
Isopropyl	36.5	60	89.8	103	196	
Allyl	44.6	71	103	114	155	
Propyl	46.4	71	102.5	92	181	147
t-Butyl	50.7	73.2	103.3	128	161	123
sec-Butyl	68	91.2	120.	108	190	30.5
Isobutyl	68.9	91	120.4	109	174	
Butyl	77.8	101.6	130	63	180	128
Neopentyl	85	109		131		118
t-Amyl	86	108	128	92		
Amyl	106	129	155	96	154	110
Hexyl	133	157	179	69	157	125
Cyclohexyl	143	165		146		
Heptyl	159	180	204	57	142	119.5
Benzyl	179.4	198	(mp 24)	117	188	104
Octyl	184	201.5		57	134	115
2-Phenylethyl	190	218		97		
1-Phenylethyl	195	205				

Aryl Halides (Liquids)

	bp(°C)	D_4^{20}	n_D^{20}	Nitro derivative Position	mp(°C)
Fluorobenzene	87.4	1.024	1.466		
Chlorobenzene	131.8	1.107	1.525	2,4	52
Bromobenzene	156.2	1.494	1.560	2,4	72
o-Chlorotoluene	159.3	1.082	1.524	3,5	63
m-Chlorotoluene	162.3	1.072	1.521	4,6	91
p-Chlorotoluene	162.4	1.071	1.521	2	38
p-Dichlorobenzene (mp 53°C)	173				
m-Dichlorobenzene	173	1.788	1.546	4,6	103
o-Dichlorobenzene	179	1.305	1.552	4,5	110
o-Bromotoluene	181.8	1.425		3,5	82
p-Bromotoluene (mp 28.5°C)	184.5	1.390		2	47
Iodobenzene	188.6	1.831	1.620	4	171
o-Chloroanisole	195	1.248	1.5480		
p-Chloroanisole	200	1.1851			
2,6-Dichlorotoluene	199				
2,4-Dichlorotoluene	200	1.249	1.549	3,5	104
o-Dibromobenzene	219.3	1.956	1.609	4,5	114
1-Chloronaphthalene	259.3	1.191	1.633	4,5	180
1-Bromonaphthalene	281.2	1.484	1.658	4	85

9.3
ALCOHOLS AND PHENOLS

In the absence of carbonyl absorption, the appearance of a medium-to-strong band in the infrared spectrum in the 3600 to 3400 cm^{-1} (2.78 to 2.94 μm) region indicates an alcohol, a phenol, or a primary or secondary amine, or possibly a wet sample (see Fig. 9.6 for the infrared spectrum of water). Amines can be distinguished from alcohols and phenols by taking advantage of their basic characters. Water-insoluble amines are soluble in dilute hydrochloric acid. The water-soluble amines have a characteristic ammoniacal odor, and their aqueous solutions are basic to litmus. Phenols are considerably more acidic than alcohols and can be differentiated from the latter by their solubility in 5% sodium hydroxide solution and by the colorations usually produced when treated with ferric chloride solution. For additional details, the classification sections on alcohols (9.3.1 and 9.3.2), phenols (9.3.3), and amines (9.14) should be consulted.

9.3.1 Alcohols

9.3.1a Classification

Once it has been established that the compound in question is an alcohol, the next logical step in a structure determination is to distinguish between primary, secondary, and tertiary alcohols. Although there are certain correlations between both the O—H and C—O stretching frequencies and the subclass of alcohols, the exact peak positions and the factors affecting the peak positions are quite variable. One is cautioned against trying to use infrared spectroscopy to determine the subclass. With experience in a given series of alcohols, infrared can be used with some degree of success (Sec. 5.2.7), but it certainly cannot be recommended for general use.

Nuclear magnetic resonance spectroscopy can be used to classify alcohols with respect to subclasses. In the most common NMR solvents, such as deuterochloroform or carbon tetrachloride, the hydroxyl hydrogen resonance is not only often obscured, but traces of acid present in the solvents catalyze proton exchange so that spin-spin splitting of the hydroxyl peaks is rarely observed. However, in dimethyl sulfoxide, the strong hydrogen bonding to the solvent shifts the hydroxyl hydrogen resonance downfield to $\delta 4$ or lower, and the rate of proton exchange slows sufficiently to permit observation of the carbinol hydrogen-hydroxyl hydrogen coupling (Fig. 9.7).[16] Because of the low-field resonance of the hydroxyl hydrogens, it is possible to use dimethyl sulfoxide rather than its deuterated

[16] O. L. Chapman and R. W. King, *J. Am. Chem. Soc.*, **86**, 1256 (1964).

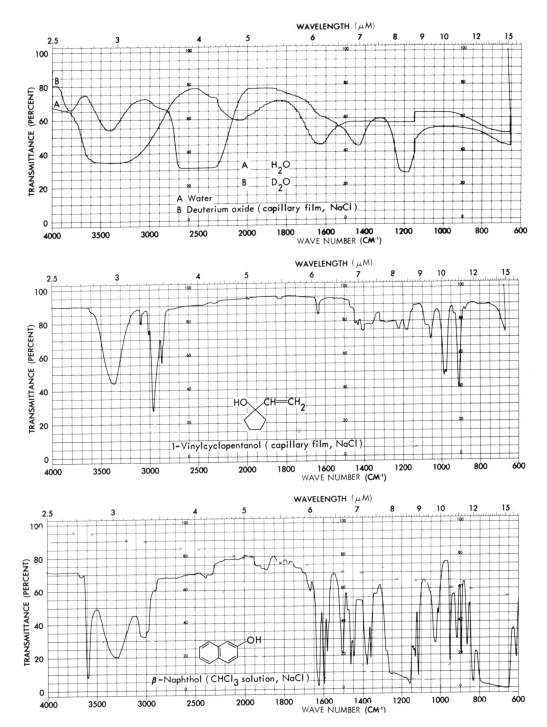

Fig. 9.6. Typical spectra of water, an alcohol, and a phenol.

analogue. The hydroxyl hydrogen in methanol gives a quartet; primary, secondary, and tertiary alcohols give clearly resolved triplets, doublets, and singlets, respectively. Polyhydroxy compounds often give separate peaks for each hydroxyl hydrogen. This method is not reliable when acidic or basic groups are present, e.g., amino alcohols, hydroxy acids, or phenols.

A solution of zinc chloride in concentrated hydrochloric acid (Lucas Reagent)

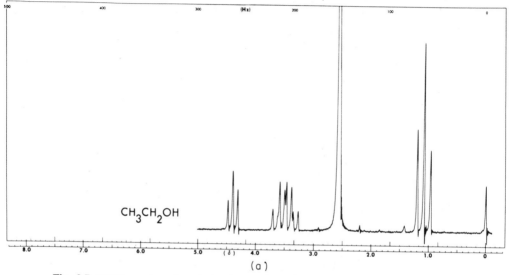

Fig. 9.7. NMR spectra of typical primary, secondary, and tertiary alcohols run in dimethyl sulfoxide. The intense resonance at δ 2.6 is due to dimethyl sulfoxide used as solvent.

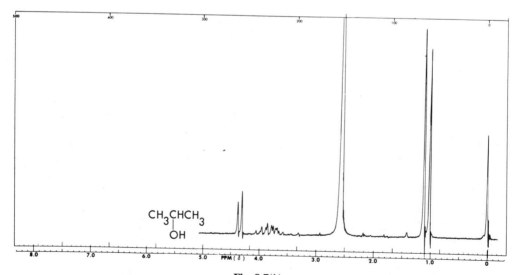

Fig. 9.7(b).

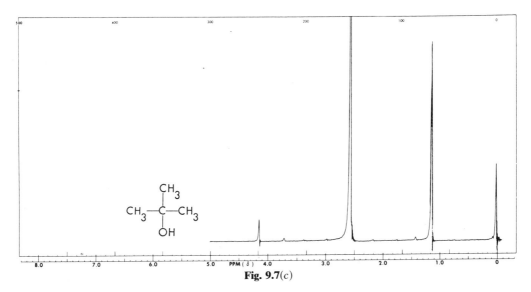

Fig. 9.7(*c*)

has been widely used as a classical reagent to differentiate between the lower primary, secondary, and tertiary alcohols. With this reagent, the order to reactivity is typical of S_N1 type reactions. The zinc chloride undoubtedly assists in the heterolytic breaking of the C—O bond as illustrated in the following equation. The test is applicable only to alcohols that dissolve in the reagent.

$$\text{ROH} + \text{ZnCl}_2 \rightleftharpoons \text{R} \overset{\delta^+}{-} \overset{\delta^-}{\text{O}} \text{---ZnCl}_2 \xrightarrow{-[\text{HOZnCl}_2]^-} \text{R}^+ \xrightarrow{\text{Cl}^-} \text{RCl}$$

9.3.1b Procedure: Lucas test

Reagent: Dissolve 16 g of anhydrous zinc chloride in 10 mL of concentrated hydrochloric acid with cooling to avoid a loss of hydrogen chloride.

Add three to four drops of the alcohol to 2 mL of the reagent in a small test tube. Shake the test tube vigorously, and then allow the mixture to stand at room temperature. Primary alcohols lower than hexyl dissolve; those higher than hexyl do not dissolve appreciably, and the aqueous phase remains clear. Secondary alcohols react to produce a cloudy solution of insoluble alkyl chloride after 2 to 5 min. With tertiary, allyl, and benzyl alcohols, there is almost immediate separation of two phases owing to the formation of the insoluble alkyl chloride. If any question remains about whether the alcohol is secondary or tertiary, the test can be repeated employing concentrated hydrochloric acid. With this reagent, tertiary alcohols react immediately to form the insoluble alkyl chloride, whereas secondary alcohols do not react.

A second method to differentiate tertiary alcohols from primary and secondary alcohols takes advantage of the inertness of tertiary alcohols toward oxidation with *chromic acid.*

$$3\,\text{H}\overset{|}{\underset{|}{-\text{C}-}}\text{OH} + 2\,\text{CrO}_3 + 6\,\text{H}^+ \longrightarrow 3\,\diagdown\text{C}{=}\text{O} + 2\,\text{Cr}^{+3} + 6\,\text{H}_2\text{O}$$

9.3.1c Procedure: Chromic acid test

Reagent: Dissolve 1 g of chromic oxide in 1 mL of concentrated sulfuric acid, and carefully dilute with 3 mL of water.

    Dissolve 20 mg or one drop of the alcohol in 1 mL of reagent grade acetone, and add one drop of the reagent. Shake the mixture. Primary and secondary alcohols react within 10 sec to give an opaque blue-green suspension. Tertiary alcohols do not react with the reagent. Other easily oxidized substances such as aldehydes, phenols, and enols also react with this reagent.

9.3.1d Characterization

Phenyl and α-naphthylurethanes

$$ROH + ArN{=}C{=}O \longrightarrow ArNH\overset{\overset{\displaystyle O}{\|}}{C}{-}OR$$

$$\text{Isocyanate} \qquad\qquad\qquad \text{Urethane}$$

The most generally applicable derivatives of primary and secondary alcohols are the urethanes, prepared by reaction of the alcohol with the appropriate isocyanate. *For the preparation of the urethanes, the alcohols must be anhydrous*; water hydrolyzes the isocyanates to give aryl amines, which combine with the reagent to produce disubstituted ureas.

$$ArNCO + H_2O \longrightarrow ArNH_2 + CO_2$$

$$ArNH_2 + ArNCO \longrightarrow ArNHCONHAr$$

These ureas, which are high melting and less soluble than the urethanes, make isolation and purification of these derivatives very difficult. It is not advisable to attempt the preparation of these urethanes from tertiary alcohols. At low temperatures the reaction is too slow; at higher temperatures the reagents cause dehydration to occur with the formation of an alkene and water, which, in turn, react to produce diaryl ureas.

9.3.1e Procedure: Urethanes

Thoroughly dry a small test tube over a flame or in an oven; cork it and allow to cool. By means of a pipette, place 0.2 mL of anhydrous alcohol and 0.2 mL of α-naphthyl or phenyl isocyanate into the tube (if the reactant is a phenol, add one drop of pyridine), and immediately replace the cork. If a spontaneous reaction does not take place, the solution should be warmed in a water bath at 60 to 70°C for 5 min. Cool in an ice bath and scratch the sides of the tube with a glass rod to induce crystallization. The urethane is purified by recrystallization from petroleum ether or carbon tetrachloride. Filter the hot solution to remove the less-soluble urea, which may form owing to traces of moisture. Cool the filtrate in an ice bath and collect the crystals. Diphenylurea, di-p-tolylurea, and di-1-naphthylurea have melting points of 241, 268, and 297°C, respectively.

Arenesulfonylurethanes. Benzenesulfonyl and p-toluenesulfonylisocyanate, which are considerably more reactive than phenyl and related isocyanates, may be used to advantage in the preparation of urethanes of tertiary and other highly hindered alcohols.[17] Extensive compilations of data are not available.

$$R-\text{⬡}-SO_2N=C=O + R_3'COH \longrightarrow R-\text{⬡}-SO_2NHCOCR_3'$$

$$R = H \text{ or } CH_3$$

Esters. Numerous esters have been used to aid in the characterization of alcohols. Among these are a wide variety of phthalic acid esters, xanthates, benzoates, and acetates. The latter two are especially useful for the characterization of glycols and polyhydroxy compounds (Sec. 9.3.2). Among the most generally useful esters are the 3,5-*dinitro* and p-*nitrobenzoates*. These esters can be prepared from the corresponding benzoyl chlorides by either of the two procedures given below in Sec. 9.3.1f. They are to be especially recommended for water-soluble alcohols that are likely to contain small amounts of moisture and hence produce trouble in the formation of the urethane derivatives. The method below employing pyridine as a solvent is one of the most useful methods for making derivatives of tertiary alcohols.

The acyl halides tend to hydrolyze on storage; it is advisable to check the melting point of the reagent prior to use [3,5-dinitrobenzoyl chloride, mp 74°C (acid, 202°C) p-nitrobenzoyl chloride, mp 75°C (acid 241°C)].

9.3.1f Procedure: 3,5-Dinitro- and p-nitrobenzoates

(a) In a small test tube place 200 mg of pure 3,5-*dinitrobenzoyl* or p-*nitrobenzoyl chloride* and 0.1 mL (two drops) of alcohol. Heat the tube for 5 min (10 min if the alcohol boils above 160°C), employing a microburner so that the melt at the bottom of the tube is maintained in the liquid state. Do not overheat. Allow the melt to cool and solidify. Pulverize the crystalline mass with a glass rod or spatula. Add 3 or 4 mL of a 2% sodium carbonate solution and thoroughly mix and grind the mixture. Warm the mixture gradually to 50 or 60°C and stir thoroughly. Collect the remaining precipitate and wash several times with small portions of water. Recrystallize from ethanol or aqueous ethanol.

---

[17] J. W. McFarland and J. B. Howard. *J. Org. Chem.*, **30**, 957 (1965); J. W. McFarland, D. E. Lenz, and D. J. Grosse, *J. Org. Chem.*, **31**, 3798 (1966); p-toluenesulfonylisocyanate is available from the Upjohn Co., Carwin Organic Chemicals, North Haven, Connecticut.

(b) In a small test tube or round-bottomed flask, place 100 mg of alcohol or phenol, 100 mg of p-nitrobenzoyl or 3,5-dinitrobenzoyl chloride, 2 mL of pyridine, and a boiling chip. Put a condenser in place and reflux gently for 1 hr or allow to stand overnight. Cool the reaction mixture and add 5 mL of water and two or three drops of sulfuric acid. Shake or stir well and collect the crystals. Suspend the crystals in 5 mL of 2% sodium hydroxide, and shake well to remove the nitrobenzoic acid. Filter, wash several times with cold water, and recrystallize the derivative from alcohol or alcohol-water mixtures.

Acetates, hydrogen phthalates, etc. of highly hindered alcohols (as well as phenols) can be prepared in excellent yield by use of the "hypernucleophilic" acylation catalyst, 4-dimethylaminopyridine.[18]

Oxidation to carbonyl compounds.   The oxidation of primary and secondary alcohols to carbonyl compounds often proves to be an exceedingly useful method of identification. From the carbonyl compounds, easily prepared derivatives can be made on a small scale, and, equally important, the spectroscopic features of the carbonyl group can provide much valuable structural information.

$$
\underset{\text{NMR spectrum complex}}{CH_3CH_2\overset{\displaystyle OH}{\underset{|}{C}}HCH_3} \quad \xrightarrow{[Ox]} \quad \underset{\text{NMR spectrum simple}}{CH_3CH_2\overset{\displaystyle O}{\overset{||}{C}}CH_3}
$$

Infrared or ultraviolet does not reveal position of hydroxyl

Conjugated ketone indicated by infrared and ultraviolet

Infrared indicates hydroxyl

Infrared reveals ketone in a five-membered ring

Numerous methods are available for the oxidation of alcohols to carbonyl compounds. Only a few are included here; they are chosen for their mild conditions and suitability for small-scale operation.

The chromic anhydride-pyridine complex (Sarett reagent)[19] is particularly useful for the oxidation of substances containing acid-sensitive groups. Alcohols, including allylic and benzylic, can be oxidized to the corresponding aldehydes or ketones.

----

[18] G. Hofle and W. Steglick, *Synthesis*, 619 (1972).

[19] G. I. Poos, G. E. Arth, R. E. Beyler, and L. H. Sarett, *J. Am. Chem. Soc.*, **75,** 422 (1963).

9.3.1g Procedure: Sarett reagent

Two mL of pyridine in a small round-bottomed flask equipped with a magnetic stirrer is cooled to 15 to 20°C. To the pyridine is added, in portions, 200 mg of chromium trioxide at such a rate that the temperature does not rise above 30°C. CAUTION: *If the pyridine is added to the chromium trioxide, the mixture will ignite.* A slurry of the yellow complex in pyridine remains after the last addition.

To the slurry is added 100 mg of the alcohol in 1 mL of pyridine. The flask is stoppered and allowed to stir overnight. The reaction mixture is poured into 20 mL of ether; the precipitated chromium salts are removed by filtration. The filtrate is placed in a separatory funnel and washed at least three times with water to remove the pyridine. The ether layer is dried over sodium sulfate and concentrated to provide the aldehyde or ketone. Low-molecular-weight, water-soluble ketones can be isolated by pouring the reaction mixture directly into water, neutralizing the pyridine by addition of acid and extraction with an appropriate solvent.

Chromic acid in acetone (Jones reagent)[20] is very convenient for the oxidation of acetone-soluble secondary alcohols to ketones. The reagent does not attack centers of unsaturation. The method is applicable on scales from 1 mmol to 1 mol. The reagent oxidizes primary alcohols and aldehydes to acids.

$$RCH_2OH \xrightarrow[]{H_2CrO_4} RCHO \xrightarrow[]{H_2CrO_4} RCOOH$$

9.3.1h Procedure: Jones reagent

Reagent: Dissolve 6.7 g of chromium trioxide (CrO_3) in 6 mL of concentrated sulfuric acid and carefully dilute with distilled water to 50 mL. One mL of this reagent is sufficient to oxidize 2 mmol of a secondary alcohol to a ketone, 2 mmol of aldehyde to an acid, or 1 mmol of primary alcohol to an acid.

The oxidation is carried out by the addition of the reagent from an addition funnel to a stirred acetone (reagent grade) solution of the alcohol maintained at 15 to 20°C. The reaction is nearly instantaneous; the mixture separates into a lower green layer of chromous salts and an upper layer that is an acetone solution of the oxidation product. The reaction mixture can be worked up by the addition of water or by other means, depending on the properties of the oxidation product. Any brown coloration, caused by an excess of the oxidizing agent remaining in the mixture, in the upper layer can be removed by a small pinch of sodium hydrogen sulfite or a few drops of methanol to destroy the excess reagent.

---

[20] K. Bowden, I. M. Heilbron, E. R. H. Jones, and B. C. L. Weedon, *J. Chem. Soc.*, 39 (1946); A. Bowers, T. G. Halsall, E. R. H. Jones, and A. J. Lemin, *J. Chem. Soc.*, 2555 (1943).

Pyridinium chlorochromate ($C_5H_6N^+$ $ClCrO_3^-$), a stable crystalline reagent that can be readily synthesized or purchased, oxidizes alcohols to the corresponding aldehydes or ketones in high yield under mild conditions.[21]

9.3.1i Procedure: Pyridinium chlorochromate oxidation

Reagent: To 18.4 mL of 6 N hydrochloric acid (0.11 mol) add 10 g (0.1 mol) of chromium trioxide rapidly with stirring. After 5 min, cool the homogeneous solution to 0°C and add dropwise, over a period of 10 min, 7.9 g (0.1 mol) of pyridine. Recool to 0°C and collect the yellow-orange pyridinium chlorochromate in a sintered glass funnel. After drying for 1 hr in vacuo, the reagent can be stored.

Pyridinium chlorochromate (0.32 g, 1.5 mmol) is suspended in 2 to 3 mL of dichloromethane in a 15 cm test tube, and the alcohol (1 mmol) in 1 mL of dichloromethane is added. After 1 to 2 hr, the oxidation is complete and black reduced reagent has precipitated. The reaction mixture is diluted with 10 to 12 mL of diethyl ether, the solids are removed by filtration, and the product is obtained by evaporation of the solvent. If the alcohol contains acid-labile functions (e.g., esters, etc.), sodium acetate (25 mg, 0.3 mmol) should be added to the reagent suspension as a buffer prior to the addition of the alcohol.

Ceric ammonium nitrate. Cerium (IV) is a very convenient oxidant for the conversion of benzyl alcohols to benzaldehydes.[22]

9.3.1j Procedure: Cerium(IV) oxidation of benzyl alcohols

Add a slight excess of 50% aqueous acetic acid solution containing 0.5 M ceric ammonium nitrate to the benzyl alcohol. Then warm the solution on the steam bath for a few minutes if necessary. After the orange cerium (IV) solution has turned to a pale yellow cerium (III) solution, extract the mixture with ether. Wash the ether solution with 1.5 N potassium hydroxide, and dry. Removal of the ether by distillation provides pure aldehyde in greater than 90% yield.

Active manganese dioxide[23] is often used to oxidize allylic and benzylic alcohols to aldehydes and ketones. Active manganese dioxide (as opposed to

---

[21] E. J. Corey and J. W. Suggs, *Tetrahedron Lett.*, 2647 (1975).

[22] W. S. Trahanovsky and L. B. Young, *J. Chem. Soc.*, 5777 (1965).

[23] S. Ball, T. W. Goodwin, and R. A. Morton, *Biochem. J.*, **42,** 516 (1948); J. Attenburrow, A. F. B. Cameron, J. H. Chapman, R. M. Evans, B. A. Hems, A. B. A. Jansen, and T. Walker, *J. Chem. Soc.*, 1094 (1952).

ordinary manganese dioxide) is obtained by the mixing of aqueous solutions of manganous sulfate, sodium hydroxide, and potassium permanganate. The oxidation is accomplished by shaking a suspension of the active manganese dioxide with the alcohol in an inert solvent, usually petroleum ether or chloroform. The oxidation of alcohols with this reagent has been used as an indication that the alcohols were allylic or benzylic. Caution should be used in applying this criterion since other alcohols are occasionally attacked by the reagent.

9.3.2 Polyhydric Alcohols

9.3.2a Classification

The polyhydric alcohols are viscous, high-boiling liquids to high-melting solids: the simpler ones are quite water-soluble; many tend to be more or less insoluble in nonpolar solvents. They can be detected and characterized by the methods already outlined under alcohols. Polyhydric alcohols often give separate peaks in the NMR for each chemically different hydroxyl when dimethyl sulfoxide is employed as solvent.

Compounds containing hydroxyl groups on adjacent carbons have characteristic reactions that are noteworthy. The simplest of these compounds are the 1,2-diols (1,2-glycols). The characteristic reactions of the 1,2-glycols are shared by the 1,2,3-triol, etc., but the situation is often more complicated (see also Sec. 9.9 on carbohydrates).

The *borax text* for 1,2-diols depends on the reversible formation of cyclic borate esters, which are much stronger acids than boric acid.

9.3.2b Procedure: Borax test

Reagent: To a 1% solution of borax (sodium borate) in water, add sufficient phenophthalein to produce a pink coloration.

Add one or two drops of the polyhydric alcohol to 0.5 mL of the reagent. A 1,2-diol causes the pink color to disappear. It reappears on warming and vanishes again upon cooling.

Periodic acid and 1,2-glycols react to produce carbonyl compounds and iodate (see discussion in Sec. 9.3.2d). The qualitative test depends upon the fact that silver iodate is only sparingly soluble in dilute nitric acid, whereas silver periodate is very soluble. It is important that exact amounts of reagents and nitric acid be employed; if too much nitric acid is present, the silver iodate will not precipitate.

9.3.2c Procedure: Periodate test for glycols

Reagent: Dissolve 0.5 g of paraperiodic acid (H_5IO_6) in 100 mL of distilled water.
    To a test tube containing 2 mL of the reagent, add one drop of concentrated nitric acid and mix thoroughly. Add one drop or a similar amount of the compound to be tested (for water-insoluble compounds, acetic acid as a co-solvent can be employed). Shake the mixture for 15 sec and add one to two drops of 5% silver nitrate solution. The instantaneous formation of a white precipitate of silver iodate constitutes a positive test.

9.3.2d Characterization

Periodic acid is a selective oxidant capable of cleavage of the carbon-carbon bond of 1,2-glycols, β-amino-alcohols, α-amino and α-hydroxy aldehydes and ketones, and α-diketones. In certain cases, α-hydroxy acids also react with the reagent.

$$R{-}CH{-}OH$$
$$\underset{R{-}CH{-}OH}{\big|} + HIO_4 \longrightarrow 2\,RCHO + H_2O + HIO_3$$

$$R'{-}C{=}O$$
$$\underset{R{-}CH{-}OH}{\big|} + HIO_4 \longrightarrow RCHO + R'COOH + HIO_3$$

$$R{-}CH{-}OH$$
$$\underset{R'{-}CH{-}NH_2}{\big|} + HIO_4 \longrightarrow RCHO + R'CHO + NH_3 + HIO_3$$

The reaction proceeds via the rapid and reversible formation of a cyclic periodate ester; the hydroxyl groups must be situated so that formation of the cyclic ester is sterically possible. In the case of carbonyl compounds, the ester is apparently formed from the hydrated carbonyl compounds, i.e.,

$$
\begin{array}{c}
\text{OH OH}\\
|\ \ |\\
{-}C{-}C{-}\\
|\ \ |\\
\text{OH H}
\end{array}
$$

The periodate reaction can be used as a qualitative test (see above) or in quantitative determinations. In the quantitative procedures, excess standard periodate is allowed to react with the compound. The excess periodate is reduced by standard sodium arsenite with iodide catalyst and controlled pH. The arsenite is then back-titrated with standard iodine solution. Detailed procedures are available in numerous standard text and reference books. For example, in a quantitative assay, it can be determined that 1 mol of glyceraldehyde will consume 2 mol of periodate and produce 2 mol of formic acid and 1 mol of formaldehyde. Similar reactions occur with lead tetraacetate.

$$\begin{array}{c} \text{CHO} \\ | \\ \text{H--C--OH} + 2\,\text{IO}_4^- \longrightarrow 2\,\text{HCOOH} + \text{CH}_2\text{O} + 2\,\text{IO}_3^- \\ | \\ \text{CH}_2\text{OH} \end{array}$$

The classic *acid catalyzed rearrangement of a 1,2-glycol* is the pinacol-pinacolone rearrangement.

$$\begin{array}{c} \text{OH OH} \\ |\;\;\; | \\ \text{CH}_3\text{--C--C--CH}_3 \\ |\;\;\; | \\ \text{H}_3\text{C}\;\;\text{CH}_3 \end{array} \xrightarrow{\;H^+\;} \begin{array}{c} \text{H}_3\text{C}\;\;\text{O} \\ |\;\;\; \| \\ \text{CH}_3\text{--C--C--CH}_3 \\ | \\ \text{CH}_3 \end{array}$$

Pinacol Pinacolone

In such rearrangements, hydrogen "migrates" in preference to an alkyl group.

$$\text{RCHOHCH}_2\text{OH} \xrightarrow{\;H^+\;} \text{RCH}_2\text{CHO}$$
$$\text{RCHOHCHOHR} \xrightarrow{\;H^+\;} \text{RCH}_2\text{COR}$$

The rearrangement, which can be carried out by warming the glycol with any strong acid, has utility in the detection and characterization of glycols. Upon heating with a little potassium hydrogen sulfate, ethylene glycol yields acetaldehyde; glycerol gives acrolein ($\text{CH}_2\text{=CHCHO}$), which is responsible for the characteristic odor of burning fat. The standard carbonyl derivatives are useful for the identification of the products of such rearrangements.

The 1,2-glycols and 1,3-glycols react with aldehydes or ketones to yield *cyclic acetals* or *ketals*, which have occasional utility in separation and identification procedures.

$$\begin{array}{c} \text{R}\;\;\text{OH} \\ \diagdown\diagup \\ \text{CH} \\ | \\ \text{CH} \\ \diagup\diagdown \\ \text{R}\;\;\text{OH} \end{array} + \begin{array}{c} \text{O} \\ \diagdown \\ \text{C--C}_6\text{H}_5 \\ \diagup \\ \text{H} \end{array} \xrightarrow{\;H^+\;} \begin{array}{c} \text{R}\quad\text{O} \\ \diagdown\diagup \\ \text{CH} \\ |\quad\quad \text{CHC}_6\text{H}_5 + \text{H}_2\text{O} \\ \text{CH} \\ \diagup\diagdown \\ \text{R}\quad\text{O} \end{array}$$

Benzylidene derivative

$$\underset{\substack{\text{R} \\ \text{CH} \\ | \\ \text{CH}_2 \\ | \\ \text{CH} \\ \text{R} \quad \text{OH}}}{\overset{\text{R} \quad \text{OH}}{}} + \underset{\text{O}}{\text{CH}_3-\overset{\|}{\text{C}}-\text{CH}_3} \xrightarrow{\text{H}^+} \underset{\substack{\text{CH}-\text{O} \\ \text{R}}}{\overset{\text{R}}{\text{H}_2\text{C}}} \underset{\text{CH}-\text{O}}{} \overset{\text{CH}_3}{\underset{\text{CH}_3}{\text{C}}}$$

Isopropylidene derivative
(acetonide)

Acetonides are readily made employing acetone or 2,2-dimethoxypropane.

9.3.2e Procedure: Acetonides

(a) Dissolve 100 mg of the diol in 2 mL of 2,2-dimethoxypropane, and add 5 mg of *p*-toluenesulfonic acid. Allow the mixture to stand for 30 min to 2 hr. Strip off the excess reactant and methanol, and purify the product. In some cases it is advantageous to drive the equilibrium towards the desired ketal by the addition of toluene to the reaction mixture and the removal of the methanol by distillation of the toluene methanol azeotrope.

(b) Dissolve 0.5 g of the diol in 50 mL of acetone. Add 0.5 mL of 70% perchloric acid, and allow the mixture to stand for 2 hr to 1 day. Add 1 g of solid sodium carbonate. Filter and remove the excess acetone to obtain the crude acetonide.

Because of their volatility, ketals, acetates, and trimethylsilyl ethers are very useful derivatives of polyhydroxy compounds for use in glc and/or mass spectral analysis.

Acetates can be made using acetic acid, acetic anhydride, and sodium acetate mixture, or with acetic anhydride containing a small amount of sulfuric acid or pyridine as catalyst. The following procedure gives better yields under milder conditions.

9.3.2f Procedure: Acetates

The polyhydroxy compound (0.5 g) is added to 5 mL of pyridine. Acetic anhydride (2 g) is added dropwise with shaking. After the initial reaction has subsided, the solution is refluxed for 3 to 5 min. The mixture is cooled and poured into 15 mL of ice water. The derivative is collected and washed with cold 2% hydrochloric acid and water. The product is recrystallized from ethanol.

The *benzoates* are often very suitable crystalline derivatives of polyhydroxy compounds.

9.3.2g Procedure: Benzoates

(a) Gradually add 0.5 mL of benzoyl chloride to a solution of 0.1 g of alcohol in 1 mL of pyridine. Warm the mixture over a low flame for one or two min; then pour into ice water with stirring. Collect the precipitate and suspend it in 1 mL of 5% sodium carbonate solution. Recollect and recrystallize from ethanol.

(b) In a test tube, place 100 mg of polyhydroxy compound and 0.5 mL of benzoyl chloride. Add 5 mL of 10% sodium hydroxide. Place a rubber stopper on the tube and shake vigorously for 1 min and then occasionally over a 5 min period. Allow to stand in a cold bath for 30 min. Collect the precipitate and wash thoroughly with water. Dry and recrystallize from ethanol.

The *trimethylsilylation* reaction proceeds smoothly, quickly, and often quantitatively to provide derivatives of mono- and polyhydric alcohols, which show high volatility and often are easily separated from closely related compounds by glc.

$$2 \text{ ROH} + (CH_3)_3\text{SiNHSi}(CH_3)_3 \longrightarrow 2 \text{ ROSi}(CH_3)_3 + NH_3$$

9.3.2h Procedure: Trimethylsilyl derivatives

Reagent: Mix anhydrous pyridine (10 mL), hexamethyldisilazane (2 mL), and trimethylchlorosilane (1 mL). (A reagent of similar composition is marketed by the Pierce Chemical Company under the name TRI-SIL.)

Place 10 mg or less of the sample in a small vial. Add 1 mL of the reagent, place the top on the vial, and shake vigorously for 30 sec. Warm if necessary to effect solution. Allow the mixture to stand for 5 min. The mixture can be injected directly into the gas chromatograph.

Table 9.4 lists the physical properties of alcohols and their derivatives.

9.3.3 Phenols

9.3.3a Classification

Phenols are compounds of acidity intermediate between that of carboxylic acids and alcohols. Alcohols do not show acid properties in aqueous systems, whereas acids and phenols react with and are soluble in 5% aqueous sodium hydroxide solution (exceptions are the highly hindered phenols, such as *o*-di-*t*-butylphenol, which are insoluble in alkali). Acids and phenols can be differentiated on the basis of the insolubility of phenols in 5% aqueous sodium hydrogen carbonate solution. Phenols that contain highly electronegative substituents, such as 2,4-dinitrophenol and 2,4,6-tribromophenol, show increased acidity and are soluble in sodium hydrogen carbonate solution.

Most phenols yield intense red, blue, purple, or green colorations in the *ferric chloride test.* All phenols do not produce color with this reagent; a negative test must not be taken as confirming the absence of the phenol grouping without additional supporting evidence. Other functional groups also produce color changes with ferric chloride: aliphatic acids give a yellow solution; aromatic acids may produce a tan precipitate; enols give a red-tan to red-violet color; oximes, hydroxamic acids, and sulfinic acids give red to red-violet colorations.

Table 9.4. Physical Properties of Alcohols and Derivatives

Liquids

Compound	bp(°C)*	α-Naphthyl-urethan	Phenyl-urethan	3,5-Dinitro-benzoate	p-Nitro-benzoate
		mp(°C) of Derivatives			
Methanol	64.65	124	47	108	96
Ethanol	78.32	79	52	93	57
2-Propanol	82.4	106	75	123	110
2-Methyl-2-propanol (mp 25.5°C)	82.5				44
3-Buten-2-ol	94.6				
2-Propen-1-ol (Allyl alcohol)	97.1	108	70	49	28
1-Propanol	97.15	80	57	74	35
2-Butanol	99.5	97	64.5	76	26
2-Methyl-2-butanol	102.35	72	42	116	85
2-Methyl-1-propanol	108.1	104	86	87	69
3-Buten-1-ol	112.5				
3-Methyl-2-butanol	114	109	68		
3-Pentanol	116.1	95	49	101	17
1-Butanol	117.7	71	61	64	36
2,3-Dimethyl-2-butanol	120.5	101	66	111	
2-Methyl-2-pentanol	121		239	72	
3-Methyl-3-pentanol	123	83.5	43.5	96.5	
2-Methyl-1-butanol	128.9	82	31	70	
4-Methyl-2-pentanol	132	88	143	65	26
3-Methyl-2-pentanol	134/749 mm	72		43.5	
3-Hexanol	136			77	
2,2-Dimethyl-1-butanol	136.7	81	66	51	
1-Pentanol	138	68	46	46.4	
2-Hexanol	138–9/745 mm	60.5		38.5	
Cyclopentanol	140.8	118	132		
3-Ethyl-3-pentanol	142				
2,3-Dimethyl-1-butanol	145		29	51.5	
2-Methyl-1-pentanol	148	76		50.5	
3-Methyl-1-pentanol (mp 8°C)	151–2	58		38	
3-Heptanol	156				
4-Heptanol	156	80		64	35
2-Heptanol	158.7	54		49	
1-Hexanol	157.5	59	42	58.4	
Cyclohexanol	161.1				
Furfuryl alcohol	172	130	45	81	76
1-Heptanol	176.8	62	60	47	10
Tetrahydrofurfuryl alcohol	178/743 mm		61	83–4	48
2-Octanol	179	64		32	28
Cyclohexylcarbinol	182				
2,3-Butanediol	182.5	201 (bis)			
1-Phenylethanol (mp 20°C)	202	106	92	95	43
Benzyl alcohol	205.5	134	77	113	85

Table 9.4 (Continued)

Liquids

Compound	bp(°C)*	α-Naphthyl-urethan	Phenyl-urethan	3,5-Dinitro-benzoate	p-Nitro-benzoate
			mp(°C) of Derivatives		
1-Nonanol	215	65.5	60	52	10
1-Phenylpropanol	219	102			60
1,4-Butanediol (mp 19.5°C)	230	199 (bis)	183.5 (bis)		175 (bis)
3-Phenylpropanol	237.4		45	92	47
1-Undecanol (mp 15.85°C)	243		62		
Cinnamyl alcohol (mp 33°C)	257	114	91.5	121	78
2,3-Dimethyl-2,3-butanediol (mp 43°C)	173				
L-Menthol (mp 44°C)	216	119 (126)	112	153	62
1,2-Diphenylethanol	167/10 mm				
Diphenylmethanol (mp 68°C)	180/20 mm	139	140	141	132
o-Nitrobenzyl alcohol (mp 74°C)	270				
2-Hydroxy-1-phenyl-1-ethanone (mp 86°C)	120/11 mm				128.6
Cholesterol (mp 148.5°C)	360d†	176	168		185
Triphenylmethanol (mp 162°C)					

* Bp at 760 mm Hg pressure unless otherwise noted.
† d denotes decomposition occurs.

9.3.3b Procedure: Ferric chloride test

To 1 mL of a dilute aqueous solution (0.1 to 0.3%) of the compound in question, add several drops of a 2.5% aqueous solution of ferric chloride. Compare the color produced with that of pure water containing an equivalent amount of the ferric chloride solution. The color produced may not be permanent, thus the observation should be made at the time of addition.

Certain phenols do not produce coloration with the foregoing procedure. As an alternative procedure, dissolve or suspend 20 mg of a solid or one drop of a liquid in 1 mL of dichloromethane, and add two drops of a saturated solution of anhydrous ferric chloride in dichloromethane. To the test solution add one drop of pyridine and observe the color change.

Phenols rapidly react with *bromine water* to produce insoluble substitution products; all available positions ortho and para to the hydroxyl are brominated. Advantage can be taken of this reaction both as a qualitative test for the presence of the phenolic grouping and for the preparation of solid derivatives. The tests should be applied with some discrimination since aniline and its substituted derivatives also react rapidly with bromine water to produce insoluble precipitates.

9.3.3c Procedure: Bromine water test

To a 1% aqueous solution of the suspected phenol, add a saturated solution of bromine water drop by drop, until bromine color is no longer discharged. A positive test is indicated by the precipitation of the sparingly soluble bromine substitution product and the production of a very strongly acidic reaction mixture. In the case of phenol, the product is 2,4,6-tribromophenol.

In the ultraviolet region, ionization of a phenol by a base increases both the wavelengths and the intensities of the absorption bands.

$$\lambda_{max}^{H_2O} \ 211, 270 \text{ nm} \qquad \lambda_{max}^{0.1N \ NaOH} \ 235, 287 \text{ nm}$$
$$(6200, 1450) \qquad\qquad (9400, 2600)$$

This shift can be visually observed in the case of *p*-nitrophenol; *p*-nitrophenol is yellow, whereas sodium *p*-nitrophenolate is red.

9.3.3d Characterization

Phenols, like alcohols, react with isocyanates to produce urethanes. The α-naphthylurethanes are the derivatives generally used for the identification of phenols. The procedure employed is that given for alcohols (Sec. 9.3.1). For highly hindered phenols, benzenesulfonylisocyanate is the reagent of choice.

The melting points of a large number of 3,5-dinitrobenzoates of phenols have also been recorded. These derivatives can be prepared using the pyridine method discussed under alcohols (Sec. 9.3.1).

The preparation of *brominated phenols* is an exceedingly simple procedure, and the bromo-substituted phenols are very useful derivatives. Although saturated bromine-water can be used satisfactorily for this bromination, the following procedure usually yields better results on a preparative scale.

9.3.3e Procedure: Bromination

In a test tube or an Erlenmeyer flask, dissolve 1 g of potassium bromide in 6 mL of water. Carefully add 0.6 g of bromine. In a second test tube, place 100 mg of the phenol, 1 mL of methanol, and 1 mL of water. Add 1 mL of the prepared bromine solution and shake. Continue the addition of bromine solution in small portions until the mixture attains a yellow color after shaking. Add 3 to 4 mL of cold water and shake vigorously. Filter the bromophenol and wash the precipitate well with water. Dissolve the crystals in hot methanol, filter the solution, and add water dropwise to the filtrate until a permanent cloudiness results. Allow the mixture to cool to complete crystallization.

The phenoxide ions, produced by dissolution of phenols in aqueous alkali, react readily with chloracetic acid to give *aryloxyacetic acids*. These derivatives are very useful; they crystallize well from water, have well-defined melting points, and can be characterized by reference to their neutralization equivalents.

$$\text{G}\text{-}\underset{\text{G}}{\bigcirc}\text{-OH} + \text{Cl}\text{-CH}_2\text{COOH} \xrightarrow[\text{(2) } H_3O^+]{\text{(1) NaOH}} \underset{\text{G}}{\bigcirc}\text{-O}\text{-CH}_2\text{COOH}$$

9.3.3f Procedure: Aryloxyacetic acids

Approximately 200 mg of the phenol is dissolved in 1 mL of 6 N sodium hydroxide in a small test tube. Additional water should be added if necessary to completely dissolve the sodium phenoxide. To the solution, 0.5 mL of a 50% aqueous solution of chloroacetic acid is added. The tube is provided with a microcondenser, and the reaction is heated on a water bath at 90 to 100°C for 1 hr. The solution is cooled, 2 mL of water is added, and the solution is acidified with dilute hydrochloric acid. The mixture is extracted several times with small portions of ether. The ether extract is washed with 2 mL of water and then extracted with 5% sodium carbonate solution. The sodium carbonate extract is acidified with dilute HCl to precipitate the aryloxy-acetic acid. The derivatives should be recrystallized from hot water. The melting point, along with the neutralization equivalent, if necessary, is determined.

Table 9.5 lists the physical properties of phenols and their derivatives.

9.4
ETHERS

[See also oxiranes (epoxides) (Sec. 9.5) and acetals and ketals (Sec. 9.6).]

9.4.1a Classification

The identification of the ether grouping in the infrared is complicated by the fact that other functional groups contain the C—O bond and, consequently, have bands in the same region. As a general guide, a relatively strong band in the 1250 to 1100 cm^{-1} (8.0 to 9.9 μm) region and the absence of C=O and O—H bands are good indications of an ether (Fig. 9.8). The aliphatic ethers absorb at the lower end of the range; conjugation raises the frequency.

Like hydrocarbons, ethers are quite unreactive, but they can be chemically distinguished from saturated hydrocarbons by the iodine charge-transfer test (Sec. 9.1.1b) and by their solubility in sulfuric acid (except diaryl ethers). Dialkyl ethers are also soluble in concentrated hydrochloric acid, whereas others are not.

In the absence of hydroxyl and carbonyl absorption in the infrared, a methyl (especially with characteristic ringing) in NMR spectra near $\delta 4$ or a methine or

Table 9.5. Physical Properties of Phenols and Derivatives

Compound	mp(°C)	bp(°C)	α-Naph-thyl-urethan	3,5-Dinitro-benzoate	N-Phenyl-urethan	Bromo Derivative	Aryloxy-acetic Acid
						mp(°C) of Derivatives	
o-Chlorophenol	7	175.6	120	143	121	48–9 (mono) 76 (di)	145
Phenol	41.8	182					
o-Cresol	31	191–2					
o-Bromophenol	5	195	129			95	143
Salicylaldehyde		197			133		
p-Cresol	36	202					
m-Cresol	12	203	128	165.4	125	84 (tri)	103
o-Ethylphenol		207		108	143.5		141
2,4-Dimethylphenol	27	211.5					
o-Hydroxyacetophenone	28	215					
m-Ethylphenol		217			137		77
p-Propylphenol	22	232			129		
p-Isobutylphenol		236					
m-Methoxyphenol		243	129			104 (tri)	118
p-Butylphenol	22	248			115		81
2-Methoxy-4-(2-propenyl)phenol	19	254.8	122	130.8	95.5	118 (tetra)	81(100)
2-Methoxy-4-(1-propenyl)phenol		267.5	150	158.4	118 (cis) 152 (trans)		94(116)
2,4-Dibromophenol	36	238–9				95	153
p-Chlorophenol	45	217	166	186	148.5		156
2,4-Dichlorophenol	43	209				68	141(135)
o-Nitrophenol	45		113	155		177 (di)	158
p-Ethylphenol	47	219	128	133	120		97
2,6-Dimethylphenol	49	203	176.5	158.8	133	79	139.5
2-Isopropyl-5-methylphenol	49.7	233	160	103.2	107	55	149
o-Phenylphenol	57.5	275					
3,4-Dimethylphenol	62.5	225	141–2	181.6	120	171 (di)	162.5
p-Bromophenol	64		169	191		95 (di)	160
2,4,6-Trichlorophenol	68	245					
3,5-Dimethylphenol	63	220	195.4	148		166 (tri)	111
2,5-Dimethylphenol	74.5	212	173	137.2	161(166)	178 (di)	118
2,3-Dimethylphenol	75						187
4-Hydroxy-3-methoxy-benzaldehyde	81						187
4-Chloro-2-nitrophenol	86						
1-Naphthol	94	280	152	217.4	178	105(2,4-di)	193.5
p-Iodophenol	94				148		156
2,4,6-Tribromophenol	95		153	174			
p-tert-Butylphenol	100	237	110		148.5	50	86
m-Hydroxybenzaldehyde	102						148
1,3-Benzenediol	110	280.8		201	164	112(4,6-di)	195
p-Nitrophenol	114		151	186		142(2,6-di)	187
p-Hydroxybenzaldehyde	116						198
2,4,6-Trinitrophenol	122						
2-Naphthol	123	286	157	210.2	156	84	154
1,2,3-Benzenetriol	133			205 (tri)	173 (tri)	158 (di)	198
o-Aminophenol	174						
p-Aminophenol	184						

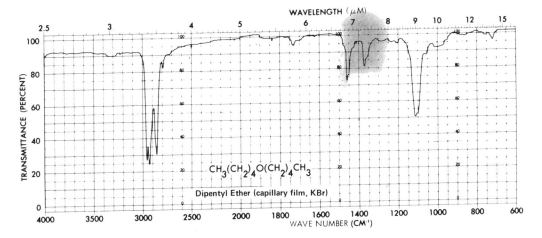

Fig. 9.8. Typical infrared spectrum of an ether.

methylene at slightly lower field should alert the investigator to the possibility of the presence of an alkyl ether.

CAUTION: *Ethers tend to form highly explosive peroxides on standing, particularly when exposed to air and light.* If peroxides are present, they accumulate in the pot during distillation and may lead to a violent explosion. The presence of peroxides can be detected by the starch-iodide test under acidic conditions (Sec. 1.15.8a). A positive test is indicated by the blue starch-iodine color. Peroxides (as well as water and alcohols) can be removed from ethers by filtering the ether through a short column of highly active alumina (Woelm basic alumina, activity grade 1, is recommended). Hydroperoxides (but not water-insoluble dialkyl or diacyl peroxides) can be removed from organic materials by treatment with ferrous sulfate; wash 10 mL of peroxide containing ether with 3 to 5 mL of 1% ferrous sulfate solution acidified with one drop of sulfuric acid.

9.4.1b Characterization

Aliphatic ethers. Although ethers are less reactive than most other functions, cleavage and conversion to other more easily characterized functions can be achieved under relatively mild conditions. The cleavage of ethers can be effected by acid chlorides or anhydrides in the presence of a Lewis acid. The reactions proceed by formation of an acyl oxonium ion, which generally fragments *via* an S_N1 mechanism.

$$R-O-R + R'-C\overset{O}{\underset{X}{\big\langle}} \xrightarrow{\text{Lewis acid}} \overset{R'}{\underset{R-O-R}{\underset{+}{\overset{|}{C}=O}}} \longrightarrow R-X + R'CO_2R$$

$$X = Cl, OCOR'$$

9.4.1c Procedure: Cleavage with Acetic Anhydride[24]

A mixture of 10 mmol of the ether and 0.5 g of anhydrous ferric chloride in 7 mL of acetic anhydride is heated for 17 hr on a steam bath. (For very volatile ethers, the reaction mixture may have to be sealed in a tube to avoid loss of the ether.) Pour the reaction mixture into 25 mL of water and extract with three portions (15 mL) of hexane. Wash the hexane extract with saturated sodium bicarbonate until no carbon dioxide is evolved. Distill or allow the hexane to evaporate. (Sometimes some unreacted ether will accompany the acetate.)

Aliphatic ethers can be cleaved with 3,5-dinitrobenzoyl chloride and $ZnCl_2$ as catalyst to yield 3,5-*dinitrobenzoates*.

The method, however, is only useful for symmetrical ethers and requires at least 0.5 mL of sample. With a low-boiling aliphatic ether, the reaction may fail owing to loss of the ether before the cleavage is effected. In these cases it may be necessary to use a sealed-tube method to acquire sufficient derivative.

9.4.1d Procedure: 3,5-Dinitrobenzoates

For best results, be sure the equipment is flame dried. Improved results are often obtained with freshly fused zinc chloride. Into a microflask, place 400 to 500 mg of zinc chloride, 0.5 mL of the ether, and 250 mg of 3,5-dinitrobenzoyl chloride. Immediately insert a condenser equipped with a drying tube. Reflux the mixture for 1 to 2 hr. Remove the condenser and allow the volatile material to distill off. Pulverize the residue. Add 5 mL of 5% sodium carbonate, and stir and warm the reaction mixture to 60 to 70°C (ca. 1 min on a steam bath). Collect the precipitate by suction filtration. Wash the precipitate with 2 mL of 5% sodium carbonate, and follow by several washings with small volumes of water and allow to dry. The crude ester can be recrystallized from aqueous ethanol. The yield may be as low as 5 mg or even lower.

Cleavage of an dialkyl or alkyl aryl ether can also be effected by boron tribromide[25] or trimethylsilyl iodide.[26]

---

[24] B. Ganem and V. R. Small, Jr., *J. Org. Chem.*, **39**, 3728 (1974).

[25] F. L. Benton and T. E. Dillon, *J. Am. Chem. Soc.*, **64**, 1128 (1942); J. F. W. McOmie and M. L. Watts, *Chem. Ind.* (London), 1658 (1963).

[26] M. G. Voronkov, V. E. Puzanova, S. F. Pavlov, and E. I. Dubinskaya, *Izv. Akad. Nauk SSR, Ser. Khim.*, 448 (1975); U. Kruerke, *Chem. Ber.*, **95**, 174 (1962); M. E. Jung and M. A. Lyster, *J. Org. Chem.*, **42**, 3761 (1977); T.-L. Ho and G. A. Olah, *Angew. Chem. Int. Ed. Engl.*, **15**, 774 (1976).

Dialkyl and alkyl aryl ethers are cleaved by hydriodic acid.

$$R_2O + 2\,HI \longrightarrow 2\,RI + H_2O$$

$$ArOR + HI \longrightarrow ArOH + RI$$

To obtain sufficient product, especially for characterization of the alkyl iodides, it is necessary to use a sample of 4 to 5 g. The alkyl iodides and phenols can be transformed into suitable derivatives.

Vinyl ethers are readily cleaved with dilute mineral acids to alcohols and carbonyl compounds.

Aromatic ethers. The derivatives of aromatic ethers most frequently employed are those obtained by electrophilic aromatic substitution.

The extent of *bromination* of aromatic ethers depends on the groups already present.

9.4.1e Procedure: Bromination

The ether (5 to 100 mg) is dissolved in glacial acetic acid and placed in an ice bath; a slight excess of bromine is added with cooling (liquid bromine can be used, though for small-scale operations, a solution of bromine in glacial acetic acid is recommended). The reaction mixture is allowed to stand for a short while in the ice bath. It is then removed from the ice bath to stand at room temperature for 10 to 15 min. The bromo compound is separated by the addition of water. The crude product is recrystallized from dilute ethanol or petroleum ether.

Aromatic ethers are readily chlorosulfonated with chlorosulfonic acid. The intermediate sulfonyl chlorides can be isolated, but it is usually more convenient to convert them directly to *sulfonamides.*

9.4.1f Procedure: Sulfonamides

A solution of 0.25 g, or 0.25 mL of aromatic ether in 2 mL of chloroform, in a test tube is cooled in an ice bath. About 1 g of chlorosulfonic acid is added, drop by drop, over a 5 min period. The tube is removed from the ice bath and allowed to stand for 30 min. The reaction mixture is then poured into a small separatory funnel containing 5 mL of ice water. The chloroform layer is separated and washed with water. This layer is then added, with stirring, to 3 mL of concentrated ammonia solution (or ammonia gas is bubbled into the chloroform solution). The solution is stirred for 10 min and then the chloroform is evaporated off. The residue is dissolved in 3 mL of 5% sodium hydroxide, and the solution is filtered to remove insoluble material. The filtrate is acidified with dilute hydrochloric acid and cooled in an ice bath. The sulfonamide is then collected and recrystallized from dilute ethanol.

Nitration can be accomplished by any of the procedures outlined for aromatic hydrocarbons (Sec. 9.1.4). Again the extent of nitration depends on the conditions employed and the groups already present.

Aromatic ethers form *molecular complexes with picric acid.*

9.4.1g Procedure: Picric acid complexes

Dissolve 1 mmol of the aromatic ether in a small volume of boiling chloroform and 1.05 mmol (0.241 g) of picric acid in 10 mL of boiling chloroform. The hot solutions are mixed and allowed to cool. The crystals are collected, dried rapidly between filter papers, and the melting point is immediately determined. Although a number of such picrates decompose on standing, some are stable enough to be recrystallized (from a minimum volume of chloroform).

Aromatic ethers containing an alkyl side chain can be oxidized to substituted benzoic acids with potassium permanganate (see procedure under aromatic hydrocarbons, Sec. 9.1.4). As noted before under aliphatic ethers, alkyl aryl ethers are cleaved by hydriodic acid and boron tribromide. Allyl aryl ethers undergo the Claisen rearrangement, upon heating, to give *o*-allylphenols.

Table 9.6 lists the physical properties of ethers and their derivatives.

Table 9.6. Physical Properties of Ethers

Compound	bp(°C)	D_4^{20*}	n_D^{20*}
Furan (mp −85.6°C)	31.27	0.9366	1.42157
Diethyl ether (mp −116.3°C)	34.60	0.71425	1.3526
1,2-Epoxypropane	35	0.830	1.466
Ethyl vinyl ether	36	0.760	
1,2-Epoxy-2-methylpropane	56		
cis-2,3-Epoxybutane	52–53		
trans-2,3-Epoxybutane	58–59		
1,2-Epoxybutane	61–2	0.837^{17}	1.385^{17}
Tetrahydrofuran	65	0.889	1.407
Allyl ethyl ether	65		
Diisopropyl ether	67.5	0.726	1.3688
Butyl methyl ether	70	0.774	1.374
Dimethoxyethane	85	0.867	1.37965
Dihydropyran	86	0.923	1.440
Tetrahydropyran	88	0.881	1.421
Dipropyl Ether (mp −122°C)	90.1	0.74698	1.3885
Butyl ethyl ether (mp −124°C)	92.3	0.7506	1.3820
Methyl pentyl ether	99	0.761	1.387
1,4-Dioxane (mp +11.8°C)	101.4	1.03361	1.4232
Diethoxyethane	121	0.841	
Di-sec-butyl ether	121	0.760	
Diisobutyl ether	123	0.762	
Hexyl methyl ether	126	0.772	1.397
Hexyl ethyl ether	142	0.772	1.401
Dibutyl ether (mp −98°C)	142.4	0.76829	1.400
Methoxybenzene (mp −37.5°C)	153.8	0.99393	1.52211
Benzyl methyl ether	171	0.9649	1.5008
2-Methoxytoluene	171	0.9853	1.505
Ethoxybenzene (mp −33°C)	172	0.9666	1.5080
Diisopentyl ether	172.5	0.778	1.409
4-Methoxytoluene	173	0.970	1.512
3-Methoxytoluene	173	0.972	1.513
Benzyl ethyl ether	184–6	0.9478	1.4958
Bis(2-ethoxyethyl)ether	188	0.906	1.411
Phenyl propyl ether	188	0.949	1.501
2-Methoxyphenol	205		
Butyl phenyl ether	206		
1,2-Dimethoxybenzene (mp 22.5°C)	207	1.080	
Dihexyl ether	229		
Diphenyl ether (mp 28°C)	259	1.073	
Dibenzyl ether (mp 3.6°C)	298	1.0428	

* Values determined at 20°C unless otherwise noted by superscript temperatures.

9.5
OXIRANES (EPOXIDES)

9.5.1 General Characterization

9.5.1a Classification

Oxiranes are three-membered cyclic ethers that differ greatly in their chemical reactivity compared with larger-ring cyclic and acyclic ethers. Oxiranes, which are often volatile liquids, display broad, moderately intense bands in the infrared at 1250 cm^{-1} (8.0 μm) and in the 900 to 800 cm^{-1} (11 to 12.5 μm) region (Fig. 9.9). The NMR spectra of oxiranes are quite complex.

The *periodate test* for oxiranes is a modification of the test used for the detection of glycols (Sec. 9.3.2c).

$$R-\overset{O}{\overset{\diagup \diagdown}{CH-CHR}} \xrightarrow{H_3O^+} R-\overset{OH}{\underset{|}{CH}}-\overset{OH}{\underset{|}{CH}}-R \xrightarrow[Ag^+]{IO_4^-} 2 \text{ RCHO} + AgIO_3 \downarrow$$

9.5.1b Procedure: Periodate test for oxiranes

Add two drops of concentrated nitric acid to 2 mL of 0.5% solution of periodic acid, and then add one or two drops of the oxirane. Water-insoluble compounds can be dissolved in acetic acid. Shake the mixture vigorously, and add two drops of 5% silver nitrate solution. A positive test is indicated by the appearance of a white precipitate of silver iodate. It is advisable to run a blank.

Most oxiranes *react with concentrated hydrochloric acid* to produce chlorohydrins that are insoluble in the reagent. This test should be run only after it has been

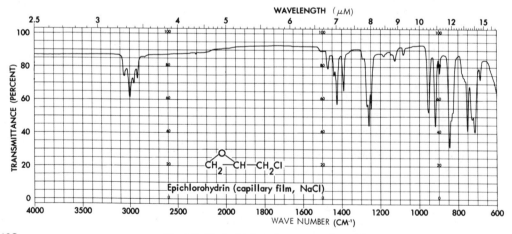

Fig. 9.9. Typical infrared spectrum of an oxirane.

established that the unknown is not an alcohol (see Lucas test, Sec. 9.3.1b). Most aliphatic ethers are soluble in, but are not cleaved by, concentrated hydrochloric acid.

$$RCH\!-\!CHR + HCl \longrightarrow R\!-\!CH\!-\!CH\!-\!R$$
$$\underset{O}{\diagdown\!\diagup} \qquad\qquad \underset{OH \quad Cl}{}$$

9.5.1c Procedure: Chlorohydrins

To 0.5 mL of cold concentrated hydrochloric acid in a small test tube, add two drops of the unknown. If the unknown is not immediately soluble in the hydrochloric acid, add a little 1,2-dimethoxyethane to produce a homogeneous solution. Allow the reaction mixture to come to room temperature and stand for 10 to 15 min. A positive test is indicated by the formation of an insoluble oil (chlorohydrin). The chlorohydrins of less than four carbons are soluble.

9.5.1d Characterization

Oxiranes are partially characterized by their boiling points and refractive indices (see Table 9.6 for physical properties of oxiranes).

Since it is generally not possible to form suitable derivatives directly from oxiranes, chemical conversions to other functional groupings are usually necessary. The chemistry of oxiranes is characterized by high reactivity toward ring-opening reactions, which fall into two categories: nucleophilic and electrophilic. Nucleophilic attack occurs at the least substituted carbon atom to give a β-substituted alcohol. Electrophilic ring-opening reactions can occur without rearrangement to give a β-substituted alcohol (note that a different compound is formed), or with rearrangement to give a carbonyl derivative via the more stable intermediate carbonium ion.

Nucleophilic ring opening

$$R_2C\!-\!CHR + N^- \longrightarrow R_2\overset{\overset{\textstyle OH}{|}}{C}\!-\!\underset{\underset{\textstyle N}{|}}{C}H\!-\!R$$

Electrophilic ring opening

$$R_2C\!-\!CH\!-\!R + HN \longrightarrow R_2\overset{\overset{\textstyle OH}{|}}{C}\!-\!\underset{\underset{\textstyle N}{|}}{C}H\!-\!R$$

Rearrangements

$$R_2C\!-\!CHR \xrightarrow{H^+} R_2\overset{+}{C}\!-\!\underset{\underset{\textstyle OH}{|}}{C}HR \longrightarrow R_2CH\!-\!\underset{\underset{\textstyle O}{\|}}{C}\!-\!R$$

The two most useful reactions for conversion of oxiranes to another functional group for purposes of derivation are the sulfuric or Lewis acid-catalyzed rearrangements (see rearrangement reactions of glycols under polyhydric alcohols, Sec. 9.3.2d) and the *lithium aluminum hydride reduction*. The latter is preferred due to fewer side reactions.

$$R_2C\overset{\displaystyle\diagdown}{\underset{\displaystyle O}{\diagup}}CH_2 \xrightarrow[\text{(2) } H_2O]{\text{(1) LiAlH}_4} R_2\underset{\displaystyle OH}{\overset{\displaystyle |}{C}}-CH_3$$

9.5.1e Procedure: Lithium aluminum hydride reduction

To a solution of a 50% excess of lithium aluminum hydride (based on an anticipated molecular weight of the oxirane) in 20 mL of absolute ether is slowly added 300 mg of the unknown oxirane dissolved in 5 mL of ether. The reaction mixture is refluxed for 30 min, and the excess hydride is carefully decomposed by the dropwise addition of saturated aqueous sodium sulfate solution or water (avoid a great excess of water). Sometimes it may be convenient to dissolve the aluminum hydroxide by addition of sodium potassium tartrate, and then to extract the mixture with ether. The reaction mixture is filtered to remove the insoluble inorganic residue. The ether solution is dried over magnesium sulfate, and the ether is removed by distillation. The alcohol can be characterized in the usual manner.

9.6
ACETALS AND KETALS

(See Sec. 9.19.2 on sulfides for dithio- and monothioacetals and ketals.)

9.6.1a Classification

A large portion of acetals and ketals are liquids with reasonably pleasant odors. An acetal or ketal should be suspected if three bands [1190 to 1160, 1195 to 1125, and 1098 to 1063 cm^{-1} (8.40 to 8.62, 8.37 to 8.89, and 9.11 to 9.41 μm, respectively)] appear in the C—O stretching region of the infrared spectrum, especially in the absence of carbonyl and hydroxyl absorptions (Fig. 9.10). In addition, the spectrum of an acetal contains a characteristic strong C—H bending deformation band in the 1116 to 1103 cm^{-1} (8.96 to 9.02 μm) region which, however, may be obscured by C—O peaks.

A compound suspected of being an acetal or ketal should be subjected to hydrolysis by dilute acid, and the hydrolysate should be tested for the presence of an aldehyde or ketone. A simple way to perform such a test is to add several drops of the compound to 2,4-dinitrophenylhydrazine reagent and then to warm the mixture. The acid contained in the reagent is usually sufficient to cause hydrolysis.

$$\underset{\diagup\diagdown OR}{\overset{OR}{\diagup}}C \xrightarrow{H_3O^+} 2\,ROH + \diagdown_{\diagup}C{=}O$$

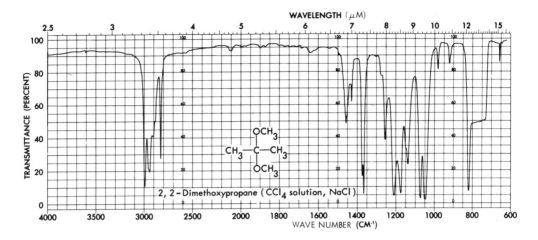

Fig. 9.10. Infrared spectrum of a simple ketal.

CAUTION: *Acetals, like ethers, form highly explosive peroxides on exposure to air and light.*

9.6.1b Characterization

The preparation of derivatives of acetals and ketals is based on their hydrolysis to the component alcohol and carbonyl compound. The hydrolysis can usually be accomplished by heating with dilute mineral acid for 10 to 30 min with a co-solvent such as acetic acid, if needed to effect solution. The hydrolysate should be divided into two portions. One portion should be used to prepare the 3,5-dinitrobenzoate or other suitable derivative of the alcohol. (In some cases, derivatization of the alcohol component may be unnecessary since that portion of the molecule can be unambiguously identified by NMR or mass spectral data obtained on the original acetal or ketal.) The second portion is used to characterize the aldehyde or ketone as the 2,4-dinitrophenylhydrazone or semicarbazone.

9.7
ALDEHYDES AND KETONES

9.7.1 General Characterization

9.7.1a Classification

The presence of an intense band in the 1850 to 1650 cm^{-1} (5.40 to 6.06 μm) region of the infrared indicates a carbonyl compound. The lack of absorption in the 3600 to 3300 cm^{-1} (2.78 to 3.03 μm) region eliminates acids and primary and secondary amides as possibilites. The differentiation between aldehydes and ketones and the other remaining carbonyl compounds (esters, anhydrides, tertiary amides, etc.) can be accomplished by the 2,4-dinitrophenylhydrazine test. From a cursory inspection of infrared spectra, the most likely compounds to be confused with aldehydes and ketones are the esters. Compare the spectra of aldehydes (Fig. 9.11) and ketones (Fig. 9.12) with those of esters (Fig. 9.19). Note the two very intense C—O bands in the 1250 to 1050 cm^{-1} (8.0 to 9.52 μm) region in the ester spectrum, which are absent in the aldehydes and ketones. With the latter, the carbonyl peak is usually the most intense in the spectrum.

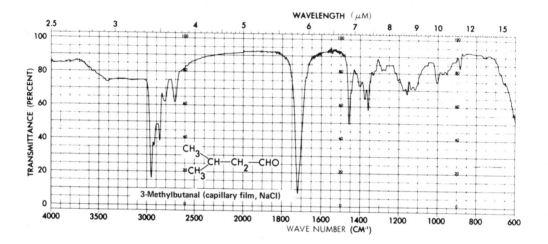

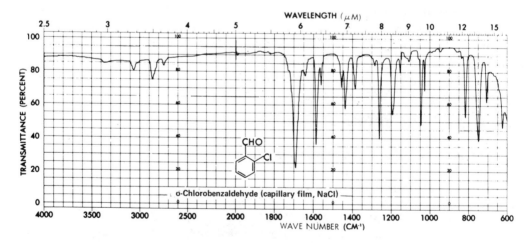

Fig. 9.11. Typical infrared spectra of aldehydes.

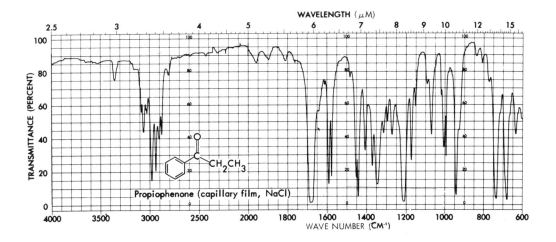

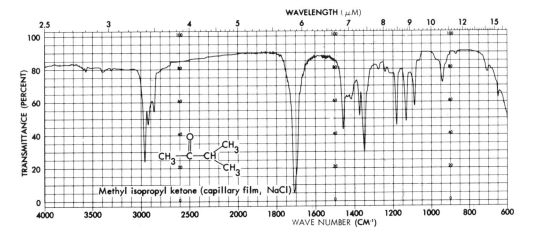

Fig. 9.12. Typical infrared spectra of ketones.

9.7.1b 2,4-Dinitrophenylhydrazine test

The utility of 2,4-dinitrophenylhydrazine lies in the fact that almost all aldehydes and ketones readily yield insoluble, solid 2,4-dinitrophenylhydrazones. There are a few exceptions among the long chain aliphatic ketones that yield oils.

R and R′ = alkyl, aryl, or hydrogen

It is wise to apply the test with some discrimination, i.e., only to compounds that give infrared evidence of being an aldehyde or ketone. The reagent can react with anhydrides or highly reactive esters; it is capable of oxidizing certain allylic and benzylic alcohols to aldehydes and ketones, which then give positive tests. The reagent can also give insoluble charge transfer complexes with amines, phenols, etc. These, however, are usually considerably more soluble than the phenylhydrazones and can be distinguished on this basis. Indiscriminate use of the reagent may provide misleading positive tests owing to the hydrolysis of compounds such as ketals or trace impurities of carbonyl compounds.

The color of a 2,4-dinitrophenylhydrazone can provide a qualitative indication about conjugation, or lack thereof, in the starting carbonyl compound. Dinitrophenylhydrazones of saturated aldehydes and ketones are typically yellow. Conjugation with a carbon-carbon double bond or with an aromatic ring changes the color from yellow to orange-red. It should be noted that 2,4-dinitrophenylhydrazone itself is orange-red, and sometimes under the reaction conditions this material precipitates or contaminates the dinitrophenylhydrazone. Any orange-red precipitate that melts near 2,4-dinitrophenylhydrazine (198°C with decomposition) should be looked at with due suspicion.

9.7.1c Procedure: 2,4-Dinitrophenylhydrazine test

Reagent: Dissolve 3 g of 2,4-dinitrophenylhydrazine in 15 mL of concentrated sulfuric acid. Add this solution with stirring to 20 mL of water and 70 mL of 95% ethanol. Mix this solution thoroughly and filter.

Place 1 mL of the 2,4-dinitrophenylhydrazine reagent in a small test tube. Add one drop of the liquid carbonyl compound to be tested or an equivalent amount of solid compound dissolved in a minimum amount of ethanol. Shake the tube vigorously; if no precipitate forms immediately, allow the solution to stand for 15 min. The formation of a yellow to orange-red precipitate is considered a positive test.

9.7.2 Differentiation of Aldehydes and Ketones

9.7.2a Spectral methods

Infrared spectra. The infrared spectra of ketones exhibit carbonyl bands in the 1780 to 1665 cm^{-1} (5.62 to 6.01 μm) region. Aldehyde carbonyl bands appear in the more restricted 1725 to 1685 cm^{-1} (5.80 to 5.94 μm) region. In the aldehyde spectra, additional bands that are not present in the spectra of ketones appear near 2820 and 2700 cm^{-1} (3.55 and 3.70 μm) (Fig. 9.11). Regrettably, these bands are often weak and/or poorly-resolved.

NMR spectra. The very low field resonance of the aldehydic hydrogen ($\delta 9.5$ to 10.1) is very diagnostic of an aldehyde. Hydrogens on saturated carbon adjacent to the carbonyl of aldehydes and ketones usually fall in the $\delta 2$ to 3 region. In the case of aldehydes, these hydrogens are coupled with the aldehydic hydrogen with a relatively small coupling constant (1 to 3 Hz). Alkyl methyl ketones exhibit sharp methyl singlets near $\delta 2.2$; aryl methyl ketones appear nearer $\delta 2.6$.

Mass spectra. If the mass spectrum exhibits a discernible parent peak and a significant parent -1 peak, the spectrum should be examined for other evidence of an aldehyde. In the mass spectrum of a ketone, the molecular ion is usually pronounced. Major fragmentation peaks for acyclic ketones result from alpha bond cleavage to give acylium ions. In the alkyl aryl ketones, the $ArCO^+$ fragment is usually the base peak.

9.7.2b Chemical methods

The lability of aldehydes toward oxidation to acids forms the basis for a number of tests to differentiate between aldehydes and ketones.

Most aldehydes reduce silver ion from *ammoniacal silver nitrate solution* (*Tollen's reagent*) to metallic silver.

$$RCHO + 2\,Ag(NH_3)_2OH \longrightarrow 2\,Ag \downarrow + RCOO^-NH_4^+ + H_2O + NH_3$$

Positive tests are also given by hydroxyketones, reducing sugars and certain amino compounds such as hydroxylamines. Normal ketones do not react. The test should be run only after it has been established that the unknown in question is an aldehyde or ketone.

9.7.2c Procedure: Tollen's test

1ml ↓   1 drop

Reagent: Solution A: Dissolve 3 g of $AgNO_3$ in 30 mL of water. *Solution B:* 10% sodium hydroxide. When the reagent is required, mix 1 mL of solution *A* with 1 mL of solution *B* in a *clean* test tube, and add dilute ammonia solution dropwise until the silver oxide is just dissolved.

CAUTION: *Prepare the reagent only immediately prior to use.* Do not heat the reagent during its preparation or allow the prepared reagent to stand, since the *very explosive* silver fulminate may be formed. Wash any residue down the sink with a large quantity of water. Rinse the test tube with dilute HNO_3 when the test is completed.

Add a few drops of a dilute solution of the compound to 2 to 3 mL of the prepared reagent. In a positive test, a silver mirror is deposited on the walls of the tube either in the cold or upon warming in a beaker of hot water.

Chromic acid in acetone rapidly oxidizes aldehydes to acids (the reagent also attacks primary and secondary alcohols, Sec. 9.3.1), whereas ketones are attacked slowly or not at all. Aliphatic aldehydes are oxidized somewhat faster than aromatic aldehydes; this rate difference has been used to distinguish between aliphatic and aromatic aldehydes.[27]

9.7.2d Procedure: Chromic acid test

Reagent: Dissolve 1 g of chromium trioxide in 1 mL of concentrated sulfuric acid, and dilute with 3 mL of water.

Dissolve one drop or 10 mg of the compound in 1 mL of reagent grade acetone (or better, in acetone that has been distilled from permanganate). Add several drops of chromic acid reagent. A positive test is indicated by the formation of a green precipitate of chromous salts. With aliphatic aldehydes, the solution turns cloudy within 5 sec, and a precipitate appears within 30 sec. Aromatic aldehydes generally require 30 to 90 sec for the formation of a precipitate.

With aliphatic aldehydes (but not aromatic) and α-hydroxyketones, *Benedict's reagent* (*a citrate complex of cupric ion*) gives a yellow-to-red precipitate. A yellow suspension in the blue solution of the reagent appears green. In certain cases, the exact composition of the precipitate is unknown; it is usually thought to be cuprous oxide. This is the classical test that has been used to distinguish between aliphatic and aromatic aldehydes.

$$RCHO + 2 Cu^{++}(citrate)_2 + 4 OH^- \longrightarrow RCOOH + Cu_2O \downarrow + H_2O$$

9.7.2e Procedure: Benedict's test

Reagent: Dissolve 86.5 g of hydrated sodium citrate and 50 g of anhydrous sodium carbonate in about 350 mL of water. Add a solution of 8.65 g of copper sulfate in 50 mL water with stirring. Dilute the resulting solution to 500 mL; filter if necessary. The solution does not deteriorate significantly on storage.

Into a test tube containing 4 mL of Benedict's solution, add two to three drops of liquid unknown or an equivalent amount of solid in a small volume of ethanol or water. Heat the mixture to boiling. A positive test is indicated by the formation of a yellow-to-red precipitate.

Aldehydes react with methone to give dimethone derivatives. Ketones do not react, and the method can serve to differentiate between the two (see next section).

9.7.3 Aldehydes

9.7.3a Characterization

The most frequently employed derivatives for aldehydes include the 2,4-dinitro-, the *p*-nitro-, and the unsubstituted phenylhydrazones, the semicarbazones, and

---

[27] J. D. Morrison, *J. Chem. Ed.*, **42**, 554 (1965).

the oximes. Discussion and procedures for these derivatives, as well as a method for reduction to a primary alcohol, will be presented in Sec. 9.7.5.

The *oxidation of an aldehyde to the corresponding acid* is a particularly useful method of identification, especially in the case of aromatic aldehydes that yield solid acids (characterized by their melting point and neutralization equivalent).

9.7.3b Procedure: Oxidation to acids

(a) The oxidation can be carried out by the chromic acid-in-acetone method outlined under alcohols (Sec. 9.3.1).

(b) In a test tube place 0.5 g of aldehyde, 5 to 10 mL of water, and several drops of 10% sodium hydroxide. Add saturated aqueous potassium permanganate several drops at a time. Shake the mixture vigorously. Add additional permanganate until a definite purple color remains. Acidify the mixture with dilute sulfuric acid, and add sodium bisulfite solution until the excess permanganate and manganese dioxide have been converted to the colorless soluble manganese sulfate. Remove the acid by filtration or extraction with dichloromethane or ether. The acid can be recrystallized from water or water-alcohol, or it can be purified by sublimation.

The equations following this paragraph illustrate the reaction between 5,5-dimethylcyclohexane-1,3-dione (methone, dimethyldihydroresorcinol) and an aldehyde. One mole of the aldehyde condenses with 2 mol of the reagent to give the *dimethone*, sometimes called a *dimedone derivative*. As ketones do not yield derivatives, the reaction can serve as another method of differentiating between aldehydes and ketones. Aldehydes can be made to react in the presence of ketones. Also, the reaction can serve as a method for the quantitative determination of formaldehyde. The dimethones are generally more suitable derivatives for low-molecular-weight aldehydes than are 2,4-dinitrophenylhydrazones.

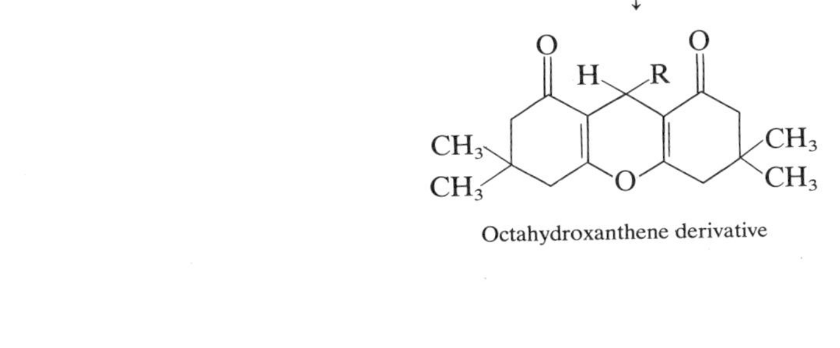

Methone   Dimethone derivative

Octahydroxanthene derivative

9.7.3c Procedure: Dimethone derivatives

To a solution of 100 mg of the aldehyde in 4 mL of 50% ethanol, add 0.3 g of methone and one drop of piperidine. Boil the mixture gently for 5 min. If the solution is clear at this point, add water dropwise to the cloud point. Cool the mixture in an ice bath until crystallization is complete. Crystallization is often slow, and one should allow, if necessary, 3 to 4 hr. Collect the crystals by vacuum filtration, and wash with a minimum amount of cold 50% ethanol. If necessary, the derivative can be recrystallized from mixtures of alcohol and water.

If the dimedone derivatives are heated with acetic anhydride or with alcohol containing a small amount of hydrochloric acid, cyclization to the *octahydroxan-thene* (*menthone anhydride*) occurs. The cyclization is rapid and quantitative, and the xanthenes can serve as a second derivative.

9.7.3d Procedure: Octahydroxanthenes

The foregoing dimedone derivative can be converted to the octahydroxanthenes by boiling a solution of 100 mg of the derivative in a small volume of 80% ethanol to which one drop of concentrated hydrochloric acid has been added. The cyclization is complete in 5 min; water is added to the cloud point, and the solution is cooled to induce crystallization. The product is usually pure, and further recrystallization is not necessary.

9.7.4 Ketones

9.7.4a Characterization

The methods described under general derivatives of aldehydes and ketones in the next section are those most often employed to obtain solid derivatives of ketones. In general, low-molecular-weight ketones are best characterized by 2,4-dinitrophenylhydrazones, *p*-nitrophenylhydrazones, or semicarbazones. Hydrazones, phenylhydrazones, and oximes are usually more suitable for the higher-molecular-weight ketones.

Other reactions sometimes useful in the structural determination of ketones include:

1. Active hydrogen determination by deuterium exchange or other means (Sec. 8.3.3).

$$RCH_2-\overset{\overset{\displaystyle O}{\|}}{C}-CH_3 \xrightarrow[Na_2CO_3]{D_2O} RCD_2-\overset{\overset{\displaystyle O}{\|}}{C}-CD_3 \begin{cases} \text{mass spectrum} \\ \text{NMR} \\ \text{microanalysis} \end{cases}$$

2. The formation of benzylidene derivatives (Sec. 8.3.3).

3. The Beckman rearrangement of oximes to amides.[28]

4. The Baeyer-Villiger oxidation of ketones to esters or lactones.[29]

5. The *haloform reaction* is useful in classification and derivatization of methyl ketones. The halogenation of aldehydes and ketones is catalyzed by acids and bases. With acid catalysis, the rate-determining step is enol formation, and with base catalysis, it is the formation of the enolate. Hence the reaction is independent of the concentration or kind of halogen. With base catalysis, the initial reaction occurs at the least-substituted carbon.

$$R-\overset{\underset{\|}{O}}{C}-CH_3 \xrightarrow{B^-} R-\overset{\underset{O^-}{|}}{C}=CH_2 \xrightarrow{X_2} R-\overset{\underset{\|}{O}}{C}-CH_2X$$

Since halogen is highly electronegative, successive hydrogens are replaced more readily, and unsymmetrical polysubstitution occurs.

$$R-\overset{\underset{\|}{O}}{C}-CH_2X \xrightarrow[X_2]{B^-} R-\overset{\underset{\|}{O}}{C}-CX_3$$

With methyl ketones or acetaldehyde, a trihalogenated product is formed. These trihalomethylcarbonyl compounds are readily cleaved by base to give a haloform and a salt of a carboxylic acid.

$$R\overset{\underset{\|}{O}}{C}-CX_3 \xrightarrow{OH^-} R\overset{\underset{\underset{OH}{|}}{C}}{\overset{\underset{O^-}{|}}{C}}-CX_3 \longrightarrow RCOH + CX_3^- \longrightarrow RCO^- + HCX_3$$

[28] L. G. Donaruma and W. Z. Heldt, *Org. Reactions*, **11**, 1 (1960).
[29] C. H. Hassall, *Org. Reactions*, **9**, 73 (1957).

The haloform reaction will proceed with aldehydes or ketones that contain the groupings $CH_3CO—$, $CH_2XCO—$, $CHX_2—CO$, or $CX_3CO—$. The reaction will also proceed with compounds that will react with the reagent to give a derivative containing one of the necessary groupings, e.g.,

$$R—\underset{\underset{OH}{|}}{C}H—CH_3 \xrightarrow[NaOH]{X_2} R—\underset{\underset{O}{\|}}{C}—CH_3 \xrightarrow[NaOH]{X_2} RCOO^- + HCX_3$$

Acetaldehyde is the only aldehyde, and ethyl alcohol is the only primary alcohol, that produces haloform. The principal types of compounds that give the haloform reaction are methyl ketones and alkylmethyl carbinols

$$CH_3—\underset{\underset{O}{\|}}{C}—R \quad \text{and} \quad CH_3—\underset{\underset{OH}{|}}{C}H—R$$

R = alkyl, vinyl, or aryl

and compounds that have the structures

$$R—\underset{\underset{O}{\|}}{C}—CH_2—\underset{\underset{O}{\|}}{C}—R, \quad R—\underset{\underset{OH}{|}}{C}H—CH_2—\underset{\underset{OH}{|}}{C}H—R, \quad \text{etc.,}$$

which are, or which upon oxidation yield, ketoalcohols or diketones that are cleaved by alkali to methyl ketones.

$$R—\underset{\underset{O}{\|}}{C}—CH_2—\underset{\underset{OH}{|}}{C}H—R' \xrightarrow{OH^-} R—\underset{\underset{O}{\|}}{C}—CH_3 + R'CHO$$

Acetic acid and its esters, and compounds such as

$$CH_3—\underset{\underset{O}{\|}}{C}—CH_2—COOR, \quad CH_3—\underset{\underset{O}{\|}}{C}—CH_2—CN$$

which substitute on the methylene rather than the methyl, do not produce haloforms.

As a method of preparation of acids from methyl ketones and related substances, sodium hypochlorite (bleaching powder or bleaching solution) is often used. As a qualitative test, the reaction is usually run with alkali and iodine, since iodoform is a water-insoluble crystalline solid readily identified by its melting point (*iodoform test*).

9.7.4b Procedure: Iodoform test

Reagent: Dissolve 20 g of potassium iodide and 10 g of iodine in 100 mL of water.

Dissolve 0.1 g or four to five drops of compound in 2 mL of water; if necessary, add 1,2-dimethoxyethane to produce a homogeneous solution. Add 1 mL of 10% sodium hydroxide and the potassium iodide-iodine reagent dropwise, with shaking, until a definite dark color of iodine persists. Allow the mixture to stand for several minutes; if no iodoform separates at room temperature, warm the tube in a 60°C water bath. If the color disappears, continue to add the iodine reagent dropwise until its color does not disappear after 2 min of heating. Remove the excess iodine by the addition of a few drops of sodium hydroxide solution. Dilute the reaction mixture with water and allow to stand for 15 min. A positive test is indicated by the formation of a yellow precipitate. Collect the precipitate; dry on filter paper. Note the characteristic medicinal odor, and determine the melting point; iodoform melts at 119 to 121°C.

9.7.5 General Derivatives of Aldehydes and Ketones

9.7.5a 2,4-Dinitrophenylhydrazones and p-nitrophenylhydrazones

$$ArNHNH_2 + RCOR' \longrightarrow ArNHN{=}CRR'$$

Ar = 2,4-dinitrophenyl, or p-nitrophenyl
R and R' = alkyl, aryl, or hydrogen

The 2,4-dinitrophenylhydrazones are excellent derivatives for small-scale operations. It is often possible to obtain sufficient amounts of the derivative for characterization from the precipitate formed in the classification test described earlier (Sec. 9.7.1b). There are certain limitations to the use of nitrophenylhydrazones as derivatives. From derivative tables one will observe that, in many cases, the melting points of the nitrophenylhydrazones are too high for practical utility; hence, the use of other derivatives such as phenylhydrazones, semicarbazones, or oximes is recommended. In general, mixtures of (E)- and (Z)-isomers about the C=N are obtained. This aspect was not recognized by earlier chemists, and the melting points of derivatives are often those of mixtures. A single recrystallization does not result in the separation of the isomers, and the melting point thus obtained generally corresponds closely with the literature value. Repeated careful fractional recrystallization often results in the separation and isolation of a single stereoisomer whose melting point does not agree with that reported. (In Tables 9.7 and 9.8, more than a single value for the melting point of a derivative is often given due to these complications.) Determination of the configuration of oximes can be accomplished by ^{13}C NMR.[30] In addition to E-,Z-isomerism about the C=N, the 2,4-dinitrophenylhydrazones tend to form several crystalline modifications with different melting points. One may find it useful to redetermine the melting points after allowing the melt to resolidify.

---

[30] G. E. Hawkes, K. Herwig, and J. D. Roberts, *J. Org. Chem.*, **39**, 1017 (1974); G. C. Levy and G. L. Nelson, *J. Am. Chem. Soc.*, **94**, 4897 (1972).

9.7.5b Procedure: 2,4-Dinitro- and p-nitrophenylhydrazones

Place 100 mg of 2,4-dinitro- or p-nitrophenylhydrazine in a test tube or Erlenmeyer flask containing 10 mL of methanol. Cautiously add five drops of concentrated hydrochloric acid and warm the solution on the steam bath, if necessary, to complete solution. Dissolve approximately 1 mmol of the carbonyl compound in 1 mL of methanol, and add this to the reagent. Warm the mixture on a water or steam bath for 1 to 2 min and allow to stand for 15 to 30 min. Most of the derivatives precipitate out on cooling, but for complete precipitation, it is advisable to add water to cloudiness. The derivative can be purified by crystallization from alcohol-water mixtures. In the case of less soluble materials, ethyl acetate is often found to be a more suitable solvent. This procedure is generally suitable for the preparation of both dinitrophenylhydrazones and nitrophenylhydrazones.

9.7.5c Semicarbazones

Semicarbazones prepared from carbonyl compounds and semicarbazide hydrochloride are excellent derivatives for ketones and aldehydes above five carbon atoms because of their ease of formation, highly crystalline properties, sharp melting points, and ease of recrystallization. The lower aldehydes react slowly and/or give soluble derivatives. As in the case of most carbonyl derivatives, semicarbazones from unsymmetrical carbonyl compounds are capable of existing in two isomeric forms. A number of substituted semicarbazones and thiosemicarbazones are occasionally used.

$$\begin{array}{c} R \\ \diagdown \\ C{=}O + NH_2NHCONH_2 \longrightarrow \\ \diagup \\ R \end{array} \qquad \begin{array}{c} R \\ \diagdown \\ C{=}NNHCONH_2 + H_2O \\ \diagup \\ R' \end{array}$$

Semicarbazide   Semicarbazone

9.7.5d Procedure: Semicarbazones

In a 8-in. test tube, place 100 mg of semicarbazide hydrochloride, 150 mg of sodium acetate, 1 mL of water, and 1 mL of alcohol. Add 100 mg (0.1 mL) of the aldehyde or ketone. If the mixture is turbid, add alcohol until a clear solution is obtained. Shake the mixture for a few minutes and allow to stand. The semicarbazone typically will crystallize from the cold solution on standing, the time varying from a few minutes to several hours. In the case of less reactive carbonyl compounds, the reaction can be accelerated by warming the mixture on a water bath for 10 min and then cooling in ice water. The crystals are filtered off and washed with a little cold water; they can usually be recrystallized from methyl or ethyl alcohol alone or mixed with water.

9.7.5e Oximes

Oximes do not have as wide a utility as other carbonyl derivatives for the identification of aldehydes and ketones. They are somewhat more difficult to obtain crystalline and are more likely to exist as mixtures of geometrical isomers. However, they do occasionally have utility and should be considered.

$$\begin{matrix} R \\ \diagdown \\ C{=}O + NH_2OH \\ \diagup \\ R' \end{matrix}$$

$$\longrightarrow \quad \begin{matrix} R \\ \diagdown \\ C{=}N \\ \diagup \qquad \diagdown \\ R' \qquad OH \end{matrix} \quad \text{and/or} \quad \begin{matrix} R \qquad OH \\ \diagdown \quad \diagup \\ C{=}N \\ \diagup \\ R' \end{matrix} \quad + H_2O$$

(E)- and (Z)- isomeric oximes

9.7.5f Procedure: Oximes

(a) The method described for preparation of semicarbazones can be employed. One substitutes hydroxylamine hydrochloride for the semicarbazide hydrochloride. The heating period is usually necessary.

(b) Reflux a mixture of 0.1 g of the aldehyde or ketone, 0.1 g of hydroxylamine hydrochloride, 2 mL of ethanol, and 0.5 mL of pyridine on a water bath for 15 to 60 min. Remove the solvent at reduced pressure or by evaporation with a current of air in a hood. Add several milliliters of cold water, and triturate thoroughly. Collect the oxime and recrystallize from alcohol, alcohol-water, or toluene.

9.7.5g Phenylhydrazones

Phenylhydrazones are recommended particularly for aryl carbonyl compounds whenever the dinitrophenylhydrazone is not suitable or adequate for identification. The manipulation of these derivatives should be as rapid as possible since they undergo slow decomposition in air.

9.7.5h Procedure: Phenylhydrazones

CAUTION: *Phenylhydrazine is a suspected carcinogen.*

In a test tube place 100 mg of aldehyde or ketone, 4 mL of methanol, and four drops of phenylhydrazine. Boil the mixture for 1 min, add one drop of glacial acetic acid, and boil gently for 3 min. Add cold water dropwise until a permanent cloudiness results, cool, collect the crystals, and wash with 1 mL of water containing one drop of acetic acid. Recrystallize the product immediately by dissolving in hot methanol, add water to the solution until cloudiness appears, cool, and scratch the sides of the tube if necessary to crystallize. Collect the crystals, wash with a few drops of dilute methanol, and determine the melting point as soon as possible. Dry the crystals by pressing between filter paper.

9.7.5i Reduction to alcohols

$$NaBH_4 + 4\ RR'C{=}O \xrightarrow[(2)\ NaOH,H_2O]{} 4RR'CHOH + NaBO_2$$

Alcohols are likely to be liquids, oils, or low-melting solids; however, there are occasions when the alcohols formed by reduction of aldehydes or ketones are well-defined crystalline solids suitable for identification purposes. Metal hydride reductions of carbonyl compounds to the corresponding alcohols generally proceed smoothly and give a high yield of the desired product. Sodium borohydride is less reactive than other hydrides such as lithium aluminum hydride, but it has the advantage of being stable in the presence of air and moisture. Sodium borohydride can be dissolved in cold water without extensive decomposition, whereas lithium aluminum hydride decomposes explosively on contact with water, and one must use an ether solvent for the reduction medium.

A commonly used method of reduction is to dissolve the organic compound in ethanol and to add an aqueous solution of sodium borohydride. Sodium borohydride, although stable in water, reacts at an appreciable rate with ethanol. It is therefore necessary to use an excess of the borohydride in order to offset losses sustained through solvolysis. Excess borohydride remaining in the reaction is destroyed by boiling the reaction mixture for a few minutes. The borate ester formed in the reduction is hydrolyzed during the boiling step by the addition of aqueous base.

9.7.5j Procedure: Sodium borohydride reduction

In a small test tube, dissolve 2 mmol of the carbonyl compounds in 3 to 5 mL of ethanol. In a second test tube, place 50 mg of sodium borohydride and dissolve in five drops of distilled water. Add the ethanol solution of the carbonyl compound to the borohydride solution. Shake the reaction mixture several times during a 15-min period. Add 0.5 mL of 6 N aqueous sodium hydroxide solution and a boiling chip to the reaction mixture. Boil it gently on the steam bath for 5 min. Pour the reaction mixture onto a small amount of ice, and collect the resulting precipitate by vacuum filtration. If the alcohol does not crystallize, it can be recovered by extraction with ether and then purified.

9.7.6 Special Methods of Separation and Purification of Aldehydes and Ketones

Most aldehydes, alkyl methyl ketones, and unhindered cyclic ketones react with saturated sodium hydrogen sulfite (bisulfite) solution to form crystalline *bisulfite addition compounds.* Other aliphatic and aromatic ketones do not yield addition compounds.

The reaction is reversible, and the aldehyde or ketone can be recovered by reaction of the addition compound with aqueous sodium carbonate or dilute hydrochloric acid. The sequence of formation and decomposition of the addition compounds can be used for the purification and/or separation of aldehydes and certain ketones from other materials.

9.7.6a Procedure: Bisulfite addition compounds

Shake or stir thoroughly the aldehyde- or ketone-containing mixture with a saturated solution of sodium hydrogen sulfite. The temperature will rise as the exothermic addition reaction takes place. Extract the residual organic material into ether and recover. Collect the crystalline precipitate, wash it with a little ethanol followed by ether, and allow it to dry.

To decompose the addition compound, use 10% sodium carbonate solution or dilute hydrochloric acid.

The reagent will undergo Michael addition to α,β-unsaturated carbonyl compounds to yield salts of sulfonic acids.

$$\text{C=C-C-} \xrightarrow{\text{NaHSO}_3} \text{Na}^+ \, ^-\text{O-S-C-C-C-}$$

Another chemical method for the separation of aldehydes and ketones from other neutral, water-insoluble compounds involves the use of *Girard's reagent T.* The reagent reacts with carbonyl compounds to yield water-soluble quaternary ammonium hydrazones. The water-insoluble compounds are then separated by ether extraction. The aldehyde or ketone is regenerated by hydrolysis with dilute acid.

$$(\text{CH}_3)_3\overset{+}{\text{N}}\text{CH}_2\text{CONHNH}_2 \ \text{Cl}^- + \ \text{C=O} \ \longrightarrow$$

Girard's reagent T

$$\text{C=NNHCOCH}_2\overset{+}{\text{N}}(\text{CH}_3)_3 \ \text{Cl}^- \xrightarrow[\Delta]{\text{HCl}} \ \text{C=O}$$

water soluble

9.7.6b Procedure: Girard's reagent T

A solution of 0.5 g of impure aldehyde or ketone, 0.5 g (or slight excess) of Girard-T reagent, and 0.5 mL of acetic acid in 5 mL of 95% ethanol is refluxed for 30 min. Cool and pour the reaction mixture into a separatory funnel. Add water, ether, and sufficient saturated sodium chloride solution to avoid emulsification. Shake well. Separate the layers and recover the ether-soluble material. Treat the aqueous layer with 1 mL of concentrated hydrochloric acid and heat on the steam bath to effect hydrolysis of the derivative back to the aldehyde or ketone. Isolate the aldehyde or ketone by extraction.

Table 9.7. Physical Properties of Aldehydes and Derivatives

Liquids

Compound	bp(°C)†	Semi-carbazone	2,4-Dinitro-phenyl-hydrazone	p-Nitro-phenyl-hydrazone	Phenyl-hydra-zone	Oxime	Methone
Acetaldehyde (Ethanal)	20.2	162	148(168)	128.5	57(63, 99)	47	140
Propionaldehyde (Propanal)	49.0	89	155	125	(oil)	40	155
Glyoxal (mp 15°C)	50	270	328		180	178	228
Acrolein	52.4	171	165	151	52 (pyra-zoline)		192
2-Methylpropanal	64	125–6	187	131	(oil)		154
Butanal	74.7	106(96)	123	95(87)			135(142)
2,2-Dimethylpropanal	75	190	209	119		41	
3-Methylbutanal	92.5	107	123	111	(oil)	48.5	154
2-Methylbutanal	92–3	103	120				
Chloral	98	90d	131			56(40)	
Pentanal	103		107(98)			52	104.5
Hexanal	131	106	104(107)			51	108.5
Tetrahydrofurfural	144/740 mm	166					123
Heptanal	153	109	108	73		57	135(103)
Furfural	161.7 (90/65 mm)	202	230(214)	154(127)	97	92(76)	160
Cyclohexanecarboxyaldehyde	162	173				91	
Octanal	171 (81/32 mm)	101(98)	106	80		60	90
Benzaldehyde	179	222	237	192	158(155)	35	193
Nonanal	185	100(84)	100			64(69)	86
Salicylaldehyde	197	231	252	227	142	57(63)	208
m-Tolualdehyde	199	204(224)		157	91(84)	60	
o-Tolualdehyde	200	218(212)	194	222	101(106)	49	
p-Tolualdehyde	204	234(215)	234	200.5	114(121)	80	
n-Decanal	209	102	104			69	91.7
Methoxybenzaldehyde	248	210	253–4d	160–1	120–1	133	145
Cinnamaldehyde (trans)	252	215(208)	255d	195	168	138	213

Solids

mp(°C)

Compound	mp(°C)	Semi-carbazone	2,4-Dinitro-phenyl-hydrazone	p-Nitro-phenyl-hydrazone	Phenyl-hydra-zone	Oxime	Methone
1-Naphthaldehyde (bp 292°C)	34	221		234	80	90	
Phenylacetaldehyde (bp 195°C)	34	156	121	151	63(58,102)	99(103)	165
o-Nitrobenzaldehyde	44	256	250(265)	263	156	102(154)	
p-Chlorobenzaldehyde (bp 214.5–216.5°C)	47	230	265	237(220)	127	110(146)	
Phthalaldehyde	56				191(di)		
2-Naphthaldehyde	60	245	270	230	206(218)	156	
p-Aminobenzaldehyde	72	153			156	124	
p-Dimethylaminobenzaldehyde	74	222	325	182	148	185	
m-Hydroxybenzaldehyde	102	198	260d	221–2 147	131	90	
p-Nitrobenzaldehyde	106	221	320	249	159	133	

† bp at 760 mm Hg pressure unless otherwise noted.
* "d" denotes decomposition.

Table 9.8. Physical Properties of Ketones and Derivatives

Liquids

Compound	bp(°C)	Semi-carbazone	2,4-Dinitro-phenyl-hydrazone	p-Nitro-phenyl-hydrazone	Phenyl-hydrazone	Oxime
			mp(°C) of Derivatives			
2-Propanone (Acetone)	56.11	190	126	149	42	59
2-Butanone	80	135	117	129		(bp 152°C)
3-Buten-2-one	81	141				
Butane-2,3-dione	88	279(di)	315(di)	230(mono)	243(di)	76(mono)
						245–6(di)
2-Methyl-3-butanone	94.3	114	120	109	(oil)	(oil)
3-Pentanone	102	138–9	156	144(139)	(oil)	(bp 165°C)
2-Pentanone	102.3	112	144	117	(oil)	(bp 167°C)
3,3-Dimethyl-2-butanone	106	158	125		(oil)	75(79)
4-Methyl-2-pentanone	116.8	132(134)	95			(bp 176°C)
2,4-Dimethyl-3-pentanone	124	160	88(98)			
2-Hexanone	128	125(122)	110	88	(oil)	49
4-Methylpent-3-en-2-one	130	164(133)	200	134	142	49
Cyclopentanone	130.7	210	146	154	55(50)	56.5
2,4-Pentanedione	139	122(mono)				149(di)
		209(di)				
4-Heptanone	144	132	75			(bp 193°C)
3-Hydroxy-2-butanone	145	185(202)	318			
1-Hydroxy-2-propanone	146	196	128.5	173	103	71
2-Heptanone	151.2	123	89		207	
Cyclohexanone	156	167	162	147	81	91
3,5-Dimethyl-4-heptanone	162(173)	84				
2-Methylcyclohexanone	165	197d	137	132		43
2,6-Dimethyl-4-heptanone	168	122(126)	66(92)			
3-Methylcyclohexanone	170	179	155	119	94	
4-Methylcyclohexanone	171	203		128	110	39
2-Octanone	173	122	58	93		
Cycloheptanone	181	163	148	137		23
Ethyl acetoacetate	181	133(129d)	93			
5-Nonanone	187	90				
2,5-Hexanedione	194	185d(mono)	257(di)	212(di)	120(di)	137(di)
		224(di)				
Acetophenone (mp 20°C)	202	198–9(203)	240(250)	184–5	106	60
1-(2-Methylphenyl)ethanone	214	203	159			
1-(4-Methylphenyl)ethanone	214	203–5	159			
Isophorone	215	199.5d			68	79.5
1-Phenyl-2-propanone (mp 27°C)	216					
Propiophenone (mp 20°C)	218	182(174)	191			54
n-Butyrophenone (mp 13°C)	221	188				50
Isobutyrophenone	222	181	163		73	94
Isovalerophenone	226	210				72(64.5)
1-Phenyl-2-butanone	226	135(146)				
1-Phenyl-1-butanone	230	191	190			50

Table 9.8 (Continued)

Liquids

Compound	bp(°C)	Semi-carbazone	2,4-Dinitro-phenyl-hydrazone	p-Nitro-phenyl-hydrazone	Phenyl-hydrazone	Oxime
				mp(°C) of Derivatives		

Solids

Compound	mp(°C)	Semi-carbazone	2,4-Dinitro-phenyl-hydrazone	p-Nitro-phenyl-hydrazone	Phenyl-hydrazone	Oxime
1,3-Diphenyl-2-propanone (bp 330°C)	34	146(126)	100		121(129)	125
α-Naphthyl methyl ketone (bp 302°C)	34	229			146	139
4-Phenyl-3-buten-2-one (bp 262°C)	42	188	227	165–7	159	117
1-Indanone	42	233	258	234–5	130–1	146
Benzophenone (bp 360°C)	48	167	238–9	154–5	137–8	144
Phenacyl bromide	50	146				89.5(97.0)
Benzalacetophenone	58	168	245		120	140(116)
4-Methylbenzophenone	60	122	202.4		109	154(115)
Desoxybenzoin (bp 320°C)	60	148	204	163	116	98
1-Phenyl-1,3-butanedione	61					
4-Methoxybenzophenone	62		180	199	132(90)	147(138) 115–6
1,3-Diphenyl-1,3-propanedione	78	205				
Fluorenone	83		284	269	152	196
Benzil	95	175(182) (mono) 243–4(di)	189	290	134(mono) 255(di)	138(mono) 237(di)
Benzoquinone	116	243(di) 166(mono) 178(mono)	231(di) 185.6(mono)		152	240
1,4-Naphthoquinone	125	247(mono)	278(mono)	279(mono)	206(mono)	198(mono)
Benzoin	137	206	245(234)		159(syn) 106(anti)	99(syn) 152(anti)
Anthraquinone	286				183	

Girard-P reagent can be employed in a like manner.

Girard's reagent P

Although not recommended as a general procedure, it is sometimes advantageous to isolate a carbonyl compound as one of its derivatives. Procedures have been developed for regenerating the original carbonyl compound by derivative exchange reactions with levulinic, pyruvic, and other keto acids,[31] or acetone.[32]

[31] E. Hershberg, *J. Org. Chem.*, **13**, 542 (1948); C. H. DePuy and B. W. Ponder, *J. Am. Chem. Soc.*, **81**, 4629 (1959).

[32] S. R. Maynez, L. Pelavin, and G. Erker, *J. Org. Chem.*, **40**, 3302 (1975).

9.7.6c Procedure: Derivative Exchange with
          Acetone

One gram of the derivative is dissolved in 5 mL of acetone and is refluxed for 24 hr. The acetone is evaporated and 30 mL of pentane is added to the residue. The precipitate is removed by filtration, and the pentane is removed from the filtrate by distillation or evaporation. The resulting carbonyl compound is purified by recrystallization or distillation. (Oximes require longer reflux times.)

Ketones can be conveniently regenerated from 2,4-dinitrophenylhydrazones by reaction with aqueous titanium trichloride.[33]

Tables 9.7 and 9.8 list the physical properties of aldehydes and ketones, respectively, and their derivatives.

9.8
QUINONES

9.8.1a Classification

Quinones are colored crystalline compounds (most are yellow) with a penetrating odor. Most *o*- and *p*-quinones have a carbonyl band in the infrared near 1670 cm^{-1} (6.0 μm) (Fig. 9.13). Quinones that have carbonyls in different rings, for example, 2,6-naphthoquinone, appear near 1645 cm^{-1} (6.08 μm). Some *p*-benzoquinones exhibit doublet carbonyl absorptions. The ultraviolet spectrum is often useful in establishing the presence and type of quinone as well as its substitution pattern (Sec. 4.3.3).

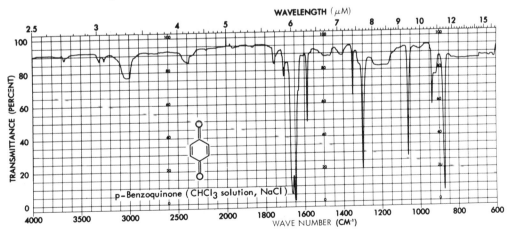

Fig. 9.13. Infrared spectrum of *p*-benzoquinone.

---

[33] J. E. Murray and M. Silvestri, *J. Org. Chem.*, **40**, 1502 (1976).

Most quinones liberate iodine from acidified potassium iodide solutions.

$$\text{(quinone)} + 2\,HI \longrightarrow \text{(hydroquinone)} + I_2$$

9.8.1b Characterization

Quinones form oximes and semicarbazones, usually from an aqueous alcohol medium. The carbonyl derivatives may not be of the usual structure, e.g.,

$$O_2N\text{—}\langle\,\rangle\text{—NHNH}_2 + O{=}\langle\,\rangle{=}O \longrightarrow O_2N\text{—}\langle\,\rangle\text{—N}{=}\text{N—}\langle\,\rangle\text{—OH}$$

Quinones can be readily *reduced to hydroquinones* with a variety of reducing reagents, e.g., zinc and dilute hydrochloric acid, or sodium dithionate.

$$\text{(naphthoquinone)} + Zn/HCl \text{ or } Na_2S_2O_4 \longrightarrow \text{(naphthohydroquinone)}$$

9.8.1c Procedure: Reduction to hydroquinones

(a) Suspend the quinone in dilute hydrochloric acid and add a little zinc dust. When the solution has turned colorless, neutralize with sodium hydrogen carbonate and extract the hydroquinone into ether. Evaporate the solvent and identify the hydroquinone (phenol-type derivatives can be made if necessary).

(b) Dissolve or suspend the quinone (0.5 g) in 5 mL of benzene or ether in a small separatory funnel. Add a solution of 1 g of sodium dithionate ($Na_2S_2O_4$) in 10 mL of 1 N sodium hydroxide. Shake the mixture until the characteristic color of the quinone has disappeared. Separate the aqueous alkaline solution, cool in an ice bath, and carefully acidify with concentrated hydrochloric acid. Collect the product (extract with ether if necessary), and recrystallize from dilute ethanol.

Like other 1,2-diketones, *o*-quinones react with *o*-phenylenediamine to afford *quinoxalines*.

$$\text{(o-phenylenediamine)} + \text{(o-quinone)} \longrightarrow \text{(quinoxaline)}$$

9.8.1d Procedure: Quinoxalines

Dissolve the *o*-quinone in alcohol or glacial acetic acid. Add an equivalent amount of *o*-phenylenediamine in alcohol. Warm the mixture on the steam bath for 15 to 20 min, cool, and dilute with water to crystallize. Recrystallize from aqueous ethanol.

The physical properties of some quinones appear in Table 9.8.

9.9
CARBOHYDRATES

*9.9.1*a Classification

Mono- and disaccharides are colorless solids or syrupy liquids. They are readily soluble in water but are almost completely insoluble in most organic solvents. The solids typically melt at high temperatures (over 200°C) with decomposition (browning and producing a characteristic caramel-like odor), and they char when treated with concentrated sulfuric acid. The polysaccharides possess similar properties but are generally insoluble or only slightly soluble in water.

The infrared spectra of carbohydrates obtained as mulls or potassium bromide pellets exhibit prominent hydroxyl absorption with correspondingly strong absorption in the C—O stretching region. Almost all sugars of four or more carbons exist in cyclic hemiacetal, acetal, hemiketal, or ketal forms, and hence exhibit no carbonyl band in the infrared (Fig. 9.14).

Sugars containing an aldehyde or α-hydroxy ketone grouping present in the free or hemiacetal (but not acetal) form are oxidized by Benedict's and Tollen's reagents (Sec. 9.7.2c). Such sugars are referred to as "reducing sugars."

Pentoses and hexoses are dehydrated by concentrated sulfuric acid to form furfural or hydroxymethylfurfural, respectively. In the *Molisch test*, these furfurals condense with 1-naphthol to give colored products.

9.9.1b Procedure: Molisch test

Place 5 mg of the substance in a test tube containing 0.5 mL of water. Add two drops of a 10% solution of 1-naphthol in ethanol. By means of a dropper, allow 1 mL of concentrated sulfuric acid to flow down the side of the inclined tube so that the

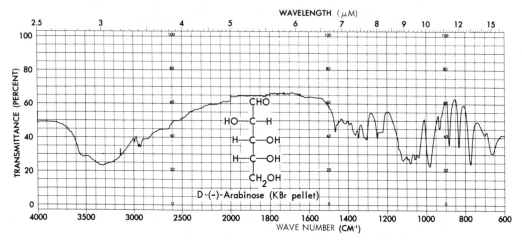

Fig. 9.14. Typical carbohydrate infrared spectrum. Note the absence of a carbonyl band.

heavier acid forms a bottom layer. If a carbohydrate is present, a red ring appears at the interface of the two liquids. A violet solution is formed on mixing. Allow the mixed solution to stand for 2 min; then dilute with 5 mL of water. A dull-violet precipitate will appear.

9.9.1c Characterization

Since sugars decompose on heating, they do not have well-defined characteristic melting points. The same is true of some of their derivatives, e.g., osazones. Fortunately, the number of sugars normally encountered in identification work is relatively small. Authentic samples of most are readily available for comparative purposes.

Thin-layer and paper chromatographic comparison[34] of an unknown with authentic samples provides an excellent means for tentative identification. For simple free sugars, paper chromatography is generally superior to thin-layer chromatography. Recommended solvent systems for paper chromatography of sugars include:

1. 1-Butanol, ethanol, water (40, 11, 19 parts by volume).
2. 1-Butanol, pyridine, water (9, 5, 4).
3. 2-Propanol, pyridine, water (7, 7, 5).
4. 1-Butanol, acetic acid, water (2, 1, 1).

---

[34] E. Heftmann, *Chromatography*, 2nd ed., Reinhold Publishing Corporation, New York, 1967; R. J. Block, E. L. Durram, and G. Zweig, *Paper Chromatography and Paper Electrophoresis*, Academic Press, New York, 1958; E. Lederer and M. Lederer, *Chromatography*, 2nd ed., Elsevier Publishing Company, Amsterdam, 1957.

For the *detection of reducing sugars* on paper chromatograms, the authors suggest the following spray.

9.9.1d Procedure: Spray for reducing sugars

The dried chromatogram is sprayed with a 3% solution of *p*-anisidine hydrochloride in 1-butanol and heated to 100°C for 3 to 10 min. Aldo- and ketohexoses, as well as other sugars, give different colored spots with this reagent.

The structure and stereochemistry of a number of sugars and sugar derivatives have been examined by NMR and mass spectrometry. The primary literature should be consulted for details. The relationship between dihedral angle and coupling constants between the hydrogens has been of significant utility in determining the stereochemical relationships of the various ring hydrogens in cyclic sugar derivatives.

The specific rotation of sugars and derivatives is a useful means of identification. Rotations must be measured under specified conditions of concentration, solvent, and temperature, employing pure samples.

9.9.1e Reaction of sugars with phenylhydrazine

Sugars containing an aldehyde or keto group (as the hemiacetal or hemiketal) react with an equivalent of phenylhydrazine in the cold to produce the corresponding phenylhydrazones. These derivatives are water-soluble and do not precipitate. Heating these sugars with excess (3 to 4 equivalents) phenylhydrazine produces *osazones* and *polyzones*.[35]

$$
\begin{array}{c}
\text{H} \\
| \\
\text{C}=\text{O} \\
| \\
\text{H}-\text{C}-\text{OH} \\
| \\
\text{R}
\end{array}
+ 3\ C_6H_5NHNH_2 \longrightarrow
\begin{array}{c}
\text{H} \\
| \\
\text{C}=\text{N}-\text{NHC}_6\text{H}_5 \\
| \\
\text{C}=\text{N}-\text{NHC}_6\text{H}_5 \\
| \\
\text{R}
\end{array}
+ C_6H_5NH_2 + 2\ H_2O + NH_3
$$

An osazone

It should be noted that, in the formation of osazones, one carbinol grouping is oxidized, and hence a number of isomeric sugars give the same osazone, for example, D-glucose, D-mannose, and D-fructose all produce the same osazone. The osazones exist in chelated structures in equilibrium with a nonchelated isomer.[35]

9.9.1f Procedure: Osazones

CAUTION: *Phenylhydrazine is a suspected carcinogen.*

Place a 0.1-g sample of the unknown sugar and a 0.2-g sample of a known sugar (suspected to be the unknown) in separate test tubes. To each sample add 0.2 g of phenylhydrazine

[35] O. L. Chapman, R. W. King, W. J. Welstead, Jr., and T. J. Murphy, *J. Am. Chem. Soc.*, **86**, 4968 (1964); O. L. Chapman, *Tetrahedron Letters*, 2599 (1966).

Table 9.9.   Physical Properties of Carbohydrates and their Phenylosazones

Compound		mp(°C)	$[\alpha]_D^{20}$ (water)*	Phenylosazone mp(°C)
D-Erythrose		Syrup	$+1 \rightarrow -14.5°(c = 11)$	164
L-Erythrulose		Syrup	$+11.4°(c = 2.4)$	164
α-L-Rhamnose(H$_2$O)		82–92	$-7.7 \rightarrow +8.9°$	222
D-Ribose		87(95)	$-19.7°(c = 4)$	164
2-Deoxy-D-ribose		90	$+2.88 \rightarrow -56.2°$	
β-Maltose(H$_2$O)		102–3	$+111.7 \rightarrow +130.4°(c = 4)$	206
D-Fructose		102–4	$-132.2 \rightarrow -92°(c = 2)$	210
D-Altrose		105	$+32.6(c = 7.6)$	178
α-D-Lyxose		106–7	$+5.5 \rightarrow -14.0°(c = 0.8)$	164
D-Threose		126–32	$? \rightarrow -12.3°(c = 4)$	164
β-D-Allose		128	$+0.58 \rightarrow +14.4°(c = 5)$	178
α-D-Mannose		133	$+29.3 \rightarrow +14.2°(c = 4)$	210
β-D-Mannose		132	$-17.0 \rightarrow +14.2°(c = 4)$	210
D-Xylose		144	$+92 \rightarrow +18.6°(c = 10)$	164
α-D-Glucose		146(anh.)	$+112.2 \rightarrow +52.7°(c = 10)$	210
β-D-Glucose		148–50	$+18.7 \rightarrow +52.7°(c = 10)$	210
β-L-Arabinose		158–9	$+173 \rightarrow +105(c = 3)$	163
L-Sorbose		165	$-43.7°(c = 5)$	156
α-D-Galactose		167	$+150.7 \rightarrow +80.2°$	196
Sucrose		169–70	$+66.5°(c = 2)$	
Lactose	α-form	223(anh.)	$+92.6 \rightarrow +52.3°(c = 4.5)$	200
	β-form	252(anh.)	$+34 \rightarrow 52.3°(c = 4.5)$	200

*When two rotations are shown, the first is the initial rotation and the second is that observed at equilibrium (mutarotation).

hydrochloride, 0.3 g of sodium acetate, and 2 mL of distilled water. Stopper the test tubes with vented corks, and place them together in a beaker of boiling water. Note the time of immersion and the time of precipitation of each osazone. Shake the tubes occasionally. The time required for the formation of the osazone can be used as evidence for the identification of the unknown sugar.

Under these conditions, fructose osazone precipitates in 2 min, glucose in 4 to 5 min, xylose in 7 min, arabinose in 10 min, and galactose in 15 to 19 min. Lactose and maltose osazones are soluble in hot water.

After 30 min, remove the tubes from the hot water and allow them to cool. Carefully collect the crystals, and compare the unknown with the known crystals under a low-power microscope. The melting points (decomposition) of osazones depend on the rate of heating and lie too close together to be of value.

Osazones can be converted to osotriazoles, which have sharp melting points.[36]

---

[36] W. T. Haskins, R. M. Hahn, and C. S. Hudson, *J. Am. Chem. Soc.*, **69**, 1461 (1947) and previous papers.

$$
\begin{array}{c}
\text{H}-\text{C}=\text{NNHC}_6\text{H}_5 \\
\mid \\
\text{C}=\text{NNHC}_6\text{H}_5 \\
\mid \\
\text{R}
\end{array}
\xrightarrow{\text{Cu}^{++}}
\begin{array}{c}
\text{H}-\text{C}=\text{N} \\
\mid \qquad\quad \text{N}-\text{C}_6\text{H}_5 \\
\text{C}=\text{N} \\
\mid \\
\text{R}
\end{array}
$$

An osotriazole

Other derivatives of value in identification of sugars include acetates, and acetates of thioacetals, benzoates, acetonides, and benzylidene derivatives. The trimethylsilyl ethers are useful for mass spectral and glc analysis. Several procedures are given under polyhydric alcohols (Sec. 9.3.2).

Table 9.9 lists the physical properties of carbohydrates and their derivatives.

9.10
CARBOXYLIC ACIDS

9.10.1a Classification

Acidic compounds containing only carbon, hydrogen, and oxygen are either carboxylic acids, phenols, or possibly enols. An indication of whether a water-insoluble compound is an acid or a phenol can be obtained from simple solubility tests. Both classes of compounds are soluble in sodium hydroxide, but only carboxylic acids are soluble in 5% sodium hydrogen carbonate with the liberation of carbon dioxide (for exceptions, see Sec. 8.4.4). Water-soluble acids also liberate carbon dioxide from sodium hydrogen carbonate solution. Other classes of compounds that liberate carbon dioxide from sodium hydrogen carbonate solution include salts of primary and secondary amines (these can be differentiated readily on the basis of the liberation of free amines, their melting point behavior, and their elemental analysis); sulfinic and sulfonic acids (differentiated on the basis of elemental analysis, infrared spectra, and their more acidic character); and a variety of substances such as acid halides and acid anhydrides which are readily hydrolyzed to acidic materials. Phenols and enols give positive color tests with ferric chloride solution. Fortunately, all these categories of acidic compounds can be readily differentiated from carboxylic acids on the basis of the extremely characteristic infrared spectrum of a carboxylic acid. Carboxylic acids give rise to a very broad and characteristic O—H band and a carbonyl band near 1700 cm^{-1} ($5.88 \ \mu\text{m}$) (Fig. 9.15). The chemist with even minimum spectroscopic experience will soon learn to distinguish other compounds from carboxylic acids by infrared spectroscopy.

9.10.1b Characterization

One of the simplest and most informative ways to characterize a carboxylic acid is to determine its *neutralization equivalent* (Sec. 8.6). The neutralization equivalent of an acid is its equivalent weight; the molecular weight can be determined from

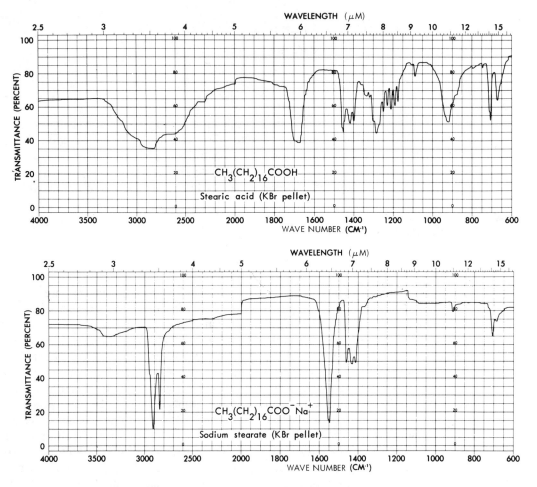

Fig. 9.15. Infrared spectra of a typical acid and its sodium salt. Note the shift in the carbonyl frequency.

the neutralization equivalent by multiplying that value by the number of acidic groups in the molecule.

Since the pK_a's of both the organic acid and the indicator are sensitive to solvent changes, one should employ only enough ethanol to dissolve the organic acid. With high concentrations of ethanol, sharp endpoints are not obtained with phenolphthalein. If it is necessary to employ absolute or 95% ethanol, brom-thymol blue should be used as the indicator.

For accurate results, a blank should always be run on the solvent, and one should take care that the neutralization equivalent is determined from a substance that is pure and anhydrous. Neutralization equivalents can be obtained with an accuracy of 1% or less.

With good technique, one can obtain an accurate neutralization equivalent with samples as small as 5 mg, employing more dilute standard alkali in burettes designed for greater accuracy with small volumes.

9.10.1c Procedure: Neutralization equivalents

Approximately 200 mg of the acid is accurately weighed and dissolved in 50 to 100 mL of water or aqueous ethanol. This solution is titrated with standard sodium hydroxide solution (approximately 0.1 N), employing phenolphthalein as the indicator or employing a pH meter.

$$\text{Neutralization equivalent} = \frac{\text{weight of sample in milligrams}}{\text{milliters of base} \cdot \text{normality}}$$

If an acid has been well characterized, it is often sufficient for identification purposes to obtain the neutralization equivalent and the melting point of a carefully chosen derivative. More than 70 types of derivatives have been suggested at various times for the identification of carboxylic acids. The majority of these derivatives fall in the categories of amides, esters, and salts of organic bases. Representative and recommended examples from each of these categories will be discussed in the following paragraphs.

9.10.1d Amides and substituted amides

Acids or salts of acids can be converted directly to acid chlorides by the action of thionyl chloride. The acid chlorides are converted to the *amides* and *substituted amides* by reaction with excess ammonia or an amine, as illustrated below.

$$RCOOH + SOCl_2 \longrightarrow RCOCl + SO_2 + HCl$$
$$RCOCl + H_2NR' \longrightarrow RCONHR' + HCl$$

9.10.1e Procedure: Amides from acids

(a) *Acid chloride.* In a 25-mL round-bottomed flask fitted with a condenser and a calcium chloride tube or a cotton plug in the top of the condenser to exclude moisture, place 0.5 to 1.0 g of the anhydrous acid or anhydrous sodium salt and add 2.5 to 5 mL of thionyl chloride. Reflux the mixture gently for 30 min. Arrange for distillation, and distill off the excess thionyl chloride (bp 78°C). For acids below four carbon atoms, the bp of the acid chloride may be too near that of thionyl chloride to afford practical separation by distillation. In this event, the excess of the reagent can be destroyed by the addition of formic acid.

$$HCOOH + SOCl_2 \longrightarrow CO + SO_2 + 2HCl$$

(b) *Primary amides.* For the preparation of primary amides, it is unnecessary to distill off the excess thionyl chloride. The entire reaction mixture can be cautiously poured into 15 mL of ice-cold concentrated ammonia. The precipitated amide is collected by vacuum filtration and purified by recrystallization from water or aqueous ethanol.

As an alternate method, the acid chloride is dissolved in 5 to 10 mL of dry toluene and the excess ammonia gas is passed through the solution. If the amide does not precipitate, it can be recovered by evaporation of the solvent.

(c) *Anilides and substituted anilides.* Dissolve the acid chloride in 5 mL of toluene and add a solution of 2 g of pure aniline, *p*-bromoaniline, or *p*-toluidine in 15 mL of toluene. It may be convenient to run this reaction in a small separatory funnel. Shake the reaction mixture with 5 mL of dilute hydrochloric acid to remove the excess aniline, wash the toluene layer with 5 mL of water, evaporate the solvent, and recrystallize the anilide from water or aqueous alcohol.

Anilides and *p-toluides* can also be prepared directly from the acids or from alkali metal salts of the acids by heating them directly with the aniline or the hydrochloride salt of the aniline.

$$RCOOH + C_6H_5NH_2 \longrightarrow RCOO^{-\,+}NH_3C_6H_5 \xrightarrow{\Delta} RCONHC_6H_5 + H_2O$$

$$RCOONa + C_6H_5NH_3^+Cl^- \xrightarrow{\Delta} RCONHC_6H_5 + NaCl + H_2O$$

9.10.1f Procedure: Anilides and *p*-toluides

Place 0.5 g of the acid and 1 g of aniline or *p*-toluidine in a small, dry, round-bottomed flask. Attach a short air condenser and heat the mixture in an oil bath at 140 to 160°C for 2 hr. Caution should be used to avoid heating the mixture too vigorously, thus causing loss of the acid by distillation or sublimation. If the material is available as the sodium salt, use 0.5 g of the salt and 1 g of the amine hydrochloride. If there is evidence that the substance is a diacid, use double the quantity of the amine and increase the reaction temperature to 180 to 200°C. At the end of the reaction time, cool the reaction mixture and triturate it with 20 to 30 mL of 10% hydrochloric acid, or dissolve the residue in an appropriate solvent and wash it with hydrochloric acid, dilute sodium hydroxide, and water. Then evaporate the solvent. The amides can usually be recrystallized from aqueous ethanol.

9.10.1g Esters

Solid esters form a second series of compounds useful in the identification and characterization of acids. The most commonly used are the *p*-nitrobenzyl and *p*-bromophenacyl esters, although the phenacyl, *p*-chlorophenacyl, and *p*-phenylphenacyl are also occasionally used. The halides corresponding to these esters, for example, the *p*-nitrobenzyl halides and the phenacyl halides, all undergo facile S_N2 displacement reactions. These esters are prepared in aqueous alcoholic solution by displacement of the corresponding halide with the weakly nucleophilic carboxylate anions, as illustrated in the following equation.

$$RCOO^- Na^+ + BrCH_2\overset{\overset{O}{\|}}{C}-\!\!\!\left\langle\right\rangle\!\!\!-Br \longrightarrow RCOOCH_2CO-\!\!\!\left\langle\right\rangle\!\!\!-Br + NaBr$$

The method is particularly advantageous because it does not require the acid to be anhydrous, and it can be run with equal facility with the alkali salts of the acids.

Although the phenacyl and benzyl esters are extremely useful for making derivatives of acids that cannot be easily obtained in anhydrous conditions (e.g., from saponification of esters of lower carboxylic acids) or directly from alkali metal salts, there are certain disadvantages that make them undesirable for routine use in identification. *The benzyl and phenacyl halides all have severe lachrymatory and blistering properties.* The formation of the esters is generally slow, and any unreacted halide is often difficult to separate from the ester. For this reason, less than one equivalent of the halide should be used, and the reaction should be continued for 1.5 to 2 hr to ensure completeness. The traces of the halides remaining with the esters impart irritating properties to the esters and severely depress the melting point.

9.10.1h Procedure: Phenacyl and
p-nitrobenzyl esters

In a small test tube or a small, round-bottomed flask equipped with a reflux condenser, place 1 mmol of the acid and 1 mL of water. Add one drop of phenol-phthalein, and carefully neutralize by the dropwise addition of 10% sodium hydroxide solution until the color of the solution is just pink. Add one or two drops of dilute hydrochloric acid to discharge the pink color of the indicator. Add an alcoholic solution of 0.9 mmol of the halide dissolved in 5 to 8 mL of alcohol. Reflux the solution for 1.5 to 2.5 hr, cool, add 1 mL of water, and scratch the sides of the tube. When precipitation is complete, collect the ester by filtration; wash with a small amount of 5% sodium carbonate solution and then several times with small quantities of cold water. The esters can usually be recrystallized from aqueous ethanol. A frequently used procedure is to dissolve the crystals in hot alcohol, filter, and add water to the hot filtrate until a cloudiness appears. Rewarm the solution until the cloudiness disappears, and then cool. Scratch the sides of the test tube, if necessary to induce crystallization. *Avoid handling or contact of the crystals with the skin.*

If the original acid is available as an alkali salt, dissolve one equivalent of it in a minimum amount of water, add a drop of phenolphthalein, and adjust the acidity as above.

Methyl esters are also occasionally used as derivatives. In certain cases these are solids, but in the majority of the cases they appear as oils. When the methyl esters are relatively high-melting solids, they are often very respectable derivatives for the purposes of melting-point determinations. Frequently one may wish to make the methyl ester for purposes of vapor phase chromatographic comparison of retention times of unknown esters with known methyl esters. The methyl esters have the advantage in that they are often more volatile, more easily separated, and they elute as sharper peaks from the vapor phase chromatograph than do the corresponding acids. Methyl esters can be made quantitatively on a very small scale, employing diazomethane. They can also be made conveniently on a small scale employing 2,2-dimethoxypropane with a small amount of p-toluenesulfonic or concentrated hydrochloric acid.

$$RCOOH + CH_2N_2 \longrightarrow RCOOCH_3 + N_2$$

$$RCOOH + CH_3\overset{\displaystyle OCH_3}{\underset{\displaystyle OCH_3}{\overset{\displaystyle |}{\underset{\displaystyle |}{C}}}}CH_3 \xrightarrow{\;H^+\;} RCOOCH_3 + CH_3OH + CH_3COCH_3$$

9.10.1i Salts

Alkylthiuronium halides prepared by heating alkyl halides with thiourea react with the sodium or potassium salts of carboxylic acids in aqueous ethanol to yield the corresponding alkyl thiuronium carboxylates. These salts are easily made and are formed in good yield and high purity. Regrettably, they possess melting points that are very close together. The most commonly used is the S-benzylthiuronium salt; sometimes the S-p-bromo- and S-p-chlorobenzylthiuronium, and S-1-naphthylmethylthiuronium salts are made.

$$C_6H_5CH_2-S-C\overset{\overset{+}{N}H_2}{\underset{NH_2}{\big\langle}}\; Cl^- + RCOONa \longrightarrow C_6H_5CH_2-S-C\overset{\overset{+}{N}H_2}{\underset{NH_2}{\big\langle}}\; RCOO^- + NaCl$$

9.10.1j Procedure: S-Benzylthiuronium salts

CAUTION: *Thiourea is a suspected carcinogen.*

Reagent: S-Benzylthiuronium chloride, although commercially available, can be easily prepared by refluxing a mixture of 2 g of benzyl chloride, 1.2 g of thiourea, and 3 mL of methanol for 30 min. Cool in an ice bath. Collect the product and wash several times with small portions of ethyl acetate.

 Dissolve or suspend 0.25 g of the acid or its sodium or potassium salt in 5 mL of water. Adjust the pH to phenolphthalein endpoint with 1 N sodium hydroxide. Add several drops of 0.1 N hydrochloric acid until the solution shows a very pale pink coloration (almost neutral). Add a solution of 1 g of S-benzylthiuronium chloride in 5 mL of water or 10 mL of methanol. Cool in an ice bath. Recrystallize the salt from dilute ethanol.

 A second series of salts that have had some utility in the identification of organic acids are the phenylhydrazonium salts. Salts of phenylhydrazine are obtained from the stronger aliphatic acids such as α-chlorocarboxylic acids, from sulfonic acids, and from aliphatic dibasic acids when they are warmed with a toluene solution of phenylhydrazine. Phenylhydrazine can also be used to convert acids directly into phenylhydrazides. At the boiling point of phenylhydrazine (243°C), simple aliphatic acids form phenylhydrazides.

$$RCOOH + C_6H_5NHNH_2 \longrightarrow \underset{\text{Phenylhydrazonium salt}}{C_6H_5NH\overset{+}{N}H_3RCOO^-}$$

$$RCOOH + C_6H_5NHNH_2 \xrightarrow{\Delta} \underset{\text{Phenylhydrazide}}{RCONHNHC_6H_5}$$

9.10.2 Salts of Carboxylic Acids

9.10.2a Classification

Metallic salts of carboxylic acids—the most commonly encountered are the sodium and potassium salts—will generally be suspected from their solubility, melting point, and ignition-test behavior. They are quite water-soluble, providing slightly alkaline solutions from which free acids sometimes precipitate upon neutralization with mineral acid. They are insoluble in most organic solvents. The salts melt or decompose only at very high temperatures. They leave a light-colored alkaline residue of oxide or carbonate upon ignition. If necessary, the identity of the metal can be made by the standard tests outlined in inorganic qualitative analysis texts or by flame photometry.

Occasionally one encounters the carboxylic acid salts of ammonia or amines. The ammonia or amine is liberated from the salt on dissolution in aqueous alkali. Place 20 to 30 mg of the sample in a small test tube containing 0.5 mL of dilute sodium hydroxide. Ammonia or a low-molecular-weight amine can be detected by placing a piece of dampened litmus paper in the vapor produced by warming the tube with a small flame. For a more sensitive test, dampen filter paper with a 10% solution of copper sulfate and hold the paper in the vapors. A positive test is indicated by the blue color of the ammonia or amine complex of copper sulfate. The amine can be isolated by extraction of a basic solution of the salt and identified as described in Sec. 9.14.

The infrared spectra of acid salts, usually taken as Nujol mulls, show strong carbonyl bands at 1610 to 1550 cm^{-1} (6.21 to 6.45 μm) and near 1400 cm^{-1} (7.14 μm) (Fig. 9.15).

9.10.2b Characterization

For characterization of carboxylic acid salts, it is necessary to rely on the chemistry and physical constants of the acid. Addition of sufuric or other strong mineral acids to an aqueous solution of a salt of a carboxylic acid liberates the free acid, which can be isolated by extraction, steam distillation, or filtration. Suitable derivatives are then made from the free acid. The sodium or potassium

Table 9.10. **Physical Properties of Carboxylic Acids and Derivatives**

Name of acid	bp(°C)*	mp(°C)	Amide	mp(°C) of Derivatives Anilide	p-Toluide
Acetic	118	16	82	114	147
Acrylic	140	13	85	105	141
Propionic	141		81	106	126
Propynoic	144d	18		87	62
2-Methylpropanoic	154		128	105	107
Methacrylic	161	16	106		
Butyric	162.5		115	96	75
3-Methylbutyric	176			137	107
Pentanoic	186		106	63	74
Dichloroacetic	194		98	118	153
Hexanoic	205		82	95	75
Heptanoic	223		96	70	81
Octanoic	239		110	57	70
Nonanoic	255		99	57	84
Oleic	216/5 mm		76	41	42
Undecanoic	284	30	103(99)	71	80
4-Oxopentanoic (levulinic)	246	35	108d	102	109
Decanoic	270	31	108(98)	70	78
2,2-Dimethylpropanoic	164	35	178	129	120
Dodecanoic (lauric)	299	44	110(102)	78	87
trans-9,10-Octadecenoic (elaidic)	234/15 mm	45	94		
(Z)-2-Methyl-2-butenoic (angelic)	185	46	127–8	126	
3-Phenylpropanoic	280/754 mm	48	105	98	135
4-Phenylbutanoic	290	52	84		
Tetradecanoic	202/16 mm	54	107(103)	84	93
Trichloroacetic	197	58	141	97	113
Hexadecanoic	222/16 mm	63	107	90	98
Chloroacetic	189	63	121	134	162
Phenylacetic	265	76.5	156	118	136
Glutaric	200/20 mm	97	137	224	218
Phenoxyacetic		100	148	99	
Oxalic (dihydrate)		101	419d(di) 210(mono)	254(di) 148–9(mono)	268
o-Toluic		104	143	125	144
m-Toluic		113	97	126	118
D,L-Mandelic		118 133(D or L)	132	152	172
Benzoic		122	110	163	158
Maleic		130	266(di)	187(di)	142(di)
Furancarboxylic		134	143	123.5	107.5
Cinnamic (trans)		133	148	153	168
Malonic		135	50(mono) 170(di)	132(mono) 230(di)	86(mono) 235(di)
o-Nitrobenzoic		146	176	155	
Diphenylacetic		148	168	180	173

Table 9.10 (Continued)

Name of acid	bp(°C)*	mp(°C)	Amide	mp(°C) of Derivatives Anilide	p-Toluide
α-Naphthoic		162	202	163	
4-Nitrophthalic		165	200d	192	172(mono)
D-Tartaric		171	196d	180d(mono)	
				264d(di)	
p-Toluic	275	180	160	148	160(165)
Acetylenedicarboxylic		179	294d(di)		
4-Methoxybenzoic		186	167	169–71	186
2-Naphthoic		185.5	195	171	192
Butanedioic (succinic)	235d	186	157(mono)	148(mono)	180(mono)
		185	260(di)	230(di)	255(di)
p-Aminobenzoic		188	193(di)		
Terephthalic		300		237	
Fumaric		302	266d(di)	314	

*bp at 760 mm Hg pressure unless otherwise noted.

salts can be converted directly to thiuronium salts (Sec. 9.10.1h) or phenacyl esters (Sec. 9.10.1i).

Table 9.10 lists the physical properties of carboxylic acids and their derivatives.

9.11 ACID ANHYDRIDES AND ACID HALIDES

9.11.1a Classification

The combination of high reactivity and unique spectral properties greatly facilitates the classification and identification of acid halides and acid anhydrides.

Acid anhydrides characteristically exhibit two bands in the carbonyl region of the infrared spectrum (Fig. 9.16); in acyclic aliphatic anhydrides, these bands appear near 1820 and 1760 cm^{-1} (5.49 and 5.68 μm). They show the usual variation with unsaturation and ring size. The relative intensity of the two bands is variable; the higher-wave number band is stronger in acyclic anhydrides, and the lower-wave number band is usually stronger in cyclic anhydrides. Diacyl peroxides also exhibit double carbonyl bands in the infrared (Sec. 9.12).

Acid chlorides, which are by far the most common of the acid halides, have a strong infrared band near 1800 cm^{-1} (5.56 μm) (Fig. 9.17). The band is only slightly shifted to lower wave number on conjugation; in aroyl halides, a prominent shoulder usually appears on the lower-wave number side of the carbonyl band. As would be predicted, acid fluorides are shifted to higher wave numbers, whereas the bromides and iodides are at lower wave numbers. The presence of

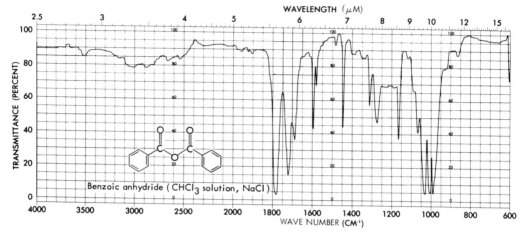

Fig. 9.16. Infrared spectrum of benzoic anhydride. Note the benzoic acid impurity that shows up as a broad weak OH band and a carbonyl band near 1690 cm⁻¹.

this strong carbonyl band and the absence of bands in the O—H, N—H, and C—O regions of the spectrum, the presence of reactive halogen (silver nitrate test, Sec. 9.2.1b), and the high reactivity of the substance with water, alcohols, and amines provide sufficient diagnosis for the acid halide grouping.

9.11.1b Characterization

The most commonly encountered anhydrides are the simple symmetrical acetic and benzoic anhydrides and the cyclic anhydrides of succinic, maleic, and phthalic acids. Mixed anhydrides formed from two carboxylic acids or from a carboxylic acid and a sulfonic acid are known, but are seldom encountered.

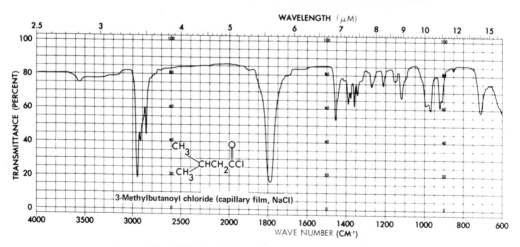

Fig. 9.17. Typical infrared spectrum of an acid chloride.

Table 9.11. Physical Properties of Acid Anhydrides and Acid Chlorides

Acid Anhydrides	bp(°C)
Trifluoroacetic	39
Acetic	140
Propanoic	167
2-Methylpropanoic	182
Butanoic	198
2-Methyl-2-butenedioic	214
Pentanoic	218
trans-2-Butenoic	248
Hexanoic	254
Benzoic (mp 42°C)	360
Maleic (mp 52°C)	200(82/14 mm)
Glutaric (mp 56°C)	
Phenylacetic (mp 72°C)	

Acid Chlorides	bp(°C)
Acetyl	51
Oxalyl	64
Propanoyl	80
2-Methylpropanoyl	92
Butanoyl	101
2,2-Dimethylpropanoyl	105
Chloroacetyl	108
Dichloroacetyl	108
3-Methylbutanoyl	115
Pentanoyl	126
Succinoyl	190d
Benzoyl	197
Phenylacetyl	210
Phthaloyl (mp 15°C)	281
3,5-Dinitrobenzoyl (mp 69°C)	
p-Nitrobenzoyl (mp 75°C)	150/15 mm

* bp at 760 mm Hg pressure unless otherwise noted.

Acid halides and anhydrides are most often converted directly to amides for identification purposes. Recall that, in the identification of a carboxylic acid, the acids are usually converted to the acid chlorides in order to prepare amides. The anilides are highly recommended (Sec. 9.10.1d).

$$RCOCl + 2 NH_2C_6H_5 \longrightarrow RCONHC_6H_5 + C_6H_5\overset{+}{N}H_3Cl^-$$

$$(RCO)_2O + 2 NH_2C_6H_5 \longrightarrow RCONHC_6H_5 + C_6H_5\overset{+}{N}H_3CH_3COO^-$$

Cyclic anhydrides, on reaction with amines, may give the monamide or the imide, depending upon the conditions of the reaction.

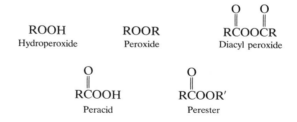

Acid halides and anhydrides react with alcohols and phenols to produce esters. With acid halides, the reaction is often rapid and exothermic. These esters can serve for identification purposes if the alcohol or phenol is chosen so that the product is an appropriate melting solid. In the case of cyclic anhydrides, acid esters are produced; phthalic anhydride reacts with alcohols to give alkyl hydrogen phthalates.

The acid halides and anhydrides are hydrolyzed with water to produce the corresponding acids which, if solid, serve as excellent derivatives. In many cases, some warming may be necessary to complete the reaction. The hydrolysis can be effected by a quantitative reaction employing standard alkali.

Table 9.11 lists physical properties of acid chlorides and anhydrides.

9.12
PEROXIDES

9.12.1a Classification

All organic peroxides should be regarded as being potentially explosive. Safety glasses, gloves, and shields should be used at all times when working with organic peroxides.

<table>
<tr><td></td><td></td><td>O   O</td></tr>
<tr><td></td><td></td><td>‖   ‖</td></tr>
<tr><td>ROOH</td><td>ROOR</td><td>RCOOCR</td></tr>
<tr><td>Hydroperoxide</td><td>Peroxide</td><td>Diacyl peroxide</td></tr>
</table>

<table>
<tr><td>O</td><td>O</td></tr>
<tr><td>‖</td><td>‖</td></tr>
<tr><td>RCOOH</td><td>RCOOR′</td></tr>
<tr><td>Peracid</td><td>Perester</td></tr>
</table>

Most organic peroxides[37] can be detected by acidified starch-iodide paper. The dialkyl peroxides can be detected only after they have been hydrolyzed to hydroperoxides by strong acid.

In the infrared region, the —O—O— stretching frequency appears near 877 cm^{-1} (11.4 μm). Because of the symmetry of the grouping, the absorption is generally weak. It is often obscured by, or confused with, other skeletal vibrations that occur in the same region. The O—H absorption in hydroperoxides is very

---

[37] A. G. Davies, *Organic Peroxides*, Butterworths, London, 1961.

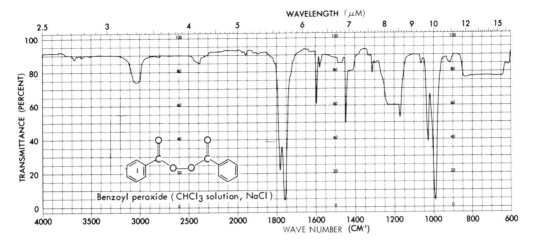

Fig. 9.18. Infrared spectrum of benzoyl peroxide. Compare with the spectrum of benzoic anhydride (Fig. 9.16).

similar to that in alcohols. The peracids exhibit an intramolecularly hydrogen-bonded O—H band at 3280 cm^{-1} ($3.05 \mu\text{m}$) and a carbonyl near 1750 cm^{-1} ($5.71 \mu\text{m}$). In aliphatic peresters, the carbonyl band appears near 1770 cm^{-1} ($5.65 \mu\text{m}$), somewhat higher than that of normal esters. The symmetrical diacyl peroxides exhibit a doublet carbonyl in the infrared. The aliphatic diacyl peroxides absorb near 1815 and 1790 cm^{-1} (5.51 and $5.59 \mu\text{m}$); the aromatic compounds absorb at slightly lower wave numbers (Fig. 9.18). In linear diacylperoxides, the lower-wave number band is stronger and sharper (see anhydrides, Sec. 9.11).

9.12.1b Characterization

Identification of peroxides is usually accomplished by quantitative reduction procedures. The reduced products, acids, alcohols, etc., can be characterized by the standard methods. Iodometric methods are usually employed for the *quantitative estimation of peroxides.*

$$\text{ROOH} + 2\text{I}^- + \text{H}^+ \longrightarrow \text{ROH} + \text{H}_2\text{O} + \text{I}_2$$

9.12.1c Procedure: Titration of peroxides

The peracid or hydroperoxide (0.1 to 0.2 g) is accurately weighed in an iodine flask. The sample is dissolved in 10 mL of dichloromethane and the flask is flushed with nitrogen. Saturated sodium iodide solution (2 mL) is added, followed by 15 mL of glacial acetic acid. The flask is stoppered and permitted to stand in the dark for 5 to 10 min. Water (50 mL) is added, and the solution is titrated to starch endpoint with $0.1 N$ thiosulfate. A blank determination should be run on the reagents.

This procedure can be adapted for use with peresters and diacyl peroxides, provided the acetic acid employed contains 0.002% ferric chloride hexahydrate as a catalyst for the reduction.[38]

Dialkyl peroxides can be estimated if they are treated with hydrogen iodide in acetic acid at 60°C for 45 to 60 min.

$$\text{Percent "active oxygen"} = \frac{\text{milliliters} \cdot N \cdot 0.8}{\text{weight of sample in grams}}$$

9.13
ESTERS

9.13.1a Classification

The majority of esters are liquids or low-melting solids, many with characteristic flowery or fruity odors. The presence of the ester function can usually be established by infrared spectroscopy. Esters have strong carbonyl bands in the infrared region in the 1780 to 1720 cm^{-1} (5.62 to 5.81 μm) region accompanied by two strong C—O absorptions in the 1300 to 1050 cm^{-1} (7.69 to 9.52 μm) region (Fig. 9.19). The higher-wave number C—O band is usually stronger and broader than the carbonyl band. As a general rule, in the spectra of aldehydes, ketones, and amides, the carbonyl band is the strongest in the spectrum (Figs. 9.11, 9.12, 9.19 and 9.23).

Carbonyl compounds whose infrared spectra have the foregoing characteristics, which lack NH and OH absorption, and which do not give a positive

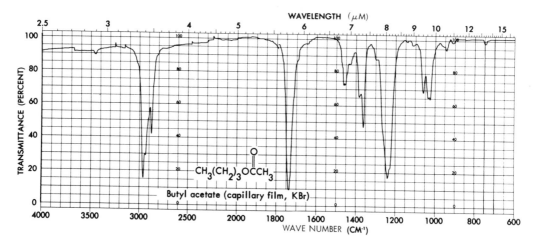

Fig. 9.19. Typical infrared spectrum of an ester.

---

[38] L. S. Silbert and D. Swern, *Anal. Chem.* **30**, 385 (1958); L. S. Silbert and D. Swern, *J. Am. Chem. Soc.*, **81**, 2364 (1959).

2,4-dinitrophenylhydrazine test, are most likely esters. An ester spectrum is most likely to be mistaken for that of a ketone. If the NMR spectrum of the suspected ester is available, evidence for an ester can be provided by the chemical shifts and coupling patterns of aliphatic hydrogens attached to the ethereal oxygen (near $\delta4$) and hydrogens *alpha* to the carbonyl group (near $\delta2$). Esters, like other carbonyl compounds, exhibit very characteristic fragmentation patterns in mass spectrometry.

9.13.1b Characterization

The principal procedure for the characterization of an ester involves the identification of the parent acid and alcohol. Because it is often difficult to separate and purify the hydrolysis products of esters on a small scale, it is usually advantageous to prepare derivatives of the acid and alcohol portion by reactions on the original ester. The sample should be apportioned accordingly. Fortunately, a large number of the esters encountered are derivatives of simple acids (e.g., acetates) or alcohols (e.g., methyl or ethyl esters); it is often possible to unequivocally establish the presence of an acetyl or methyl or ethyl grouping by spectroscopic means.

Aromatic esters can sometimes be identified through solid derivatives prepared directly by aromatic substitution (nitration, halogenation, etc.), thus eliminating the necessity of preparing separate derivatives of both the alcohol and acid portion.

The *saponification equivalent* (the molecular weight divided by the number of ester moieties in the molecule) is determined on a known weight of an ester by hydrolysis with an excess of standard alkali; the excess alkali is then back-titrated with standard acid. The method is exceedingly useful in limiting the number of structural possibilities for an unknown; regrettably, it is tedious and time-consuming.

$$\text{Saponification equivalent} = \frac{\text{weight of ester in milligrams}}{(\text{milliliters} \cdot \text{normality})_{\text{alkali}} - (\text{milliliters} \cdot \text{normality})_{\text{acid}}}$$

Of the two methods given here for determination of the saponification equivalent, the diethylene glycol method is often more advantageous because of faster reaction times, a more stable standard solution, and less likelihood of loss of esters because of transesterification to the more volatile ethyl esters. However, the viscous glycol solution is somewhat more difficult to transfer with accuracy. With either procedure, ground-glass equipment should be employed. If such equipment is not available, rubber stoppers that have been boiled with alkali and then thoroughly washed with distilled water can be employed.

9.13.1c Procedure: Saponification equivalent (diethylene glycol)

Reagent: Dissolve 6.0 g of potassium hydroxide pellets in 30 mL of diethylene glycol with the aid of heat. Use a thermometer as a stirring rod, and keep the temperature below 130°C to prevent discoloration of the solution. Transfer the solution to a glass-stoppered bottle, and add 70 mL of diethylene glycol. Mix the solution thoroughly, and allow it to cool. Standardize with 0.5 N hydrochloric acid. Pipette 10 mL of the reagent (ca. 1 N) into a 125-mL Erlenmeyer flask, add 15 mL of water, and titrate to phenolphthalein endpoint. Because of the viscosity of the solution, a large-bore pipette is desirable. Alternatively, the standardization and saponification can be conducted with accurately weighed quantities of the glycol solution.

Procedure: For the determination of the saponification equivalent of the ester, transfer approximately 10 mL of the standardized reagent, accurately weighed or measured, into a 50-mL, glass-stoppered Erlenmeyer flask. Quantitatively transfer about 0.5 g of the ester, accurately weighed, into the flask. While holding the stopper in place with swirling, heat the mixture in an oil bath that has been preheated to 120 to 130°C. After 2 or 3 min, remove the flask from the bath, shake well, allow to drain, and carefully remove the stopper to allow the internal pressure to escape. Replace the stopper and continue heating in the oil bath until the temperature of the reaction mixture is about 120°C. If the ester is high-boiling, the stopper can be removed and a thermometer inserted. After several minutes at this temperature, cool the flask to 80 to 90°C; remove the stopper and wash it well with distilled water, allowing the rinsing to drain into the flask. Add 10 to 15 mL of distilled water to the flask and several drops of phenolphthalein indicator, and titrate with standard hydrochloric or sulfuric acid.

9.13.d Procedure: Saponification equivalent (ethanol)

Reagent:- Place 8 g of potassium hydroxide in 250 mL of ethyl alcohol (95%), and shake to effect dissolution. Allow the mixture to stand for 24 hr; then decant or filter the clear solution from the residue of potassium carbonate, It is necessary to standardize the ethanolic potassium hydroxide solution (approximately 0.5 N) immediately before use. Either standard hydrochloric or sulfuric acid, using phenolphthalein as the indicator, is suitable.

Procedure: Place an accurately weighed sample of the ester (about 0.5 g) in a 100-mL, round-bottomed flask. Add 25 mL of standard 0.5 N alcoholic potassium hydroxide by means of a pipette (CAUTION: *Be sure to use a bulb pipette control*). Attach an efficient reflux condenser and reflux gently on a steam bath for 1.5 to 2 hr, or until the hydrolysis is complete. Allow the mixture to cool. Introduce about 25 mL of water through the condenser. Remove the condenser, add two to three drops of phenolphthalein, and titrate with standard acid.

9.13.2 Derivatives of the Acyl Moiety

9.13.2a Hydrolysis to acids

If the acid obtained by hydrolysis of the ester is a solid, it will serve as an excellent derivative (mp and neutralization equivalent). The acid can be obtained by quantitative or qualitative application of the procedures for the determination of saponification equivalents, or by *hydrolysis with aqueous sodium hydroxide.* The potassium hydroxide/diethylene glycol procedure is especially effective for water-insoluble esters boiling above 200°C.

9.13.2b Procedure: Hydrolysis of esters

Into a small, round-bottomed flask place 0.2 to 1 g of the ester; add 2 to 10 mL of 25% aqueous sodium hydroxide and a boiling chip. Attach a condenser and reflux for 30 min for esters boiling below 110°C, 1 to 2 hr for esters boiling between 110 to 200°C, or until the oily layer and/or the characteristic ester odor disappears. Cool the flask and acidify with dilute acid. Phosphoric is recommended because of the high solubility of the sodium phosphate. Recover the acid by filtration or extraction, and purify by suitable means.

Alkyl esters can be dealkylated under mild conditions by treatment with trimethylsilyl iodide.[39]

$$R'COOR + (CH_3)_3SiI \longrightarrow R'COOSi(CH_3)_3 + R—I$$

$$\downarrow H_2O$$

$$R'COOH + (CH_3)_3SiOH$$

9.13.2c Conversion to amides

Esters can be heated with aqueous or alcoholic ammonia to produce amides. Some simple esters react on standing or at reflux; however, most must be heated under pressure.

$$RCOOCH_3 + NH_3 \longrightarrow RCONH_2 + CH_3OH$$

Many esters can be converted to crystalline *N-benzylamides* by refluxing with benzylamine in the presence of an acid catalyst. This if often the most preferred method of making a derivative of the acyl moiety.

$$RCOOR' + CH_3OH \xrightarrow[\text{or HCl}]{CH_3O^-} RCOOCH_3 + R'OH$$

$$RCOOCH_3 + C_6H_5CH_2NH_2 \xrightarrow{NH_4Cl} RCONHCH_2C_6H_5 + CH_3OH$$

9.13.2d Procedure: N-Benzylamides from esters

Into a test tube or a 10-mL round-bottomed flask equipped with a reflux condenser, place 100 mg of ammonium chloride, 1 mL or 1 g of the ester, 3 mL of benzylamine, and a boiling stone. Reflux gently for 1 hr. Cool and wash the reaction mixture with water to remove the excess benzylamine and soluble salts. If crystallization does not occur, add one or two drops of dilute hydrochloric acid. If crystallization still does not occur, it may be due to excess ester. It is often possible to remove the ester by

---

[39] M. E. Jung and M. A. Lyster, *J. Am. Chem. Soc.*, **99**, 968 (1977).

Table 9.12.   Physical Properties of Esters

Name of compound	mp(°C)	bp(°C)	D_4^{20}	n_D^{20}
Ethyl formate	−79.4	54.2	0.92247	1.35975
Ethyl acetate	−83.6	77.15	0.90055	1.372
Methyl propanoate	−87.5	79.9	0.9151	1.3779
Methyl acrylate		80.3	0.961	1.3984
Isopropyl acetate	−73.4	91	0.873	1.377
Methyl 2-methylpropanoate	−84.7	92.6	0.8906	1.3840
t-Butyl acetate		97.8	0.867	1.386
Ethyl propanoate	−73.9	99.1	0.8889	1.3853
Ethyl acrylate		101	0.9136	1.4059
Methyl methacrylate	−50	101	0.936	1.413
Propyl acetate	−95	101	0.8834	1.38468
Allyl acetate		104	0.9276	1.40488
sec-Butyl acetate		112	0.872	1.3865
Isobutyl acetate		117.2	0.8747	1.39008
Ethyl butanoate	−100.8	121.6	0.87917	1.40002
Butyl acetate	−73.5	126.1	0.881	1.3947
Isoamyl acetate		142	0.8674	1.40034
Amyl acetate	−70.8	148.8	0.8756	1.4031
Methyl hexanoate	−71.0	151.2	0.88464	1.405
Cyclohexyl acetate		175	0.970	1.442
Butyl chloroacetate		175	1.081	
Ethyl acetoacetate		181	1.025	1.41976
Dimethyl malonate	−62	181.5	1.1539	1.41398
Dimethyl succinate	18.2	196.0	1.1192	1.41965
Phenyl acetate		196.7	1.078	1.503
Methyl benzoate	−12.5	199.5	1.089	1.5164
Ethyl benzoate	−34.2	213.2	1.0465	1.506
Benzyl acetate		217	1.055	1.5200
Diethyl succinate	−21	217.7	1.0398	1.41975
Diethyl fumarate	0.2	218.4	1.052	1.44103
Methyl phenylacetate		220	1.068	1.507
Diethyl maleate	−17	222.7	1.066	1.44156
Allyl benzoate		230	1.067	
Ethyl cinnamate	6.5	271	1.0490	1.55982
Diethyl phthalate		283.8	1.191	1.5138
Benzyl benzoate	21	324	1.1224	1.5681

boiling the reaction mixture for a few minutes in an evaporating dish. Collect the solid N-benzylamide by fitration, wash it with a little hexane, and recrystallize from aqueous alcohol or acetone.

The aminolysis of esters higher than ethyl is usually quite slow. In these cases, it is best to convert the ester to the methyl ester before attempting to make the benzylamide. This can be done by refluxing 1 g of ester with 5 mL of absolute methanol in which about 0.1 g of sodium has been dissolved, or by refluxing in 3% methanolic hydrogen chloride. Remove the methanol by distillation, and treat the residue as described above.

$$O_2N\text{—}\underset{O_2N}{\overset{}{\bigcirc}}\text{—COOH} + R'COOR \xrightarrow{\;H_2SO_4\;} O_2N\text{—}\underset{O_2N}{\overset{}{\bigcirc}}\text{—COOR} + R'COOH$$

9.13.3 Derivatives of the Alcohol Moiety

Occasionally the alcohol derived from the ester by hydrolysis is a solid, in which case it and subsequent derivatives can serve to identify that portion of the ester. In general, alcohols above four carbons can be recovered from aqueous hydrolysates by extraction procedures. In many cases, it is much more convenient to identify alcohol moiety by a transesterification reaction with 3,5-dinitrobenzoic acid to produce the *alkyl 3,5-dinitrobenzoate.*

9.13.3a Procedure: 3,5-Dinitrobenzoates from esters

Mix about 0.5 g of the ester with 0.4 g of 3,5-dinitrobenzoic acid, and add one drop of concentrated sulfuric acid. If the ester boils below 150°C, heat to reflux with the condenser in place. Otherwise run the reaction in an open test tube placed in an oil bath at 150°C. Stir the mixture occasionally. If the acid dissolves within 15 min, heat the mixture for 30 min; otherwise continue the heating for 1 hr. Cool the reaction mixture, dissolve it in 20 mL of ether, and extract thoroughly with 10 mL of 5% sodium carbonate solution. Wash with water and remove the ether. Dissolve the crystalline or oily residue in 1 to 2 mL of hot ethanol and add water slowly until the 3,5-dinitrobenzoate begins to crystallize. Cool, collect, and recrystallize from aqueous ethanol if necessary.

Table 9.12 lists physical properties of esters.

9.14 AMINES

9.14.1a General classification

Most simple amines are readily recognized by their solubility in dilute mineral acids. However, many substituted aromatic amines (e.g., diphenylamine) and aromatic nitrogen heterocycles, which might formally be classified as amines, fail to dissolve in dilute acids. (The parent ring system of the latter can often be determined from their characteristic ultraviolet spectra.) Water-soluble amines can be detected by their basic reaction to litmus or other indicators, or by the *copper ion test.*

9.14.1b Procedure: Copper ion test

Add 10 mg or a small drop of the unknown to 0.5 mL of a 10% solution of copper sulfate. A blue to blue-green coloration or precipitate is indicative of an amine. The test can be run as a spot test on filter paper that has been treated with the copper sulfate solution. Ammonia also gives a positive test.

Infrared spectroscopy can be very useful in the recognition and classification of amines (Figs. 9.20 and 9.21). Primary amines, both aliphatic and aromatic, exhibit a weak but recognizable doublet in the 3500 to 3300 cm^{-1} (2.86 to 3.03 μm) region (symmetric and asymmetric NH stretch) and strong absorption due to NH bending in the 1640 to 1560 cm^{-1} (6.10 to 6.41 μm) region (Figs. 5.12 and 9.20). The 3500 to 3000 cm^{-1} (2.8 to 3.3 μm) bands in aliphatic amines are quite broad due to hydrogen bonding, but are reasonably sharp in aromatic amines. Secondary amines exhibit a single band in the 3450 to 3310 cm^{-1} (2.90 to 3.02 μm) region; aromatic amines absorb nearer the higher end of the range, and the aliphatics absorb at the lower end. The NH bending band in secondary amines in the 1580 to 1490 cm^{-1} (6.33 to 6.71 μm) region is weak and generally of no diagnostic value. Tertiary amines have no generally useful characteristic absorptions. (However, the absence of characteristic NH absorptions is a helpful observation in the classification of a compound known to be an amine.) The presence of a tertiary amine grouping can be established by infrared examination of the hydrochloride or other proton-acid salt of the amine; $\overset{+}{\geq}$NH absorption occurs in the 2700 to 2250 cm^{-1} (3.70 to 4.45 μm) region.

Color reactions with sodium nitroprusside have been recommended as classification tests to differentiate between the various classes of amines.[40]

9.14.1c Hinsberg test

This test is based on the fact that primary and secondary amines react with arenesulfonyl halides to give N-substituted sulfonamides. The tertiary amines do not give derivatives. The sulfonamides from primary amines react with alkali to form soluble salts (the sodium salts of certain primary alicyclic amines and some long chain alkyl amines are relatively insoluble and give confusing results).

Primary amines

$$\text{ArSO}_2\text{Cl} + \text{RNH}_2 + 2\,\text{NaOH} \longrightarrow \underset{\text{Soluble}}{[\text{ArSO}_2\text{NR}]^-\text{Na}^+} + \text{NaCl} + 2\,\text{H}_2\text{O}$$

$$[\text{ArSO}_2\text{NR}]^-\text{Na}^+ + \text{HCl} \longrightarrow \underset{\text{Insoluble}}{\text{ArSO}_2\text{NHR}} + \text{NaCl}$$

---

[40] R. L. Baumgarten, C. M. Dougherty, and O. N. Nercissian, *J. Chem. Ed.*, **54**, 189 (1977).

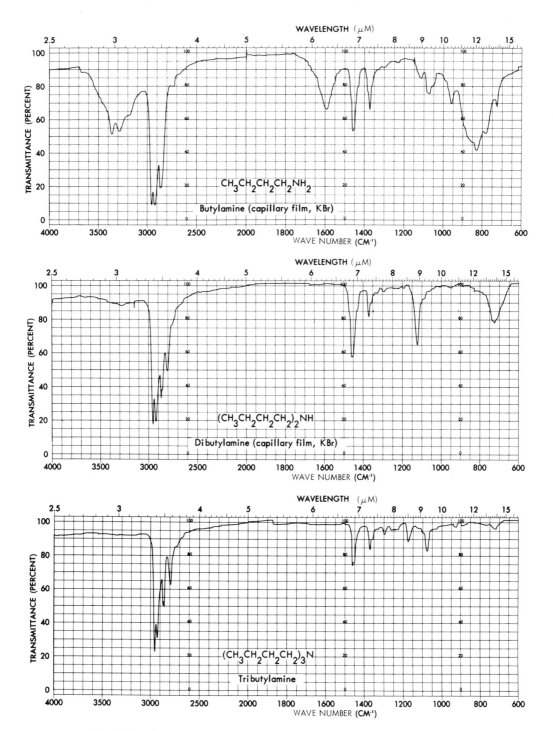

Fig. 9.20. Infrared spectra of a primary, secondary, and tertiary amine. In the spectrum of the primary amine note the strong NH stretching, bending, and deformation bands. The NH stretching and bending bands of secondary amines are frequently very weak and may not be observed at the usual cell thickness.

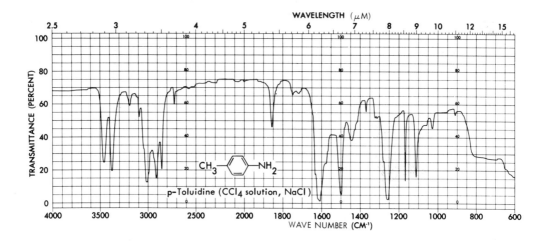

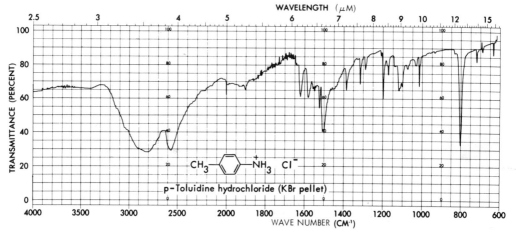

Fig. 9.21. Typical infrared spectra of an aromatic amine and its hydrochloride. In the spectrum of *p*-toluidine note the strong 1,4-disubstitution bands in the 2000 to 1660 cm^{-1} (5 to 6 μm) region.

Secondary amines

$$ArSO_2Cl + R_2NH + NaOH \longrightarrow \underset{\text{Insoluble}}{ArSO_2NR_2} + H_2O + NaCl$$

Tertiary amines

$$ArSO_2Cl + R_3N + NaOH \longrightarrow \text{tertiary amine recovered}$$

9.14.1d Procedure: Hinsberg test

To 0.2 g or 0.2 mL of the amine and 5 mL of 10% KOH is added 0.4 mL of benzenesulfonyl chloride. Stopper the mixture and shake, with cooling if necessary, until the odor of the benzenesulfonyl chloride is gone. At this point, the mixture should still be strongly basic. If not, add small volumes of the base until basic.

If the mixture has formed two layers, separate the two layers and test the organic layer for solubility in 10% HCl. Tertiary amines will be soluble. Some higher-molecular-weight amines form benzenesulfonamides that have low solubility in aqueous base. They may give the same solubility characteristics as the benzenesulfonamides of secondary amines. In order to differentiate between these possibilities, the separated aqueous phase is brought to pH 4. A precipitate indicates that a benzenesulfonamide of a primary amine was formed.

If the original mixture did not separate into two layers, a soluble salt of a benzenesulfonamide derived from a primary amine was formed. This can be substantiated by bringing the pH to 4, in which case the benzenesulfonamide of the primary amine will precipitate.

This procedure can be scaled up and used to separate mixtures of primary, secondary, and tertiary amines. Unreacted tertiary amines can be recovered by solvent extraction or steam distillation. The sulfonamides of primary and secondary amines can be separated by taking advantage of the solubility of the primary derivative in dilute sodium hydroxide. The original primary and secondary amines can be recovered by hydrolysis of the sulfonamides with 10 parts of 25% hydrochloric acid, followed by neutralization and extraction. The sulfonamides of primary amines require 24 to 36 hr of reflux; those of secondary amines require 10 to 12 hr for hydrolysis.

For a discussion of the reaction of benzenesulfonyl chloride with tertiary amines, see C. R. Gambill, T. D. Roberts, and H. Shechter, *J. Chem. Ed.*, **49**, 287 (1972).

9.14.1e Nitrous acid

A number of N-nitroso amines (R₂N—NO) are toxic and/or carcinogenic! Because of the high risk involved, we do not recommend treating amines of unknown structure with nitrous acid.

The following equations illustrate the reactions of nitrous acid with amines. The reactions are the basis for a classical method of distinguishing between various types of amines. The details of test procedures can be found in most organic texts and organic qualitative analysis texts.

Primary amines

$$RNH_2 \xrightarrow{\text{HONO}} R-N_2^+ \xrightarrow{-N_2} R^+ \longrightarrow \text{Carbonium ion products}$$

$$ArNH_2 \xrightarrow{\text{HONO}} ArN_2^+ \xrightarrow[\beta\text{-naphthol}]{OH^-}$$

—N=N—Ar

Red azo dye

Secondary amines

$$R_2NH \xrightarrow{HONO} R_2N-NO$$
(yellow oils or solids)

Tertiary amines

$$R_3N \xrightarrow{HONO} R_3\overset{+}{N}-NO \xrightarrow{H_2O} R_2NNO + \text{aldehydes, ketones}$$

The greatest utility is the demonstration of the presence of primary aliphatic amino group (by the loss of N_2) and primary aromatic amine group (through coupling with reactive aromatic nuclei such as β-naphthol to form azo dyes).

9.14.1f Characterization

Note: A number of the procedures for the preparation of derivatives of amines are quick and simple methods that can be run on a very small scale. For this reason, it is often expedient to use certain of these reactions as a combination classification-derivation procedure.

In many cases, with simple amines the combination of boiling or melting point, solubility in acid, and infrared spectrum provides sufficient information to allow the investigator to choose an appropriate derivative. In other cases, it is wise to examine selected classification tests and to obtain additional physical (e.g., pK_b) and spectral (NMR or ultraviolet) data before proceeding. See Sec. 6.4.1b for NH chemical shifts. In some cases the NH proton signals are broadened by quadruple interactions with the nitrogen to the extent that the signal can barely be differentiated from the base line. In other cases, the signal is much sharper. If there is any doubt in the assignment of the peak due to protons on nitrogen, the investigator may find it helpful to exchange the nitrogen protons for deuterium (add a drop of D_2O to the NMR tube and shake) or to examine the spectrum of a salt of the amine. A simple technique for the latter is to use trifluoroacetic acid as solvent; with this solvent and tertiary amines, the $\geq\overset{+}{N}H$ shows up between $\delta 2.3$ and 2.9.

In ultraviolet spectra of aromatic amines, a dramatic change is observed in going from the free base to the protonated salt.

λ_{max}230, 280 nm λ_{max}203, 254 nm

9.14.2 Derivatives of Primary and Secondary Amines

9.14.2a Amides and related derivatives

The most commonly employed derivatives are the *benzene-* and *p-toluenesulfonamides*. These can be made by the procedure outlined earlier for the Hinsberg test. The product should be recrystallized from 95% ethanol.

Benzyl- (*α-toluene-*), *p-bromobenzene-*, *m-nitrobenzene-*, *α-naphthalene-*, and *methanesulfonamides* are also occasionally used.

9.14.2b Procedure: Sulfonamides

Reflux for 5 to 10 min a mixture of 1 mmol of the sulfonyl chloride and 2.1 mmol of the amine in 4 mL of toluene. Allow the mixture to cool. Filter off the amine hydrochloride that precipitates. Evaporate the toluene filtrate to obtain the crude sulfonamide. Recrystallize from 95% ethanol.

Primary and secondary amines react with isothiocyanates to yield *substituted thioureas.* The most useful derivatives are obtained from phenyl- and 2-naphthylisothiocyanate; these reagents have the advantage of being insensitive to water and alcohols.

$$RNH_2 + ArN{=}C{=}S \longrightarrow ArNH\overset{\overset{\displaystyle S}{\|}}{-}C{-}NHR$$

9.14.2c Procedure: Thioureas

(a) Reflux a solution of 1 mmol of the isothiocyanate and 1.2 mmol of the amine in 2 mL of alcohol for 5 to 10 min. Add water dropwise to the hot mixture until a permanent cloudiness occurs. Cool, and scratch with a glass rod if necessary, to induce crystallization. Collect the solid. Wash with 1 to 2 mL of hexane. Recrystallize from hot alcohol.

(b) Mix equal amounts of the amine and isothiocyanate in a test tube and shake for 2 min. If no reaction occurs, heat the mixture gently over a small flame for 1 to 2 min. Cool in an ice bath, and purify by recrystallization.

Primary and secondary amines react with aryl isocyanates to produce substituted ureas. In contrast to the isothiocyanates, the isocyanates react rapidly with water and alcohols. Traces of moisture in the amines result in the formation of diaryl ureas, which are often difficult to separate from the desired product.

$$R_2NH + ArN{=}C{=}O \longrightarrow ArNHCONR_2$$

$$2\,ArN{=}C{=}O + H_2O \longrightarrow ArNHCONHAr + CO_2$$

Both aliphatic and aromatic primary and secondary amines readily react with acid anhydrides or acid chlorides to form substituted amides. The reaction of benzoyl chloride with amines under Schotten-Baumann conditions (aqueous sodium hydroxide) produces *benzamides,* which are generally excellent derivatives. The *p-nitro-* and *3,5-dinitrobenzamides* are also useful derivatives.

$$\text{C}_6\text{H}_5-\text{COCl} + \text{R}_2\text{NH} \xrightarrow{\text{NaOH}} \text{C}_6\text{H}_5-\text{CONR}_2$$

9.14.2d Procedure: Benzamides

CAUTION: *Benzoyl chloride is a lacrymator and should be used and stored in a hood.*

(a) About 1 mmol of the amine is suspended in 1 mL of 10% sodium hydroxide, and 0.3 to 0.4 mL of benzoyl chloride is added dropwise, with vigorous shaking and cooling. After about 5 to 10 min, the reaction mixture is carefully neutralized to about pH 8 (pH paper). The *N*-substituted benzamide is collected, washed with water, and recrystallized from ethanol-water.

(b) Benzoyl, *p*-nitrobenzoyl, or 3,5-dinitrobenzoyl chloride (2 to 3 mmol) is dissolved in 2 to 3 mL of toluene; 1 mmol of the amine is added, followed by 1 mL of 10% sodium hydroxide. The mixture is shaken for 10 to 15 min. The crude amide is obtained by evaporation of the toluene layer. Alternatively the amine and aroyl chlorides are mixed with 2 mL of pyridine, and the mixture is refluxed for 30 min. The reaction mixture is poured into ice water, and the derivative is collected and recrystallized from ethanol-water.

Most water-insoluble primary and secondary amines form crystalline *acetamides* upon reaction with acetic anhydride.

$$\text{RNH}_2 + (\text{CH}_3\text{CO})_2\text{O} \xrightarrow{\text{NaOH}} \text{CH}_3\text{CONHR}$$

9.14.2e Procedure: Acetamides

Dissolve about 0.2 g of water-insoluble amine in 10 mL of 5% hydrochloric acid. Add sodium hydroxide (5%) with a dropping pipette until the mixture becomes cloudy, and remove the turbidity by the addition of a few drops of 5% hydrochloric acid. Add a few chips of ice and 1 mL of acetic anhydride. Swirl the mixture and add 1 g of sodium acetate (trihydrate) dissolved in 2 mL of water. Cool in an ice bath and collect the solid; recrystallize from ethanol-water.

Primary and secondary amines react with 3-nitrophthalic anhydride to provide phthalamic acids; those from primary amines dehydrate when heated to 150°C to form *N*-alkyl 3-nitrophthalimides.

9.14.2f Salts

The amine salts of a number of proton acids are useful for purposes of identification or purification. Some salts have reproducible melting or decomposition points; others may be useful only as solid derivatives that can be carefully purified for purposes of microanalysis, spectroscopic studies, X-ray crystallography, etc.

Hydrochlorides can be prepared by passing dry hydrogen chloride (from a tank or from a sodium chloride/sulfuric acid generator) into an ether, toluene, or isopropyl alcohol solution of primary, secondary, or tertiary amines. The hydrobromides and hydroiodides are usually more hygroscopic than the hydrochlorides.

The *fluoroborates* (BF_4^-) and *perchlorates* (ClO_4^-) are often well-defined crystalline substances. *All perchlorates should be handled as potentially explosive substances.* Melting points of perchlorates should be taken with very small samples and, in general, should not be determined for perchlorates melting over 230°C.

Picrates are frequently employed as derivatives of the tertiary amines. Picric acid forms addition compounds with primary and secondary amines as well, but extensive data are available only for the tertiary amines. (Picric acid also forms crystalline complexes with amine oxides, certain aromatic hydrocarbons, and aryl ethers.)

9.14.2g Procedure: Picrates

Dissolve 1 mmol of the amine in 5 mL of 95% ethanol. Add 5 mL of a saturated solution of picric acid in 95% ethanol. Heat the solution to boiling and allow to cool slowly. The yellow crystals of the picrate can be recrystallized from ethanol or methanol if necessary.

Ammonium tetraphenylborates are quite water-insoluble. The addition of an aqueous solution of sodium tetraphenylborate to a solution of an amine salt (usually the hydrochloride) causes the immediate precipitation of the amine as the tetraphenylborate salt. The method can be adapted as a means of detection of amines and amine salts, including quaternary ammonium salts. The melting points of a number of ammonium tetraphenylborates have been recorded.[41]

$$RNH_2 \xrightarrow{HCl} R\overset{+}{N}H_3 \ Cl^- \xrightarrow{NaB(C_6H_5)_4} R\overset{+}{N}H_3 \ B(C_6H_5)_4^-$$

$$R_4N^+ X^- \xrightarrow{NaB(C_6H_5)_4} R_4N^+ \ B(C_6H_5)_4^-$$

9.14.2h Procedure: Tetraphenylborates

Dissolve or suspend the amine (0.1 g) in 5 to 10 mL of water. Add sufficient hydrochloric acid to dissolve the amine and to adjust the solution to pH 2 or 3. Slowly add, with stirring, an aqueous solution of sodium tetraphenylborate until precipitation of the salt is complete. Collect the white precipitate, wash well with distilled water, and dry below 60°C.

---

[41] F. E. Crane, Jr., *Anal. Chem.*, **28**, 1794 (1956).

Table 9.13. Physical Properties of Primary and Secondary Amines and Derivatives

Name of Compound	mp(°C)	bp(°C)	Acetamide	Benzamide	Benzene-sulfon-amide	p-Toluene-sulfon-amide	Phenyl thiourea
Isopropylamine		33		71	26		101
t-Butylamine		46		134			120
Propylamine		49		84	36	52	63
Methylisopropylamine		50		(oil)			120
Diethylamine		56		42	42		34
Allylamine		58			39	64	98
sec-Butylamine		63		76	70	55	101
Butylamine		77					65
Diisopropylamine		86					
Pyrrolidine		89		(oil)		123	69
Pentylamine		104					
Piperidine		106	(oil)	48	94	96	101
Dipropylamine		109	(oil)		51		69
Diallylamine		112					
Ethylenediamine		116	172(di) 51(mono)	244(di) (249)	168(di)	360(di) 123(mono)	102
Hexylamine		129		40	96		77
Morpholine		130		75	118	147	136
Cyclohexylamine		134	104	149	89		148
Diisobutylamine		139	101		57		113
Heptylamine		155·	86				75
Dibutylamine		159					86
Octylamine		180					
Aniline		184	114	160	112	103	54
Benzylamine		185	60	105	88	116	156
N-Methylaniline		196	102	63	79	94	87
β-Phenylethylamine		198	51	116	69		135
o-Toluidine		200	112	43(146)	124	108	136
m-Toluidine		203	65	125	95	114	
N-Ethylaniline		205	54	60	(oil)	87	89
o-Ethylaniline		216	111	147			
p-Ethylaniline		216	94	151		104	
2,4-Dimethylaniline		217	133	192	130	81	152(133)
3,5-Dimethylaniline		220	144	136			153
N-Propylaniline		222	47		54		104
o-Methoxyaniline		225	88	60(84)	89	127	136
Dibenzylamine		300		112	68	159	
				Solids			
N-Benzylaniline	37	298	58	107	119	149	103
p-Toluidine	45	200	147	158	120	118	141
1-Naphthylamine*	50		159	160	167	157(147)	165
Indole	52	253	158	68	254		
4-Aminobiphenyl*	53		171	230			
Diphenylamine	53–4		101	180	124	141	152
p-Anisidine	58	240	130	157	95	114	157(171)
2-Aminopyridine	60	204	71	165			
o-Nitroaniline	71		94	110(98)	104	142	

Table 9.13 (Continued)

Name of Compound	mp(°C)	bp(°C)	Acetamide	Benzamide	Benzene-sulfon-amide	p-Toluene-sulfon-amide	Phenyl thiourea
					mp(°C) of Derivatives		
p-Chloroaniline	72		179(172)	192	122	95(119)	152
o-Phenylenediamine	102		185(di)	301(di)	185	260(di)	
2-Naphthylamine*	112		132	162	102	133	129
Benzidine*	127		317(di)	252(di)	232(di)	243(di)	
			199(mono)	205(mono)			
o-Toluidine	129		314	265			
p-Nitroaniline	147		215	199	139	191	
o-Aminophenol	174		209(201)	167	141	146	
			124(di)				
p-Aminophenol	184s†		150(di)	217			150
			168(mono)				
p-Aminobenzoic acid	188		251d	278	212		

* *These compounds have been shown to be carcinogenic!* The use and handling of these compounds have been restricted by the federal government.
Extreme caution should be exercised in the handling of all suspected amines.
† Sublimes.

Table 9.13 lists physical properties of primary amines, secondary amines, and their derivatives.

9.14.3 Derivatives of Tertiary Amines

For solid derivatives of tertiary amines, salts are usually prepared. Quaternary ammonium salts, formed by alkylation of the tertiary amine with methyl iodide, benzyl chloride, or methyl p-toluenesulfonate, have been used. The most extensive data are available on the *methiodides*.

$$R_3N + CH_3I \longrightarrow R_3\overset{+}{N}CH_3\ I^-$$

9.14.3a Procedure: Methiodides

CAUTION: *Methyl iodide can cause serious toxic effects on the central nervous system. Methyl iodide should be used in an efficient hood, and care should be taken to avoid exposure of the skin to liquid methyl iodide.*

(a) A mixture of 0.5 g of the amine and 0.5 g of methyl iodide is warmed in a test tube over a low flame or in a water bath for a few minutes and then cooled in an ice bath. The tube is scratched with a glass rod, if necessary, to induce crystallization. Recrystallize from alcohol or ethyl acetate.

(b) A tertiary amine-methyl iodide mixture dissolved in toluene is refluxed until precipitation of the methiodide is complete.

The salts formed between tertiary amine and protonic acids form an important category of derivatives. By far the most important are the *picrates*. Extensive data are also available on the *hydrochlorides* and *chloroplatinates*. The procedures are those given under derivatives of primary and secondary amines.

For structure determinations of complex tertiary amines, it is often necessary to employ degradation methods. The following equations illustrate the most commonly used reactions.

Von Braun reaction[42]

$$R_2NR' + BrCN \longrightarrow R_2NCN + R'Br$$

Amine oxide pyrolysis[43]

$$R_2NCH_2CH_2R \xrightarrow{H_2O_2} R_2\overset{+}{\underset{}{N}}CH_2CH_2R \xrightarrow{\Delta} R_2NOH + CH_2{=}CHR$$
(with $\overset{O^-}{\underset{}{|}}$ on the nitrogen)

Hoffman elimination[43]

$$R_2NCH_2CH_2R \xrightarrow[2)\ Ag_2O]{1)\ CH_3I} R_2\overset{+}{N}{-}CH_2CH_2R \xrightarrow{\Delta} R_2NCH_3 + CH_2{=}CHR$$
(with CH_3 above N and ^-OH below)

A special category of tertiary amines are the vinyl amines or *enamines*.[44] Primary and secondary vinyl amines are unstable and rearrange to imines. The enamines are important synthetic intermediates whose reactivity closely resembles that of enolates.

Enamines form stable iminium salts by *C*-protonation; the perchlorates are usually nicely crystalline salts.[45]

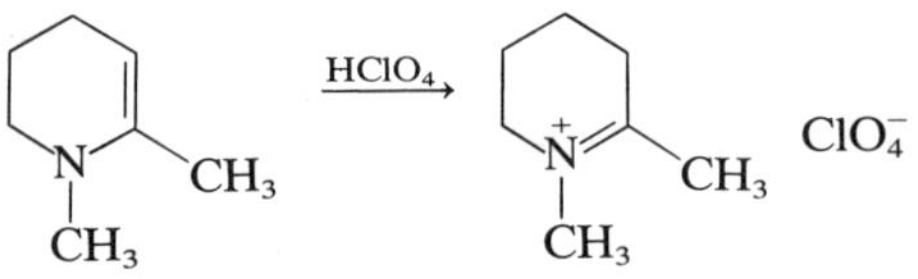

Most enamines can be hydrolyzed to secondary amines and carbonyl compounds, which can be separated and identified by the usual methods.[46]

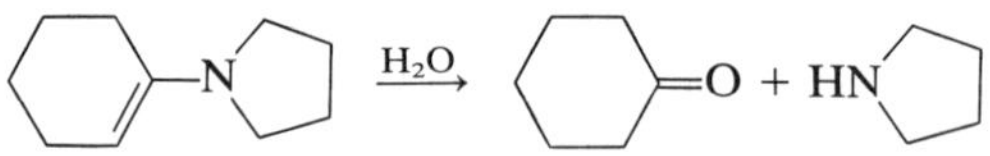

---

[42] H. A. Hageman, *Org. Reactions*, **7**, 198 (1953).

[43] A. C. Cope and E. R. Trumbull, *Org. Reactions*, **11**, 317 (1960).

[44] J. Szmuszkovicz, *Advan. Org. Chem.*, **4**, 1 (1963).

[45] N. J. Leonard and K. Jann, *J. Am. Chem. Soc.*, **84**, 4806 (1962).

[46] J. L. Johnson, M. E. Herr, J. C. Babcock, A. E. Fonken, J. E. Stafford, and F. W. Heyl, *J. Am. Chem. Soc.*, **78**, 430 (1956); J. Joska, J. Fajkos, and F. Sorm, *Collection Czech. Chem. Comm.*, **26**, 1646 (1961).

Table 9.14. Physical Properties of Tertiary Amines and Derivatives

Name of Compound	bp(°C)*	Picrate	Quaternary Methiodide
Trimethylamine	3	216	230d†
Triethylamine	89	173	
Pyridine	116	167	117
2-Methylpyridine	129	169	230
2,6-Dimethylpyridine	143	168	233
3-Methylpyridine	143	150	
4-Methylpyridine	143.1	167	
Tripropylamine	156.5	116	208
2,4-Dimethylpyridine	159	183	
N,N-Dimethylbenzylamine	181	93	
N,N-Dimethylaniline (mp 1.5–2.5°C)	196	163	228d
Tributylamine	216.5 (93–5/15 mm)	106	180
N,N-Diethylaniline	218 (93–5/10 mm)	142	102
Quinoline (mp 15.6°C)	239	203	133
Isoquinoline (mp 26.5°C)	243		
2-Phenylpyridine	268–9	175	
N,N-Dibenzylaniline (mp 72°C)		131d	135
Tribenzylamine (mp 91°C)		190	184
Triphenylamine (mp 127°C)			

* bp at 760 mm Hg pressure unless otherwise noted.
† "d" denotes decomposition occurs at that temperature.

Enamines are reduced to the saturated tertiary amines by catalytic hydrogenation or formic acid.[47]

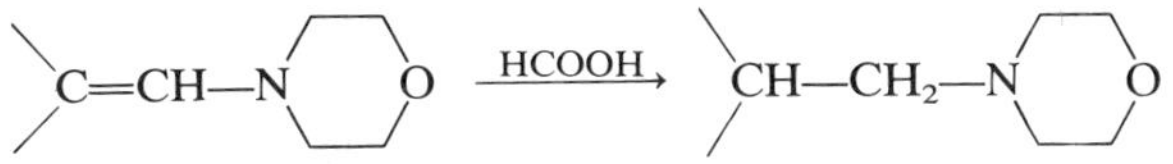

$$\begin{array}{c}\diagup\\C=CH-N\\\diagup\end{array}\bigcirc O \xrightarrow{\ HCOOH\ } \begin{array}{c}\diagup\\CH-CH_2-N\\\diagup\end{array}\bigcirc O$$

Table 9.14 lists the physical properties of tertiary amines and their methiodides.

9.14.4 Amine Salts

Salts of amines fall into two categories—the salts of primary, secondary, and tertiary amines with proton acids, and the quaternary ammonium salts. Both types are generally water-soluble and insoluble in nonpolar solvents such as ethers and hydrocarbons. Many are soluble in alcohols, methylene chloride, and chloroform.

---

[47] P. L. DeBenneville and J. H. Macartney, *J. Am. Chem. Soc.*, **72**, 3073 (1950).

See Sec. 5.4.3 for characteristic infrared absorptions of $-\overset{+}{N}H_3$, $-\overset{+}{N}H_2$, and $-\overset{+}{N}H$. There are no characteristic bands for quaternary ammonium salts. Inorganic anions can be identified by the standard methods of inorganic qualitative analysis.

The *salts of proton acids* usually give slightly acidic solutions; exceptions are the salts of weak acids such as acetic, propionic, etc. Upon treatment with dilute alkali, the amine will separate if it is insoluble or sparingly soluble in water. Even if separation does not occur, the presence of the free amine will be recognized by its characteristic odor. It is often convenient to recover the free amine from the alkaline solution by extraction with ether. In other cases, it may be more convenient to adjust the pH of the solution and make a sulfonamide, benzamide, or acetamide according to the directions given under amines.

Quaternary ammonium salts form neutral-to-basic solutions. The hydroxides and alkoxides are very strong bases. These salts can be converted to other salts that may be suitable for identification purposes by anion exchange reactions, as illustrated in the following equations.

$$R_4N^+I^- + AgBF_4 \longrightarrow R_4N^+ BF_4^- + AgI$$

$$R_4N^+ Cl^- + Na\ B(C_6H_5)_4 \longrightarrow R_4N^+ B(C_6H_5)_4^- + NaCl$$

$$R_4N^+ SO_4H^- \xrightarrow[\text{column}]{\text{Ion exchange}} R_4N^+ OH^- \xrightarrow{HClO_4} R_4N^+ ClO_4^-$$

They can also be degraded to tertiary amine by the Hoffman elimination reaction illustrated for tertiary amines.

9.14.5 α-Amino Acids

9.14.5a Classification

Amino acids are commonly encountered in identification work as the free amino acids, the hydrochlorides, or peptides. The free amino acids exist as the internal salts or zwitter ions, e.g.,

$$\begin{array}{c} R-CH-COO^- \\ | \\ \overset{+}{N}H_3 \end{array}$$

They are very slightly soluble in nonpolar organic solvents, sparingly soluble in ethanol, and very soluble in water to give neutral solutions. They show increased solubility in both acidic and basic solutions, and they have melting points or decomposition points between 120 and 300°C, which depend on the rate of heating.

The principal infrared absorption bands of free amino acids are those arising from the carboxylate [asymmetric stretching near $1600\ cm^{-1}$ ($6.25\ \mu m$) and symmetric stretching near $1400\ cm^{-1}$ ($7.05\ \mu m$)] and the ammonium group [$\overset{+}{N}H_3$ stretching bands overlapping CH bands near $3000\ cm^{-1}$ ($3.33\ \mu m$); overtones are

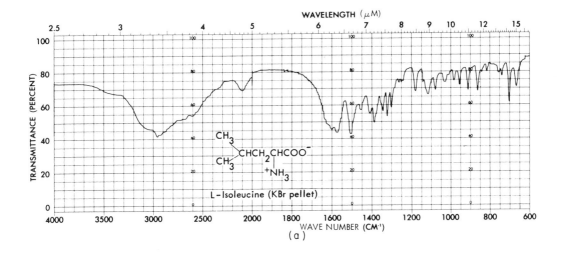

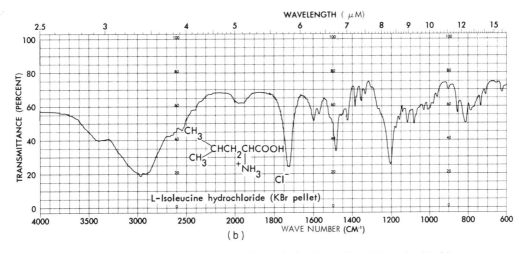

Fig. 9.22. Infrared spectra of a typical amino acid and its hydrochloride.

sometimes present in the 2600 to 1900 cm^{-1} (3.85 to 5.26 μm) region, asymmetric bending near 1600 cm^{-1} (6.25 μm), and symmetric bending near 1500 cm^{-1} (6.67 μm)]. (Fig. 9.22.) In the hydrochloride salts, the very broad —OH band of the carboxylic acid obscures the —$\overset{+}{\text{N}}$H$_3$ and CH stretching region. The other —$\overset{+}{\text{N}}$H$_3$ bands appear as above, along with carboxylic carbonyl near 1720 cm^{-1} (5.81 μm) (Fig. 9.22). Because of low solubility, the free acids and salts are usually run as potassium bromide discs or as mulls. Under these conditions, the spectra of the racemic and optically active forms may show considerable differences. The solution spectra, however, are identical.

Upon warming an aqueous solution of an α-amino acid with a few drops of 0.25% *ninhydrin* (indane-1,2,3-trione hydrate), a blue-to-violet color appears. Proline and hydroxyproline give a yellow color. (Ammonium salts also give a positive test.)

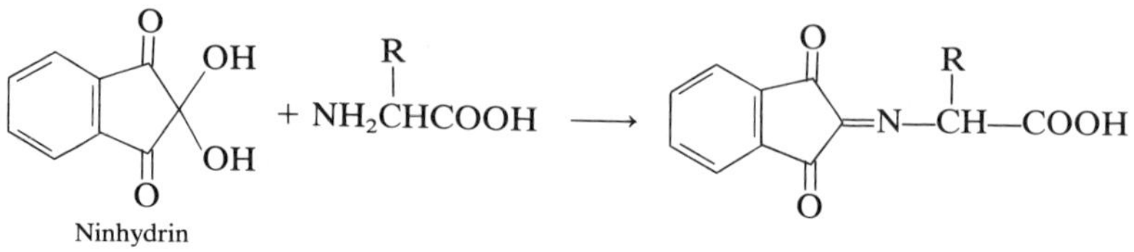

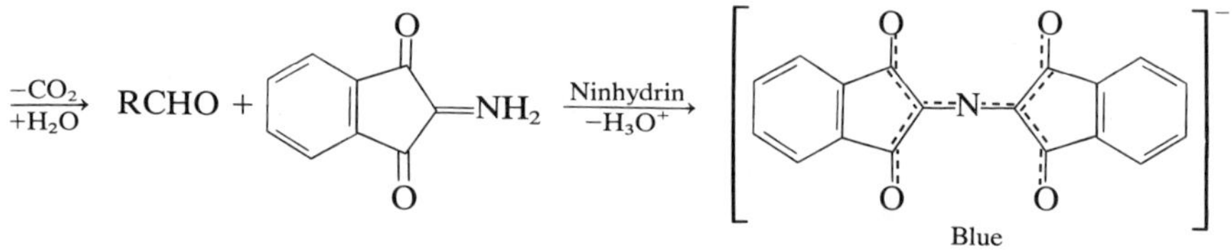

Ninhydrin

Blue

Addition of an aqueous solution of an α-amino acid to a *copper sulfate* solution produces a deep blue coloration. (Ammonia and amines also give a blue color with copper sulfate.)

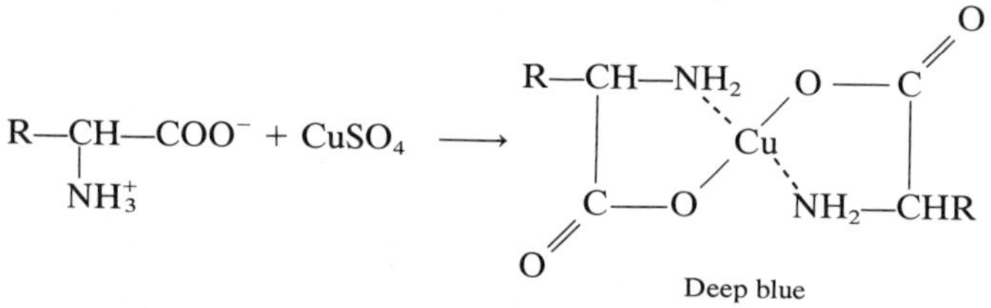

Deep blue

Amino acids yield hydroxy acids and nitrogen when treated with *nitrous acid.* The amount of nitrogen present in amino acids or peptides as primary amino groups can be quantitatively determined by measurement of the volume of nitrogen evolved.

9.14.5b Characterization

The twenty common amino acids found as constituents of proteins are optically active (except glycine) and have the L configuration. A number of less common amino acids are found elsewhere in nature, some of which have the D configuration. The specific rotations are valuable in identification.

Paper and thin-layer chromatography[48] are excellent techniques for the tentative identification of amino acids. *Tables for Identification of Organic Compounds* lists the R_F values on paper using a variety of solvent systems. Whenever possible, direct comparison should be made using a known, the unknown,

---

[48] E. Stahl, *Thin-Layer Chromatography*, Springer-Verlag, Berlin, 1965.

and a mixture of the two. Two-dimensional paper chromatography has proven especially useful in the separation of amino acid mixtures.[49]

With paper or thin-layer chromatography, the spots are usually detected by developing the chromatogram with ninhydrin. The chromatogram is sprayed with a 0.25% solution of ninhydrin in acetone or alcohol, and developed in an oven at 80 to 100°C for 5 min.

An interesting technique for determining the absolute configuration of microgram quantities of amino acids has been developed. By spraying a paper chromatogram with a solution of D-amino acid oxidase (a commercially available enzyme isolated from pig kidneys) in a pyrophosphate buffer and incubating, only D-amino acids are destroyed;[50] the α-keto acids, thus formed, can be detected with an acidified 2,4-dinitrophenylhydrazine spray.[51]

The qualitative and quantitative composition of amino acid mixtures (as obtained from the hydrolysis of peptides) can be achieved by commercially available automated amino acid analyzers. These instruments employ buffer solutions to elute the amino acids from ion-exchange columns. The effluent from the column is mixed with ninhydrin, and the intensity of the blue color is measured photoelectrically and plotted as a function of time.

Crystalline derivatives of the amino acids are usually obtained by reactions at the amino group by treatment with conventional amine reagents in alkaline solution. Extensive data are available on the *N-acetyl, N-benzoyl* and *N-p-toluenesulfonyl* derivatives.

$$R\!-\!\underset{\underset{NH_3^+}{|}}{CH}\!-\!COO^- + NaOH \longrightarrow R\!-\!\underset{\underset{NH_2}{|}}{CH}\!-\!COO^-Na^+ + H_2O$$

$$R\!-\!\underset{\underset{NH_2}{|}}{CH}\!-\!COO^-Na^+ \xrightarrow[\text{(2) HCl}]{\text{(1) R'COCl}} R\!-\!\underset{\underset{NHCOR'}{|}}{CH}\!-\!COOH$$

9.14.5c Procedure: *N*-Acetyl, *N*-benzoyl, and *N-p*-toluenesulfonyl derivatives

(a) Aqueous alkaline solutions of amino acids react with acetic anhydride or benzoyl chloride to yield the acetyl or benzoyl derivatives (Secs. 9.14.2d and e). In order to precipitate the derivative, the solution should be neutralized with dilute mineral acid. In the benzoyl case, any benzoic acid that precipitates can be removed from the desired derivative by washing with a little cold ether.

(b) The *p*-toluenesulfonyl derivatives can be conveniently made by stirring an alkaline solution of 1 equivalent of amino acid with an ether solution of a slight excess of *p*-toluenesulfonyl chloride for 3 to 4 hr. The aqueous layer is acidified to Congo red with dilute hydrochloric acid and cooled if necessary. The crude sulfonamide can be recrystallized from dilute ethanol.

---

[49] R. J. Block, E. L. Durrum, and G. Zweig, *Paper Chromatography and Paper Electrophoresis*, 2nd ed., Academic Press, New York, 1958.

[50] T. S. G. Jones, *Biochem. J.*, **42**, LIX (1948).

[51] J. L. Auclair and R. L. Patton, *Rev. Can. Biol.*, **9**, 3 (1950).

The 2,4-*dinitrophenyl derivatives* are yellow crystalline materials of relatively sharp melting points. Mixture of these "DNP" derivatives of amino acids can be separated by thin-layer chromatography on silica gel. This method has been used for *N*-terminal amino acid analysis in proteins and peptides; the free amino groups on the end of proteins or peptides are reacted with the DNP reagent prior to hydrolysis.[52] The DNP derivatives of simple primary and secondary amines are sometimes useful in identification work.

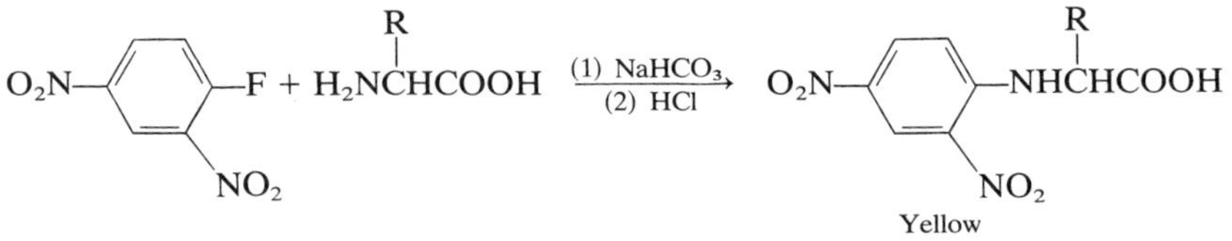
R
O_2N
F + $H_2NCHCOOH$
(1) $NaHCO_3$
(2) HCl
R
O_2N
NHCHCOOH
NO_2
NO_2
Yellow

9.14.5d Procedure: *N*-2,4-Dinitrophenyl derivatives

Dissolve or suspend 0.25 g of the amino acid in 5 mL of water containing 0.5 g of sodium hydrogen carbonate. Add a solution of 0.4 g of 2,4-dinitrofluorobenzene in 3 mL of ethanol. Stir the mixture for 1 hr. Add 3 mL of saturated salt solution and extract several times with small volumes of ether to remove any unchanged reagent. Pour the aqueous layer into 12 mL of cold 5% hydrochloric acid, with stirring. The mixture should be distinctly acid to Congo red. The derivative can be recrystallized from 50% ethanol. These derivatives are light-sensitive; the reaction should be run in the dark (cover flask with aluminum foil), and the products should be kept in a dark place or in dark containers. As a substitute for the 2,4-dinitrofluorobenzene, the chloro compound can be used; in this case, however, the reaction mixture should be refluxed for 4 hr.

Table 9.15 lists the physical properties of amino acids and their derivatives.

9.15 AMIDES AND RELATED COMPOUNDS

9.15.1a Classification

Amides, imides, ureas, and urethanes are nitrogen-containing compounds that exhibit a carbonyl band in the infrared and do not give a positive 2,4-dinitrophenylhydrazine test or show evidence for other nitrogen functions such as nitrile, amino, and nitro. Almost all are colorless crystalline solids; notable exceptions are certain *N*-alkylformamides and tetraalkylureas, which are high-boiling liquids.

---

[52] F. Sanger, *Biochem. J.*, **45,** 563 (1949) and previous papers.

Table 9.15. Physical Properties of Amino Acids and Derivatives

Name of Compound	mp(°C)	Acetyl	Benzoyl	p-Toluene sulfonyl	Dinitro-phenyl Derivative
L-(+)-Valine	96	156	127	147	
N-Phenylglycine	127	194	63		
L-Ornithine	140		240(mono) 189(di)		
o-Aminobenzoic acid	147	185	182	217	
D-(−)-Valine	157	156			
D,L-Glutamic acid	199	187.5	155	117	
β-Alanine	200		120		
D,L-Proline(monohyd., mp 191°C)	203				
L-Arginine	207d		298(mono) 235(di)		252
L-Glutamic acid	213	199		131	
N-Methylglycine	213	135	104		
L-Proline	222d		156(mono)	133	137
L-Lysine	225d		150(di)		
L-Aspargine	227		189	175	
D,L-Threonine	229		145		152
L-Serine	228d				
Glycine	232	206	187.5	150	195
D,L-Arginine	238		230(di,anh)		
D,L-Serine	246d		150	213	199
D,L-Isoserine	248d		151		
D,L-Threonine	252		148		
L-Cystine	260d		181(di)	205	109
D,L-Aspartic acid	270d		119(hyd) 177(anh)		196
L-Aspartic acid	270		185	140	
L-Hydroxyproline	274		100(mono) 92(di)	153	
D,L-Tryptophane	275–82			176	
D,L-Methionine	281	114	145	105	117
L-Methionine	283d	99			
D,L-Phenylalanine	283d(320d)		146	165	
L-(+)-Isoleucine	283d		117	132	
L-Histidine	288(253)		230d(mono)	202–4d	
L-Tryptophane	290		183	176	175
D,L-Isoleucine	292		118	141	166
D,L-Leucine	293–5d		141		
D,L-Alanine	295	137	166	139	
D,L-Norleucine	300			124	
D,L-Valine	298d		132	110	
Creatine	303	165(di)			
Tyrosine	314–8d (290–5d)	148 172(di)	167 212(di)	188 114(di)	

Table 9.15 (Continued)

Name of Compound	mp(°C)	mp(°C) of Derivatives†			
		Acetyl	Benzoyl	p-Toluene sulfonyl	Dinitro-phenyl Derivative
D,L-Tyrosine	340 290–5		197	226	
D,L-Ornithine			288(mono) 188(di)	188(mono)	
D,L-Lysine			249(mono) 146(di)		

* "d" denotes decomposition.
† "anh" denotes anhydrous; "hyd" denotes a hydrate.

Infrared spectra can be of great diagnostic value in the classification of amides (Fig. 9.23). Attention should be given to the presence of NH stretch, the exact position of the carbonyl band (amide I band), and the presence and position of the NH deformation (amide II band). Refer to Chapter 5 on infrared spectroscopy for details.

Occasionally the investigator may have difficulty in deciding from infrared analysis whether a compound in question is an amide. In such cases it may be useful to test for liberation of an amine by treatment of a small sample of the unknown with refluxing 20% sodium hydroxide or fusion with powdered sodium

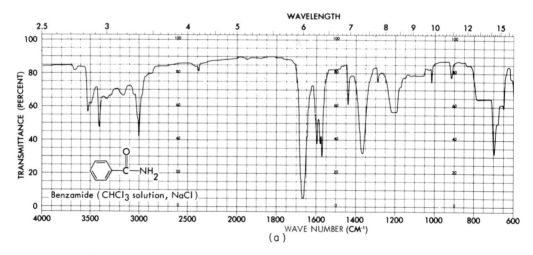

Fig. 9.23. Infrared spectra of typical amides. Note the absence of the amide II band in the spectrum of the lactam.

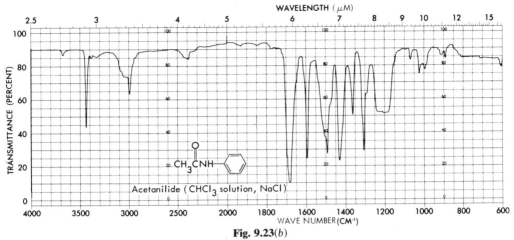

Fig. 9.23(*b*)

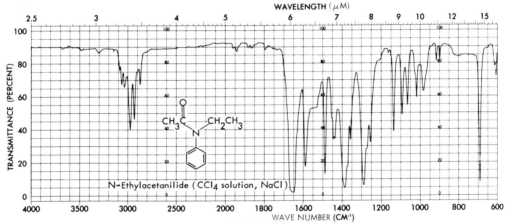

Fig. 9.23(*c*)

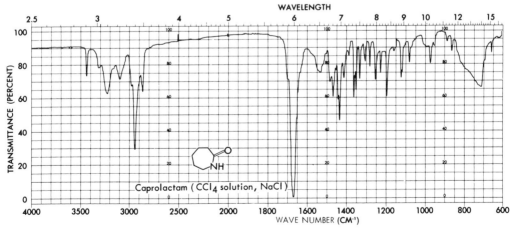

Fig. 9.23(*d*)

hydroxide. A number of color tests have been developed for various classes of amides, ureas, etc.[53]

In the NMR, amide hydrogens exhibit broad absorptions in the region of $\delta 1.5$ to 5.0. The spectra of tertiary amides indicate magnetic nonequivalence of the N-alkyl groups due to the partial double-bond character of the C—N bond.

9.15.1b Characterization

The only completely general method for the preparation of derivatives of amides and related compounds is *hydrolysis* and identification of the products.

$$R'CONR_2 \xrightarrow{H_2O} R'COOH + R_2NH$$
Amides

Lactams → $NH_2—(C)_n—COOH$

Imides → $\begin{array}{c} CO_2H \\ CO_2H \end{array}$ + RNH_2

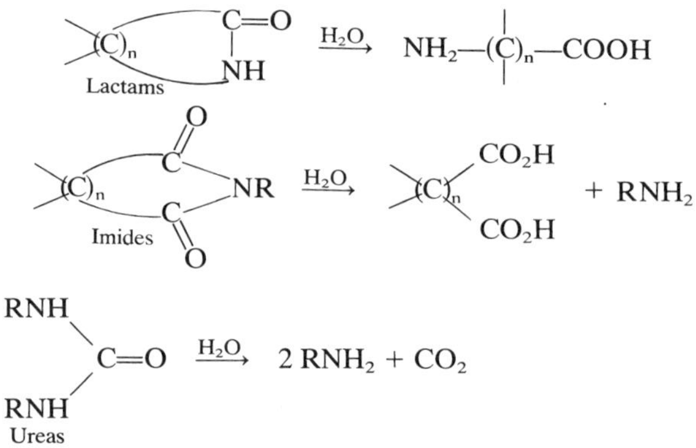

Ureas → $2\,RNH_2 + CO_2$

$$\overset{O}{\overset{\|}{R'OCNHR}} \xrightarrow{H_2O} R'OH + RNH_2 + CO_2$$
Urethanes

Alkaline hydrolysis produces the free amine and the salt of the carboxylic acid; acidic hydrolysis yields the free acid and a salt of the amine. A special case is presented by the lactams, which yield amino acids upon hydrolysis; these amino acids can be converted to derivatives utilizing the general principles discussed in Sec. 9.14.5 for the α-amino acids.

The hydrolysis can be effected by refluxing with 6 N hydrochloric acid or 10 to 20% sodium hydroxide. The latter is usually faster. Resistant amides can be hydrolyzed by boiling with 100% phosphoric acid (prepared by dissolving 400 mg of phosphorous pentoxide in 1 g of 85% phosphoric acid), or by heating at 200°C in a 20% solution of potassium hydroxide in glycerol. Tertiary amides are readily hydrolyzed with anhydrous hydroxide/t-butoxide in ether.[54]

---

[53] N. D. Cheronis, J. B. Entrikin, and E. M. Hodnett, *Semimicro Qualitative Organic Analysis*, 3rd ed., Interscience Publishers, Inc., New York, 1965; F. Feigl, *Spot Tests in Organic Analysis*, Elsevier, Amsterdam, 1960.

[54] P. G. Gassman, P. K. G. Hodgson, and R. J. Balchunis, *J. Am. Chem. Soc.*, **98**, 1275 (1976).

Table 9.16.   Melting Points of Amides

Name of Compound	mp(°C)
N,N-Dimethylformamide (bp 153°C)	
N,N-Diethylformamide (bp 176°C)	
Formamide (bp 196°C)	
1-Methyl-2-pyrrolidinone (bp 202°C)	−24
Propanamide	81
Acetamide	82
4-Phenylbutanamide	84
Acrylamide	85
Propiolanilide	87
m-Toluamide	97
Heptanamide	96
Octanamide	99
Acetanilide	114
Butanamide	115
Methacrylamide	116
Phenylacetanilide	118
Adipamide	130(mono)
Succinimide	126
Nicotinamide	128
Isobutyramide	129
Benzamide	130
o-Toluamide	143
Phthalamide	149(mono)
Succinimide	157(mono)
p-Toluamide	160
Maleimide	170(190)
Benzanilide	175
Phthalamide	220(di)

Primary amides can be converted to acids by reaction with nitrous acid.

$$RCONH_2 + HONO \longrightarrow RCOOH + N_2 + H_2O$$

The Hoffman degradation is of occasional value in structural work. Under appropriate conditions, urethanes or symmetrical ureas can be obtained from the intermediate isocyanates.

$$RCONH_2 \xrightarrow{Br_2, NaOH} RNCO \xrightarrow{H_2O} RNH_2 + CO_2$$

Table 9.16 lists melting points of amides.

9.16
NITRILES

9.16.1a Classification

Aliphatic nitriles are usually liquids; the simpler aromatic nitriles are liquids or low-melting solids. Both types have a characteristic odor reminiscent of cyanide.

In the infrared, the nitrile absorption appears as a weak band in the 2260 to

$2210 \, \text{cm}^{-1}$ (4.42 to 4.52 μm) region. This region somewhat overlaps that of an isocyanate, but these can be readily differentiated on the basis of their reactivity with water, alcohols, etc., and the fact that the isocyanate absorption is much more intense than that of a nitrile (Fig. 9.25). Further preliminary tests for the nitrile grouping are usually not necessary. Other texts have recommended the conversion of the nitrile to a hydroxamic acid by reaction with hydroxylamine, whose presence can be indicated by color test with ferric chloride. Alternatively, a small amount of the supposed nitrile can be heated with potassium hydroxide in glycerol, with the evolution of ammonia being detected by odor or by the litmus or copper sulfate test (see Sec. 9.14 on amines). Both of these methods, however, depend on the prior establishment of the absence of other functional groups such as esters and amides.

9.16.1b Characterization

Nitriles can be *hydrolyzed to acids* under acidic or alkaline conditions. If the resulting acid is solid, it can be employed as such a derivative; if it is a liquid or water-soluble, it is usually converted directly to a solid derivative.

$$\text{RCN} + 2 \, \text{H}_2\text{O} \longrightarrow \text{RCOOH} + \text{NH}_3$$

9.16.1c Procedure: Hydrolysis to acids

(a) *Acid hydrolysis.* In a small flask equipped with a reflux condenser, place 4 mL of 85% phosphoric acid, 1 mL of 75% sulfuric acid, and 0.2 to 0.5 g of nitrile. Add a boiling chip and gently reflux the mixture for 1 hr. Cool the mixture and pour onto a small amount of crushed ice. If the acid is a solid, it may precipitate at this point. It should be collected on a filter. If amide impurities are present, the precipitate can be taken up in base, filtered, and reacidified to recover the acid. If the acid does not precipitate, the reaction mixture should be extracted several times with ether, the ether evaporating to recover solid acid. If the acid is a liquid, it can be converted to conventional acid derivatives such as the *p*-toluidide, *p*-bromobenzyl ester, or *S*-benzylthiuronium salt.

(b) *Alkaline hydrolysis.* In a small flask equipped with a reflux condenser, place 2 g of potassium hydroxide, 4 g of glycerol or ethylene glycol, and 250 to 500 mg of nitrile. Reflux for 1 hr. Dilute with 5 mL of water, cool, and add several milliliters of ether. Shake and allow the layers to separate. Decant the ether layer and discard. Acidify the aqueous solution by slow addition of 6 N hydrochloric or sulfuric acid. Extract the acid with several portions of ether, and handle the extract as indicated above under acid hydrolysis.

For those nitriles that yield solid, water-insoluble amides, partial *hydrolysis to amides* is often a very satisfactory method of derivatization.

$$\text{RCN} \xrightarrow[\text{H}_2\text{O}]{\text{H}_2\text{SO}_4} \text{RCONH}_2$$

9.16.1d Procedure: Hydrolysis to amides

A solution of the nitrile (0.1 g) in 1 mL of concentrated sulfuric acid is warmed on the steam bath for several minutes. The mixture is cooled and poured into cold water. The precipitated crude amide is collected and suspended in a small volume of bicarbonate solution to remove any acid formed. The insoluble amide is recollected and recrystallized (aqueous alcohol). The amide can be further converted to the acid by action of nitrous acid.

Nitriles, especially aromatic nitriles, can be smoothly converted to amides by *alkaline hydrogen peroxide.*

$$\text{Ph–C}\equiv\text{N} \xrightarrow[\text{H}_2\text{O}]{\text{HOO}^-} \overset{\text{OOH}}{\underset{}{\text{Ph–C=NH}}} \xrightarrow{\text{HOOH}} \overset{\text{O}}{\underset{}{\text{Ph–C–NH}_2}} + \text{O}_2 + \text{H}_2\text{O}$$

9.16.1e Procedure: Conversion to amides by alkaline hydrogen peroxide

In a small test tube, place 0.1 g of nitrile, 1 mL of ethanol, and 1 mL of 1 N sodium hydroxide. To this mixture add dropwise (cooling if the reaction foams too vigorously) 1 mL of 12% hydrogen peroxide. After addition is complete, maintain the solution at 50 to 60°C in a water bath for an additional 30 min to 1 hr. Dilute the reaction mixture with cold water, and collect the solid amide. Recrystallize from aqueous alcohol.

Nitriles can be *reduced to primary amines* by lithium aluminum hydride or by sodium dissolving in alcohol. The conversion can also be made catalytically, but secondary amine by-products may result; addition of ammonium carbamate suppresses the formation of secondary amine. The amines obtained by reduction are usually liquids and require further transformation to a solid derivative.

$$\text{RCN} \xrightarrow[\text{or Na/ROH}]{\text{LiAlH}_4} \text{RCH}_2\text{NH}_2$$

$$\text{RCN} \xrightarrow{\text{H}_2\text{ cat.}} \text{RCH}_2\text{NH}_2 \xrightarrow{\text{RCN}} \overset{\text{NH}}{\underset{}{\text{RCH}_2\text{NHC–R}}}$$

$$\xrightarrow{\text{H}_2\text{ cat.}} (\text{RCH}_2)_2\text{NH} + \text{NH}_3$$

9.16.1f Procedure: Reduction to amines

(a) *Reduction with lithium aluminum hydride.* In a small dry flask equipped with a dropping funnel and magnetic stirrer, place 0.1 g of lithium aluminum hydride and 10 mL of anhydrous ether. To the hydride solution add dropwise 0.2 g of the nitrile in 10 mL of anhydrous ether. When the reaction has subsided, add water dropwise and cautiously with cooling, until hydrogen evolution ceases. Several drops of water should be sufficient. Then add to the reaction mixture 10 mL of a 20% solution of sodium potassium tartrate. Transfer the mixture to a separatory funnel. Separate the ether layer and extract the aqueous layer one or more times with small volumes of ether. The ether should be removed, and amine derivatives should be made in the usual manner. The phenylthiourea is a particularly suitable derivative for small scale reactions.

(b) *Reduction with sodium in alcohol.* Place 0.5 g of the nitrile and 10 mL of absolute ethanol in a dry 100-mL round-bottomed flask equipped with a reflux condenser. Add through the top of the condenser approximately 0.75 g of clean, finely cut sodium at such a rate that the reaction remains under control. When the sodium has dissolved, cool the reaction mixture in an ice bath and cautiously add 10 mL of concentrated hydrochloric acid dropwise through the condenser. The solution should be acid to litmus. Transfer the reaction mixture to a beaker and place it on the steam bath to remove the ethanol. Cool the reaction mixture in an ice bath, and cautiously make the mixture basic with concentrated sodium hydroxide. Extract the amine with ether, and use it to make an amide or thiourea derivative. If the amine is particularly volatile, distill the amine directly from the solution made basic in the distillation flask into water (the tip of the delivery tube should be under the water surface—however, be sure to remove the delivery tube before reducing the heat to the distillation pot!) Add 20 drops of phenyl isothiocyanate and shake the reaction mixture vigorously for several minutes. If necessary, cool and scratch the flask to induce crystallization.

The transformation of nitriles to *aldehydes by partial reduction and hydrolysis,* or to ketones by reaction with Grignard reagents, is a particularly suitable identification method for lower-molecular-weight nitriles, which would yield highly volatile amines on reduction, liquid acids on hydrolysis, or water-soluble amides on partial hydrolysis. Solid derivatives of the aldehydes or ketones are readily made. Sometimes it may be necessary to isolate the aldehydes or ketones by use of the Grignard *T* reagent (Sec. 9.7.5).

In the Grignard reaction, the best yields are obtained by the use of phenyl magnesium bromide in large excess.[55] The reaction can easily be adapted to small-scale use.

$$\text{RC} \equiv \text{N} + \text{C}_6\text{H}_5\text{MgBr} \longrightarrow R-\overset{\overset{\displaystyle \text{NMgBr}}{\|}}{C}-\text{C}_6\text{H}_5 \xrightarrow{\text{H}_3\text{O}^+} R-\overset{\overset{\displaystyle \text{O}}{\|}}{C}-\text{C}_6\text{H}_5$$

Aldehydes can be readily prepared from nitriles by use of complex hydride reducing agents. Reagents of particular utility for this reduction are diisobutylaluminum hydride[56] and lithium triethoxyaluminium hydride.[57]

$$\text{RC} \equiv \text{N} \xrightarrow{\text{(IsoBu)}_2\text{AlH}} R-\overset{\overset{\displaystyle \text{N}-\text{Al(IsoBu)}_2}{\|}}{C}-H \xrightarrow{\text{H}_2\text{O}} R-C\overset{\displaystyle O}{\underset{\displaystyle H}{\diagdown}}$$

Table 9.17 lists physical properties of nitriles.

---

[55] R. L. Shriner and T. A. Turner, *J. Am. Chem. Soc.,* **52,** 1267 (1930).

[56] A. I. Zakhaikin and I. M. Kharlina, *Polk. A Rad. Nauk SSSR,* **116,** 422 (1957); *Chem. Abstracts,* **52,** 8040 (1958).

[57] H. C. Brown and C. P. Garg, *J. Am. Chem. Soc.,* **86,** 1085 (1964).

Table 9.17.   **Physical Properties of Nitriles**

Name of Compound	bp(°C)	mp(°C)
Acrylonitrile	78	
Acetonitrile	81.6	
Propanonitrile	97.3	
2-Methylpropanonitrile	104	
Butanonitrile	117.4	
trans-2-Butenonitrile	119	
3-Methylbutanonitrile	130	
Pentanonitrile	141	
Hexanonitrile	165	
Benzonitrile	190.2	
o-Toluonitrile	205	
Ethylcyanoacetate	207	
m-Toluonitrile	212	
p-Toluonitrile	217	27
Phenylacetonitrile	234	
Cinnamonitrile	256	20
Adiponitrile	295(di)	21
Malononitrile	219	31
1-Naphthonitrile		32
o-Chlorobenzonitrile	232	43
Succinonitrile		57
Cyanoacetic acid		66
p-Chlorobenzonitrile		96
o-Nitrobenzonitrile		110
p-Nitrobenzonitrile		147

9.17
NITRO COMPOUNDS

9.17.1a Classification

Most nitroalkanes are liquids that are colorless when pure, but they develop a yellow tinge on storage. The liquid nitro compounds have characteristic odors; they are insoluble in water and have densities greater than unity. Many of the aromatic compounds are crystalline solids. The aromatic nitro compounds are generally yellow, the color increasing markedly on polynitration.

When the nitro group is the primary functionality in the molecule, its presence is readily revealed on casual inspection of the infrared spectrum; strong bands appear near 1560 and 1350 cm^{-1} (6.41 and 7.41 μm) (Fig. 9.24). More frequently, the group appears as a substituent in other classes of compounds, and its presence is revealed by the physical constants of the compound and its derivation and/or by more careful inspection of the infrared spectrum. Like halogen substituents, when the nitro group is present along with a more reactive functionality,

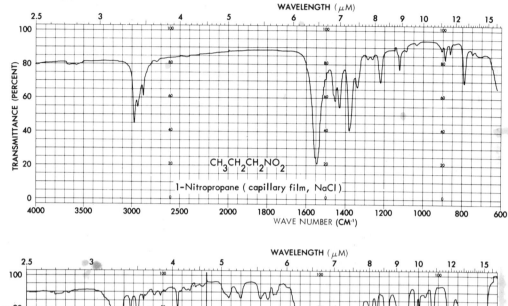

Fig. 9.24. Typical infrared spectra of nitro compounds.

for example, carbonyl, hydroxyl, etc., derivatization is made at the more reactive site.

The high electronegativity of the nitro group results in extensive deshielding of adjacent aliphatic hydrogens in NMR spectra; nitromethane appears at $\delta 5.72$, RCH_2NO_2 at $\delta 5.6$, and R_2CHNO_2 at $\delta 5.3$. The characteristic chemical shift, along with coupling patterns, allows distinction to be made between primary and secondary aliphatic nitroalkanes.

For the classification of a material as a nitroalkane or nitroarene, it is usually not necessary to resort to chemical methods; however, the following simple tests may have occasional utility.

Most nitro compounds (exceptions are nitroethane and 1-nitropropane) will oxidize *ferrous hydroxide* to ferric hydroxide, as will nitroso compounds, quinones, hydroxylamines, nitrates, and nitrites.

$$RNO_2 + 4\,H_2O + 6\,Fe(OH)_2 \longrightarrow RNH_2 + 6\,Fe(OH)_3$$

9.17.1b Procedure: Ferrous hydroxide test

In a small test tube, mix about 10 to 20 mg of the compound with 1.5 mL of freshly prepared 5% solution of ferrous ammonium sulfate. Add one drop of 3 N sulfuric acid and 1 mL of 2 N potassium hydroxide in methanol. Stopper the tube quickly and shake well. In a positive test, a red-brown precipitate of ferric hydroxide appears within 1 min.

Nitrous acid can be used to differentiate between primary, secondary, and tertiary aliphatic nitro compounds.

Nitrolic acid   red

Pseudonitrole (blue)

$$R_3CNO_2 \xrightarrow{\text{HONO}} \text{No Reaction}$$

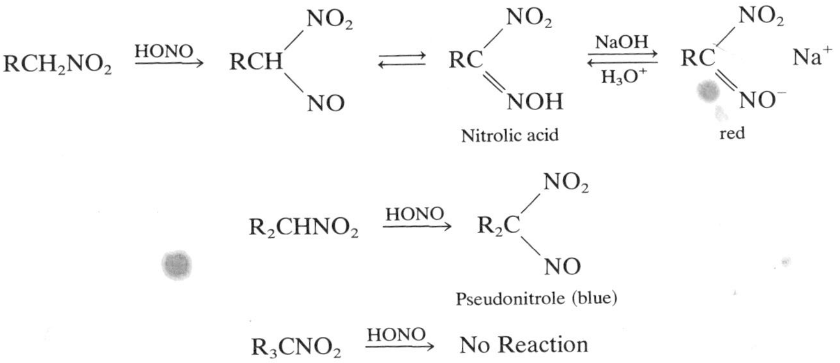

9.17.1c Characterization

Many of the simple nitroalkanes can be identified by reference to physical and spectral properties. Relatively few vinyl nitro compounds are known; nitroethylene and other lower nitroalkenes readily polymerize.

Both aromatic and aliphatic nitro compounds can be converted to *primary amines by reduction with tin and hydrochloric acid*. The reduction can also be effected by catalytic hydrogenation. The amine thus obtained is identified by standard procedures. This provides the only general method for the conversion of nitroalkanes to solid derivatives.

$$RNO_2 \xrightarrow{\text{Sn/HCl}} RNH_2$$

9.17.1d Procedure: Reduction with tin and hydrochloric acid

In a 50-mL round-bottomed flask fitted with a reflux condenser, place 1 g of the nitro (nitroso, azo, azoxy, or hydrazo) compound, and 2 to 3 g of granulated tin. If the compound is highly insoluble in water, add 5 mL of ethanol. Add 20 mL of 10%

hydrochloric acid in small portions, shaking after each addition. After the addition is complete, warm the mixture on the steam bath for 10 min. Decant the hot solution into 10 mL of water and add sufficient 40% sodium hydroxide to dissolve the tin hydroxide. Extract the mixture several times with ether. Dry the ether over magnesium sulfate; remove the ether and identify the amine by conversion to one or more crystalline derivatives.

For highly volatile amines, the procedure must be modified. As the solution is made alkaline, the amine can be distilled into dilute hydrochloric acid.

With base catalysis, primary and secondary nitroalkanes give *adducts with carbonyl and α,β-unsaturated carbonyl and related compounds*. Such adducts have only occasional use in identification.

$$\text{ArCHO} + \text{RCH}_2\text{NO}_2 \xrightarrow{\text{OH}^-} \underset{H}{\overset{Ar}{\diagdown}}C=C\underset{NO_2}{\overset{R}{\diagup}}$$

$$\underset{G}{\overset{\diagdown}{\diagup}}C=C\underset{\diagdown}{\diagup} + \text{RCH}_2\text{NO}_2 \xrightarrow{\text{OH}^-} \underset{G}{\overset{\diagdown}{\diagup}}CH-\underset{R}{\overset{|}{\underset{|}{C}}}-\text{CHNO}_2$$

$$G = -\text{COR}, -\text{CO}_2\text{R}, \text{CN}, -\text{SO}_2\text{R}$$

Primary and secondary nitroalkanes can be *oxidized to the corresponding carbonyl compounds* by potassium permanganate. Magnesium sulfate solution is added to act as a buffering agent.[58]

$$\text{R}-\text{CH}_2\text{NO}_2 + \text{KMnO}_4 \xrightarrow[\text{MgSO}_4]{} \text{RCHO}$$

Aromatic mononitro compounds can be characterized by conversion to *di-* or *trinitro derivatives* (follow procedures for aromatic hydrocarbons, Sec. 9.1.4). Many polynitro aromatics form stable *charge-transfer complexes* with naphthalene and other electron-rich aromatic compounds.[59]

Nitro derivatives of alkylbenzenes can be oxidized to the corresponding nitrobenzoic acids with dichromate (preferred) or permanganate (follow procedures for alkyl benzenes, Sec. 9.1.4). Bromination of the ring often affords a suitable derivative.

Table 9.18 lists the physical properties of nitro compounds.

---

[58] H. Shechter and F. T. Williams, Jr., *J. Org. Chem.*, **27**, 3699 (1962).
[59] O. C. Dermer and R. B. Smith, *J. Am. Chem. Soc.*, **61**, 748 (1939).

Table 9.18.   Physical Properties of Nitro Compounds

Name of Compound	bp(°C)	mp(°C)
Nitromethane	101	
Nitroethane	114	
2-Nitropropane	120.3	
1-Nitropropane	131.6	
2-Nitrobutane	140	
1-Nitrobutane	153	
2-Nitropentane	154	
1-Nitropentane	173	
Nitrobenzene	210	
o-Nitrotoluene	221.7	
o-Nitroanisole	265	10
m-Nitrobenzyl alcohol		27
o-Chloronitrobenzene		32
2,4-Dichloronitrobenzene	258	33
2-Nitrophenol		37
4-Chloro-2-nitrotoluene	240	38
p-Nitrotoluene	238	52
p-Nitroanisole		52.5
1-Nitronaphthalene		57
2,4,6-Trinitroanisole		68
2,4-Dinitrotoluene		70
p-Nitrobenzyl chloride		72
2,4-Dinitrobromobenzene		72
o-Nitrobenzyl alcohol		74
2-Nitronaphthalene		78
2,4,6-Trinitrotoluene		80.1
2,4,6-Trinitrochlorobenzene		83
p-Chloronitrobenzene		84
m-Dinitrobenzene		90
2,4-Dinitroanisole		95
4-Nitrobiphenyl*		114

* *Compound is carcinogenic.* Its use is severely restricted by the federal government.

9.18
MISCELLANEOUS NITROGEN COMPOUNDS

(See Tables 5.5 and 5.6 for characteristic infrared absorptions of miscellaneous nitrogen compounds.)

9.18.1 Isocyanates and Carbodiimides

The isocyanate group can readily be detected by infrared spectroscopy—a very strong band in the 2275 to 2250 cm^{-1} (4.40 to 4.04 μm) region—much stronger than a nitrile that also appears in the same region (Fig. 9.25). Isocyanates can also

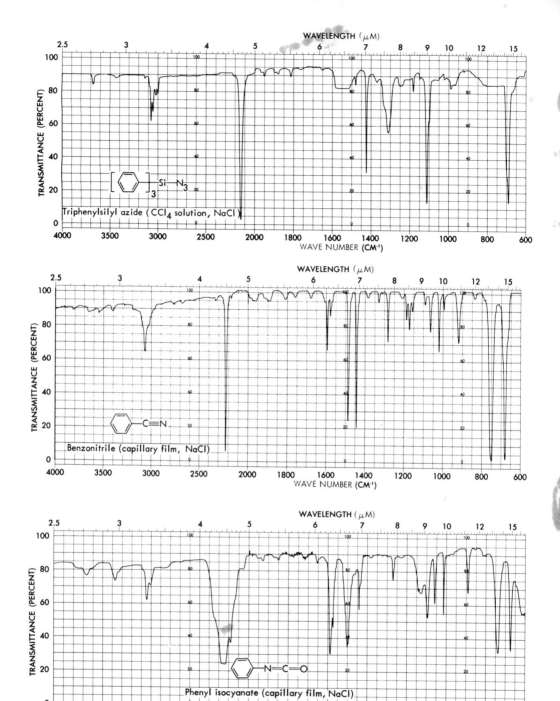

Fig. 9.25. Typical spectra of representative nitrogen compounds having bands in the 2100 to 2200 cm⁻¹ region.

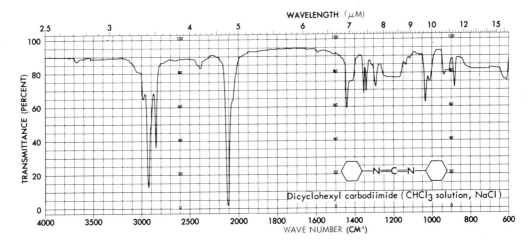

Fig. 9.25 (Continued)

be distinguished from nitriles in their ready reactivity with water, alcohols, and amines to yield urethanes and ureas, respectively, all of which can serve to identify the isocyanate.

$$2 \; RN{=}C{=}O \xrightarrow{\text{Cat.}} R{-}N{=}C{=}N{-}R + CO_2$$

An isocyanate   A carbodiimide

Carbodiimides, readily available from isocyanates, are used in peptide synthesis and other condensation reactions that involve the removal of water. They are highly reactive (with amines, alcohols, acids, etc.) and can be readily identified by their reaction with water to produce symmetrical ureas. The most commonly encountered carbodiimide is dicyclohexylcarbodiimide (Fig. 9.25).

9.18.2 Azo, Azoxy, and Hydrazo Compounds

Azo and azoxy compounds are highly colored substances. Ultraviolet and visible spectroscopy, as well as comparative thin-layer chromatography, should be extensively used as aids in their identification.

The following equation illustrates types of reactions that may be useful in preparing derivatives.

$$\text{ArN}{=}\text{NAr} \xrightleftharpoons[\text{Zn, NaOH}]{\text{H}_2\text{O}_2} \underset{\text{Azoxyarene}}{\text{ArN}{=}\overset{\overset{\text{O}^-}{|+}}{\text{N}}\text{Ar}}$$

$$\Big\Vert \overset{\text{H}_2/}{\underset{\text{cat.}}{}}\text{HgO} \quad \text{Sn, HCl}$$

$$\text{ArNHNHAr} \xrightarrow{\text{Sn, HCl}} 2\ \text{ArNH}_2$$

Hydrazoarene

Hydrazobenzene   Benzidine (*carcinogen!*)

9.18.3 Hydrazines

$$\text{RNHNH}_2 \qquad \text{R}_2\text{NNH}_2 \qquad \text{RNHNHR}$$

Hydrazines are weak-to-intermediate strength bases; they dissolve in mineral acids to form salts. The mono- and unsymmetrically disubstituted hydrazines can be detected and identified by condensation with carbonyl compounds to yield hydrazones.

The N—N bond of hydrazines can be easily cleaved by tin and hydrochloric acid (method for nitro compounds, Sec. 9.17.1d) or in some cases with aluminum amalgam.

9.18.4 Oximes, Hydrazones, and Semicarbazones

These compounds can be hydrolyzed by the action of concentrated hydrochloric acid to yield the hydrochloride of the hydroxylamine, the hydrazine, or the semicarbazine and the carbonyl compound. The original carbonyl compound may in some cases be regenerated by exchange reactions[60] (see Sec. 9.7.6c).

Semicarbazones also react with nitrous acid to regenerate aldehydes or ketones.

Oximes can be identified and their stereochemistry determined by the Beckman rearrangement to amides,[61] or by ^{13}C NMR spectroscopy.[62]

$$\underset{\text{R}-\overset{\overset{\displaystyle\text{N}-\text{OH}}{\|}}{\text{C}}-\text{R}'}{} \xrightarrow{\text{H}^+} \underset{\text{RNH}-\overset{\overset{\displaystyle\text{O}}{\|}}{\text{C}}-\text{R}'}{}$$

[60] E. B. Hershber, *J. Org. Chem.*, **13**, 542 (1948).

[61] L. G. Donaruma and W. Z. Heldt, *Org. Reactions*, **11**, 1 (1960).

[62] G. E. Hawkes, K. Herwig, and J. D. Roberts, *J. Org. Chem.*, **39**, 1017 (1974); G. C. Levy and G. L. Nelson, *J. Am. Chem. Soc.*, **94**, 4897 (1972).

9.18.5 Azides

Azides (Fig. 9.25) can be reduced to *primary amines* by sodium borohydride, lithium aluminum hydride, or by catalytic hydrogenation.

$$R—N_3 \xrightarrow{[H]} R—NH_2$$

9.18.5a Procedure: Borohydride reduction of azides

To 0.3 g of the azide in 5 mL of 2-propanol, add 0.2 to 0.3 g of sodium borohydride and reflux on a steam bath overnight. Evaporate the 2-propanol, add water, and reflux until effervescence has subsided. Isolate the amine by extraction with ether.

Alkyl and aryl azides will undergo 1,3-dipolar additions with reactive alkenes[63] and alkynes.[64]

$$RN_3 + HC{\equiv}C—COOH \longrightarrow$$

$$ArN_3 +$$

9.18.6 Isocyanides

Isocyanides are very uncommon. They have a very characteristic vile odor. They can be identified by acid hydrolysis to the corresponding primary amine and formic acid, or by reduction to secondary amines.

$$RN{=}C: \xrightarrow{H_3O^+} RNH_3^+ + HCOOH$$
$$RN{=}C: \xrightarrow{H_2} RNHCH_3$$

9.18.7 Nitramines

Primary nitramines are acidic and form salts with aqueous alkali. These salts can be alkylated with reagents such as methyl iodide to give secondary nitramines, which are neutral.

---

[63] R. Huttle, *Chem. Ber.*, **74B**, 1680 (1941).

[64] P. Scheiner, J. H. Schumaker, S. Deming, W. J. Libbey, and G. P. Nowack, *J. Am. Chem. Soc.*, **87**, 306 (1965).

$$R{\diagdown}N{-}NO_2 \xrightarrow{\text{NaOH}} R{-}N{=}\overset{+}{N}\diagup\overset{O^-}{\underset{Na^+}{\diagdown}}O^- \xrightarrow{\text{CH}_3\text{I}} \underset{CH_3}{\overset{R}{\diagdown}}N{-}NO_2$$

Most alkylnitramines eliminate the nitro group upon treatment with acid. Arylnitramines, like the nitrosoamines, undergo isomerization in the presence of mineral acids, e.g., phenyl nitramine gives a mixture of *o*- and *p*-nitroanilines. Nitramines can be reduced to the corresponding primary amines.

9.18.8 C-Nitroso Compounds

The simple nitrosoalkanes and arenes generally exist as dimers that are white crystalline solids, but that assume the blue color of the monomer when melted or vaporized.

Nitroso compounds are reduced to primary amines by zinc, iron, or tin in acid solution, or by catalytic hydrogenation. Nitrosobenzene and its derivatives readily condense with aniline to form azo compounds.[65] Aliphatic nitroso compounds that contain the partial structure ${\diagdown}$CH${-}$N${=}$O undergo tautomerization to oximes.

$$X{\diagdown}\text{⬡}{-}N{=}O + H_2N{-}\text{⬡} \xrightarrow[25°]{\text{HOAc}} X{\diagdown}\text{⬡}{-}N{=}N{-}\text{⬡}$$

9.18.9 N-Nitroso Compounds

CAUTION: *Many N-nitrosoamines are carcinogenic.*

Only the *N*-nitroso derivatives of secondary amines are stable; they are neutral to weakly basic compounds. Arylnitrosoamines rearrange to the *p*-nitrosoarylamines in the presence of mineral acid. Nitrosoamines can be reduced to hydrazines by zinc and acetic acid, or to secondary amines by tin and hydrochloric acid. The *N*-nitrosoamides yield diazo compounds upon treatment with sodium hydroxide.

9.18.10 Nitrites and Nitrates

$$R{-}O{-}N{=}O \qquad\qquad R{-}O{-}\overset{O}{\underset{O^-}{\overset{\|}{\underset{}{N^+}}}}$$

An alkyl nitrite   An alkyl nitrate

These materials are esters of inorganic acids. *They are toxic and should be handled with care.* Both nitrites and nitrates give a positive ferrous hydroxide test.

---

[65] H. D. Anspon, *Org. Syn.*, **Coll. Vol. 3,** 711 (1955); C. M. Atkinson, *J. Chem. Soc.*, 2023 (1954).

They can be hydrolyzed in alkaline solution to yield alcohols and salts of nitric or nitrous acid. The common nitrites are ethyl, *n*-butyl, and isoamyl.

9.18.11 Amine Oxides

$$R_3N \xrightarrow{H_2O_2} R_3N^+\!\!-\!O^-$$

Amine oxides, formed by the oxidation of tertiary amines by peroxides, are usually highly hygroscopic, viscous syrups. Some form crystalline hydrates.

Amine oxides show an intense N—O stretching band at 970 to 950 cm^{-1} (10.31 to 10.53 μm). In heteroaromatic amine oxides, for example, pyridine *N*-oxide, this band appears in the 1300 to 1200 cm^{-1} (7.69 to 8.33 μm) region. These bands are shifted to lower frequencies upon hydrogen bonding. Amine oxides form crystalline picrates when treated with alcoholic or aqueous solutions of picric acid.

Reduction to the tertiary amine can be accomplished with a variety of reagents including triphenylphosphine, triethyl phosphite, and metal-acid mixtures. The former is especially useful for aromatic amine oxides. The reactants are mixed neat and heated until the tertiary amine distills out.[66]

$$R_3N^+\!\!-\!O^- + (C_6H_5)_3P \longrightarrow R_3N + (C_6H_5)_3P\!\!=\!\!O$$

Amine oxides with β-hydrogens that are sterically accessible to the oxide are easily pyrolyzed to hydroxylamines and alkenes.[67]

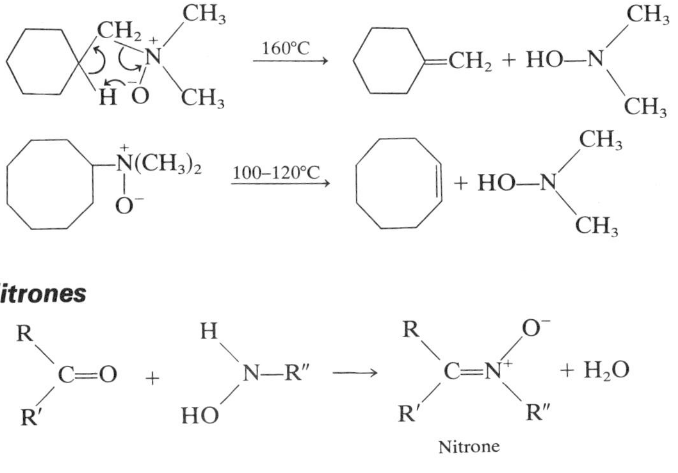

9.18.12 Nitrones

Nitrone

Nitrones,[68] formed by the condensation of aldehydes or ketones with *N*-substituted hydroxylamines, can be solids or liquids. Unless aryl groups are

---

[66] L. Horner and H. Hoffman, *Angew. Chem.*, **68**, 480 (1956). E. Howard, Jr. and W. F. Olszewski, *J. Am. Chem. Soc.*, **81**, 1483 (1959).

[67] A. C. Cope and E. R. Trumbull, *Org. Reactions*, **11**, 317 (1960).

[68] G. R. Delpierre and M. Lamchen, *Quart. Revs.*, **19**, 329 (1965).

present, nitrones tend to be quite water-soluble. The nitrones exhibit a strong band in the infrared in the 1615 to 1560 cm^{-1} (6.20 to 6.40 μm) region. Aliphatic nitrones have a characteristic ultraviolet absorption near 229 to 235 nm (ε = 9,000), which moves to longer wavelengths on conjugation.

Nitrones can be reduced to hydroxylamines with sodium borohydride or lithium aluminum hydride, and to secondary amines with zinc and mineral acid or catalytic hydrogenation.

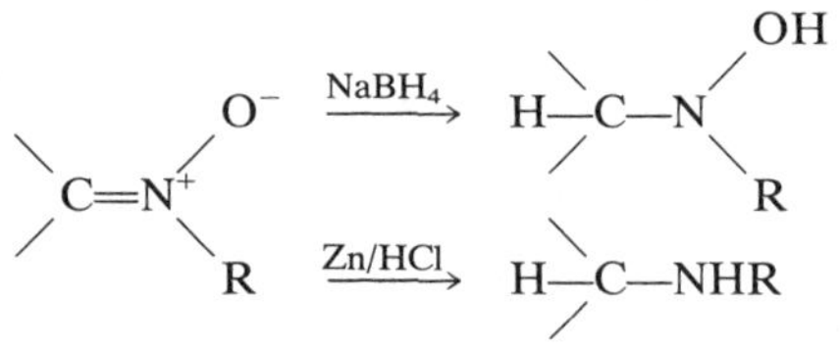

With alkenes, nitrones undergo 1,3-cycloaddition reactions to provide isoxazolidines. The ease of such reactions is considerably enhanced by phenyl or carbonyl substituents on the alkene.

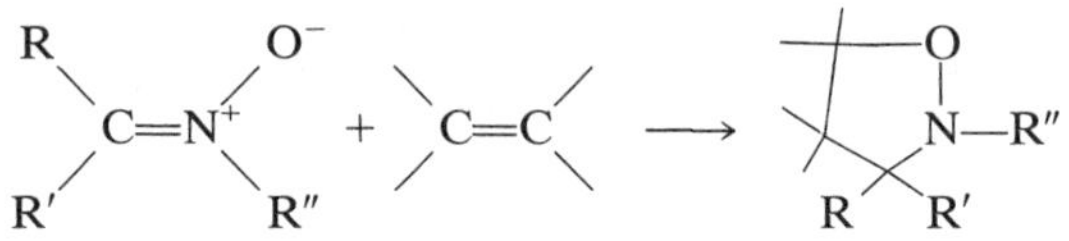

9.19 SULFUR COMPOUNDS

A comprehensive discussion of the methods of detection and identification of the many types of organic sulfur compounds is far beyond the scope of this text. In general, organic sulfur chemistry follows along the lines of organic oxygen chemistry; many of the same general types of functional groups occur, but the variety is greatly increased by the variable oxidation states of sulfur, as indicated by the following examples.

Oxygen series

ROH	Alcohol	ROR	Ether	ROOR	Peroxide

Sulfur series

RSH	Thiol	RSR	Thioether (sulfide)	RSSR	Disulfides
RSO_2H	Sulfinic Acid and Derivatives	RS(=O)R	Sulfoxide		
RSO_3H	Sulfonic Acid and Derivatives	RS(=O)(=O)R	Sulfones		

In this chapter, the most commonly occurring organic sulfur functions will be discussed in some detail; many others will be mentioned only briefly, perhaps only to make the reader aware of their existence. The most comprehensive listing of the properties of organic bivalent sulfur compounds is found in the series by Reid.[69]

9.19.1 Thiols (Mercaptans)

The classification of a compound as a thiol seldom awaits the interpretation of the infrared spectrum. For the uninitiated, most thiols have an objectionable odor reminiscent of hydrogen sulfide. Thiols are generally insoluble in water, but they react with aqueous alkali to form soluble salts.

$$RSH + OH^- \longrightarrow RS^- + H_2O$$

In the infrared, the S—H band appears in the 2600 to 2550 cm^{-1} (3.85 to 3.92 μm) region (Fig. 9.26). This is a relatively weak band and in some cases may be overlooked, especially if the spectrum is run as a dilute solution. This may result in misassignment of the compound as a sulfide. Sulfides and thiols can be differentiated on the basis of solubility of the latter in dilute alkali and by the *lead mercaptide test*; the former is insoluble in base and does not precipitate in the lead mercaptide test.

$$2\,RSH + Pb^{++} + 2\,H_2O \longrightarrow Pb(SR)_2 \downarrow + 2\,H_3O^+$$

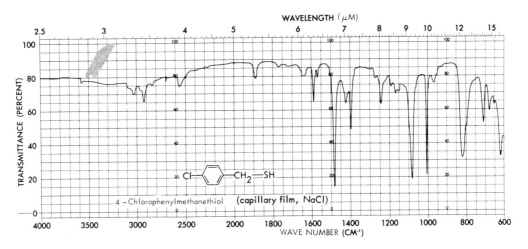

Fig. 9.26. Typical infrared spectrum of a thiol. Note the very weak SH band at 2560 cm^{-1}.

---

[69] E. E. Reid, *Organic Chemistry of Bivalent Sulfur*, Vols. I–V, Chemical Publishing Co., Inc., New York.

9.19.1a Procedure: Lead mercaptide test

Add two drops of the thiol to a saturated solution of lead acetate in ethanol; yellow lead mercaptide precipitates.

Perhaps the most useful derivatives of the alkyl and aryl thiols are the 2,4-*dinitrophenyl sulfides* and the corresponding sulfones.

$$RSH \xrightarrow{\text{NaOH}} RSNa$$

$$RSNa + Cl-\underset{O_2N}{\bigcirc}-NO_2 \longrightarrow RS-\underset{O_2N}{\bigcirc}-NO_2 + NaCl$$

9.19.1b Procedure: 2,4-Dinitrophenyl sulfides

In a test tube dissolve 250 mg of the thiol in 5 mL of methanol. Add 1 mL of 10% sodium hydroxide solution. Add this solution to 3 mL of methanol containing 0.5 g of 2,4-dinitrochlorobenzene in a small flask arranged for reflux. Reflux the mixture for 10 min, cool, and recrystallize from methanol or ethanol.

Occasionally it may be expedient to identify thiols by conversion to simple sulfides or sulfones. The methyl and benzyl sulfides, produced by alkylation of the sodium thiolates with methyl iodide or benzyl bromide in alcohol, are the most useful. These sulfides, as well as the 2,4-dinitrophenyl sulfides, can be oxidized to sulfones by procedures outlined under sulfides in the next section. The 2,4-dinitrophenylsulfones are particularly valuable because they exhibit a wide range of melting points.

The *3,5-dinitrothiobenzoates* can be easily prepared by reaction of the thiol and 3,5-dinitrobenzoyl chloride in the presence of pyridine as catalyst. However, many have low melting points that fall within a narrow range.

$$\underset{O_2N}{\overset{O_2N}{\bigcirc}}-COCl + RSH \xrightarrow{\text{Pyridine}} \underset{O_2N}{\overset{O_2N}{\bigcirc}}-COSR$$

9.19.1c Procedure: 3,5-Dinitrothiobenzoates

In a dry test tube, place 200 mg of 3,5-dinitrobenzoyl chloride, 100 mg (five or six drops) of thiol, and three drops of pyridine. Heat the mixture in a beaker of boiling water for 10 min. Add 2 mL of water and a few drops of pyridine to destroy any remaining reagent. Stir with a glass rod to induce crystallization. Filter, wash with water, suspend the crude powder in several milliliters of sodium hydrogen carbonate solution, and warm if necessary to remove any 3,5-dinitrobenzoic acid. Collect, wash again with water, and recrystallize from dilute ethanol.

In those cases where the *disulfide* is a crystalline solid, disulfides of thiols can be readily made by oxidation with sulfoxides.[70]

$$2\ RSH + CH_2\overset{\overset{\displaystyle O}{\|}}{S}CH_3 \longrightarrow RSSR + CH_3SCH_3 + H_2O$$

With thiophenols, the reaction can be carried out at room temperature for 24 hr. With aliphatic thiols, heating up to 160°C must be applied for 24 hr. Disulfides can also be obtained by oxidation of thiols with hypoiodite, ferricyanide, and a variety of other mild oxidants.

9.19.2 Sulfides and Disulfides

Infrared spectroscopy alone usually cannot provide sufficient information to adequately classify a compound as belonging to the sulfide or disulfide class (Fig. 9.27). Both these classes of compounds exhibit a carbon-sulfur stretching frequency in the 800 to 600 cm^{-1} (12.5 to 16.7 μm) region, which is of little utility because it is easily confused with other bands in that region and, furthermore, is frequently obscured by the solvent. The sulfur-sulfur stretching frequency is a weak band in the 550 to 450 cm^{-1} (18.2 to 22.2 μm) region, and it is thus inaccessible on most instruments.

Aliphatic and most aromatic sulfides and disulfides have unpleasant odors somewhat resembling that of hydrogen sulfide. Compounds with such odors, which are not mercaptans or sulfoxides (as determined by the absence of characteristic infrared bands), should be strongly suspect as sulfides or disulfides. Most aliphatic sulfides and disulfides are liquids. The aromatic sulfides are liquids or low-melting solids; aromatic disulfides are low-melting solids. A number of polysulfides of general formula $R—S_n—R$ are also known.

Most sulfides give *crystalline complexes with mercuric chloride.*

$$R_2S + n\,HgCl_2 \longrightarrow R_2S(HgCl_2)_n \xrightarrow[NaOH]{} R_2S$$

9.19.2a Procedure: Mercuric chloride complexes

The sulfide, neat or in alcohol solution, is added dropwise to a saturated solution of mercuric chloride in alcohol. The complex, which usually precipitates immediately, varies in ratio of mercuric chloride to sulfide, depending on the structure of the sulfide and the exact conditions of preparation. The sulfides can be regenerated from the complex by aqueous sodium cyanide or dilute hydroxide. The procedure can be used as a test for sulfides or as a method for obtaining a solid derivative.

---

[70] T. J. Wallace, and J. J. Mahon, *J. Org. Chem.*, **30**, 1502 (1965).

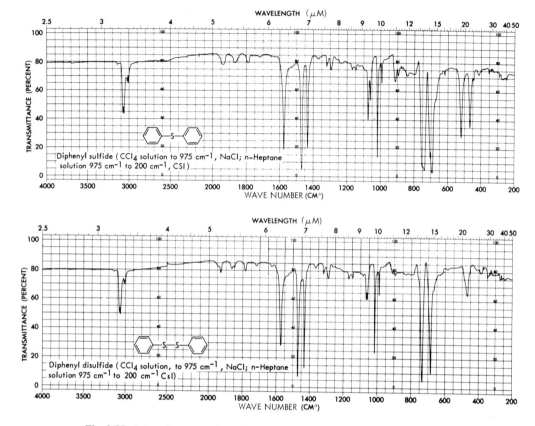

Fig. 9.27. Infrared spectra of a sulfide and the corresponding disulfide. Note the striking similarity in the two spectra.

The most useful solid derivatives of the sulfides are the *sulfones* produced by oxidation with permanganate or peroxides.

$$R\!-\!S\!-\!R \xrightarrow{\text{[Ox]}} R\!-\!\overset{\displaystyle O}{\underset{\displaystyle O}{\overset{\|}{\underset{\|}{S}}}}\!-\!R$$

9.19.2b Procedure: Oxidation to sulfones

(a) Dissolve the sulfide in a small volume of warm glacial acetic acid. Shake and add 3% potassium permanganate solution in 2- or 3-mL portions as fast as the color disappears. Continue the addition until the color persists after shaking for several minutes. Decolorize the solution by addition of sodium hydrogen sulfite. Add 2 to 3 volumes of crushed ice. Filter off the sulfone and recrystallize from ethanol.

(b) Dissolve 1 mmol of the sulfide in 3 mL of toluene and 3 mL of methanol. Add 0.5 to 1 mmol of vanadium pentoxide and heat to 50°C with stirring. Slowly add a slight excess of *t*-butyl hydroperoxide until the excess can be detected by acidified starch iodide paper. Add solid sodium hydrogen sulfite in small portions, and stir for 5 to 10 min. Filter and concentrate to a heavy oil. Add a small amount of water, extract with methylene chloride or other suitable solvent, and dry over magnesium sulfate.

(c) Dissolve the sulfide in glacial acetic acid. Add excess hydrogen peroxide and allow to stand overnight. Warm the solution to 50 to 60°C for 5 to 10 min. Pour into ice water and collect the crystalline sulfone.

Other derivatives of sulfides that may prove useful include the sulfoxides, sulfilimines, and sulfonium salts.

Sulfoxides are cleanly produced by oxidation of sulfides with aqueous or aqueous-alcoholic sodium periodate at 0°C for aliphatic sulfides and at room temperature for aromatic sulfides.[71]

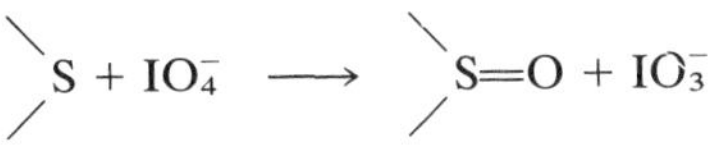

$$\diagdown\!\!\diagup S + IO_4^- \longrightarrow \diagdown\!\!\diagup S{=}O + IO_3^-$$

The *N-p*-toluenesulfonyl sulfilimines are readily prepared by shaking a suspension of the sulfide in an aqueous solution of Chloroamine-*T*.

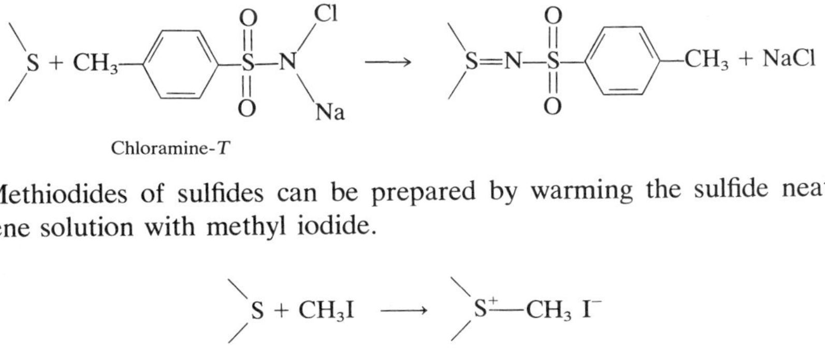

Chloramine-*T*

Methiodides of sulfides can be prepared by warming the sulfide neat or in toluene solution with methyl iodide.

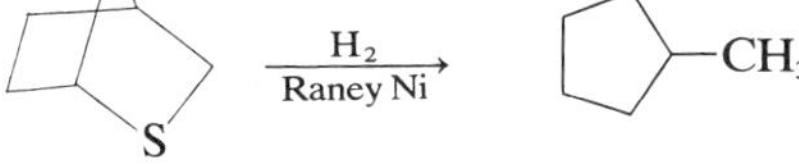

$$\diagdown\!\!\diagup S + CH_3I \longrightarrow \diagdown\!\!\diagup S^+{-}CH_3 \; I^-$$

Sulfides and disulfides (as well as thiols, sulfones, sulfoxides, and many other sulfur compounds) undergo hydrogenolysis when treated with Raney nickel catalyst.[72]

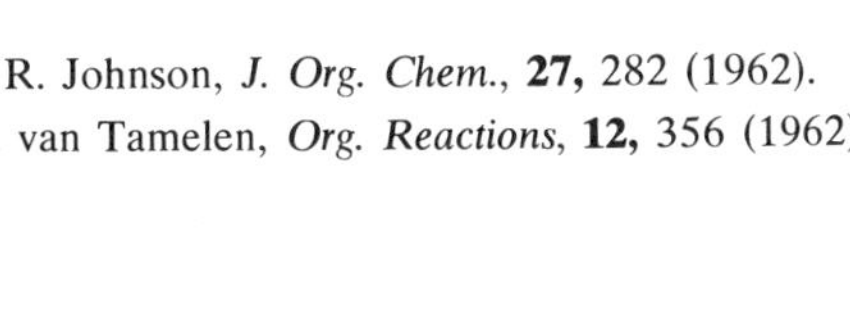

---

[71] N. J. Leonard and C. R. Johnson, *J. Org. Chem.*, **27**, 282 (1962).
[72] G. R. Pettit and E. E. van Tamelen, *Org. Reactions*, **12**, 356 (1962).

Dithio- and hemithioacetals and ketals form a special class of sulfides. The most common are the ethylene thioketals and ethylenehemithioketals. The following equations illustrate reactions that may be useful in their structure determination.

Disulfides can be reduced to thiols by the action of zinc and dilute acid, or they can be oxidized to sulfonic acids by potassium permanganate.

$$RSSR \xrightarrow[H^+]{Zn} 2\,RSH$$

$$RSSR \xrightarrow{KMnO_4} 2\,RSO_3H$$

9.19.3 Sulfoxides

Sulfoxides are usually low-melting, highly hygroscopic solids with a noticeable odor similar to that of sulfides—which, in fact, is often due to trace amounts of contaminating sulfide. The sulfoxide grouping is pyramidal and configurationally stable; hence sulfoxides can exist as enantiomeric and/or diastereomeric pairs. Sulfoxides are weak bases; isolable salts are formed with some strong acids, including nitric acid. CAUTION: *Sulfoxides react explosively on contact with perchloric acid.*

The sulfoxide grouping is readily noted by the characteristic strong infrared band at 1050 to 1000 cm^{-1} (9.52 to 10.0 μm) (Fig. 9.28). This band is quite sensitive to hydrogen bonding; the addition of methanol to the solution causes a

---

[73] G. R. Pettit and E. E. van Tamelen, *Org. Reactions*, **12**, 356 (1962).
[74] M. L. Wolfrom, *J. Am. Chem. Soc.*, **51**, 2188 (1929); H. W. Arnold and W. L. Evans, *J. Am. Chem. Soc.*, **58**, 1950 (1936); J. English, Jr., and P. H. Griswold, *J. Am. Chem. Soc.*, **67**, 2040 (1945).
[75] Pettit and van Tamelen, p. 356.

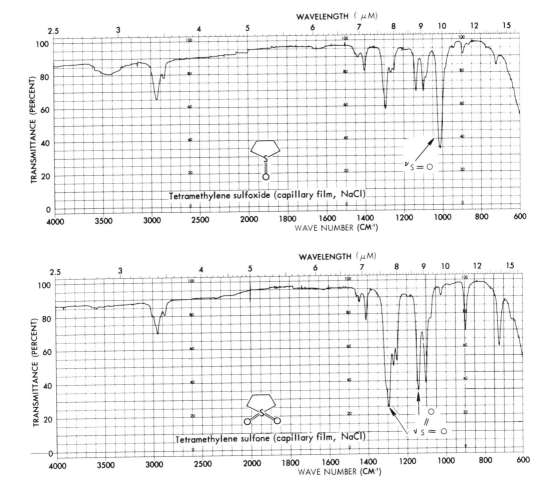

Fig. 9.28. Typical infrared spectra of a sulfoxide and sulfone.

significant shift to lower frequency. Likewise, a shift to lower frequency is observed in going from carbon tetrachloride to chloroform.

In the NMR, hydrogens alpha to sulfoxide appear in the $\delta 2.0$ to 2.5 region. Methylene protons adjacent to sulfoxide often appear as an *AB* system.

Sulfoxides can be reduced to sulfides by reduction with sodium borohydride[76] or aqueous sodium hydrogen sulfite.[77] The yields are high and the sulfides are obtained in essentially pure form.

[76] D. W. Chasar, *J. Org. Chem.*, **36**, 613 (1971).
[77] C. R. Johnson, C. C. Bacon, and J. J. Rijan, *J. Org. Chem.*, **37**, 919 (1972).

Sulfoxides can be oxidized to sulfones by the methods outlined under sulfides.

$$R_2SO \xrightarrow{[Ox]} R_2SO_2$$

Conversion to sulfilimines can be brought about by reaction with sulfonyl isocyanates[78] or N-sulfinylsulfonamines.[79]

$$G = C \text{ or } S$$

Most sulfoxides form crystalline alkoxysulfonium salts when treated with triethyloxonium fluoroborate. These salts can be characterized by their melting points and neutralization equivalents (titrate with aqueous sodium hydroxide to phenolphthalin endpoint).[80]

Like sulfides, sulfoxides also form crystalline complexes with mercuric chloride; however, these complexes are sometimes difficult to purify to constant composition.

Like amine oxides, sulfoxides that contain β-hydrogens can be pyrolyzed to yield alkenes.[81] The primary product is an unstable sulfenic acid, which usually decomposes to a variety of unisolated sulfur-containing materials.

$$R\text{—}\underset{\overset{\|}{O}}{S}\text{—}CH_2\text{—}CH_2\text{—}R' \xrightarrow{\Delta} [RSOH] + CH_2\text{=}CH\text{—}R$$

---

[78] C. King, *J. Org. Chem.*, **25**, 352 (1960).

[79] G. Schultz and G. Kresze, *Angew. Chem.*, **75**, 1022 (1963); J. Day and D. J. Cram, *J. Am. Chem. Soc.*, **87**, 4398 (1965).

[80] C. R. Johnson and W. G. Phillips, *J. Org. Chem.*, **32**, 1926 (1967).

[81] L. Bateman, M. Cain, T. Colclough, and J. I. Cunneen, *J. Chem. Soc.*, 3570 (1962).

9.19.4 Sulfones

Sulfones are colorless, odorless solids; however, a number of the dialkyl sulfones have low melting points.

When the sulfone is the primary functionality in the molecule, the two strong sulfone bands at 1325 and 1175 cm^{-1} (7.55 and 8.51 μm) are the most prominent in the infrared spectrum (Fig. 9.28).

Sulfones are relatively inert substances; the group is not readily subject to chemical alteration. Some sulfones can be reduced to sulfide by lithium aluminum hydride.[82] The sulfone group markedly increases the acidity of α-hydrogens. Sulfone anions, readily formed by action of strong bases (e.g., butyllithium), undergo the expected reactions of halogenation, alkylation, and acylation.

Infrared, nuclear magnetic resonance and mass spectral data (the SO_2 grouping is transparent down to 185 nm in the ultraviolet), along with physical constants, constitute the best method of identification.

Aromatic sulfones yield phenols when fused with solid potassium hydroxide.

$$\text{ArSO}_2\text{R} \xrightarrow[\text{(2) } H_2O^+]{\text{(1) KOH, }\Delta} \text{ArOH}$$

9.19.5 Sulfenic Acids and Derivatives

<table>
<tr><td align="center">RSOH</td><td align="center">RSCl</td></tr>
<tr><td align="center">Alkane- or arenesulfenic acid</td><td align="center">Sulfenyl chloride</td></tr>
<tr><td align="center">RSOR'</td><td align="center">R—S—NR$_2'$</td></tr>
<tr><td align="center">Alkyl sulfenate</td><td align="center">Sulfenamide</td></tr>
</table>

The free sulfenic acids are highly unstable substances; only a few have been characterized. They are produced by the pyrolysis of sulfoxides.

The sulfenyl chlorides (prepared by reaction of chlorine and disulfides) are somewhat more stable. The most commonly encountered ones are 2,4-dinitrobenzene-, benzene-, and trichloromethanesulfenyl chloride. Derivatives can be made by reaction with amines to produce amides, reaction with alkoxides in a nonpolar solvent (ether) to produce sulfenates, or by addition to alkenes to produce β-chlorosulfides.

9.19.6 Sulfinic Acids and Derivatives

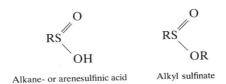

Alkane- or arenesulfinic acid   Alkyl sulfinate

---

[82] F. G. Bordwell and W. H. McKellin, *J. Am. Chem. Soc.*, **73**, 2251 (1951).

The free sulfinic acids are relatively unstable, but their sodium and potassium salts are quite stable. The salts are readily oxidized (permanganate) to sulfonic acid salts.

The sodium or potassium sulfinates can be alkylated to produce sulfones.

$$\text{ArSO}_2\text{Na} + \text{CH}_3\text{I} \longrightarrow \underset{\underset{\text{O}}{\|}}{\overset{\overset{\text{O}}{\|}}{\text{Ar}-\text{S}-\text{CH}_3}} + \text{NaI}$$

The sulfinyl chlorides are readily produced by reaction of the sodium or potassium salts with thionyl chloride. The sulfinyl chlorides undergo the expected reactions with alcohols, etc. The esters (sulfinates), which are capable of existing in optically active forms, readily undergo transesterification with alcohols in the presence of hydrogen chloride, and they react with Grignard reagents to produce sulfoxides.

9.19.7 Sulfonic Acids; RSO₃H

Sulfonic acids are strong acids, comparable to sulfuric acid. The free acids and their alkali metal salts are soluble in water and insoluble in nonpolar organic solvents (ether, etc.).

The free acids are very hygroscopic and do not give sharp melting points. The 1160 cm^{-1}, 1080 to 1000 cm^{-1}, and 700 to 610 cm^{-1} (8.00 to $8.62 \, \mu\text{m}$, 9.26 to $10.0 \, \mu\text{m}$, and 14.28 to $16.39 \, \mu\text{m}$). The first of these is the most intense.

The free acids are very hygroscopic and do not give sharp melting points. The acids are listed in *Tables of Identification of Organic Compounds* in order of increasing melting point of the sulfonamide.

The preparation of *S-benzylthiuronium sulfonates* from free sulfonic acids and their salts is highly recommended because of the ease with which they are made and because of the extensive data available.

$$\text{RSO}_3\text{Na} + \text{C}_6\text{H}_5\text{CH}_2\text{SC(NH}_2)_2^+\text{Cl}^- \longrightarrow \text{C}_6\text{H}_5\text{CH}_2\text{SC(NH}_2)_2^+ \text{ RSO}_3^- + \text{NaCl}$$

9.19.7a Procedure: S-Benzylthiuronium sulfonates

To a solution of 0.5 g of the salt in 5 mL of water, add several drops of 0.1 N hydrochloric acid. Add a cold solution of 1 g of *S*-benzylthiuronium chloride in 5 mL of water. Cool in an ice bath and collect and recrystallize the derivative from aqueous ethanol.

If the free acid is available, dissolve 0.5 g in 5 mL of water and neutralize to phenolphthalein endpoint with sodium hydroxide. Then add two to three drops of 0.1 N hydrochloric acid and proceed as before.

Probably the most commonly employed method for the derivatization of sulfonic acids and salts is the *conversion to sulfonamides* via the sulfonyl chlorides.

$$RSO_3Na + PCl_5 \longrightarrow RSO_2Cl + POCl_3 + NaCl$$

$$RSO_2Cl + 2 NH_3 \longrightarrow RSO_2NH_2 + NH_4Cl$$

9.19.7b Procedure: Sulfonamides

Place 1.0 g of the free acid or anhydrous salt and 2.5 g of phosphorous pentachloride in a small, dry flask equipped with a reflux condenser. Heat in an oil bath at 150°C for 30 min. Cool and add 20 mL of toluene; then warm and stir to effect complete extraction. (If desired, the sulfonyl chloride can be isolated by removal of the solvent. Recrystallize from petroleum ether.) Add the toluene solution slowly, with stirring, to 10 mL of concentrated ammonia solution. If the sulfonamide precipitates, isolate it by filtration. Otherwise it can be isolated by evaporation of the organic layer. Recrystallization can be accomplished from water or ethanol.

Substituted sulfonamides can be obtained by treatment of the foregoing sulfonyl chloride with amines in the usual manner.

The preparation of *arylamine salts* of sulfonic acids is represented by the following equation.

$$RSO_3^- Na^+ + CH_3 - \!\!\!\left\langle \underset{}{\bigcirc} \right\rangle\!\!\! - NH_3^+ Cl^- \longrightarrow CH_3 - \!\!\!\left\langle \underset{}{\bigcirc} \right\rangle\!\!\! - NH_3^+ RSO_3^- + NaCl$$

9.19.7c Procedure: Arylamine salts

Dissolve 0.5 g of the sulfonic acid or its salt in a minimum volume of boiling water and add a saturated aqueous solution of 1 g of *p*-toluidine hydrochloride. Cool and recrystallize from hot water containing a drop of concentrated hydrochloric acid or from dilute ethanol.

Salts of other aromatic amines can be prepared in a similar manner; those of aniline, *o*-toluidine, pyridine, and phenylhydrazine have been employed.

9.19.8 Sulfonyl Chlorides; RSO₂Cl

The simple aliphatic and a number of the aromatic sulfonyl chlorides are liquids. High-molecular-weight aliphatic and substituted aromatic sulfonyl chlorides are solids. They are insoluble in water and soluble in most organic solvents. Their vapors have a characteristic irritating odor and are lacrymatory. In the infrared, strong bands are observed at 1360 and 1170 cm^{-1} (7.35 and 8.55 μm).

Sulfonyl chlorides are easily converted to amides or substituted amides by reaction with concentrated ammonia solution or with amines in the presence of aqueous alkali (see Sec. 9.14.2 on amines).

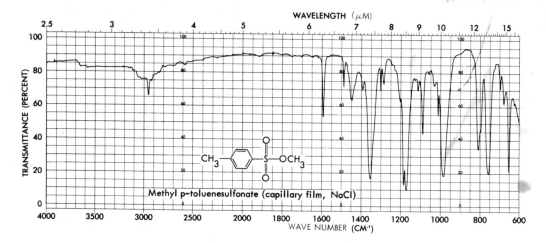

Fig. 9.29. Infrared spectrum of a sulfonate.

9.19.9 Alkyl Sulfonates; RSO₃R′

The sulfonate group is an excellent leaving group from carbon; consequently, these esters, which may be liquids or crystalline solids, are highly reactive in nucleophilic displacement reactions. Probably the most commonly encountered ones are the *p*-toluenesulfonates (*tosylates*) (Fig. 9.29).

Most sulfonates are solvolyzed by heating in water, alcohols, or acetic acid to yield the sulfonic acid and alcohols, ethers, or acetates, respectively, and alkenes. Primary and some secondary alkyl sulfonates (especially tosylates) can be reduced to corresponding hydrocarbons with lithium aluminum hydride.

$$RO_3S\!-\!\!\bigcirc\!\!-CH_3 + LiAlH_4 \longrightarrow RH + CH_3\!-\!\!\bigcirc\!\!-SO_3^-$$

The primary and secondary sulfonates can be used as general alkylating agents to form quaternary ammonium salts, sulfonium salts, etc.

9.19.10 Sulfonamides; RSO₂NH₂

Most sulfonamides are colorless crystalline solids of high melting point that are insoluble in water. The primary and secondary sulfonamides are soluble in aqueous alkali (Hinsberg test). In the infrared, sulfonamides exhibit the appropriate N—H stretching and deformation bands, as well as asymmetric [1370 to 1330 cm⁻¹ (7.30 to 7.52 μm)] and symmetric [1180 to 1160 cm⁻¹ (8.47 to 8.62 μm)] stretching of the S═O (Fig. 9.30).

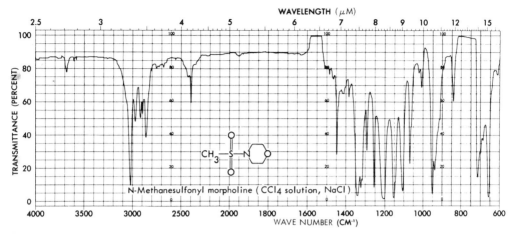

Fig. 9.30. Infrared spectrum of a sulfonamide.

A very sensitive color test for sulfonamides has been developed employing *N,N*-dimethyl-*α*-naphthylamine and nitrous acid.[83]

The only general method for characterization of sulfonamides is *hydrolysis to the sulfonic acid and amine.* The hydrolysis can be effected in a number of ways. Refluxing with 25% hydrochloric acid requires 12 to 36 hours. Eighty percent sulfuric acid or a mixture of 85% phosphoric and 80% sulfuric can also be used. The latter method is particularly efficient.

$$CH_3-\left\langle \bigcirc \right\rangle-SO_2-NHR \xrightarrow{H_3O^+} CH_3-\left\langle \bigcirc \right\rangle-SO_3H + RNH_3^+$$

9.19.10a Procedure: Hydrolysis of sulfonamides

In a test tube mix 1 mL of 80% sulfuric acid with 1 mL of 85% phosphoric acid. Add 0.5 g of the sulfonamide and heat the mixture to 160°C for 10 min or until the sulfonamide has dissolved. A dark, viscous mixture will result. Cool and add 6 mL of water. Continue cooling and, while stirring, render the solution alkaline with 25% sodium hydroxide solution. Separate the amine by extraction or distillation. Prepare an appropriate derivative of the amine (benzoyl) and of the sodium sulfonate (*S*-benzylthiuronium salt).

The sodium salts of primary and secondary sulfonamides can be alkylated by refluxing with an appropriate alkyl halide or sulfate in ethanol.

$$RSO_2NHNa + R'X \longrightarrow RSO_2NHR' + NaX$$

---

[83] C. Hackmann, *Deut. med. Wechschr.*, **72,** 71 (1947); *Chem. Abstracts,* **41,** 4824 (1947).

**Table 9.19. Physical Properties of Sulfides, Sulfones,
Sulfoxides, Thiols, and Derivatives**

Sulfide	bp(°C)	Sulfoxide (mp, °C)	Sulfone (mp, °C)
Dimethyl	38	18.4 (bp 189°C)	109 (bp 238°C)
Ethyl methyl	67		36
Thiophene	84		
Diethyl	91		71
Methyl propyl	95.5		
Ethyl propyl	118.5		
Tetrahydrothiophene	119		27 (bp 285°C)
Diisopropyl	119		36
Dipropyl	142.8		30
Dibutyl	189	34(bp 250°C)	45 (bp 287–295°C)
Methyl phenyl	188		
Ethyl phenyl	204		42
Diphenyl	296	70.4	128
	mp(°C)		
Benzyl phenyl	41	123	146
Dibenzyl	50	134.8	151.7
Di-p-tolyl	57	93	158

| | | mp(°C) of Derivatives | |
Thiol	bp(°C)	2,4-Dinitrophenyl thioether	2,4-Dinitrophenyl sulfone
Methanethiol	6	128	189.5
Ethanethiol	36	115	160
2-Propanethiol	56	94	140.5
2-Methyl-1-propanethiol	88	76	105.5
2-Propen-1-thiol	90	72	105
Butanethiol	97	66	92
Pentanethiol	126	80	83
Hexanethiol	151	74	97
Cyclohexanethiol	159	148	172
Benzenethiol	169	121	161
Phenylmethanethiol	194	130	182.5
1-Phenylethanethiol	199	89	133.4

Table 9.19 lists the physical properties of thiols, sulfoxides, sulfones, sulfonic acids, and their derivatives.

9.19.11 Miscellaneous Organic Sulfur Compounds

The most commonly encountered *thioacids* are thioacetic and thiobenzoic acid. Such acids could, in theory, exist in two forms. In the case of thioacetic acid, these would correspond to thiolacetic and thionacetic acid. However only one acid is known; its properties [carbonyl band at 1720 cm^{-1} ($5.81 \ \mu\text{m}$)] and reactions are best accounted for by the thiol structure. (Mercaptoacetic acid has the structure $HSCH_2COOH$.)

$$CH_3C \overset{O}{\underset{SH}{\big\backslash}} \qquad CH_3C \overset{S}{\underset{OH}{\big\backslash}}$$

Thiolacetic acid   Thionacetic acid

The following equations illustrate characteristic reactions that are of value in the preparation of derivatives of thioacids.

$$RCOSH + C_6H_5NH_2 \xrightarrow{25°C} RCONHC_6H_5 + H_2S$$

$$RCOSH + H_2O_2 \longrightarrow RCOOH$$

$$RCOSH + CH_2{=}CHC_6H_5 \longrightarrow RCOSCH_2CH_2C_6H_5$$

$$RCOS^- + R'X \longrightarrow RCOSR' + X^-$$

Thiolesters are readily hydrolyzed by acid or base to give the carboxylic acid and the thiol.

$$RC \overset{O}{\underset{SR'}{\big\backslash}} \xrightarrow[H^+ \text{ or } OH^-]{H_2O} RCOOH + R'SH$$

The carbonyl band of aliphatic thiolesters appears near 1690 cm^{-1} ($5.92 \ \mu\text{m}$). Thiolesters are oxidized to sulfonic acids by hydrogen peroxide.[84]

$$CH_3C \overset{O}{\underset{SR}{\big\backslash}} \xrightarrow{H_2O_2} CH_3COOH + RSO_3H$$

Thionesters hydrolyze to give an acid, an alcohol, and hydrogen sulfide. Relatively few of these esters are known.

$$RC \overset{S}{\underset{OR'}{\big\backslash}} \xrightarrow{H_2O} RC \overset{O}{\underset{OH}{\big\backslash}} + R'OH + H_2S$$

Because of their instability, *dithioacids* are not well known. The *dithioesters* are reddish-to-yellow oils, many of which are thermochromic. They are very sensitive to atmospheric oxygen, and they are readily desulfurized by Raney nickel.

$$C_6H_5CH_2{-}\overset{\overset{S}{\|}}{C}{-}SCH_3 \xrightarrow{H_2(Ni)} C_6H_5CH_2CH_3$$

---

[84] J. S. Showell, J. R. Russell, and D. Swern, *J. Org. Chem.*, **27**, 2853 (1962).

Thioamides are among the more common and more stable of the compounds containing a thiocarbonyl group. They form addition compounds with mercuric chloride, are alkylated on sulfur to produce salts (analogous to the *S*-alkylthiuronium salts), and react with many oxidants to produce the corresponding amides.

$$
\text{(pyrrolidine ring)} {=}\!S + CH_3I \longrightarrow \text{(pyrrolidine ring)} {-}SCH_3 \quad I^-
$$

$$
C_6H_5CH_2C\!\!\underset{NH_2}{\overset{S}{<}} + PbO \longrightarrow C_6H_5C\!\!\underset{NH_2}{\overset{O}{<}} + PbS
$$

Relatively few *thials* and *thiones* are known. Most are very unstable with respect to dimerization, trimerization, or polymerization.

$$
3\ CH_2S \longrightarrow \text{(1,3,5-trithiane ring)}
$$

Thiobenzophenone is a fairly stable blue compound with a thiocarbonyl band at $1220\ cm^{-1}$ ($8.20\ \mu m$).

The nitrile band in aliphatic *thiocyanates* appears as a very strong absorption near $2140\ cm^{-1}$ ($4.67\ \mu m$); the aromatic derivatives appear at slightly higher wave numbers.

$$
RSCN + R'OH + HCl \longrightarrow RSC\!\!\underset{OR'}{\overset{\overset{+}{N}H_2\ Cl^-}{<}} \longrightarrow RS\overset{O}{\overset{\|}{C}}NH_2
$$

$$
RSCN + [Ox] \longrightarrow RSO_3H + HCN
$$

Isothiocyanates have an extremely intense absorption in the infrared in the 2140 to $1990\ cm^{-1}$ (4.67 to $5.26\ \mu m$) region. The majority appear at the higher-wave number end of the range.

$$
R{-}N{=}C{=}S + R'NH_2 \longrightarrow R{-}NH{-}\overset{S}{\overset{\|}{C}}{-}NH{-}R'
$$

10

Searching
the Literature

The location of specific information in the chemical literature is a challenging task for the chemist. The results of chemical investigations are usually published in journals and/or described in patents. Since many countries issue patents, and the number of journals that accept articles on chemistry is in excess of 10,000, the location of specific chemical data would be almost impossible if not for abstracting services. These abstracting services summarize articles and patents and systematically index chemical information. Articles and other scientific documents dealing with chemistry now appear at an annual total of about 500,000.

The degree of difficulty of completing a thorough literature search depends on the amount and the kind of data available to the searcher, and on the kind of information desired. Specific information on the physical and chemical properties of a specific compound can usually be found with relatively little difficulty. Knowledge of only the molecular formula and/or functional class requires considerable culling of the literature. Literature reviews covering chemical reactions are quite involved, entailing a search of the functional groups involved, reagents used, general chemical processes involved, and the types of products formed. It should be pointed out that the ability to efficiently locate a specific compound in the literature requires a knowledge of systematic nomenclature. It may often be necessary to search under several names of the compound because changes in nomenclature may have occurred. The reader is advised to consult basic texts on organic chemistry for a review of systematic nomenclature.

Various abstract journals and compilations of data are outlined in the following sections. The extent of the discussions are, of necessity, limited, and the reader is referred to the more comprehensive literature guides listed at the end of this chapter. In general, the more recent literature is more likely to contain the most reliable data, and the reader is advised to begin there.

10.1
ABSTRACT JOURNALS

The two most important abstract journals are *Chemical Abstracts* and *Chemisches Zentralblatt*. These two journals provide nearly the same data, the only minor differences appearing in the methods of indexing. Specific compound information is most readily located by use of the subject and formula indexes. Author and patent indexes are also provided, the former being particularly useful in locating or tracing a particular scientist's research.

10.1.1 Chemical Abstracts Service

Chemical Abstracts (*CA*) published by Chemical Abstracts Service, a division of the American Chemical Society, first appeared in 1907. Yearly subject and author indexes are continuous since 1907, with yearly formula and patent indexes being initiated in 1920 and 1935, respectively. Since 1960, semiannual indexes have been published. Decennial indexes appeared in 1916, 1926, 1936, 1946, and 1956, and five-year indexes began appearing in 1961. The current *Chemical Abstracts* indexes are a *Chemical Formula Index*, *Chemical Substance Index*, *General Subject Index*, *Author Index*, and *Patent Index and Concordance*.

Knowledge of *CA* nomenclature practices is required to find chemical compounds in the *Chemical Substance Index*. Nomenclature used by *CA* has evolved so that the same molecular structure may be represented by several names over the years. Since the molecular formula of a chemical substance is invariant, the *Formula Index* provides the quickest reference to an individual compound and gives the name used in the *Chemical Substance Index*. With this name, the *Chemical Substance Index* can be searched for all *CA* references to this compound, as well as to its derivatives, which appear nearby in alphabetical order. Entries for each compound include information to indicate briefly in what context the compound was cited, for example, physical characterization, derivatization, participation in chemical reactions, or use.

The formula indexes are arranged according to the Hill system. Molecular formulas are arranged alphabetically, except when the compound contains carbon; in this case, the carbon is listed first, followed by hydrogen, if present. In compounds containing carbon and hydrogen, the remaining elements are listed in alphabetical order. For example, ammonium silver chloroaurate appears as $Ag_2Au_3Cl_{17}H_{24}N_6$, and 5-bromo-4-methyl-2-thiazolecarboxylic acid is listed as $C_5H_5BrNO_2S$. Under each formula, the preferred names of the compounds included are arranged in alphabetical order. Following the names are reference numbers locating the abstract.

Chemical Abstracts Service Registry Numbers are an integral part of the *Formula* and *Chemical Substance Indexes*. They serve to tie together the printed indexes with a growing computer-produced and computer-readable data base. They represent computer-file addresses at which are collected unique chemical structures and all names (both systematic and nonsystematic) previously employed for a given substance. A *Registry Handbook-Number Section* has been issued. It lists in Registry Number order all of the chemical substances that have been registered in the Registry System, together with their molecular formulas and *CA* index names. Between the inception of the Registry System in 1965 and the end of 1977, about 4 million unique structures had been recorded. New structures are being entered at the rate of about 350,000 per year. Registry Numbers can have up to nine digits and are recognized by the format YYYYXX-XX-X, where the Y's represent optional digits contiguous with those represented by X's. For example, the Registry Number of (*E*)-2-hexenal is 6728-26-3.

Chemical Abstracts Service has published the *Parent Compound Handbook*; supplements are issued every two months. This is a major reference tool containing information about all the ring systems and natural products presently in the Chemical Registry System. The *Parent Compound Handbook* includes the following for a parent compound: (1) chemical structural diagram illustrating the nomenclature locant numbering system; (2) Registry Number; (3) current *CA* index name used in the *Chemical Substance Index*; (4) molecular formula; (5) Wiswesser line notation (a linear abbreviation of the structural formula); (6) ring data (number of rings, size of rings, and elemental composition of rings); (7) *CA* reference [if parent compound was referenced in a *CA* abstract after Volume 78 (1973)].

Author indexes are used when a specific work or series of articles in an area by a given scientist is to be located. Patent indexes list only patent numbers and reference numbers to the abstract. The Concordance gives cross references to the same patent issued in more than one country.

The abstracts provided by *Chemical Abstracts* generally contain only a portion of the essential material contained in the original paper. If at all possible, the original literature should be consulted for the desired information.

10.1.2 Chemisches Zentralblatt

Chemisches Zentralblatt is the oldest abstract journal, first appearing in 1830. A number of minor changes in the title and in the numbering of yearly volumes have occurred.

The information provided by *Chemisches Zentralblatt* is nearly identical to that provided by *Chemical Abstracts*. The main difference is in the organization of the formula indexes, which employ the Richter system of classification of carbon-containing compounds, beginning with the sixth collective index (1922 to 1924).

Compounds are tabulated according to their molecular formula, beginning with those compounds containing one carbon atom. The sequence of elements in combination with carbon are given in the sequence H, O, N, Cl, Br, I, F, S, and P, the remaining elements of the periodic table following in alphabetical order. A "guide number" is assigned to each formula. This guide number is composed of an Arabic numeral referring to the number of carbon atoms present and a Roman numeral referring to the number of other elements. For example, 5-bromo-4-methyl-2-thiazolecarboxylic acid would be assigned a guide number 5V. One then simply looks at the top of the pages in the formula index until the guide number is located. The compound can then be located on that page.

10.1.3 Miscellaneous Abstract Journals

An abstracting and indexing service is provided by the Russian language publication, *Referativnyi Zhurnal, Khimiya*. Several other abstract journals are available. Many of these cover special, and thus more limited, areas. However, owing to their more limited nature, they can well cover highly specialized journals, which in some cases can be rather obscure. The abstract journals include *Journal of the Chemical Society*, 1871–1925: *Journal of the Society of Chemical Industry*, 1882–1925; *British Chemical Abstracts*, 1926–1953; *Bulletin de la Societe chimique de France*, 1863–1946; *Berichte der deutschen chemischen Gesellschaft (Referate)*, 1880–1895; *Nuclear Science Abstracts*, beginning in 1948 and continuing to the present, and *Science Abstracts*, beginning in 1898 and continuing to the present.

10.2
CURRENT LITERATURE COVERAGE SERVICES

The indexes of the various abstract journals usually lag significantly behind the publication of the abstracts. To browse the abstract journals requires considerable time. In order to help keep the scientist abreast of the current literature, current literature surveying journals have been introduced.

Chemical Titles, published by the Chemical Abstracts Service of the American Chemical Society, is a bi-monthly publication that publishes the titles of all the articles appearing in approximately 700 selected journals of pure and applied chemistry and chemical engineering. It also contains author and keyword indexes in each volume.

Current Contents: Physical and Chemical Sciences, a weekly publication of the Institute for Scientific Information, provides a reproduction of the tables of contents of chemical journals appearing that week. An author index listing authors' addresses is provided; a subject index is an available option.

Current Abstracts of Chemistry and Index Chemicus, published weekly by the Institute for Scientific Information, provides a search system to locate new organic compounds, syntheses, and reactions reported in chemistry journals. Article abstracts along with key structural formulas usually appear within 45 days after their original journal publication.

Individual chemists can subscribe to various customized computer search services based on a personal listing of subjects and authors of special interest. Personalized computer reports are provided from a data base consisting of titles, authors, and, in some cases, key words taken from current chemical journals.

Using data bases provided by Chemical Abstracts Service interactive on-line computer searching of *Chemical Abstracts* (since 1970) is possible. The data base is updated biweekly. Searches can be made by key-word phases, bibliographic data (authors' names, locations, journal titles, dates, country of origin of patent, etc.), *Chemical Abstracts* nomenclature (full name or substructure name, e.g., 5-nitro-), general subject index entries, and registry numbers. DIALOG® (a service of Lockheed Information Systems) allows interactive searching of *Chemical Abstracts* data bases using a computer located in Palo Alto, California, accessible to subscribers through national and international (satellite) telephone networks.

10.3
COMPILATIONS OF CHEMICAL
AND/OR PHYSICAL DATA

Numerous attempts have been made to tabulate the physical data and, in many cases, the chemical data of specific compounds. Many of these compilations are included in Fig. 10.1, illustrating their time-period coverage.

Several of the most important of these compilations will be discussed individually.

10.3.1 Beilstein

Perhaps the most important compilation of chemical and physical data on individual compounds, and hence frequently referred to as the "organic chemists' bible," is Beilstein's *Handbuch der organischen Chemie*. Beilstein is composed of a main series (29 volumes), which covers all organic compounds through 1909, and four supplementary series covering the years 1910 to 1919, 1920 to 1929, 1930 to 1949, and 1950 to 1959 (the last was not complete at the time of publication of this text).

The classification of compounds in Beilstein is complex and will not be discussed in this text. A detailed description of the classification system is given on pp. 1 to 46 of Vol. 1 of the main series. *The Beilstein Guide* by O. Weissback (Springer-Verlag, New York and Heidelberg, 1976) describes the organization of Beilstein with precision and succinctness. Many helpful examples are included, as well as a German–English–French glossary, which enables the reader who knows little or no German to make effective use of Beilstein. *A Brief Introduction of*

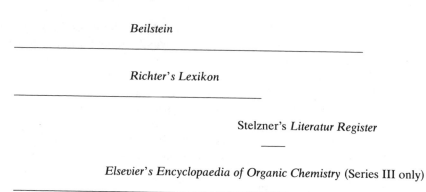

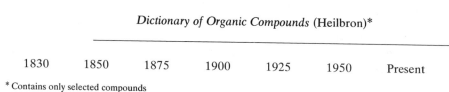

| 1830 | 1850 | 1875 | 1900 | 1925 | 1950 | Present |

* Contains only selected compounds

Fig. 10.1 Compilations of data on individual compounds

Beilstein's Handbuch der organischem Chemie by E. H. Huntress (John Wiley & Sons, Inc., New York, 1938) is also useful.

Locating a desired compound in Beilstein is greatly facilitated by the availability of a compound formula index covering the main series and the first two supplementary series. The formula index gives the names of various isomers possessing a given formula, and it gives a reference number. The reference number is composed of a Roman numeral indicating the volume (band) number and an Arabic numeral that indicates on what page of the main series the compound reference will be found. Occasionally a compound will not be found in the main series; the Beilstein reference number then indicates the volume and page number on which the compound would appear if contained in the main series. The supplementary series contain the Beilstein reference number at the top center of each page; the student then simply searches for the reference number to locate a given compound in the supplementary series.

The citation for each compound includes bibliographical citations, mode of preparation or occurrence, physical properties, chemical reactions, physiological properties, uses, methods of analysis for, and derivatives of.

10.3.2 Richter's Lexikon

Richter's *Lexikon* contains a compilation of all known organic compounds up to the date of its publication in 1910. The compounds are classified according to the Richter system. The four volumes of Richter's *Lexikon* are classified as follows: Volume I, C_1 to C_9 (1I to 9III): Volume II, C_9 to C_{13} (9III to 13IV); Volume III, C_{13} to C_{20} (13IV to 20II); and Volume IV, C_{20} and higher (20II to 1039IV). Richter's *Lexikon* contains melting- and boiling-point data and a few references to the original literature.

10.3.3 Stelzner's Literatur-Register

Stelzner's *Literatur-Register* is a compilation of physical data on compounds covering the period of 1910 to 1921. The accompanying references are more extensive than those contained in Richter's *Lexikon*. The entries are classified according to the Richter system. The five volumes of this work are divided on the basis of time covered and not on compound classification, as is Richter's *Lexikon*: Volume I covers the period 1910 to 1911; Volume II, 1912 to 1913; Volume III, 1914 to 1915; Volume IV, 1916 to 1918; and Volume V, 1919 to 1921.

10.3.4 Compilations by Mulliken and Huntress

A series of volumes prepared by Mulliken and/or Huntress have appeared tabulating physical properties for a variety of compounds. The original volume in this series by Mulliken, entitled *A Method for the Identification of Pure Organic Compounds*, Vol. 1 (1904), also describes certain techniques of structure determination.

The compounds appearing in these volumes are classified according to the types of atoms present, possession of color and functional groups, and their physical properties. Compounds are divided into *orders* on the basis of the types of atoms present and into two *suborders*, depending on whether the compound is colored or colorless. Further subdivision into *genus* is based on the types of functional groups present. The final classifications by *divisions* and *sections* are dependent on the physical state (solid or liquid) and solubility or density data, respectively. In the respective sections, compounds are listed in order of increasing melting or boiling points. In the first two titles listed below, location numbers are assigned to each section (see the introduction section of these two volumes), and the compounds are located in the text by use of these location numbers. These volumes provide data on physical properties, derivatives, literature references, and frequency on the chemical properties.

1. HUNTRESS, E. H., *Organic Chlorine Compounds*, (order III containing chlorine, in addition to C, H, and C, H, and O). New York; John Wiley & Sons, Inc., 1948.

2. HUNTRESS, E. H., and MULLIKEN, S. P., *Identification of Organic Compounds of Order I*, (containing C, H or C, H, and O). New York; John Wiley & Sons, Inc., 1941.

3. MULLIKEN, S. P., *Identification of Pure Organic Compounds*, Vol. II. New York; John Wiley & Sons, Inc., 1916.

4. MULLIKEN, S. P., *Identification of Pure Organic Compounds*, Vol. IV. New York; John Wiley & Sons, Inc., 1922.

10.3.5 Melting-Point Tables of Organic Compounds

Melting-Point Tables of Organic Compounds [second supplemented edition by W. Utermark and W. Schicke, and published by Interscience Publishers, Inc., New York, N.Y. (1963)] is a tabulation of physical and chemical properties of pure organic compounds. The tables present the melting point, boiling point, crystalline form, other physical properties, chemical reactions, and the Beilstein reference number. Although the compounds are listed in order of increasing melting point, a formula index facilitates locating a specific compound.

Another compilation essentially identical to the one just described is *Schmelzpunktstabellen* by R. Kempf and F. Kutter (Druck und Verlag von Friedr. Vieweg & Sohn Akt.-Ges., Braunschweig, Germany, 1928).

10.3.6 Elsevier's Encyclopaedia of Organic Compounds

Elsevier's *Encyclopaedia of Organic Compounds* (Elsevier Publishing Company, Amsterdam) was originally designed "to give complete information on all chemical and physical properties and on the most important physiological properties of organic compounds." The accomplishment of such a proposed design is truly a tremendous task and has proved, apparently, to be too great. Although the area of organic chemistry was originally divided into four major series (Series I, acyclic compounds; Series II, cyclic noncondensed compounds; Series III, cyclic condensed compounds, and Series IV, heterocyclic compounds) only Volumes 12, 13, and 14 of Series III have appeared. Volume 12 covers bicyclic compounds, Volume 13 covers tricyclic compounds, and Volume 14 covers tetracyclic and higher polycyclic compounds. Several supplements have appeared in recent years, making a total of 19 individual covers.

The compounds in each category are divided according to the total number of carbon atoms and, if cyclic, to the smallest cycle first. Further classification is according to the types of functional groups present in the order hydrocarbons, halides, nitrogen-containing, hydroxy compounds, carbonyl compounds, carboxylic acids, and followed by compounds containing S, Se, Te, P, As, and other metals. The information provided for each compound includes the physical properties, mode of formation or source, references, chemical reactions, and derivatives. Each greater system (particular carbon structure system) is introduced by a detailed systematic survey covering the various modes of formation of the carbon structure.

10.3.7 Dictionary of Organic Compounds

The *Dictionary of Organic Compounds* (Oxford University Press), commonly referred to as *Heilbron*, is a compilation of physical and chemical properties, and syntheses and reactions of compounds, with references to the original literature, The first volume was published in 1934, with the most recent, tenth volume and cumulative supplement being published in 1974 (Oxford University Press, New York). The compounds are listed alphabetically according to the I.U.P.A.C. 1957 nomenclature rules.

10.3.8 Chemical Rubber Handbooks

The CRC Press, Cleveland, Ohio, publishes a series of handbooks containing data useful for the identification of organic compounds.

1. *Handbook of Tables for Organic Compound Identification*, Z. Rappaport, ed. 1967; Information on each compound includes name, boiling point or melting point, refractive index, density, and properties of up to eight derivatives.
2. *Handbook of Chemistry and Physics*, 57th ed., R. C. Weast, ed. 1976; A classic handbook of mathematical tables, physical properties of the elements, inorganic and organic compounds, and miscellaneous information.
3. *Atlas of Spectral Data and Physical Constants for Organic Compounds*, 2nd ed., J. G. Grasselli and W. M. Ritchey, eds. 1975; A six-volume atlas providing names, structures, Wiswesser line notations, CAS Registry Numbers, literature references, molecular weights, melting points, boiling points, densities, refractive indexes, solubilities, and data on infrared, Raman, ultraviolet, carbon-13 and proton magnetic resonances, and mass spectra.

10.3.9 Merck Index

Merck and Company, Rahway, N. J., periodically publishes a compilation of the physical and chemical properties of drugs and chemicals related to or used in the drug industry. This compilation is particularly useful in this rather limited area.

10.3.10 Other Sources of Physical Properties
of Compounds

The following references provide additional, but quite limited, sources of data on the physical properties of compounds. No attempt to describe the use of these references will be made, since they are all quite self-explanatory.

1. CHERONIS, N. D., ENTRIKIN, J. B., and HODNETT, E. M., *Semimicro Qualitative Organic Analysis*, 3rd ed. New York; Interscience Publishers, Inc., 1965.
2. *Lange's Handbook of Chemistry*, Handbook Publishers, Inc., Sandusky, Ohio.
3. SHRINER, R. L., FUSON, R. C., and CURTIN, D. Y., *The Systematic Identification of Organic Compounds*, 5th ed. New York: John Wiley & Sons, Inc., 1964.

10.3.11 Spectra Compilations

Extensive compilations of spectra in the various regions of the electromagnetic spectrum have been published. The references for such compilations appear at the end of the respective spectroscopy chapters. See also ref. 3, Sec. 10.3.8.

10.4
REFERENCE TEXTS ON THE USE OF THE LITERATURE

Several more extensive texts are available describing the use of library resources. These are now listed.

1. CRANE, E. J., PATTERSON, A. M., and MARR, E. B., *A Guide to the Literature of Chemistry*, 2nd ed. New York; John Wiley & Sons, Inc., 1957.
2. MELLON, M. G., *Chemical Publications—Their Nature and Their Use*. New York; McGraw-Hill Book Company, 1958.
3. SOULE, B. A., *Library Guide for the Chemist*. New York; McGraw-Hill Book Company, 1938.

A series of articles by Hancock [*J. Chem. Ed.*, **45,** 193, 260, 336 (1968)] are highly recommended for further reading.

11

Introduction to Structural Problems

11.1
INTRODUCTION

The successful solution of a structure problem requires recognition of the various partial structures indicated to be present in the molecule by the various bits of chemical and physical (primarily spectral) data. It is the assembling of these bits of data that is the most difficult part of the problem. It is necessary to be able to recognize certain common features (atoms or groups of atoms) that may appear in two or more of the partial structures that can be used as cement in putting the various fragments together. There are certain techniques that are very useful in solving structure problems, which will be outlined by the use of specific examples.

Sample Problem 1. The IR and NMR spectra of Compound **A** are given in Figs. 11.1 and 11.2. Compound **A**, $C_{10}H_{14}O$, reacts with the Lucas reagent to produce an insoluble oil.

Oxidation of **A**, with chromic acid produces **B**, $C_{10}H_{12}O$, which produces an orange-red precipitate when treated with 2,4-dinitrophenylhydrazine. The IR and NMR spectra of **B** are provided in Figs. 11.3 and 11.4.

Deduce the structure of **A** and **B**.

The first objective is to determine the number of sites of unsaturation in **A** and **B** by use of Eq. (8.1) (Sec. 8.3.1). Compounds **A** and **B** have four and five sites of unsaturation, respectively.

The infrared spectrum of **A** indicates the presence of a hydroxyl group (3300 cm^{-1}). The NMR spectrum shows a singlet at $\delta 7.13$ with a relative intensity

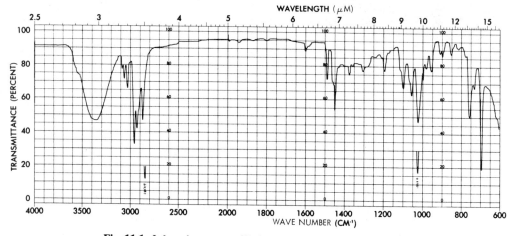

Fig. 11.1. Infrared spectrum (film) of unknown **A**, sample problem 1.

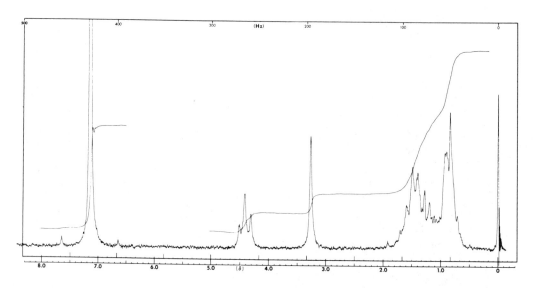

Fig. 11.2. NMR spectrum of unknown **A**, sample problem 1.

of 5, consistent with a monoalkyl substituted benzene; a benzene ring would account for the four sites of unsaturation. The slightly distorted triplet at $\delta 4.11$ (relative intensity 1) is substantially deshielded and is adjacent to a methylene (CH_2) group. This system probably represents the partial structure

$$
\begin{array}{c}
\text{H} \\
| \\
-\text{C}-\text{OH} \\
| \\
-\text{CH}_2
\end{array}
$$

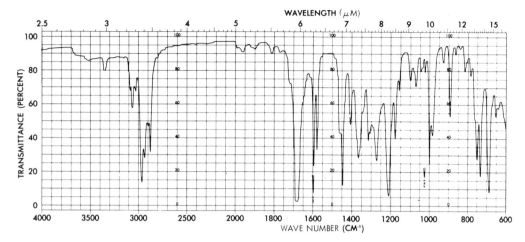

Fig. 11.3. Infrared spectrum (film) of unknown **B**, sample problem 1.

The positive Lucas test indicates the presence of an alcohol (or other functional group capable of reacting with the Lucas reagent such as an epoxide) that reacts via a S_N^1 mechanism (tertiary, allylic, or benzylic). This chemical test suggests that a benzene ring is attached to the carbinol carbon (Note: This is confirmed by the IR data of **B**). The NMR spectrum displays a highly distorted system at δ0.83 (relative intensity of 3), which probably represents one methyl bonded to a methylene group (this pattern is typical of an A_2B_3 spin system).

The IR spectrum of **B** indicates the presence of an aryl ketone (1680 cm^{-1}) and is consistent with the partial structure of **A** indicated above. The NMR spectrum

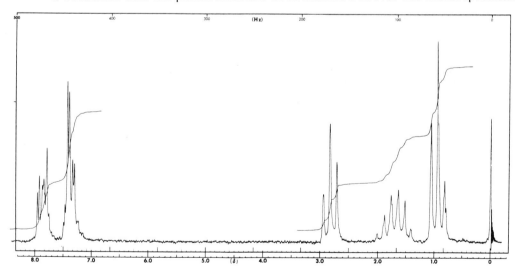

Fig. 11.4. NMR spectrum of unknown **B**, problem 1.

of **B** readily clarifies the nature of the remaining portion of the molecule. The triplet at δ2.81 (relative intensity of 2) represents the methylene group adjacent to the carbonyl group. The high field resonance at δ0.98 represents a methyl attached to methylene of not too different chemical shift. The multiplet at δ1.72 (relative intensity of 2) must represent the methylene group bonded to the methyl group and to the methylene attached to the carbonyl group.

The structures of **A** and **B** must therefore be

$$\underset{\textbf{A}}{\overset{\overset{\displaystyle OH}{|}}{C_6H_5CHCH_2CH_2CH_3}} \qquad \underset{\textbf{B}}{\overset{\overset{\displaystyle O}{\parallel}}{C_6H_5CCH_2CH_2CH_3}}$$

Sample Problem 2. Sarkomycin, $C_7H_8O_3$, is a very unstable acid isolated from a certain type of bacteria. Sarkomycin displays characteristic peaks in the infrared region at 3450 (broad), 1710 (intense and broad), 1639, and 886 cm^{-1}. Sarkomycin adsorbs 1 mol of hydrogen, giving a new acid, which displays peaks in the infrared region at 3450 (broad), 1739, and 1710 cm^{-1}. Ozonolysis of sarkomycin gives formaldehyde and an acid (not isolated), which on oxidation with potassium permanganate gives succinic acid. Devise suitable structures for sarkomycin and all degradation products.

Sarkomycin contains four sites of unsaturation of which one site is present in the carboxyl group, which must be present in the molecule.

The infrared spectrum of sarkomycin indicates the presence of a bonded —OH which, in conjunction with the carbonyl absorption at 1710 cm^{-1}, represents the —COOH group. The bands at 1639 and 886 cm^{-1} indicate the presence of a carbon-carbon double bond that is probably a terminal methylene group that we will represent as =CH$_2$. This is confirmed by the formation of formaldehyde on ozonolysis.

Hydrogenation of sarkomycin produces a dihydroderivative, which still displays —COOH absorption at 3450 and 1710 cm^{-1}. However, a new band has been generated in the carbonyl region at 1739 cm^{-1}, which is characteristic of either a five-membered cyclic ketone or a six-membered or acyclic ester. The latter possibility is ruled out on the basis that an insufficient number of oxygen atoms are present to accommodate both a —COOH group and a —COO— group of an ester. The production of the 1739 cm^{-1} band on hydrogenation requires the conversion of an α,β-unsaturated five-membered ring ketone (1710 cm^{-1}) to a saturated system. Therefore the partial structure **1** must be present in sarkomycin. It is therefore only necessary to locate the position of the —COOH group.

Ozonolysis of sarkomycin must give partial structure **2**, which on further oxidation gives succinic acid. This oxidation involves the loss of two carbon atoms, probably as carbon dioxide. Only two possibilities exist for the placement of the —COOH group in **2**, either adjacent to the carbonyl group (**3**), or the 4-position (**4**). Permanganate oxidation of **4**, in the enol form in which it probably exists, would give **5**, which on decarboxylation would give succinic acid. Oxidation of **3**, in its expected preferred enol form, would give succinic acid directly.

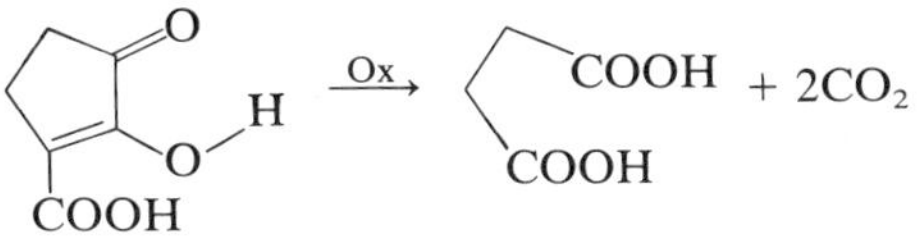

3

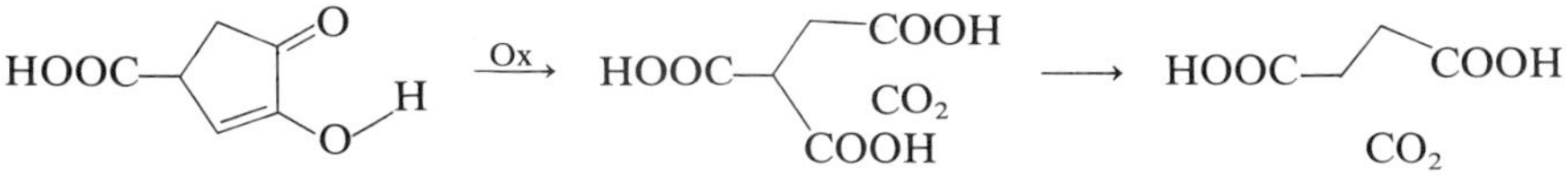

4   **5**

It is obvious that no unique structure can be proposed for Sarkomycin strictly on the basis of the information provided. Three structures are possible and are given below (**6, 7,** and **8**). The indicated instability of sarkomycin would tend to favor **6**, in which a β-ketoacid is present and may undergo decarboxylation.

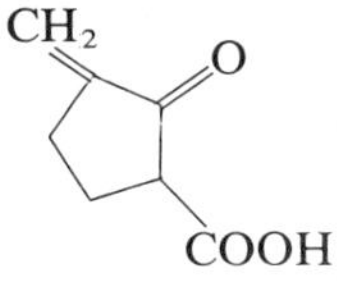

6 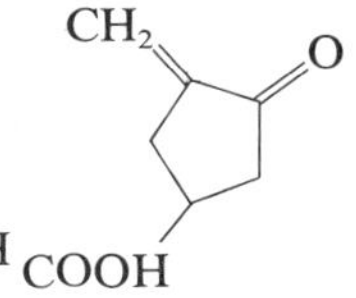   **7** 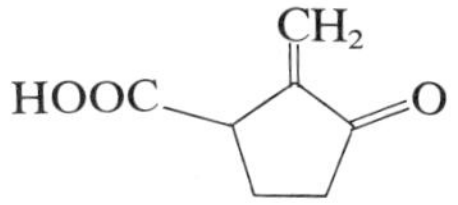   **8**

However, no pertinent data concerning this point is given. The availability of the nuclear magnetic resonance spectrum would readily indicate the correct structure. The student is encouraged to consider the expected differences in the resonance spectra of **6, 7,** and **8**.

In cases in which no unique structure can be derived the student should consider what additional information may be needed and how to derive that information (either chemical or physical). An important part of any structure problem is recognizing the limitations of the data and what additional data are needed.

11.2
PROBLEMS

In the following problems infrared data will be given only in cm^{-1}. The NMR data will be given as δ. Immediately following the δ value the multiplicity,[1] coupling

---

[1] Multiplicity is indicated by s(singlet), d(doublet), t(triplet), and m(complex multiplet).

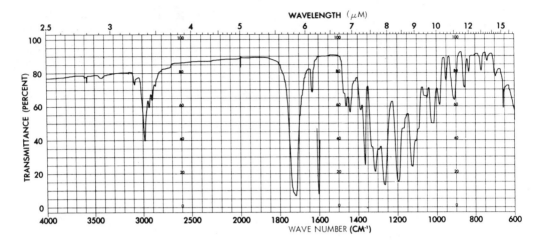

Fig. 11.5. Infrared spectrum (film) of unknown **A**, problem 1.

constant, and relative integrated intensity will be given in parentheses, e.g., $\delta 2.09$ $(d, J = 6.0\ \text{Hz};\ 3)$.

1. Reaction of commercially available 1,4-dibromo-2-butene with diethyl malonate in ethanol containing two equivalents of sodium ethoxide produces compound **A** in high yield. The high resolution mass spectrum of **A** indicates a molecular formula of $C_{11}H_{16}O_4$. The infrared spectrum of **A** is shown in Fig. 11.5. At 450°C compound **A** is cleanly isomerized to **B**. The NMR spectrum of **B** is reproduced in Fig. 11.6. Suggest structures for **A** and **B**.

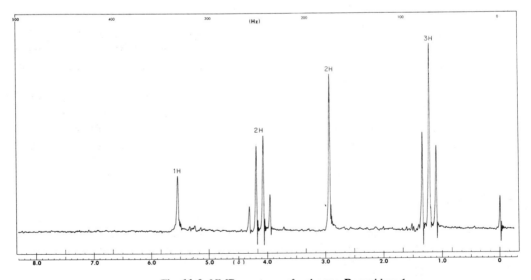

Fig. 11.6. NMR spectrum of unknown **B**, problem 1.

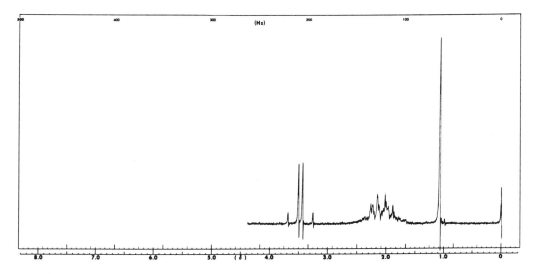

Fig. 11.7. Complete NMR spectrum of unknown **B**, problem 2.

2. Reaction of cyclobutanone with isopropenyl magnesium chloride in tetrahydrofuran, followed by treatment with aqueous ammonium chloride, provides compound **X**. When **X** is reacted with *t*-butyl hypochlorite in chloroform solution in the dark, compound **Y** is produced. The IR spectrum of **Y** exhibits no significant absorption above 3000 cm^{-1} but does contain a very strong band at 1743 cm^{-1}. The complete NMR spectrum of **Y** is shown in Fig. 11.7. Suggest structures for **X** and **Y**.

3. Beginning with enamine **I** the following reaction sequence was achieved:

$$
\begin{array}{c}
\overset{\displaystyle \text{N(CH}_3)_2}{\underset{|}{}} \\
\text{C}_6\text{H}_5\!-\!\text{C}\!=\!\text{CH}_2 + \text{CH}_3\text{SO}_2\text{Cl} \xrightarrow[\text{Benzene}]{\text{Triethylamine}} \text{C}_{11}\text{H}_{15}\text{NO}_2\text{S} \\
\mathbf{I} \qquad\qquad\qquad\qquad\qquad\qquad\qquad \mathbf{II}
\end{array}
$$

$$
\downarrow \begin{array}{l} (1)\ \text{H}_2\text{O}_2 \\ (2)\ \Delta \end{array}
$$

$$
\text{CH}_3\text{SO}_2\text{CH}_3 + \mathbf{V} \xleftarrow[\Delta]{\text{OH}^-} \mathbf{IV} \xleftarrow{\text{OH}^-} \text{C}_9\text{H}_8\text{O}_2\text{S}
$$
$$
\mathbf{III}
$$

Compound **III** IR: 1600, 1310 (*s*), 1123 (*s*) cm^{-1}.
NMR: $\delta 4.77$ (*s*, 2), 6.92 (*s*, 1), 7.42 (*s*, 5).

Compound **IV** IR: 1675 (*s*), 1325 (*s*), 1151 (*s*) cm^{-1}.
NMR: $\delta 3.12$ (*s*, 3), 4.16 (*s*, 2), 7.8 (*m*, 5).

Compound **V** mp 121°C; neut. equiv. 121

Suggest structures for **II** through **V**.

4. Compound **A** was isomerized to **B** ($\nu_{max}^{C=O}$ 1712 cm^{-1}) by boiling in water under a nitrogen atmosphere. Treatment of **B** with hydrogen peroxide gave **C**. Titration of **C**

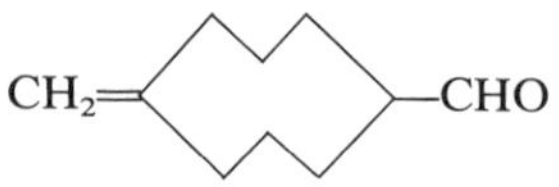

$$\text{CH}_2 = \langle\rangle\langle\rangle - \text{CHO}$$

A

with 0.01 N sodium hydroxide indicated a neutralization equivalent of 167. The NMR of **C** included a singlet (3H) at δ1.1 and no absorptions due to olefinic hydrogens. Suggest structures of **B** and **C**.

5. Compound **I**, $C_{11}H_{16}O$, displays absorption in the infrared region at 1683 and 1600 cm^{-1}. The ultraviolet spectrum displays maxima at 239 (6500) and 320 nm (75). Kuhn-Roth determination indicates the presence of 0.8 C—CH$_3$.

Treatment of **I** with methyl magnesium iodide produces an alcohol **II**, which on dehydration yields as the predominant product **III**, $C_{12}H_{18}$ with λ_{max}^{EtOH} 283 nm (16,000). Dehydrogenation of **III** with selenium produces 1,3-dimethylnaphthalene. Propose structures for **I**, **II**, and **III**.

6. Extraction of black Hemiptera bugs produces a colorless oil **I** possessing a characteristic nauseating odor. Microanalysis (*Anal.* Found: C, 75.02; H, 10.62) indicates the absence of nitrogen, sulfur, and halogen. Treatment of **I** with 2,4-dinitrophenylhydrazine produces a yellow precipitate. Compound **I** gives positive tests with both Tollen's and Fehling's solutions. The infrared spectrum of **I** displays characteristic peaks at 1725, 1380, and 975 cm^{-1}.

Synthetic compound **I** can be produced by thermal rearrangement of 3-(1-pentenyl) vinyl ether.

Propose a structure for compound **I**.

7. Pyrolysis of 6.7-dimethylbicyclo[3.2.0]hept-6-en-2-ol at 405°C produces two compounds **A** and **B** with molecular formula $C_9H_{14}O$. Compounds **A** and **B** display the following spectral properties:

Compound **A**   IR: 1670 and 1650 cm^{-1}.
UV: λ_{max}^{EtOH} 238 nm (11,300).
NMR: δ1.18 (d, J = 6.8 Hz, 3), 1.85 (m, 4), 1.91. (d, J = 1.5 Hz, 3), 2.50 (m, 3), and 5.72 (m, 1).

Compound **B**   IR: 1705 and 1661 cm^{-1}.
UV: λ_{max}^{EtOH} 295 nm (450).
NMR: δ1.80 (s, 6), 2.00 (m, 2), 2.40 (m, 4), and 3.02 (s, 2).

Suggest structures for **A** and **B**.

APPENDICES

Appendix I

Vapor Pressure-Temperature Nomograph

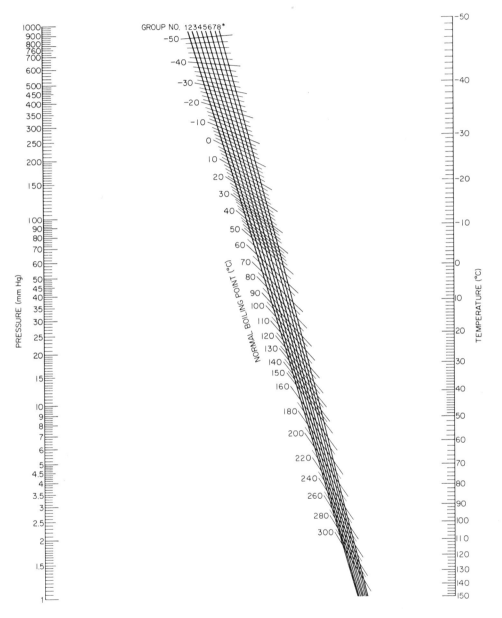

From *Industrial and Engineering Chemistry*, Vol. 38, p. 320, 1946. Copyright 1946 by the American Chemical Society and reproduced by permission of the copyright owner.

* See Table A.1.

Table A.1. Groups of Compounds Represented in Nomographs

Group 1	Group 3	Group 5
Anthracene	Acetaldehyde	Ammonia
Anthraquinone	Acetone	Benzyl alcohol
Butylethylene	Amines	Methylamine
Carbon disulfide	Chloroanilines	Phenol
Phenanthrene	Cyanogen chloride	Propionic acid
Sulfur monochloride	Esters	
Trichloroethylene	Ethylene oxide	
	Formic acid	Group 6
Group 2	Hydrogen cyanide	Acetic anhydride
Benzaldehyde	Mercuric chloride	Isobutyric acid
Benzonitrile	Methyl benzoate	Water
Benzophenone	Methyl ether	
Camphor	Methyl ethyl ether	
Carbon suboxide	Naphthols	Group 7
Carbon sulfoselenide	Nitrobenzene	Benzoic acid
Chlorohydrocarbons	Nitromethane	Butyric acid
Dibenzyl ketone	Tetranitromethane	Ethylene glycol
Dimethylsilicane		Heptanoic acid
Ethers	Group 4	Isocaproic acid
Halogenated hydrocarbons	Acetic acid	Methyl alcohol
Hydrocarbons	Acetophenone	Valeric acid
Hydrogen fluoride	Cresols	
Methyl ethyl ketone	Cyanogen	
Methyl salicylate	Dimethylamine	Group 8
Nitrotoluenes	Dimethyl oxalate	n-Amyl alcohol
Nitrotoluidines	Ethylamine	Ethyl alcohol
Phosgene	Glycol diacetate	Isoamyl alcohol
Phthalic anhydride	Methyl formate	Isobutyl alcohol
Quinoline	Nitrosyl chloride	Mercurous chloride
Sulfides	Sulfur dioxide	n-Propyl alcohol

Vapor Pressure-Temperature Nomograph

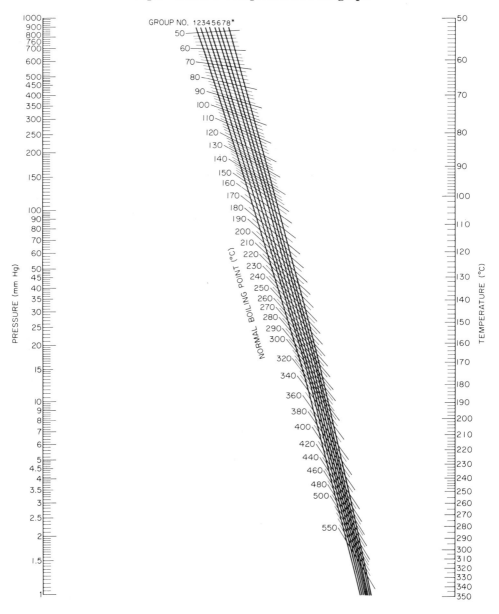

From *Industrial and Engineering Chemistry*, Vol. 38, p. 320, 1946. Copyright 1946 by the American Chemical Society and reproduced by permission of the copyright owner.

523   * See Table A.1.

Appendix II

Wave-number Wavelength Conversion Table

Wavelength(μm)	Wave number(cm⁻¹) 0	1	2	3	4	5	6	7	8	9
2.0	5000	4975	4950	4926	4902	4878	4854	4831	4808	4785
2.1	4762	4739	4717	4695	4673	4651	4630	4608	4587	4566
2.2	4545	4525	4505	4484	4464	4444	4425	4405	4386	4367
2.3	4348	4329	4310	4292	4274	4255	4237	4219	4202	4184
2.4	4167	4149	4132	4115	4098	4082	4065	4049	4032	4016
2.5	4000	3984	3968	3953	3937	3922	3906	3891	3876	3861
2.6	3846	3831	3817	3802	3788	3774	3759	3745	3731	3717
2.7	3704	3690	3676	3663	3650	3636	3623	3610	3597	3584
2.8	3571	3559	3546	3534	3521	3509	3497	3484	3472	3460
2.9	3448	3436	3425	3413	3401	3390	3378	3367	3356	3344
3.0	3333	3322	3311	3300	3289	3279	3268	3257	3247	3236
3.1	3226	3215	3205	3195	3185	3175	3165	3155	3145	3135
3.2	3125	3115	3106	3096	3086	3077	3067	3058	3049	3040
3.3	3030	3021	3012	3003	2994	2985	2976	2967	2959	2950
3.4	2941	2933	2924	2915	2907	2899	2890	2882	2874	2865
3.5	2857	2849	2841	2833	2825	2817	2809	2801	2793	2786
3.6	2778	2770	2762	2755	2747	2740	2732	2725	2717	2710
3.7	2703	2695	2688	2681	2674	2667	2660	2653	2646	2639
3.8	2632	2625	2618	2611	2604	2597	2591	2584	2577	2571
3.9	2564	2558	2551	2545	2538	2532	2525	2519	2513	2506
4.0	2500	2494	2488	2481	2475	2469	2463	2457	2451	2445
4.1	2439	2433	2427	2421	2415	2410	2404	2398	2392	2387
4.2	2381	2375	2370	2364	2358	2353	2347	2342	2336	2331
4.3	2326	2320	2315	2309	2304	2299	2294	2288	2283	2278
4.4	2273	2268	2262	2257	2252	2247	2242	2237	2232	2227
4.5	2222	2217	2212	2208	2203	2198	2193	2188	2183	2179
4.6	2174	2169	2165	2160	2155	2151	2146	2141	2137	2132
4.7	2128	2123	2119	2114	2110	2105	2101	2096	2092	2088
4.8	2083	2079	2075	2070	2066	2062	2058	2053	2049	2045
4.9	2041	2037	2033	2028	2024	2020	2016	2012	2008	2004
5.0	2000	1996	1992	1988	1984	1980	1976	1972	1969	1965
5.1	1961	1957	1953	1949	1946	1942	1938	1934	1931	1927
5.2	1923	1919	1916	1912	1908	1905	1901	1898	1894	1890
5.3	1887	1883	1880	1876	1873	1869	1866	1862	1859	1855
5.4	1852	1848	1845	1842	1838	1835	1832	1828	1825	1821
5.5	1818	1815	1812	1808	1805	1802	1799	1795	1792	1789
5.6	1786	1783	1779	1776	1773	1770	1767	1764	1761	1757
5.7	1754	1751	1748	1745	1742	1739	1736	1733	1730	1727
5.8	1724	1721	1718	1715	1712	1709	1706	1704	1701	1698
5.9	1695	1692	1689	1686	1684	1681	1678	1675	1672	1669
	0	1	2	3	4	5	6	7	8	9

Wavelength(μm)	Wave number(cm^{-1})									
	0	1	2	3	4	5	6	7	8	9
6.0	1667	1664	1661	1658	1656	1653	1650	1647	1645	1642
6.1	1639	1637	1634	1631	1629	1626	1623	1621	1618	1616
6.2	1613	1610	1608	1605	1603	1600	1597	1595	1592	1590
6.3	1587	1585	1582	1580	1577	1575	1572	1570	1567	1565
6.4	1563	1560	1558	1555	1553	1550	1548	1546	1543	1541
6.5	1538	1536	1534	1531	1529	1527	1524	1522	1520	1517
6.6	1515	1513	1511	1508	1506	1504	1502	1499	1497	1495
6.7	1493	1490	1488	1486	1484	1481	1479	1477	1475	1473
6.8	1471	1468	1466	1464	1462	1460	1458	1456	1453	1451
6.9	1449	1447	1445	1443	1441	1439	1437	1435	1433	1431
7.0	1429	1427	1425	1422	1420	1418	1416	1414	1412	1410
7.1	1408	1406	1404	1403	1401	1399	1397	1395	1393	1391
7.2	1389	1387	1385	1383	1381	1379	1377	1376	1374	1372
7.3	1370	1368	1366	1364	1362	1361	1359	1357	1355	1353
7.4	1351	1350	1348	1346	1344	1342	1340	1339	1337	1335
7.5	1333	1332	1330	1328	1326	1325	1323	1321	1319	1318
7.6	1316	1314	1312	1311	1309	1307	1305	1304	1302	1300
7.7	1299	1297	1295	1294	1292	1290	1289	1287	1285	1284
7.8	1282	1280	1279	1277	1276	1274	1272	1271	1269	1267
7.9	1266	1264	1263	1261	1259	1258	1256	1255	1253	1252
8.0	1250	1248	1247	1245	1244	1242	1241	1239	1238	1236
8.1	1235	1233	1232	1230	1229	1227	1225	1224	1222	1221
8.2	1220	1218	1217	1215	1214	1212	1211	1209	1208	1206
8.3	1205	1203	1202	1200	1199	1198	1196	1195	1193	1192
8.4	1190	1189	1188	1186	1185	1183	1182	1181	1179	1178
8.5	1176	1175	1174	1172	1171	1170	1168	1167	1166	1164
8.6	1163	1161	1160	1159	1157	1156	1155	1153	1152	1151
8.7	1149	1148	1147	1145	1144	1143	1142	1140	1139	1138
8.8	1136	1135	1134	1133	1131	1130	1129	1127	1126	1125
8.9	1124	1122	1121	1120	1119	1117	1116	1115	1114	1112
9.0	1111	1110	1109	1107	1106	1105	1104	1103	1101	1100
9.1	1099	1098	1096	1095	1094	1093	1092	1091	1089	1088
9.2	1087	1086	1085	1083	1082	1081	1080	1079	1078	1076
9.3	1075	1074	1073	1072	1071	1070	1068	1067	1066	1065
9.4	1064	1063	1062	1060	1059	1058	1057	1056	1055	1054
9.5	1053	1052	1050	1049	1048	1047	1046	1045	1044	1043
9.6	1042	1041	1040	1038	1037	1036	1035	1034	1033	1032
9.7	1031	1030	1029	1028	1027	1026	1025	1024	1022	1021
9.8	1020	1019	1018	1017	1016	1015	1014	1013	1012	1011
9.9	1010	1009	1008	1007	1006	1005	1004	1003	1002	1001
	0	1	2	3	4	5	6	7	8	9

Wavelength(μm)	Wave number(cm⁻¹)									
	0	1	2	3	4	5	6	7	8	9
10.0	1000.0	999.0	998.0	997.0	996.0	995.0	994.0	993.0	992.1	991.1
10.1	990.1	989.1	988.1	987.2	986.2	985.2	984.3	983.3	982.3	981.4
10.2	980.4	979.4	978.5	977.5	976.6	975.6	974.7	973.7	972.8	971.8
10.3	970.9	969.9	969.0	968.1	967.1	966.2	965.3	964.3	963.4	962.5
10.4	961.5	960.6	959.7	958.8	957.9	956.9	956.0	955.1	954.2	953.3
10.5	952.4	951.5	950.6	949.7	948.8	947.9	947.0	946.1	945.2	944.3
10.6	943.4	942.5	941.6	940.7	939.8	939.0	938.1	937.2	936.3	935.5
10.7	934.6	933.7	932.8	932.0	931.1	930.2	929.4	928.5	927.6	926.8
10.8	925.9	925.1	924.2	923.4	922.5	921.7	920.8	920.0	919.1	918.3
10.9	917.4	916.6	915.8	914.9	914.1	913.2	912.4	911.6	910.7	909.9
11.0	909.1	908.3	907.4	906.6	905.8	905.0	904.2	903.3	902.5	901.7
11.1	900.9	900.1	899.3	898.5	897.7	896.9	896.1	895.3	894.5	893.7
11.2	892.9	892.1	891.3	890.5	889.7	888.9	888.1	887.3	886.5	885.7
11.3	885.0	884.2	883.4	882.6	881.8	881.1	880.3	879.5	878.7	878.0
11.4	877.2	876.4	875.7	874.9	874.1	873.4	872.6	871.8	871.1	870.3
11.5	869.6	868.8	868.1	867.3	866.6	865.8	865.1	864.3	863.6	862.8
11.6	862.1	861.3	860.6	859.8	859.1	858.4	857.6	856.9	856.2	855.4
11.7	854.7	854.0	853.2	852.5	851.8	851.1	850.3	849.6	848.9	848.2
11.8	847.5	846.7	846.0	845.3	844.6	843.9	843.2	842.5	841.8	841.0
11.9	840.3	839.6	838.9	838.2	837.5	836.8	836.1	835.4	834.7	834.0
12.0	833.3	832.6	831.9	831.3	830.6	829.9	829.2	828.5	827.8	827.1
12.1	826.4	825.8	825.1	824.4	823.7	823.0	822.4	821.7	821.0	820.3
12.2	819.7	819.0	818.3	817.7	817.0	816.3	815.7	815.0	814.3	813.7
12.3	813.0	812.3	811.7	811.0	810.4	809.7	809.1	808.4	807.8	807.1
12.4	806.5	805.8	805.2	804.5	803.9	803.2	802.6	801.9	801.3	800.6
12.5	800.0	799.4	798.7	798.1	797.4	796.8	796.2	795.5	794.9	794.3
12.6	793.7	793.0	792.4	791.8	791.1	790.5	789.9	789.3	788.6	788.0
12.7	787.4	786.8	786.2	785.5	784.9	784.3	783.7	783.1	782.5	781.9
12.8	781.3	780.6	780.0	779.4	778.8	778.2	777.6	777.0	776.4	775.8
12.9	775.2	774.6	774.0	773.4	772.8	772.2	771.6	771.0	770.4	769.8
13.0	769.2	768.6	768.0	767.5	766.9	766.3	765.7	765.1	764.5	763.9
13.1	763.4	762.8	762.2	761.6	761.0	760.5	759.9	759.3	758.7	758.2
13.2	757.6	757.0	756.4	755.9	755.3	754.7	754.1	753.6	753.0	752.4
13.3	751.9	751.3	750.8	750.2	749.6	749.1	748.5	747.9	747.4	746.8
13.4	746.3	745.7	745.2	744.6	744.0	743.5	742.9	742.4	741.8	741.3
13.5	740.7	740.2	739.6	739.1	738.6	738.0	737.5	736.9	736.4	735.8
13.6	735.3	734.8	734.2	733.7	733.1	732.6	732.1	731.5	731.0	730.5
13.7	729.9	729.4	728.9	728.3	727.8	727.3	726.7	726.2	725.7	725.2
13.8	724.6	724.1	723.6	723.1	722.5	722.0	721.5	721.0	720.5	719.9
13.9	719.4	718.9	718.4	717.9	717.4	716.8	716.3	715.8	715.3	714.8
14.0	714.3	713.8	713.3	712.8	712.3	711.7	711.2	710.7	710.2	709.7
14.1	709.2	708.7	708.2	707.7	707.2	706.7	706.2	705.7	705.2	704.7
14.2	704.2	703.7	703.2	702.7	702.2	701.8	701.3	700.8	700.3	699.8
14.3	699.3	698.8	698.3	697.8	697.4	696.9	696.4	695.9	695.4	694.9
14.4	694.4	694.0	693.5	693.0	692.5	692.0	691.6	691.1	690.6	690.1
14.5	689.7	689.2	688.7	688.2	687.8	687.3	686.8	686.3	685.9	685.4
14.6	684.9	684.5	684.0	683.5	683.1	682.6	682.1	681.7	681.2	680.7
14.7	680.3	679.8	679.3	678.9	678.4	678.0	677.5	677.0	676.6	676.1
14.8	675.7	675.2	674.8	674.3	673.9	673.4	672.9	672.5	672.0	671.6
14.9	671.1	670.7	670.2	669.8	669.3	668.9	668.4	668.0	667.6	667.1
	0	1	2	3	4	5	6	7	8	9

| Wavelength(μm) | Wave number(cm^{-1}) | | | | | | | | | |
	0	1	2	3	4	5	6	7	8	9
15.0	666.7	666.2	665.8	665.3	664.9	664.5	664.0	663.6	663.1	662.7
15.1	662.3	661.8	661.4	660.9	660.5	660.1	659.6	659.2	658.8	658.3
15.2	657.9	657.5	657.0	656.6	656.2	655.7	655.3	654.9	654.5	654.0
15.3	653.6	653.2	652.7	652.3	651.9	651.5	651.0	650.6	650.2	649.8
15.4	649.4	648.9	648.5	648.1	647.7	647.2	646.8	646.4	646.0	645.6
15.5	645.2	644.7	644.3	643.9	643.5	643.1	642.7	642.3	641.8	641.4
15.6	641.0	640.6	640.2	639.8	639.4	639.0	638.6	638.2	637.8	637.3
15.7	636.9	636.5	636.1	635.7	635.3	634.9	634.5	634.1	633.7	633.3
15.8	632.9	632.5	632.1	631.7	631.3	630.9	630.5	630.1	629.7	629.3
15.9	628.9	628.5	628.1	627.7	627.4	627.0	626.6	626.2	625.8	625.4
16.0	625.0	624.6	624.2	623.8	623.4	623.1	622.7	622.3	621.9	621.5
16.1	621.1	620.7	620.3	620.0	619.6	619.2	618.8	618.4	618.0	617.7
16.2	617.3	616.9	616.5	616.1	615.8	615.4	615.0	614.6	614.3	613.9
16.3	613.5	613.1	612.7	612.4	612.0	611.6	611.2	610.9	610.5	610.1
16.4	609.8	609.4	609.0	608.6	608.3	607.9	607.5	607.2	606.8	606.4
16.5	606.1	605.7	605.3	605.0	604.6	604.2	603.9	603.5	603.1	602.8
16.6	602.4	602.0	601.7	601.3	601.0	600.6	600.2	599.9	599.5	599.2
16.7	598.8	598.4	598.1	597.7	597.4	597.0	596.7	596.3	595.9	595.6
16.8	595.2	594.9	594.5	594.2	593.8	593.5	593.1	592.8	592.4	592.1
16.9	591.7	591.4	591.0	590.7	590.3	590.0	589.6	589.3	588.9	588.6
17.0	588.2	587.9	587.5	587.2	586.9	586.5	586.2	585.8	585.5	585.1
17.1	584.8	584.5	584.1	583.8	583.4	583.1	582.8	582.4	582.1	581.7
17.2	581.4	581.1	580.7	580.4	580.0	579.7	579.4	579.0	578.7	578.4
17.3	578.0	577.7	577.4	577.0	576.7	576.4	576.0	575.7	575.4	575.0
17.4	574.7	574.4	574.1	573.7	573.4	573.1	572.7	572.4	572.1	571.8
17.5	571.4	571.1	570.8	570.5	570.1	569.8	569.5	569.2	568.8	568.5
17.6	568.2	567.9	567.5	567.2	566.9	566.6	566.3	565.9	565.6	565.3
17.7	565.0	564.7	564.3	564.0	563.7	563.4	563.1	562.7	562.4	562.1
17.8	561.8	561.5	561.2	560.9	560.5	560.2	559.9	559.6	559.3	559.0
17.9	558.7	558.3	558.0	557.7	557.4	557.1	556.8	556.5	556.2	555.9
18.0	555.6	555.2	554.9	554.6	554.3	554.0	553.7	553.4	553.1	552.8
18.1	552.5	552.2	551.9	551.6	551.3	551.0	550.7	550.4	550.1	549.8
18.2	549.5	549.1	548.8	548.5	548.2	547.9	547.6	547.3	547.0	546.7
18.3	546.4	546.1	545.9	545.6	545.3	545.0	544.7	544.4	544.1	543.8
18.4	543.5	543.2	542.9	542.6	542.3	542.0	541.7	541.4	541.1	510.8
18.5	540.5	540.2	540.0	539.7	539.4	539.1	538.8	538.5	538.2	537.9
18.6	537.6	537.3	537.1	536.8	536.5	536.2	535.9	535.6	535.3	535.0
18.7	534.8	534.5	534.2	533.9	533.6	533.3	533.0	532.8	532.5	532.2
18.8	531.9	531.6	531.3	531.1	530.8	530.5	530.2	529.9	529.7	529.4
18.9	529.1	528.8	528.5	528.3	528.0	527.7	527.4	527.1	526.9	526.6
19.0	526.3	526.0	525.8	525.5	525.2	524.9	524.7	524.4	524.1	523.8
19.1	523.6	523.3	523.0	522.7	522.5	522.2	521.9	521.6	521.4	521.1
19.2	520.8	520.6	520.3	520.0	519.8	519.5	519.2	518.9	518.7	518.4
19.3	518.1	517.9	517.6	517.3	517.1	516.8	516.5	516.3	516.0	515.7
19.4	515.5	515.2	514.9	514.7	514.4	514.1	513.9	513.6	513.3	513.1
19.5	512.8	512.6	512.3	512.0	511.8	511.5	511.2	511.0	510.7	510.5
19.6	510.2	509.9	509.7	509.4	509.2	508.9	508.6	508.4	508.1	507.9
19.7	507.6	507.4	507.1	506.8	506.6	506.3	506.1	505.8	505.6	505.3
19.8	505.1	504.8	504.5	504.3	504.0	503.8	503.5	503.3	503.0	502.8
19.9	502.5	502.3	502.0	501.8	501.5	501.3	501.0	500.8	500.5	500.3
	0	1	2	3	4	5	6	7	8	9

Wavelength(μm)	Wave number(cm⁻¹)									
	0	1	2	3	4	5	6	7	8	9
20.0	500.0	499.8	499.5	499.3	499.0	498.8	498.5	498.3	498.0	497.8
20.1	497.5	497.3	497.0	496.8	496.5	496.3	496.0	495.8	495.5	495.3
20.2	495.0	494.8	494.6	494.3	494.1	493.8	493.6	493.3	493.1	492.9
20.3	492.6	492.4	492.1	491.9	491.6	491.4	491.2	490.9	490.7	490.4
20.4	490.2	490.0	489.7	489.5	489.2	489.0	488.8	488.5	488.3	488.0
20.5	487.8	487.6	487.3	487.1	486.9	486.6	486.4	486.1	485.9	485.7
20.6	485.4	485.2	485.0	484.7	484.5	484.3	484.0	483.8	483.6	483.3
20.7	483.1	482.9	482.6	482.4	482.2	481.9	481.7	481.5	481.2	481.0
20.8	480.8	480.5	480.3	480.1	479.8	479.6	479.4	479.2	478.9	478.7
20.9	478.5	478.2	478.0	477.8	477.6	477.3	477.1	476.9	476.6	476.4
21.0	476.2	476.0	475.7	475.5	475.3	475.1	474.8	474.6	474.4	474.2
21.1	473.9	473.7	473.5	473.3	473.0	472.8	472.6	472.4	472.1	471.9
21.2	471.7	471.5	471.3	471.0	470.8	470.6	470.4	470.1	469.9	469.7
21.3	469.5	469.3	469.0	468.8	468.6	468.4	468.2	467.9	467.7	467.5
21.4	467.3	467.1	466.9	466.6	466.4	466.2	466.0	465.8	465.5	465.3
21.5	465.1	464.9	464.7	464.5	464.3	464.0	463.8	463.6	463.4	463.2
21.6	463.0	462.7	462.5	462.3	462.1	461.9	461.7	461.5	461.3	461.0
21.7	460.8	460.6	460.4	460.2	460.0	459.8	459.6	459.3	459.1	458.9
21.8	458.7	458.5	458.3	458.1	457.9	457.7	457.5	457.2	457.0	456.8
21.9	456.6	456.4	456.2	456.0	455.8	455.6	455.4	455.2	455.0	454.8
22.0	454.5	454.3	454.1	453.9	453.7	453.5	453.3	453.1	452.9	452.7
22.1	452.5	452.3	452.1	451.9	451.7	451.5	451.3	451.1	450.9	450.7
22.2	450.5	450.2	450.0	449.8	449.6	449.4	449.2	449.0	448.8	448.6
22.3	448.4	448.2	448.0	447.8	447.6	447.4	447.2	447.0	446.8	446.6
22.4	446.4	446.2	446.0	445.8	445.6	445.4	445.2	445.0	444.8	444.6
22.5	444.4	444.2	444.0	443.9	443.7	443.5	443.3	443.1	442.9	442.7
22.6	442.5	442.3	442.1	441.9	441.7	441.5	441.3	441.1	440.9	440.7
22.7	440.5	440.3	440.1	439.9	439.8	439.6	439.4	439.2	439.0	438.8
22.8	438.6	438.4	438.2	438.0	437.8	437.6	437.4	437.3	437.1	436.9
22.9	436.7	436.5	436.3	436.1	435.9	435.7	435.5	435.4	435.2	435.0
23.0	434.8	434.6	434.4	434.2	434.0	433.8	433.7	433.5	433.3	433.1
23.1	432.9	432.7	432.5	432.3	432.2	432.0	431.8	431.6	431.4	431.2
23.2	431.0	430.8	430.7	430.5	430.3	430.1	429.9	429.7	429.6	429.4
23.3	429.2	429.0	428.8	428.6	428.4	428.3	428.1	427.9	427.7	427.5
23.4	427.4	427.2	427.0	426.8	426.6	426.4	426.3	426.1	425.9	425.7
23.5	425.5	425.4	425.2	425.0	424.8	424.6	424.4	424.3	424.1	423.9
23.6	423.7	423.5	423.4	423.2	423.0	422.8	422.7	422.5	422.3	422.1
23.7	421.9	421.8	421.6	421.4	421.2	421.1	420.9	420.7	420.5	420.3
23.8	420.2	420.0	419.8	419.6	419.5	419.3	419.1	418.9	418.8	418.6
23.9	418.4	418.2	418.1	417.9	417.7	417.5	417.4	417.2	417.0	416.8
24.0	416.7	416.5	416.3	416.1	416.0	415.8	415.6	415.5	415.3	415.1
24.1	414.9	414.8	414.6	414.4	414.3	414.1	413.9	413.7	413.6	413.4
24.2	413.2	413.1	412.9	412.7	412.5	412.4	412.2	412.0	411.9	411.7
24.3	411.5	411.4	411.2	411.0	410.8	410.7	410.5	410.3	410.2	410.0
24.4	409.8	409.7	409.5	409.3	409.2	409.0	408.8	408.7	408.5	408.3
24.5	408.2	408.0	407.8	407.7	407.5	407.3	407.2	407.0	406.8	406.7
24.6	406.5	406.3	406.0	405.8	405.8	405.7	405.5	405.4	405.2	405.0
24.7	404.9	404.7	404.5	404.4	404.2	404.0	403.9	403.7	403.6	403.4
24.8	403.2	403.1	402.9	402.7	402.6	402.4	402.3	402.1	401.9	401.8
24.9	401.6	401.4	401.3	401.1	401.0	400.8	400.6	400.5	400.3	400.2
	0	1	2	3	4	5	6	7	8	9

Answers
to Problems

Chapter 4

1. a. 310 nm   b. 232 nm   c. 243 nm   d. 227 nm   e. 234 nm   f. <200 nm
for $\pi \rightarrow \pi^*$, ~275 nm for $n \rightarrow \pi^*$ g. 244 nm   h. 219 nm   i. 312 nm
j. 227 and 253 nm

2. a.

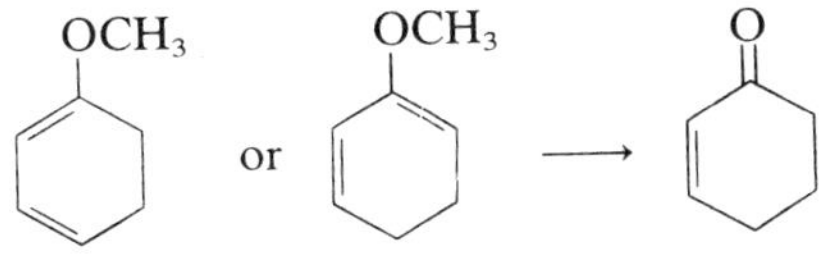

b.

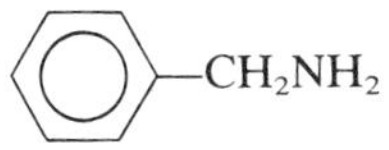

c. $CH_3CH\!=\!CHCHO$

d.

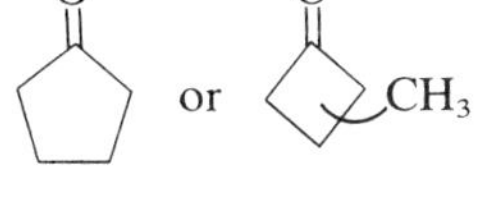

e.

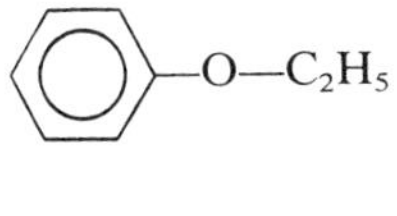

f.

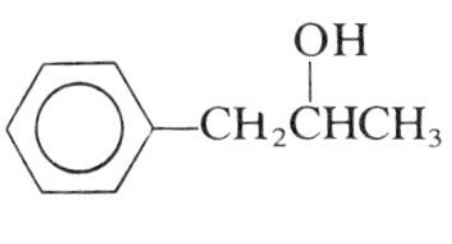

g.

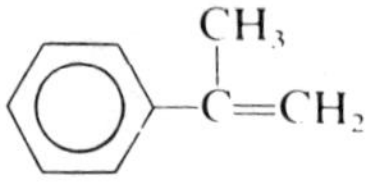

CH$_3$

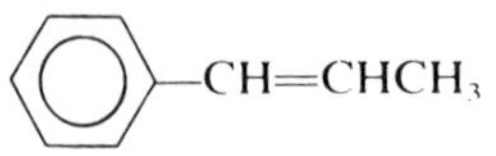

$\bigcirc$—C=CH$_2$ or $\bigcirc$—CH=CHCH$_3$

h.   O
     ‖
CH$_3$SCH$_2$CH$_3$

i. (CH$_3$)$_2$C=CH—CH=CH$_2$

$\longrightarrow$ (CH$_3$)$_2$C=CHCHO

j. 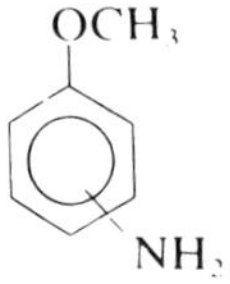   OCH$_3$

$\bigcirc$

NH$_2$

Chapter 5

1. Isopropenyl acetate

2. (*E*)-3-Penten-2-one

3. 2,5-Dimethylphenol

4. (*E*)-3-Phenyl-2-propen-1-ol

5. δ-Butanolactone

6. *m*-Nitrobenzyl alcohol

7. (*E*)-1-Bromo-2-butene

8. 2-Cyclohexenone

Chapter 6

1. H$_2$C=CHCH$_2$CH(CH$_3$)$_2$

2. C$_6$H$_5$CH$_2$CO$_2$CH$_2$CH$_3$

3.

4. 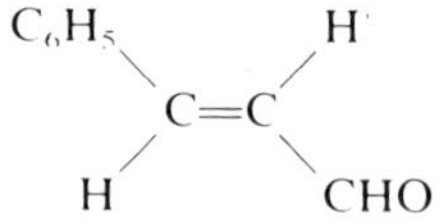 C$_6$H$_5$ H
     \   /
      C=C
     /   \
    H   CHO

5. ▷—CH$_2$OH

6.
     O
     ‖

CH$_3$ CH$_3$
    CH$_3$

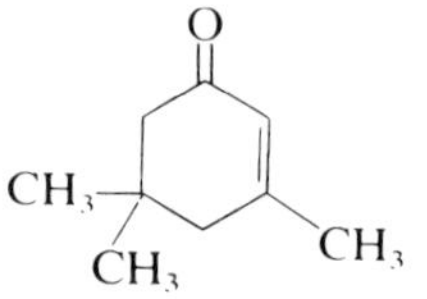

7. 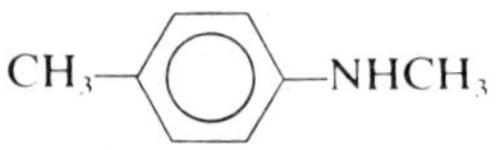 CH$_3$—$\bigcirc$—NHCH$_3$

8. 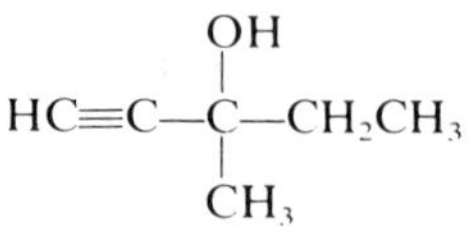   OH
              |
HC≡C—C—CH$_2$CH$_3$
              |
             CH$_3$

9. C$_6$H$_5$CH$_2$CH$_2$NH$_2$

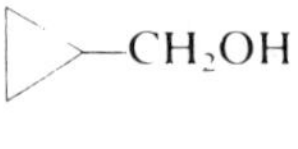

10. H$_2$C=CHCH$_2$OH

11. CH$_3$COCH(CH$_3$)$_2$

12. 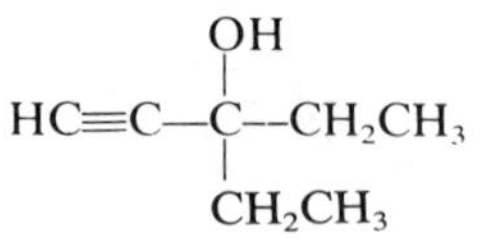   OH
               |
HC≡C—C—CH$_2$CH$_3$
               |
              CH$_2$CH$_3$

13.

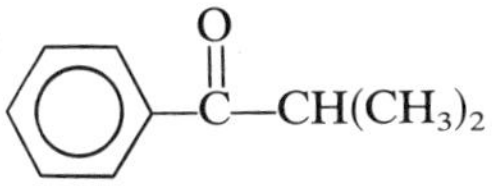

14. $CH_3CHOHCH_2CH(CH_3)_2$

15. 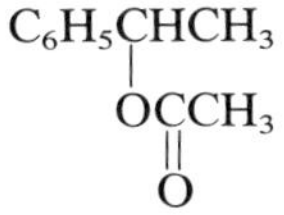 $C_6H_5CHCH_3$
      |
      $OCCH_3$
      ||
      O

16. $(CH_3CH_2CH_2CH_2)_2NH$

17. 5-Hexen-2-one

Chapter 7

1.

$$CH_3CH_2CH_2CH_2\overset{O}{\overset{||}{C}}CH_2CH_2CH_2CH_3$$

2.

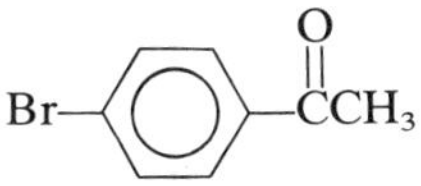

3. $C_{15}H_{31}\overset{O}{\overset{||}{C}}CH_2\overset{O}{\overset{||}{C}}OC_2H_5$

4. 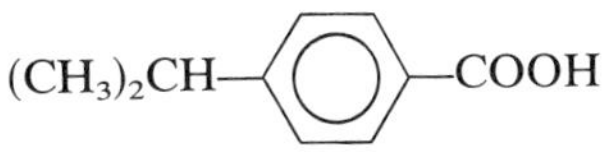 $(CH_3)_2CH\!-\!\bigcirc\!-\!COOH$

5. $CH_3\overset{O}{\overset{||}{C}}CH_2CH_2\overset{O}{\overset{||}{C}}CH_3$

6. $CH_3CH(OC_2H_5)_2$

7.

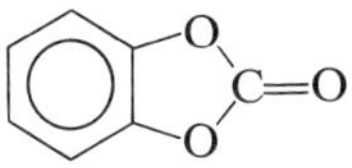

8. $(CH_3)_2C\!=\!CH\overset{O}{\overset{||}{C}}CH_3$

9. $BrCH_2CH_2C\!\equiv\!N$

10. $CH_3(CH_2)_6CH_2SH$

Chapter 11

1. 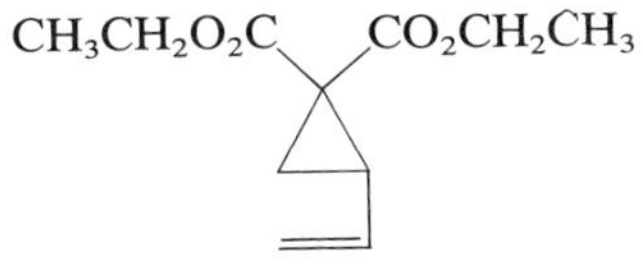 $CH_3CH_2O_2C\diagdown\diagup CO_2CH_2CH_3$

A

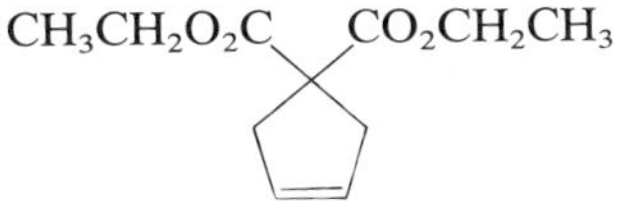

$CH_3CH_2O_2C\diagdown\diagup CO_2CH_2CH_3$

B

2.

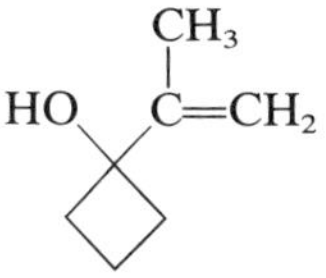

X 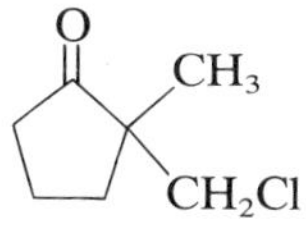   **Y**

3.

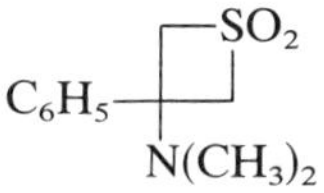

II 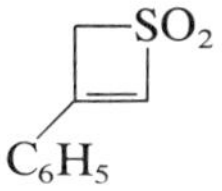   **III**   **IV**   **V**

$C_6H_5\overset{O}{\overset{||}{C}}CH_2SO_2CH_3$ C_6H_5COOH

532 Answers to Problems

4.

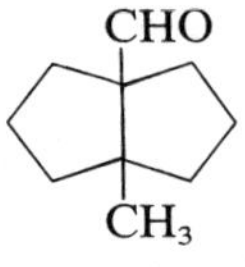

CHO

B

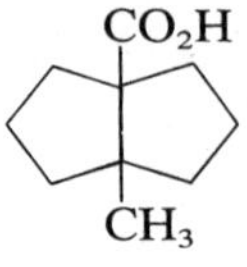

CO$_2$H

C

5.

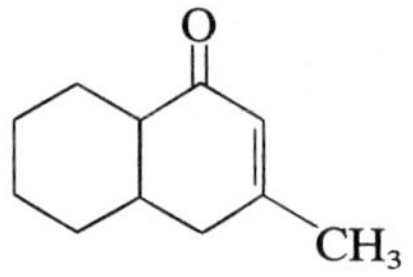

I

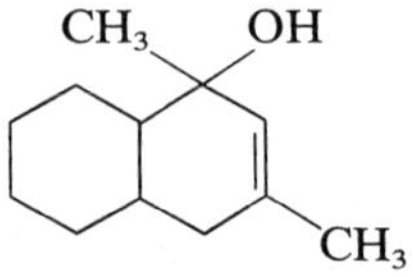

CH$_3$

II

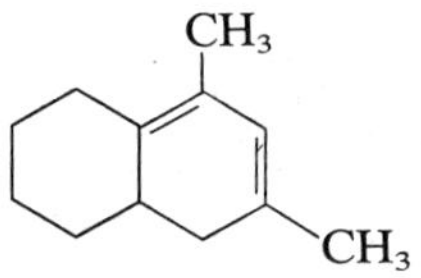

CH$_3$ OH

CH$_3$

III

CH$_3$

CH$_3$

CH$_3$

6.

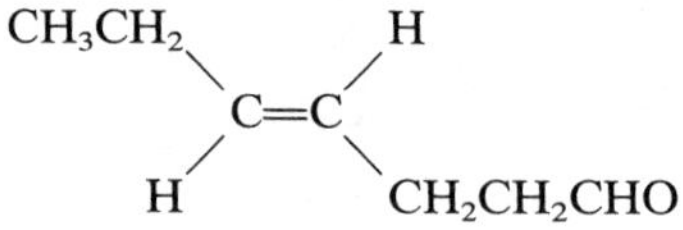

$$CH_3CH_2 \quad H$$
$$C{=}C$$
$$H \quad CH_2CH_2CHO$$

I

7.

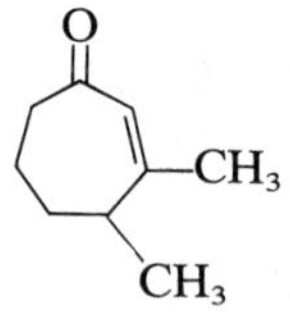

O

—CH$_3$

CH$_3$

A

O

—CH$_3$

CH$_3$

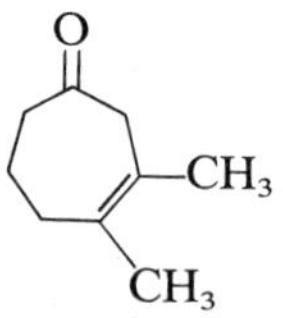

B

Index